Methods in Enzymology

Volume 106
POSTTRANSLATIONAL MODIFICATIONS
Part A

METHODS IN ENZYMOLOGY

EDITORS-IN-CHIEF

Sidney P. Colowick   Nathan O. Kaplan

Methods in Enzymology

Volume 106

Posttranslational Modifications

Part A

EDITED BY

Finn Wold
DEPARTMENT OF BIOCHEMISTRY
AND MOLECULAR BIOLOGY
UNIVERSITY OF TEXAS MEDICAL SCHOOL
AT HOUSTON
HOUSTON, TEXAS

Kivie Moldave
DEPARTMENT OF BIOLOGICAL CHEMISTRY
CALIFORNIA COLLEGE OF MEDICINE
UNIVERSITY OF CALIFORNIA
IRVINE, CALIFORNIA

1984

ACADEMIC PRESS, INC.

(Harcourt Brace Jovanovich, Publishers)

Orlando   San Diego   San Francisco   New York   London
Toronto   Montreal   Sydney   Tokyo   São Paulo

ACADEMIC PRESS, INC.
Orlando, Florida 32887

United Kingdom Edition published by
ACADEMIC PRESS, INC. (LONDON) LTD.
24/28 Oval Road, London NW1 7DX

LIBRARY OF CONGRESS CATALOG CARD NUMBER: 54-9110
ISBN 0-12-182006-8

PRINTED IN THE UNITED STATES OF AMERICA

84 85 86 87   9 8 7 6 5 4 3 2 1

Table of Contents

Section IV. Covalent Modification of the α-Amino and α-Carboxyl Groups of Proteins

Section V. Protein Alkylations/Dealkylations (Arylations)

Section VI. Protein Glycosylations; ADP-Ribosylation

Contributors to Volume 106

Article numbers are in parentheses following the names of contributors.
Affiliations listed are current.

PETER ADAMIETZ (48), *Institut für Physiologische Chemie der Universität Hamburg, 2000 Hamburg 20, Federal Republic of Germany*

ALASTAIR AITKEN (19), *Department of Pharmaceutical Chemistry, School of Pharmacy, London University, London WC1N 1AX, England*

RAFAEL ALVAREZ-GONZALES (50), *Departments of Chemistry and Biochemistry, North Texas State University, Texas College of Osteopathic Medicine, Denton, Texas 76203*

JEFFREY L. BADA (9), *Amino Acid Dating Laboratory, Scripps Institution of Oceanography, University of California, San Diego, La Jolla, California 92093*

JOHN W. BAYNES (8), *Department of Chemistry, University of South Carolina, Columbia, South Carolina 29208*

KLAUS BIEMANN (4), *Department of Chemistry, Massachusetts Institute of Technology, Cambridge, Massachusetts 02139*

SYLVAIN BLANQUET (11), *Laboratoire de Biochimie, Ecole Polytechnique, 91128 Palaiseau Cedex, France*

JAMES W. BODLEY (38), *Department of Biochemistry, University of Minnesota, Minneapolis, Minnesota 55455*

REINHARD BREDEHORST (49), *Institut für Physiologische Chemie der Universität Hamburg, 2000 Hamburg 20, Federal Republic of Germany*

HARRIS BUSCH (23), *Department of Pharmacology, Baylor College of Medicine, Houston, Texas 77030*

STEVEN A. CARR (4), *Department of Nutrition, Harvard School of Public Health, and Department of Biological Chemistry, Harvard Medical School, Boston, Massachusetts 02115*

CHRISTOPHER C. Q. CHIN (2, 6), *Department of Biochemistry and Molecular Biology, The University of Texas Medical School, Houston, Texas 77025*

STEVEN CLARKE (29, 31), *Department of Chemistry and Biochemistry and the Molecular Biology Institute, University of California, Los Angeles, Los Angeles, California 90024*

PHILIP COHEN (19), *Protein Phosphorylation Group of the Medical Research Council, Department of Biochemistry, University of Dundee, Medical Sciences Institute, Dundee DD1 4HN, Scotland*

WALDO E. COHN (1), *Biology Division, Oak Ridge National Laboratory, Oak Ridge, Tennessee 37831*

HERBERT L. COOPER (32), *Cellular and Molecular Physiology Section, Laboratory of Tumor Immunology and Biology, National Cancer Institute, National Institutes of Health, Bethesda, Maryland 20205*

PHILIPPE DESSEN (11), *Laboratoire de Biochimie, Ecole Polytechnique, 91128 Palaiseau Cedex, France*

CHARLES E. DEUTCH (18), *Department of Biology, Kalamazoo College, Kalamazoo, Michigan 49007*

PETER DIMARIA (26), *Department of Pathology, Washington University School of Medicine, St. Louis, Missouri 63110*

JACK E. DIXON (15), *Department of Biochemistry, Purdue University, West Lafayette, Indiana 47907*

PATRICIA C. DUNLOP (38), *Department of Medicine, Veteran's Administration Hospital, Minneapolis, Minnesota 55417*

MARTIN FLAVIN (22), *Laboratory of Cell Biology, National Heart, Lung, and Blood Institute, National Institutes of Health, Bethesda, Maryland 20205*

RUDOLF FLÜCKIGER (7), *Zentrum für Lehre und Forschung Diabetologie, Kantonsspital Basel, CH 4031 Basel, Switzerland*

J. FÖHLES (5), *German Wool Research Institute, Aachen Institute of Technology, D-5100 Aachen, Federal Republic of Germany*

J. E. FOLK (32), *Enzyme Chemistry Section, Laboratory of Oral Biology and Physiology, National Institute of Dental Research, National Institutes of Health, Bethesda, Maryland 20205*

PAUL M. GALLOP (7), *Laboratory of Human Biochemistry, Department of Orthopedic Surgery, Children's Hospital Corporation and Departments of Biological Chemistry, Oral Biology and Pathophysiology, Harvard Schools of Medicine and Dental Medicine, Boston, Massachusetts 02115*

ALEXANDER N. GLAZER (35), *Department of Microbiology and Immunology, University of California, Berkeley, Berkeley, California 94720*

C. G. GOFF (43), *Department of Biology, Haverford College, Haverford, Pennsylvania 19041*

DONALD J. GRAVES (41), *Department of Biochemistry and Biophysics, Iowa State University, Ames, Iowa 50011*

OSAMU HAYAISHI (47, 52), *Osaka Medical College, Takatsuki, Osaka 569, Japan*

HELMUTH HILZ (48, 49), *Institut für Physiologische Chemie der Universität Hamburg, 2000 Hamburg 20, Federal Republic of Germany*

B. L. HORECKER (17), *Roche Research Center, Roche Institute of Molecular Biology, Nutley, New Jersey 07110*

GABOR HUSZAR (27), *Department of Obstetrics and Gynecology, Yale University School of Medicine, New Haven, Connecticut 06510*

ELAINE L. JACOBSON (50), *Department of Biology, Texas Women's University, Denton, Texas 76204*

MYRON K. JACOBSON (50), *Departments of Chemistry and Biochemistry, North Texas State University/Texas College of Osteopathic Medicine, Denton, Texas 76203*

HECTOR JUAREZ-SALINAS (50), *Life Sciences Research Group, Bio-Rad Laboratories, Richmond, California 94804*

DANIEL KAHN (11), *Laboratoire de Biologie Moléculaire des Relations Plantes-Microorganismes, Institut National de la Recherche Agronomique, 31320 Castanet-Tolosan, France*

WOON KI PAIK (24, 25, 26), *Fels Research Institute, Temple University School of Medicine, Philadelphia, Pennsylvania 19140*

HIROSHI KIDO (17), *Department of Enzyme Chemistry, Institute for Enzyme Research, School of Medicine, The University of Tokushima, 3 Kuramoto-cho, Tokushima 770, Japan*

WILLIAM R. KIDWELL (53), *Cell Cycle Regulation Section, Laboratory of Pathophysiology, National Cancer Institute, National Institutes of Health, Bethesda, Maryland 20205*

SANGDUK KIM (28), *Fels Research Institute, Temple University School of Medicine, Philadelphia, Pennsylvania 19140*

H. KLOSTERMEYER (5), *Dairy Science Institute, Munich Institute of Technology, D-8050 Freising, Federal Republic of Germany*

P. E. KOLATTUKUDY (20), *Institute of Biological Chemistry, Washington State University, Pullman, Washington 99164*

DANIEL E. KOSHLAND, JR. (29, 30), *Department of Biochemistry, University of California, Berkeley, Berkeley, California 94720*

GÜNTHER KREIL (21), *Institute of Molecular Biology, Austrian Academy of Sciences, A-5020 Salzburg, Austria*

J. OLIVER LAMPEN (36), *Waksman Institute of Microbiology, Rutgers University, Piscataway, New Jersey 08854*

DEREK T. A. LAMPORT (55), *MSU-DOE Plant Research Laboratory, Michigan State University, East Lansing, Michigan 48824*

KONRAD LERCH (34), *Biochemisches Institut der Universität Zürich, CH-8057 Zürich, Switzerland*

LILLIAN L. LOU (31), *Department of Chemistry and Biochemistry, and the Molecular Biology Institute, University of California, Los Angeles, Los Angeles, California 90024*

MARY E. MARSH (33), *Department of Biology, Rice University, Houston, Texas 77251*

NANCY C. MARTIN (12), *Division of Molecular Biology, Department of Biochemistry, The University of Texas Health Science Center at Dallas, Dallas, Texas 75235*

PHILIP N. MCFADDEN (31), *Department of Chemistry and Biochemistry, and the Molecular Biology Institute, University of California, Los Angeles, Los Angeles, California 90024*

WILLIAM S. MCINTIRE (37), *Veterans Administration Medical Center, San Francisco, California 94121*

JOYCE F. MILLER (41), *Department of Neurology, University of Iowa, Iowa City, Iowa 52242*

MASANAO MIWA (46, 51), *National Cancer Center Institute, Tsukiji 5-chome, Chuoku, Tokyo 104, Japan*

J. M. MOEHRING (39), *Department of Medical Microbiology, The University of Vermont, College of Medicine, Burlington, Vermont 05405*

T. J. MOEHRING (39), *Department of Medical Microbiology, The University of Vermont, College of Medicine, Burlington, Vermont 05405*

YOSHIMASA MORINO (10), *Department of Biochemistry, Kumamoto University Medical School, Kumamoto 860, Japan*

JOEL MOSS (42, 44), *National Heart, Lung, and Blood Institute, National Institutes of Health, Bethesda, Maryland 20205*

HIROMU MUROFUSHI (22), *Department of Biophysics and Biochemistry, Faculty of Science, University of Tokyo, Bunkyo-ku, Tokyo 113, Japan*

MARTHA H. MURTIASHAW (8), *Department of Chemistry, University of South Carolina, Columbia, South Carolina 29208*

FUJIO NAGASHIMA (10), *Department of Biochemistry, Kumamoto University Medical School, Kumamoto 860, Japan*

JENNIFER B. K. NIELSEN (36), *Merck, Sharp and Dohme Research Laboratories, Rahway, New Jersey 07065*

CLARE M. O'CONNOR (31), *Department of Chemistry and Biochemistry, and the Molecular Biology Institute, University of California, Los Angeles, Los Angeles, California 90024*

NORMAN J. OPPENHEIMER (40), *Department of Pharmaceutical Chemistry, University of California, San Francisco, San Francisco, California 94143*

MYUNG HEE PARK (32), *Enzyme Chemistry Section, Laboratory of Oral Biology and Physiology, National Institute of Dental Research, National Institutes of Health, Bethesda, Maryland 20205*

D. MICHAEL PAYNE (50), *Departments of Chemistry and Biochemistry, North Texas State University/Texas College of Osteopathic Medicine, Denton, Texas 76203*

MICHAEL R. PURNELL (53), *Cell Cycle Regulation Section, Laboratory of Pathophysiology, National Cancer Institute, National Institutes of Health, Bethesda, Maryland 20205*

MURRAY RABINOWITZ (12), *Departments of Medicine and Biochemistry, University of Chicago, Chicago, Illinois 60637*

KENT L. REDMAN (16), *Department of Biochemistry, College of Medicine, University of Iowa, Iowa City, Iowa 52242*

KARIN RÖPER (5), *German Wool Research Institute, Aachen Institute of Technology, D-5100 Aachen, Federal Republic of Germany*

PETER A. RUBENSTEIN (16), *Department of Biochemistry, College of Medicine, University of Iowa, Iowa City, Iowa 52242*

FUMIO SAKIYAMA (14), *Institute for Protein Research, Osaka University, Suita, Osaka 565, Japan*

TATSURU SASAGAWA (3), *Toya Soda Co. Ltd., 2743-1 Hayakawa, Ayase-shi, Kanagawa-ken, Japan*

RONALD L. SASS (33), *Department of Biology, Rice University, Houston, Texas 77251*

STEPHEN SHARP (13), *Department of Microbiology and Molecular Genetics, California College of Medicine, University of California, Irvine, Irvine, California 92717*

JAMES L. SIMS (50), *Departments of Chemistry and Biochemistry, North Texas State University/Texas College of Osteopathic Medicine, Denton, Texas 76203*

THOMAS P. SINGER (37), *Department of Pharmaceutical Chemistry, School of Pharmacy, and Department of Biochemistry and Biophysics, School of Medicine, University of California, San Francisco, California 94143, and Veterans Administration Medical Center, San Francisco, California 94121*

MARK SMULSON (45, 54), *Department of Biochemistry, Georgetown University Schools of Medicine and Dentistry, Washington, D.C. 20007*

MARK A. SNYDER (30), *Department of Microbiology, University of California, San Francisco, San Francisco, California 94143*

GOPALAN SOMAN (41), *Department of Biochemistry and Biophysics, Iowa State University, Ames, Iowa 50011*

THOMAS S. STEWART (13), *School of Biochemistry, University of New South Wales, Kensington, New South Wales 2033, Australia*

JEFFREY B. STOCK (29, 30), *Department of Molecular Biology, Princeton University, Princeton, New Jersey 08544*

TAKASHI SUGIMURA (45, 46, 51), *National Cancer Research Institute, Tsukiji 5-chome, Chuo-ku, Tokyo 104, Japan*

SUZANNE R. THORPE (8), *Department of Chemistry, University of South Carolina, Columbia, South Carolina 29208*

SUSUMU TSUNASAWA (14), *Institute for Protein Research, Osaka University, Suita, Osaka 565, Japan*

MARTIN TUCK (25), *Fels Research Institute, Temple University School of Medicine, Philadelphia, Pennsylvania 19140*

KUNIHIRO UEDA (47, 52), *Department of Medical Chemistry, Kyoto University Faculty of Medicine, Kyoto 606, Japan*

BRIAN G. VANNESS (38), *Department of Biochemistry, University of Iowa, Iowa City, Iowa 52242*

MARTHA VAUGHAN (42, 44), *National Heart, Lung, and Blood Institute, National Institutes of Health, Bethesda, Maryland 20205*

ALBERTO VITA (17), *Laboratorio Biochemica Applicata, Università diCamerino, 62032 Camerino (Mc), Italy*

KENNETH A. WALSH (3), *Department of Biochemistry, University of Washington, Seattle, Washington 98195*

KLAUS WIELCKENS (49), *Institut für Physiologische Chemie der Universität Hamburg, 2000 Hamburg 20, Federal Republic of Germany*

FINN WOLD (6), *Department of Biochemistry and Molecular Biology, University of Texas Medical School, Houston, Texas 77225*

TERRY A. WOODFORD (15), *Division of Cell Growth and Regulation, Charles A. Dana Cancer Center, Dana-Farber Cancer Institute, Boston, Massachusetts 02115*

JINGYUAN ZHANG (52), *Shanghai Institute of Biochemistry, Academia Sinica, Shanghai, China*

Introduction: A Short Stroll through the Posttranslational Zoo

The area of posttranslational modifications of proteins is often, more or less affectionately, referred to as the "amino acid zoo," and in assembling these volumes we have encountered what we imagine might be many of the zookeepers' problems, first in classifying and then in properly housing the animals in proper relation to the other existing vivariums. The present menagerie of posttranslational modifications underwent quite a few changes from the planning stage to the present exposition. Most of the changes were derived from concessions to the fact that many of the posttranslational reactions are very much a part of other subdivisions of biochemistry, indeed, that some classes of posttranslational reactions (e.g., glycosylations, phosphorylations) have become subdivisions in their own right.

With the starting premise that duplication of other volumes in this series should be kept to a minimum, the desired goal of having Volumes 106 and 107 as complete as possible represented a dilemma. We tried to resolve this dilemma by the extensive use of cross-references to other volumes in the *Methods in Enzymology* series. Since these cross-references were to replace planned chapters in these volumes, the next question was where the cross-references should appear, and that problem was finally resolved by the inclusion of this Introduction.

The outline that follows represents our current classification as it has evolved from the original outline of the various posttranslational reactions. When appropriate, a brief statement explains how a particular section has been handled in these volumes and where the omitted components can be found in other volumes.

The meaning of "posttranslational modification" needs to be briefly considered to apprise the readers of the definition that has been used in these volumes. Several authors, when first approached to contribute their chapters, pointed out that their particular derivative or reaction was not in fact "posttranslational" but "cotranslational." Our uniform response was that for the purpose of these volumes we did not feel it desirable to distinguish between the two. In other words, our usage of the term posttranslational goes back to the original usage of "translation" as a broad description of the entire process by which the polymer of three-letter codons in the mRNA is "translated" into a polypeptide chain of 20 "primary" amino acids. This process includes (1) the synthesis of the individual aminoacyl-tRNA derivatives as the commitment step in the translation, (2) the polymerization of the amino acids on the polysome using these aminoacyl-tRNA derivatives as the monomer building blocks, and (3) the release of the completed polyamino acid product. The processing of this polymer, either by shortening or lengthening the original encoded sequence or by covalently modifying the original 20 amino acids into unique noncoded amino acid derivatives, should consequently be considered posttranslational whether the processing occurs before (at the level of aminoacyl-tRNA), during, or after polymerization. According to current usage, these three stages in protein synthesis should be designated pretranslational (presumably?), cotranslational, and posttranslational, respectively. This now well-established nomenclature has unquestionably been useful in communication among the informed, and in fact one of the arguments advanced for making the distinction between co- and posttranslational processes is that it sharpens the definition of the "sloppy" term "translation." However, implicit in this refinement to distinguish the temporal stages of protein processing is the brand-new definition of translation as being synonymous with polymerization, and the benefits of the refinement should be balanced against the requirement to replace the perfectly good chemical term polymerization with the rather vague and equivocal term translation. The choice of these terms must obviously be left to the individual readers. The important point to emphasize is that in these volumes posttranslational modification will include all processing steps by which the direct translation product is altered from the structure specified by the gene.

As the list of topics grew, it became obvious that the space limitation of a single volume would be exceeded. Consequently, the list had to be cut roughly in half for distribution into two volumes. There was no obvious place to make the division, so the arbitrary guide was simply to make Volumes 106 and 107 approximately the same size.

Posttranslational Covalent Modification of Proteins: A List of Reaction Types and General Topics

VOLUME 106

Reactions Involving the Polypeptide Backbone (Limited Proteolysis)

Such reactions have been covered extensively in several recent volumes of *Methods in Enzymology,* and consequently were not included. Major general treatments of proteases, including those involved in zymogen activation, can be found in Volumes 19, 45, and 80 on proteolytic enzymes; the more specialized ones on proteolytic processing in the transport of proteins into and through membranes in Volume 96; on blood clotting, complement, the plasmin system in Volume 80; and on peptide hormones in Volume 37.

General Aspects (Naming, Releasing, and Analyzing Posttranslationally Derivatized Amino Acids)

This section will provide an overview of most of the known amino acid derivatives found to date in proteins and general starting points for proper nomenclature, analytical procedures, and the enzymatic hydrolysis of proteins. The major and most obvious cross-references for this section can be found in Volumes 25, 26, 27, 47, 48, 49, 61, and 91 on enzyme structure.

Nonenzymatic Modifications

Many individuals question the inclusion of nonenzymatic reactions as true posttranslational reactions. We feel that any reaction taking place under physiological conditions to alter the gene-specified polymer should be included. In fact, some reactions (e.g., disulfide interchange, P-450-catalyzed oxygenations) proceed either with or without specific enzyme involvement.

Modifications at the Level of Aminoacyl-tRNA
Covalent Modification of the α-Amino and γ-Carboxyl Groups of Proteins
Reactions Involving Covalent Modification of the Side Chains of Amino Acids in Proteins
Protein Alkylations/Dealkylations (Arylations)
Protein Glycosylations; ADP-Ribosylation

Glycosylation reactions have also been treated extensively in previous volumes of *Methods in Enzymology*. The principal reference sources are Volumes 8, 28, 50, and 83 on complex carbohydrates and Volumes 96 and 98 on membrane biogenesis. Glycosylation of one posttranslational hydroxylated amino acid, hydroxylysine, has been presented in Volume 82 on collagen and elastin, but others (hydroxyproline, β-hydroxytyrosine) are included here. The main emphasis of this section is the glycosylation reactions involving ADP-ribosylation.

VOLUME 107

Reactions Involving Covalent Modification of the Side Chains of Amino Acids (continued)
Protein Acylations/Deacylations (other than phosphorylations)

A major component of the acylation reactions, protein phosphorylation, has been treated in many previous volumes (e.g., 68, 90, and 102) and a recent one, Volume 99, is devoted entirely to protein kinases. To avoid major overlaps with the previous volumes, this section consequently does not treat specific systems, but rather focuses on a broader overview of phosphorylations/dephosphorylation systems and analytical procedures. Some derivatives involving phosphate diesters through the incorporation into proteins or glycoproteins of sugar phosphoryl or nucleotidyl groups are included as special cases of phosphorylations. Other acylation reactions are included.

Oxidations/(Reductions), Hydroxylations, and Halogenations

This is a rather heterogeneous section, justified as a unit only by the common feature that an oxidation/reduction is part of the process by which these derivatives are made. Many of the reactions and products covered in this section have been treated in other volumes, which are cross-referenced in the individual contributions. Many of the posttranslational hydroxylated amino acids which should represent a major component of this section, such as HyL and HyP, have been considered as components of mammalian structural proteins in Volume 82. Hypusin is covered in Volume 106 (see Section V).

Miscellaneous Derivatives

The content of these two volumes, even with the cross-references to other volumes, obviously does not represent a complete display of all known posttranslational reactions and reaction products. [The chapter on nomenclature (Chapter 1, Volume 106) contains a tabulation of most of the posttranslational derivatives known by mid-1983.] The main purpose of compiling this heterogeneous and complex set of reactions into one unit is to emphasize the well-established fact that there is more to protein synthesis than the polymerization of twenty primary amino acids on the mRNA template transcript of the structural gene. The subsequent covalent manipulations of the genetically specified polypeptide chain represent an exciting set of current problems in biochemistry. The problems involve the common challenge of trying to understand all the different ways in which the information of the primary amino acid sequence can be translated into specificity signals for the modification reactions as well as the separate challenges of trying to understand each individual reaction in terms of its chemical mechanisms and biological functions.

FINN WOLD
KIVIE MOLDAVE

METHODS IN ENZYMOLOGY

EDITED BY

Sidney P. Colowick and Nathan O. Kaplan

VANDERBILT UNIVERSITY
SCHOOL OF MEDICINE
NASHVILLE, TENNESSEE

DEPARTMENT OF CHEMISTRY
UNIVERSITY OF CALIFORNIA
AT SAN DIEGO
LA JOLLA, CALIFORNIA

I. Preparation and Assay of Enzymes
II. Preparation and Assay of Enzymes
III. Preparation and Assay of Substrates
IV. Special Techniques for the Enzymologist
V. Preparation and Assay of Enzymes
VI. Preparation and Assay of Enzymes (*Continued*)
Preparation and Assay of Substrates
Special Techniques
VII. Cumulative Subject Index

METHODS IN ENZYMOLOGY

EDITORS-IN-CHIEF

Sidney P. Colowick   Nathan O. Kaplan

VOLUME 75. Cumulative Subject Index Volumes XXXI, XXXII, and XXXIV–LX
Edited by EDWARD A. DENNIS AND MARTHA G. DENNIS

VOLUME 76. Hemoglobins
Edited by ERALDO ANTONINI, LUIGI ROSSI-BERNARDI, AND EMILIA CHIANCONE

VOLUME 77. Detoxication and Drug Metabolism
Edited by WILLIAM B. JAKOBY

VOLUME 78. Interferons (Part A)
Edited by SIDNEY PESTKA

VOLUME 79. Interferons (Part B)
Edited by SIDNEY PESTKA

VOLUME 80. Proteolytic Enzymes (Part C)
Edited by LASZLO LORAND

VOLUME 81. Biomembranes (Part H: Visual Pigments and Purple Membranes, I)
Edited by LESTER PACKER

VOLUME 82. Structural and Contractile Proteins (Part A: Extracellular Matrix)
Edited by LEON W. CUNNINGHAM AND DIXIE W. FREDERIKSEN

VOLUME 83. Complex Carbohydrates (Part D)
Edited by VICTOR GINSBURG

VOLUME 84. Immunochemical Techniques (Part D: Selected Immunoassays)
Edited by JOHN J. LANGONE AND HELEN VAN VUNAKIS

VOLUME 85. Structural and Contractile Proteins (Part B: The Contractile Apparatus and the Cytoskeleton)
Edited by DIXIE W. FREDERIKSEN AND LEON W. CUNNINGHAM

VOLUME 86. Prostaglandins and Arachidonate Metabolites
Edited by WILLIAM E. M. LANDS AND WILLIAM L. SMITH

Methods in Enzymology

Volume 106
POSTTRANSLATIONAL MODIFICATIONS
Part A

Section I

General Aspects

[1] Nomenclature and Symbolism of α-Amino Acids

By WALDO E. COHN

The traditional and well-known trivial names of the α-amino acids commonly found in proteins and represented in the genetic code were, in general, given to them by their discoverers and bear no relationship to their chemical structures, which are given explicitly by their systematic names. Nevertheless, the nomenclature used by those who deal with α-amino acids in any biochemical context, from proteins to metabolism to bacterial cell walls to antibiotics and hormones, utilizes almost exclusively those trivial names that were common in the pre-molecular-biology days, when the composition of proteins dominated biochemical research. Thus, when the Commission on Biochemical Nomenclature (CBN), created by the International Union of Pure and Applied Chemistry (IUPAC) and the International Union of Biochemistry (IUB), met for the first time in 1964 and undertook the task of codifying "abbreviations and symbols for chemical names of special interest in biological chemistry,"[1] the section on polypeptides and proteins was based on those traditional trivial names, which had been formally approved by the IUPAC Commission on the Nomenclature of Organic Chemistry in 1949, and the symbols derived from them in 1947 by Brand and Edsall.[2] These are listed in Table I. The only changes from the original proposal were to change Try to Trp, to prevent error by accidental inversion, Nval to Nvl (and now to Avl), Ileu to Ile, CySH to Cys, Glu-NH$_2$ and Asp-NH$_2$ to Gln and Asn, and Hypro and Hylys to Hyp and Hyl—all these, except the first-named, to hold to the three-letter system, so that sequences could be aligned for comparison. CBN also added two "allo's," one "nor," two "homo's" to the initial list and a number of relatively newly discovered α-amino acids or near relatives. Since these seldom or ever figured in long sequences, where comparisons might require alignment, extra letters or numerals (locants) could be added. Subsequently, (1971[3] and 1983[4]) some of these

---

[1] IUPAC–IUB CBN "Abbreviations and Symbols for Chemical Names of Special Interest in Biological Chemistry (Revised Tentative Rules 1965)," published in *J. Biol. Chem.* **241**, 527 (1966), *Biochemistry* **5**, 1445 (1966), *Eur. J. Biochem.* **1**, 259 (1967), *Biochem. J.* **101**, 1 (1966), and in many other journals, and in four languages.

[2] E. Brand and J. T. Edsall, *Ann. Rev. Biochem.* **16**, 223 (224) (1947).

[3] IUPAC–IUB CBN, "Symbols for Amino-acid Derivatives and Peptides, (Recommendations 1971)," published in *J. Biol. Chem.* **247**, 977 (1972), *Biochemistry* **11**, 1726 (1972), *Eur. J. Biochem.* **17**, 201 (1970), *Biochem. J.* **126**, 773 (1972), and in many other journals.

[4] IUPAC–IUB JCBN, "Nomenclature and Symbolism for Amino Acids and Peptides (Recommendations 1982)" *Eur. J. Biochem.* **138**, 9 (1984).

METHODS IN ENZYMOLOGY, VOL. 106

TABLE I

SYMBOLS FOR α-AMINO ACIDS UNDER mRNA-DIRECTED CONTROL

Trivial name[a]	Symbol[b] 3-letter	Symbol[b] 1-letter[h]	Systematic name[c]
Alanine	Ala	A	2-Aminopropanoic acid
Arginine	Arg	R	2-Amino-5-guanidinopentanoic acid
Asparagine	Asn	N	2-Amino-3-carbamoylpropanoic acid
Aspartic acid	Asp	D	2-Aminobutanedioic acid
Cysteine	Cys[d]	C	2-Amino-3-mercaptopropanoic acid
Glutamine	Gln	Q	2-Amino-4-carbamoylbutanoic acid
Glutamic acid	Glu	E	2-Aminopentanedioic acid
Glycine	Gly	G	Aminoethanoic acid
Histidine	His	H	2-Amino-3-(1H-imidazol-4-yl)propanoic acid
Isoleucine	Ile	I	2-Amino-3-methylpentanoic acid
Leucine	Leu	L	2-Amino-4-methylpentanoic acid
Lysine	Lys	K	2,6-Diaminohexanoic acid
Methionine	Met	M	2-Amino-4-(methylthio)butanoic acid
Phenylalanine	Phe	F	2-Amino-3-phenylpropanoic acid
Proline	Pro	P	Pyrrolidine-2-carboxylic acid
Serine	Ser	S	2-Amino-3-hydroxypropanoic acid
Threonine	Thr	T	2-Amino-3-hydroxybutanoic acid
Tryptophan[e]	Trp[f]	W	2-Amino-3-(1H-indol-3-yl)propanoic acid
Tyrosine	Tyr	Y	2-Amino-3-(4-hydroxyphenyl)propanoic acid
Valine	Val	V	2-Amino-3-methylbutanoic acid
Unspecified	Xaa	X	
"Asn and/or Asp"	Asx	B	
"Gln and/or Glu"	Glx[g]	Z[g]	

[a] Refers to L, D, or DL forms; only L forms (except the achiral glycine) are coded for by mRNA.

[b] Without stereochemical prefix, specifies L form. Hyphen (after D or DL) may be omitted to avoid confusion with peptide bonds. The one-letter system was devised for computer usage (except for very long sequences) in which punctuation is critical (see Reference in footnote h for systematization).

[c] The fully systematic forms ethanoic, propanoic, butanoic, pentanoic, butanedioic, 3-carbamoylpropanoic, pentanedioic, and 4-carbamoylbutanoic may be called acetic, propionic, butyric, valeric, succinic, succinamic, glutaric, and glutaramic, respectively.

[d] Cystine has no unique symbol. See Table III.

[e] No "e."

[f] Not Try; pronounced "trip."

[g] Also represents precursors of the Glu found in acid hydrolyzates of peptides, such as 4-carboxyglutamate (Gla) and 5-oxoproline (pyroglutamate, Glp, or <Glu).

[h] IUPAC–IUB CBN, "A One-letter Notation for Amino Acid Sequences, 1968," published in *Biochem. J.* **113**, 1 (1969), *Biochemistry* **7**, 2703 (1968), *Biochim. Biophys. Acta* **168**, 6 (1968), *Biochimie* **50**, 1577 (1968), *Pure Appl. Chem.* **31**, 641 (1972), and in other journals.

were modified to avoid ambiguity or to permit greater flexibility in usage. In Table II, I list these and give the original (1964)[1] symbols as well as the latest (1983)[4] revisions, where there are such, and the reasons for any change.

Amino acids are usually linked by peptide bonds, represented by an arrow leading from the carbon atom (or carbonyl group) of the α-carboxyl terminus of one amino acid to the nitrogen atom (or imino group) of the α-amino group of another. If arrows are used, any orientation of the sequence is permissible. However, if the convention of "amino group to the left, carboxyl group to the right" is followed, a simple hyphen (replacing an arrow pointing to the right) suffices. Arrows may be necessary in representation of cyclic peptides (e.g., gramicidin), or in other situations where a single horizontal line of residues cannot be maintained (e.g., branched chains). The residue of H_2N-R-COOH embedded in a peptide written horizontally in a single line thus may appear as -R- (equivalent to → R →) or, if the orientation is HOOC-R-NH_2, as ← R ← (but not as -R-). Terminal NH_2 or OH groups may be represented by their symbols H and OH, if emphasis is required; ordinarily, the absence of a hyphen or arrow suffices.[1,3,4]

Clearly, then, the hyphen or arrow indicates the absence of a hydrogen atom from an α-amino group or the absence of a hydroxyl group from an α-carboxyl group—or both, if it lies between two residues. Inasmuch as each residue thus connected is substituted on the other, the same system may be used to indicate substitutions of atoms or groups other than α-amino acids on the α-amino or α-carboxyl groups (substitutions on other parts of an α-amino acid are dealt with below). In Table III appear a few examples showing how this convention may be applied to peptides and other α-amino acid compounds.

In 1974 Recommendations,[5] a joint effort by the IUPAC Commission on the Nomenclature of Organic Chemistry and CBN, is set out in detail the mode of formation of the semisystematic names (the kind of names favored by biochemists) for substituted derivatives of the α-amino acids. In systematic nomenclature, the carbon atom of the carboxyl carbon is numbered 1. In the semisystematic system, carbon atom 1 is defined as the carbon of the carboxyl group adjacent to the carbon atom carrying the amino group. This removes any ambiguity with regard to aspartic and glutamic acid, or any of the dioic acids listed in Table II. If Greek letters are used, carbon atom 2, which holds both the amino group and carbon

[5] IUPAC–IUB CBN and IUPAC CNOC, "Nomenclature of α-Amino Acids (Recommendations 1974)," published in *Biochemistry* **14**, 1449 (1975), *Biochem. J.* **149**, 1 (1975), and *Eur. J. Biochem.* **53**, 1 (1975).

TABLE II
ANALOGS AND MODIFIED α-AMINO ACIDS[a]

Trivial name	Symbol	Systematic or other names and comments
Allo's		
Allohydroxylysine[b]	aHyl; aLys(5-OH)	*cis*-5-Hydroxy-L-lysine
Allohydroxyproline[b]	—	*cis*-4-Hydroxy-L-proline
Alloisoleucine	aIle	*trans*-L-Isoleucine
Allothreonine	aThr	*cis*-L-Threonine
Nor's		
Norleucine[b]	Nle[b]	Use 2-aminohexanoic acid (Ahx); see below
Norvaline[b]	Nva[b]	Use 2-aminovaleric acid (Avl); see below
Hydroxy's		
5-Hydroxylysine	Lys(5-OH) or $\overset{5}{\underset{\mid}{\text{Lys}}}^{\text{OH}}$	*trans*-5-Hydroxy-L-lysine
4-Hydroxyproline	Pro(4-OH) or $\overset{4}{\underset{\mid}{\text{Pro}}}^{\text{OH}}$	*trans*-4-Hydroxy-L-proline
Homo's		
Homocysteine	Hcy	
Homoserine	Hse	
Cyclic derivatives		
Homoserine lactone	Hse> or Hsl	
Pyroglutamic acid	<Glu or Glp	5-Oxoproline
Higher unbranched amino acids		
β-Alanine	βAla	3-Aminopropanoic acid
2-Aminobutyric acid	Abu	
2-Aminovaleric acid	Avl	Recommended replacement for norvaline (Nva)
2-Aminohexanoic acid	Ahx	Recommended replacement for norleucine (Nle)
6-Aminohexanoic acid	εAhx	Recommended replacement for εAcp (caproic)
2-Aminoadipic acid	Aad	2-Aminohexanedioic acid
3-Aminoadipic acid	βAad	3-Aminohexanedioic acid
2-Aminopimelic acid	Apm	2-Aminoheptanedioic acid
2,3-Diaminopropanoic acid	A₂pr or Dpr	Not Dap
2,4-Diaminobutanoic acid	A₂bu or Dbu	Not Dab
2,4-Diaminovaleric acid	Orn	Ornithine
2,6-Diaminopimelic acid	A₂pm or Dpm	Not Dap
Natural modifications		
4-Carboxyglutamic acid	Gla	2-Amino-4-carboxypentanedioic acid
Cysteic acid	Cya	3-Sulfoalanine

TABLE II (*continued*)

Trivial name	Symbol	Systematic or other names and comments
Non-amino-acid residues that may be associated with amino acids		
Sugar residues	Glc, Gal, Rib, etc.; d (prefix)	See refs. 6, 7
Glycer-ol, -one, -al, -ic acid	Gro, Grn, Gra, Gri	See refs. 6, 7
Neuraminic, sialic acids etc.	GalNAc, Neu, Sia, Mur, etc.	See refs. 6, 7
Nucleic acid constituents	Ade, Ado, etc.; A, C, etc.; P; p	See refs. 6, 7
Alkyl residues (R)	Me, Et, · · · Dec, Und, Dod	See refs. 6, 7
Acyl residues (RCO-)	f^c, Ac, Gc, Pp, Bt, · · · Lau, Pam, Ste	See refs. 6, 7
Ceramide, choline, ethanolamine	Cer, Cho, Etn	See refs. 6, 7
Phosphatidyl, sphingosine (or -oid)	Ptd, Sph (or Spd)	See refs. 6, 7

[a] L assumed unless otherwise indicated by D or DL.
[b] Abandoned in 1983; see "comment" column.
[c] As in formylmethionine, fMet.

atom 1, is the α-carbon. Thus, in lysine, carbon atom 2 is the α-carbon, 3 is the β, 4 is the γ, etc.

The aralkyl and cyclic α-amino acids retain the α, β designation in the side (alanine) chain, as is customary in "conjunctive" (formerly "additive") nomenclature, and use systematic numbering for the rings.[5] Two exceptions were necessary. Proline retains the pyrrolidine numbering (the N is 1; the C holding it and the carboxyl group, not numbered, is 2). Histidine, because of the long-standing confusion as to which ring nitrogen atom is 1, uses π (for *pros* = near) for the nitrogen nearer the alanine chain and τ (for *tele* = far) for the nitrogen farther from the alanine chain.[3–5] The carbon between them is 2; the carbon holding the side chain is 4.

Substitutions

Substitutions on atoms other than the α-amino nitrogen or α-carboxyl carbon atoms may be dealt with in one of two ways. Where it is desired to hold each symbol to three letters (e.g., in aligning homologous sequences

$\begin{array}{c} H_2C\text{---}CH_2 \\ H_2C \diagdown \underset{N}{} \diagup CHCOOH \\ H \end{array}$

Proline

$-CH_2CH(NH_2)COOH$ (phenyl ring, positions 1–6, β α)

Phenylalanine

H N, HO, COOH

cis-4-Hydroxy-L-proline (allo-4-hydroxy-L-proline)*

H N, OH, COOH

trans-4-Hydroxy-L-proline

$HO-$ (phenyl ring, positions 1–6) $-CH_2CH(NH_2)COOH$ (β α)

Tyrosine

$CH_2CH(NH_2)COOH$ (β α) (indole ring, positions 1–7, N–H)

Tryptophan

H N, COOH, OH

cis-4-Hydroxy-D-proline (allo-4-hydroxy-D-proline)*

H N, COOH, HO

trans-4-Hydroxy-D-proline

$\begin{array}{c} HC \!\!=\!\! CCH_2CH(NH_2)COOH \\ \text{(}5\text{)}\ \text{(}4\ \beta\ \alpha\text{)} \\ ^\tau HN \diagdown \underset{C}{} \diagup N^\pi \\ H \\ 2 \end{array}$

Histidine

H N, COOH, OH

cis-3-Hydroxy-L-proline (allo-3-hydroxy-L-proline)*

H N, HO, COOH

trans-3-Hydroxy-L-proline

one below the other) and space is available above or below, a vertical stroke from the amino acid symbol to the substituent indicates the side-chain or branch nature of the addition. Where horizontal space is available and vertical space is not, the substituent may be placed in parenthe-

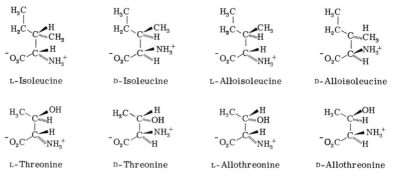

L-Isoleucine   D-Isoleucine   L-Alloisoleucine   D-Alloisoleucine

L-Threonine   D-Threonine   L-Allothreonine   D-Allothreonine

* Former names, no longer recommended.

TABLE III
FORMULATION OF PEPTIDE AND OTHER BONDS

O^1-Methyl aspartate

Asp-OMe

O^4-Methyl aspartate

Asp(OMe) or Asp or OMe

with OMe above Asp and OMe above Asp

N^2-Acetyllysine

Ac-Lys

N^6-Acetyllysine

Lys(Ac) or Lys or Ac

with A above Lys and Lys above Ac

Cystine

Cys or Cys Cys (not Cys-Cys)

with Cys above Cys

O^3-Phosphoserine

Ser(P) or Ser

with P above Ser

N^π-Methylhistidine (*pros*-Methylhistidine)[a]

His(πMe) or His

with Me (π) above His

Diiodotyrosine

Tyr(I$_2$)

Isoglutamine (glutamic 1-amide)

Glu-NH$_2$ [glutamine can be Glu(NH$_2$)]

N^2-(α-Glutamyl)lysine

Glu-Lys

N^6-(α-Glutamyl)lysine

Glu⌐ Lys

N^2-(γ-Glutamyl)lysine

Glu-Lys or Glu(-Lys)

N^6-(γ-Glutamyl)lysine

Glu Lys or Glu

with Lys above Glu

Thyroliberin (pyroglutamylhistidylprolinamide)

Glp-His-Pro-NH$_2$

Glutathione (*N*-γ-glutamyl-cysteinylglycine)

Glu or Glu(-Cys-Gly)

with Cys-Gly bracket below Glu

Vasopressin

Cys-Tyr-Phe-Gln-Asn-Cys-Pro-Arg-Gly-NH$_2$

Gramicidin S

Val-Orn-Leu-DPhe-Pro-Val-Orn-Leu-DPhe-Pro

or

cyclo(-Val-Orn-Leu-DPhe-Pro-Val-Orn-Leu-DPhe-Pro-)

or

⌐→ Val → Orn → Leu → DPhe → Pro ⌐
└ Pro ← DPhe ← Leu ← Orn ← Val ◄┘

Cyclic ester of threonylglycyl-glycylglycine

Thr-Gly-Gly-Gly

Sequence with part in unknown order

Ala-Lys-(Ala,Gly$_3$,Val$_2$)-Glu-Val

---

[a] Histidine is numbered (π = *pros*; τ = *tele*)[3-5]:

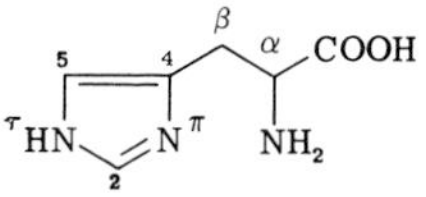

TABLE IV

MODIFICATIONS OF THE "COMMON 20" α-AMINO ACIDS FOUND IN PROTEIN THAT DO NOT INVOLVE A SECOND α-AMINO ACID[a]

A. Not Involving a Prosthetic Group

Semisystematic names	Comments or other names	Symbols[b]
L-Alanine		Ala
N-Acetyl-		Ac-Ala
N-Methyl-		Me-Ala
N,N,N-Trimethyl-	(-alaninium)	Me$_3$Ala$^+$
Didehydro-	(See cysteine, 2-Dehydro-3-demercapto-)	
L-Arginine		Arg
N$^\omega$-(ADP-ribosyl)-		Arg(ωADPR)
N$^\omega$,N$^\omega$-Dimethyl-		Arg(ω,ω-Me$_2$); m$_2^\omega$Arg
N$^\omega$,N$^{\omega'}$-Dimethyl-		Arg(ω,ω'-Me$_2$); m'mArg
N$^\omega$-Methyl-		Arg(ωMe); m$^\omega$Arg
N$^\omega$-Phospho-		Arg(ωP); p$^\omega$Arg
C^G-Deimino-C^G-oxo-	(Citrulline)	Cit
N^5-Deamidino-	(Ornithine)	Orn
L-Aspartic acid		Asp(Asx)[c]
N-Acetyl-		Ac-Asp
1-Amido-	(Isoasparagine)	Asp-NH$_2$
3-Hydroxy-		Asp(3OH)
O^4-Methyl-	(ester) ("β-methyl ester")	Asp(O^{4}Me); o^{4}Asp
L-Asparagine		Asn(Asx)[c]
N^4-Glycosyl[d]		Asn(4Sac)[d]
Iso-	(See aspartic acid, 1-amido)	
L-Cysteine		Cys
2-Dehydro-3-demer-capto-	(Didehydroalanine)	ΔCys
S-Galactosyl-		Cys(Gal)
S-Glucosyl-		Cys(Glc)
S-(γ-Glutamyl)-		Cys(γGlu)
S-(1-Glycero)-[e]		Cys(Gro)
S-Mercapto-		Cys(SSH)
Seleno-		[Se]Cys
L-Glutamic acid		Glu(Glx)[c]
O^5-(ADP-ribosyl)$_n$-		Glu(ADPR)$_n$
1-Amido-	(Isoglutamine) (Glutamide)	Glu-NH$_2$
4-Carboxy-		Gla
O^5-Methyl-		Glu(γMe)
Pyro-	(5-Oxoproline)	pGlu; <Glu

TABLE IV (*continued*)

Semisystematic names	Comments or other names	Symbols[b]
L-Glutamine		Gln(Glx)[c]
1-Amido-	(Glutaminamide)	Gln-NH$_2$
Glycine		Gly
N-Acetyl-		Ac-Gly
Amido-		Gly-NH$_2$
N-Formyl-		HCO-Gly
N-Glucuronyl-		GlcU-Gly
L-Histidine		His
1-Amido-	(Histidinamide)	His-NH$_2$
2-[(3-Carbamoyl-3-tri-methylammonio)-propyl]-	(Diphthamide)	Diph
2-[(3-Carbamoyl-3-tri-methylammonio)-propyl]-τ-(ADP-ribosyl)-		Diph(ADPR)
4-Iodo-		i^{4}His; His(4I)
π-Methyl-		m$^\pi$His; His(πMe)
π(τ)-Phospho-		p$^{\pi(\tau)}$His; His(π(τ)P)
L-Lysine		Lys
N^6-Acetyl-		Lys(6Ac)
N^6-(4-Amino-2-hydroxy-butyl)-	(Hypusine)	Hpu
N^6,N^6-Dimethyl-		m$_2^6$Lys; Lys(Me$_2$)
N^6-Fructosyl-		Lys(Fru)
5-Glycosido- (and oligoglycosido compounds)		Lys(Sac)[d]
5-Hydroxy-		o^{5}Lys; Lys(5OH)
5-Hydroxy-N^6,N^6,N^6-trimethyl-	(-inium)	o^{5}m$_3^6$Lys$^+$; Lys(6Me$_3$5OH)$^+$
N^6-Methyl-		m^{6}Lys; Lys(Me)
N^6-Phospho-		Lys(P)
N^6,N^6,N^6-Trimethyl-	(-inium)	m$_3^6$Lys$^+$; Lys(Me$_3$)$^+$
L-Methionine		Met
N-Acetyl-		Ac-Met
1-Amido-	(Methioninamide)	Met-NH$_2$
N-Formyl-		fMet
N-Methyl		Me-Met
L-Phenylalanine		Phe
1-Amido-	(Phenylalaninamide)	Phe-NH$_2$
3,4-Dihydroxy-	(Dopa)	Dopa
β-Glucosido-	(Glucosyloxy)	Phe(βGlcO)
β-Hydroxy-		o$^\beta$Phe; Phe(βOH)

(*continued*)

TABLE IV (*continued*)

Semisystematic names	Comments or other names	Symbols[b]
L-Proline		Pro
Amido-	(Prolinamide)	Pro-NH$_2$
4-Arabinosido-	(O^4-Arabinosyl-4-hydroxy-)	Pro(4AraO)
3,4-Dihydroxy-		Pro(OH)$_2$; o$_2$Pro
4-Galactosido-		Pro(4GalO)
N,N-Dimethyl-	(-inium) (stachydrin)	Me$_2$Pro$^+$; Sta
3-Hydroxy-		Pro(3OH); o^{3}Pro
4-Hydroxy-		Pro(4OH); o^{4}Pro
5-Oxo-	(See pyroglutamic acid)	o^{5}Pro
L-Serine		Ser
O^3-Glycosyl-[d]		Ser(Sac)[d]
O^3-Phospho-	(Root of many diesters)	Ser(P)
O^3-(Glycosylphospho)-		Ser(PSac)[d]
L-Threonine		Thr
O^3-Glycosyl-[d]	(Many)	Thr(Sac)[d]
O^3-Phospho-		Thr(P)
L-Tyrosine		Tyr
O^4-Adenylyl-		Tyr(pA); Tyr(AMP)
3,5-Dihalo-[e]		Tyr(Hal$_2$)[e]
β-Glycosido-[d]	(-Glycosyloxy-)	Tyr(βSac)[d]
3-Halo-[e]		Tyr(Hal)[e]
O^4-Phospho-		Tyr(P); p^{4}Tyr
O^4-Sulfo-		Tyr(S); s^{4}Tyr
O^4-Uridylyl-		Tyr(pU); Tyr(UMP)
3-Hydroxy-	(See Dihydroxyphenylalanine, dopa)	

B. Involving a Prosthetic Group

Names[f]	Symbol[g]
3β-(S-Cysteinyl)-3α,3β-dihydrophycocyanobilin; S-(3α,3β-dihydrophycocyanobilin-3β-yl)cysteine	Cys(3βPhy)
8α-(S-Cysteinyl)-8α-hydroxyriboflavin; S-(8α-hydroxyriboflavin-8α-yl)cysteine	Cys(o^{8}Flv)
6-(S-Cysteinyl)riboflavin; S-(6-riboflavinyl)cysteine	Cys(6Flv)
8α-(S-Cysteinyl)riboflavin; S-(8α-riboflavinyl)cysteine	Cys(8αFlv)
8α-($\pi(\tau)$-Histidinyl)riboflavin; $\pi(\tau)$-(8α-riboflavinyl)histidine	$\pi(\tau)$His(8αFlv)
(N^6-Lysino)biotin; N^6-(Biotinyl)lysine: biocytin	Bcy

TABLE IV (*continued*)

Semisystematic names	Comments or other names	Symbols[b]
(N^6-Lysino)lipoic acid; N^6-(lipoyl)lysine		Lys(Lip)
(N^6-Lysino)-5'-phosphopyridoxylidene; N^6-(5'-phosphopyridoxylidene)lysine[h]		Lys(=PPxd) or (PPxd=)Lys
(N^6-Lysino)retinylidene; N^6-retinylidenelysine[h]		Lys(Ret)
ω-[(O^3-Serinyl)phospho]pantetheine; O^3-[(ω-pantetheinyl)phospho]serine		Ser(PPan)
8α-(O^4-Tyrosinyl)riboflavin; O^4-(8α-riboflavinyl)tyrosine		Tyr(8αFlv)

[a] The entries in this and associated tables are taken from reviews by F. Wold [*Annu. Rev. Biochem.* **50**, 783 (1981)] and R. Uy and F. Wold [*Science* **198**, 890 (1977)], in which references to the primary literature can be found.

[b] The symbols given are either those recommended in the IUPAC/IUB CBN documents[3-6] or devised by OBN[7] in accordance with the CBN principles.

[c] The symbols Asx (and Glx) signify either Asp + Asn (Glu + Gln) or—in the case of the latter—any residue that yields Glu on acid hydrolysis (thus including pyroglutamic acid, Glp, and 4-carboxyglutamic acid, Gla).

[d] When the nature of the sugar is known, its name and symbol should replace glycosyl (or glycosido) and Sac (for saccharido).

[e] Halo (symbol Hal) for halogen should be replaced by the appropriate name (and symbol), when known.

[f] Given in terms of the prosthetic group and amino acid as principal moieties.

[g] With the exception of Pxd, all substituent symbols have been invented for this table by the author.

[h] The aldimine.

ses immediately following the main symbol (and without hyphenation, which, as noted above, would indicate a carboxylic ester). This method is used in representing short branches in carbohydrate chains.[6] In either case, locants may be added to indicate the precise point of substitution. Some examples (cystine, phosphoserine, diiodotyrosine) appear in Table III.[3,4]

Groups substituted for hydrogen or carboxyl hydroxyl may be indicated by their formulas, or by symbols constructed according to the same principles that gave rise to the symbols in Tables I, II, and III—but

[6] IUPAC–IUB CBN, "The Nomenclature of Lipids (Recommendations 1976)," published in *Lipids* **12**, 455 (1977), *Biochem. J.* **171**, 21 (1978), *Eur. J. Biochem.* **79**, 11 (1977), *Hoppe-Seyler's Z. Physiol. Chem.* **358**, 617 (1977), and elsewhere.

TABLE V

MODIFICATIONS OF "THE COMMON 20" α-AMINO ACIDS FOUND IN PROTEINS THAT INVOLVE A SECOND α-AMINO ACID[a]

A. Cross-links Involving Two of "the 20" α-Amino Acids[a]

Names	Symbols[b]
π-(3-Alaninyl)histidine; 3-(π-histidino)alanine	3Ala(πHis)
τ-(3-Alaninyl)histidine; 3-(τ-histidino)alanine	3Ala(τHis)
S-(3-Alaninyl)cysteine; 3-(S-cysteinyl)alanine	3Ala(SCys)
N^6-(β-Aspartyl)lysine; 4-(N^6-lysino)aspartic acid	βAsp(6Lys)
S-(S-Cysteinyl)cysteine; cystine	(Cys)₂
5-(S-Cysteinyl)glutamate; S-(γ-glutamyl)cysteine	SCys(γGlu)
2-(S-Cysteinyl)histidine; S-(2-histidinyl)cysteine	SCys(2His)
N^6-(S-Cysteinyl)lysine; S-(N^6-lysino)cysteine	SCys(6Lys)
N^6-(γ-Glutamyl)lysine; 5-(N^6-lysino)glutamate	γGlu(6Lys)
O^3-(γ-Glutamyl)serine; 6-(O^3-serinyl)glutamate	γGlu(3Ser)
3-(O^4-Tyrosinyl)tyrosine; isodityrosine	i(Tyr)₂
3-(3-Tyrosinyl)tyrosine; 3,3'-bityrosine	(Tyr)₂
3-[5-(3-Tyrosinyl)-3-tyrosinyl]tyrosine; 3,3':5',3"-tertyrosine	(Tyr)₃

B. Di-(α-Amino Acids) in Which at Least One Is Not a Bona Fide Member of "the 20"

Names	Symbols
1. Connected by nitrogen (the lysinonorleucines)[c]	
Lysinonorleucine[c]	Lys(6Nle)[d]; HN(6Nle)₂
Bis(6-norleucinyl)amine[c]	o⁵Lys(6Nle)
5-Hydroxy-	o⁵Lys(o⁵6Nle);
5,5'-Dihydroxy-	HN(o⁵6Nle)₂

N^6:6'-Didehydro-; 6-(Δ^6-lysino)norleucine (an aldimine) Δ^6Lys(6Nle)

N^6:6'-Didehydro-5-hydroxy- Δ^6Lys(o^{5}6Nle)

N^6:6'-Didehydro-5'-hydroxy- Δ^6o^{5}Lys(6Nle)

N^6:6'-Didehydro-5,5'-dihydroxy- Δ^6o^{5}Lys(o^{5}6Nle)

2. Connected by carbon–carbon bonds (the diallysines)[e]

Name	Structure	Trivial name	Abbreviation
Di(allysine)	$HOCHCH_2$—$(CH_2)_2CH(NH_2)COOH$ / $OCHCH$—$(CH_2)_2CH(NH_2)COOH$	Allysine aldol[f]	(All)$_2$
Didehydro-(or Δ^5-)	$CHCH_2$— / $\|$ / $OCHC$—		Δ(All)$_2$
Hydroxy-	$HOCHCHOH$ / $OCHCH$— $HOCHCH_2$ / $OCHCOH$	Syndesine[f]	o(All)$_2$; Syn
Dihydroxy-	$HOCHCHOH$ / $HOCH_2CH$— $HOCHCH_2$ / $HOCH_2CHOH$—	Syndesinol / Reduced syndesine / Syndesine alcohol	oH$_2$(All)$_2$; SynOH
Merodesmosine	CH=CH / CH_2—CH / $NH(CH_2)_2$ or structure	A ter(α-amino acid)[f]	Δ(All)$_2$Lys; Mer
Desmosine	CH_2—R / R R / N⁺ / (CH$_2$)$_2$R (pyridinium structure)	A tetrakis(α-amino acid) in 1,3,4,5-substituted-pyridinium form	(ΔAll)$_3$Lys; Des

(continued)

TABLE V (continued)

Names		Symbols[b]
Isodesmosine	A tetrakis(α-amino acid) in 1,2,3,5-substituted pyridinium form	(ΔAll)₃Lys; iDes

R as shown with ring structure, $R = (CH_2)_2CH(NH_2)COOH$

[a] All these di(α-amino acid)s are connected by bonds that do not involve the α-amino or α-carboxyl group of either component, hence each can be a member of a conventional polypeptide. Since neither component is the more important, an A-B name has no seniority over a B-A name. Hence, both are given.

[b] The symbols that constitute the abbreviation of each di(α-amino acid) can be expressed (see text) in several ways. Taking the first as an example:

$$3Ala(\pi His); \pi His(3Ala); His\ Ala; \pi His; 3Ala$$

Horizontal bonds (hyphens) are reserved for α-NH₂ and α-COOH (peptide) links, vertical strokes for bonds originating in side-chain groups [–SH, –OH, –NH, –CH, ε-NH₂, β(γ)-COOH].

[c] More complete names are: 6-(N⁶-lysino)norleucine and N⁶-(6-norleucinyl)lysine; in the newly recommended nomenclature, 2-amino-6-(N⁶-lysino)hexanoic acid and N⁶-(5-amino-5-carboxypentyl)lysine.

[d] The order can be reversed—e.g., 6Nle(Lys) is equivalent to Lys(6Nle); Lys can be expanded to N⁶Lys.

[e] Allysine (All) is 6-oxonorleucine or 2-amino-6-oxohexanoic acid.

[f] Histidine derivatives of allysine aldol, syndesine, and merodesmosine are known.

preferably not by nonevocative abbreviations. Symbols for substituents recommended by CBN or OBN appear in Tables III, IV, and V.[3,4,7]

Modifications of "the Common 20" α-Amino Acids That Have Been Found in Proteins

These have been grouped in the following way. Table IV includes modifications that do not involve a second α-amino acid, and do not, therefore, lead to cross-links between polypeptide chains. Table V includes modifications that involve a second α-amino acid. (Part A) The second amino acid is also one of "the 20," and thus, as in the case of cysteine, the modification can constitute a cross-link between two polypeptides. (Part B) The second α-amino acid is not one of "the 20," and the link is (1) via sulfur (e.g., cystathionine), (2) via a single nitrogen atom (the lysinonorleucines), or (3) carbon–carbon (syndesines) or carbon–carbon plus nitrogen–carbon (desmosines). Each entry is given a proper semisystematic name (i.e., a name based on the accepted trivial names of the components), which may differ from that given by the discoverer, but that does include the name of the (presumed) precursor α-amino acid, and an abbreviation based upon the symbols recommended by IUPAC/IUB CBN (or, if such do not exist, created according to CBN principles).

[7] NAS-NRC Office of Biochemical Nomenclature (OBN) "Symbols and Abbreviations Recommended by the IUPAC–IUB Commission on Biochemical Nomenclature (CBN)": a compilation ("OBN-6," 1975) compiled by OBN and distributed *gratis* to requestors. Most are reproduced in Instructions to Authors in *J. Biol. Chem.* **259,** 7 (1984) and in other journals.

[2] Ion-Exchange Chromatography of Some Amino Acid Derivatives Found in Proteins

By CHRISTOPHER C. Q. CHIN

By far the most common and obvious manner in which new amino acids have been recognized and eventually quantified is by ion-exchange chromatography on automatic amino acid analyzers. The common 17 (or 18 if Trp is protected by special hydrolysis conditions) amino acids that survive the standard acid hydrolysis conditions are obtained from the ion echange column in characteristic elution positions, and can be quantified quite accurately relative to reference standards through the color yield in the automated, highly reproducible ninhydrin reaction system. At this time a large number of different amino acid analyzers are in use, and with different resins, elution programs, and color development procedures, the amino acid elution patterns and color yields are not directly comparable from instrument to instrument. However, the general sequence of elution (acidic amino acids first, followed by aliphatic with increasing chain length, then aromatic and finally basic amino acids) seems to be the same for virtually all the cation exchange systems now in use. Thus it should be possible after observing a new and unique peak on one's own analyzer to at least get some idea about its identity by comparing its elution position to those of other amino acids and amino acid derivatives on another analyzer. The purpose of this chapter is to provide that type of first step information. The elution positions of a number of amino acids and amino acid derivatives known to be present in proteins have been determined on a standard amino acid analyzer and are recorded relative to the position of the common 17 amino acids present in acid hydrolysates of proteins. The elution positions of oxidation products (cysteic acid and methionine sulfone), an aminosugar (glucosamine), and an internal reference compound (norleucine) are also included.

An attempt has also been made to record the approximate elution positions of a number of derivatives discussed in other chapters in these volumes. It should be emphasized that the data for this latter group have been transferred to our elution pattern from widely different systems on the basis of their relative positions ("2 min after Gly," "between Tyr and Phe") on the other chromatograms, and that this estimated elution position on our analyzer could be off by several minutes, especially in the region of the aromatic and basic amino acids.

Procedures. The analyzer used was the Beckman Model 119 amino acid analyzer with a System AA analyzer/integrator. The elution programs followed a single column (56 × 0.9 cm, Beckman spherical resin, Type W-1) procedure, eluting with 3 buffers (A, B, C) at 56°. The buffers were prepared as follows:

Buffer A (run time 0–34 min)
    Na-Citrate · 2H$_2$O, 78.4 g
    Conc HCl, 45.0 ml
    Thioglycol, 10.0 ml
    Caprylic acid, 0.4 ml
    Water (high purity) to 4 liters, final pH 3.42 (Na$^+$ conc 0.2 M)
Buffer B (run time 34–84 min)
    Na-Citrate · 2H$_2$O, 78.4 g
    NaCl, 46.8
    Conc HCl, 28.0 ml

Thioglycol, 10.0 ml
Caprylic acid, 0.4
Water to 4 liters, final pH 4.2 (Na^+ conc 0.4 M)
Buffer C (run time 84–152 min)
Na-Citrate · $2H_2O$, 78.4 g
NaCl, 350.0 g
HCl, 1.0 ml
Caprylic acid, 0.4 ml
Water to 4 liters, final pH 6.24 (Na^+ conc 1.7 M)
Regeneration with 0.2 M NaOH (run time 152–167 min)
Shut-down at 167 min

For amino acids expected to elute after Arg, the run time for buffer C was extended by 30 min, the NaOH regeneration being started at 182 min. Most of the amino acids eluting this late give broad and rather poorly defined peaks. The program presented has been modified to optimize separation of amino sugars and is not ideally suited for the aromatic amino acid derivatives eluting late. To optimize the program for these amino acids, the amount of NaCl in buffer C should be lowered to one-half (175 g) of the value given.

AN ALPHABETICAL LISTING OF THE AMINO ACID DERIVATIVES IN FIG. 1[a]

10	Asparagine* (Asp)	24	3-(π-Histidino)alanine[15]
17	N^ε(β-Aspartyl)lysine* (Asp+Lys)[1]	23	3-(τ-Histidino)alanine[15]
20	Bityrosine[2]	22	δ-Hydroxylysine
1	β-Carboxyaspartate* (Asp)[3]	7	4-Hydroxyproline
2	γ-Carboxyglutamate* (Glu)[4,5]	31	Hypusine[16]
25	3-Chlorotyrosine[6]	9	Lanthionine[17]
11	Citrulline* (Orn)[7]	12	O^4-Methylaspartate* (Asp)[18]
30	3,5-Dibromotyrosine[6]	13	O^5-Methylglutamate* (Glu)[19,20]
16	3,4-Dihydroxyphenylalanine[8]	28	π-Methylhistidine[9–11]
33	3,5-Diiodotyrosine[6]	27	N^ε-Methyllysine[9–11]
29	N^G,N^G-Dimethylarginine[9–11]	26	Ornithine
19	Diphthine[12]	3	O-Phosphoserine* (Ser)[21]
21	(Galactosamine*)	4	O-Phosphothreonine* (Thr)[21]
8	Glutamine* (Glu)	5	O-Phosphotyrosine* (Tyr)[21]
18	N^ε-(γ-Glutamyl)lysine* (Glu + Lys)[13]	14	Selenocysteine*[22,23]
6	S-(Histidin-2-yl)cysteine[14]	15	Selenomethionine*[24]
		32	Tryptophan*

[a] The elution position of each compound is indicated by the number in front of each compound corresponding to the numbered arrows in Fig. 1. Compounds that are completely or extensively destroyed by standard acid hydrolysis procedures are indicated by an asterisk and, when known, the major hydrolysis product(s) is given in parentheses after the compound.

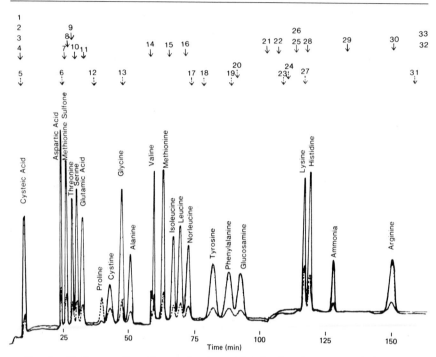

FIG. 1.   Elution position of several amino acid derivatives known to exist in proteins. The reference elution pattern represents a standard amino acid mixture of the 17 acid-stable amino acids, supplemented with cysteic acid, methionine sulfone, norleucine, and glucosamine. The solid arrows indicate the elution positions of amino acid derivatives that have been analyzed in this laboratory with the same standard analytical procedure. The dotted arrows in the lower tier represent approximate elution positions deduced from data in the literature. The compounds coded by the numbers in the figure are identified in the table. Compounds 32 (Trp) and 33 (diiodotyrosine) eluted 28 and 100 min after Arg, respectively.

To reassess the acid stability of the amino acid derivatives used in this survey, they were subjected to standard acid hydrolysis conditions (6 N HCl, 110°, 21 hr), and the hydrolyzed samples were then compared to the unhydrolyzed starting materials. In order to detect the presence of amino sugars in the acid hydrosylate of glycoproteins, we have found that milder hydrolysis conditions are required for optimal yield (2 N HCl, 110°, 4 hr). The elution positions of glucosamine and galactosamine are included in the data.

The results are summarized in Fig. 1 and the table.[1-24] Figure 1 shows as background the elution diagram of our standard amino acid (+ glucos-

[1] H. Klostermeyer, this series, Vol. 107 [14].
[2] R. Amado, R. Aschbach, and H. Neukom, this series, Vol. 107 [24].

amine) mixture. Superimposed on this standard elution pattern is a set of solid arrows representing the derivatives surveyed in this study. The second set of dotted arrows represents the derivatives reported from other laboratories and included here to indicate approximate elution positions on this particular chromatographic system. The compounds are identified in the table.

It should be noted that a number of special systems have been developed for optimal separation and quantification of different types of derivatives. As examples it can be mentioned that halogenated Tyr, His, and Trp derivatives can be separated by gel filtration,[6] methylated amino acids require special modified ion exchange systems,[10] or HPLC,[11] and the poorly resolved acidic derivatives (phospho-, sulfo-, and carboxy derivatives) can be separated by anion exchange chromatography.[3,4,25] The reader is referred to individual chapters for several such special procedures.

[3] T. H. Koch, M. R. Christy, R. M. Barkley, R. Sluski, D. Bohemier, J. J. Van Buskirk, and W. M. Kirsch, this series, Vol. 107 [38].
[4] G. Nelsestuen, this series, Vol. 107 [32].
[5] P. A. Price, this series, Vol. 91, p. 13.
[6] S. Hunt, this series, Vol. 107 [27].
[7] J. Rothnagel and G. E. Rogers, this series, Vol. 107 [44].
[8] J. H. Waite and C. V. Benedict, this series, Vol. 107 [26].
[9] W. K. Paik, this volume [24].
[10] W. K. Paik and S. Kim, Adv. Enzymol. **42**, 227 (1927).
[11] M. Elzinga and N. Alonzo, this series, Vol. 91, p. 8.
[12] J. W. Bodley, P. Dunlap, and B. G. VanNess, this volume [38].
[13] A. G. Loewy, this series, Vol. 107 [13].
[14] K. Lerch, this volume [34].
[15] R. Sass and M. E. Marsh, this volume [33].
[16] H. L. Cooper, M. H. Park, and J. E. Folk, this volume [32].
[17] P. M. Steinert, this series, Vol. 107.
[18] S. Clarke, P. N. McFadden, C. M. O'Connor, and L. L. Lou, this volume [31].
[19] J. Stock, S. Clarke, and D. E. Koshland, Jr., this volume [29].
[20] S. J. Kleene, M. L. Toews, and J. J. Adler, J. Biol. Chem. **252**, 3214 (1977).
[21] T. M. Martensen, this series, Vol. 107 [1].
[22] T. C. Stadtman, this series, Vol. 107 [39].
[23] H. E. Ganther, R. J. Kraus, and S. J. Foster, this series, Vol. 107 [40].
[24] M. X. Sliwkowski, this series, Vol. 107 [43].
[25] R. G. Spiro, this series, Vol. 28, p. 3.

[3] High-Performance Liquid Chromatography Probes for
Posttranslationally Modified Amino Acids

By KENNETH A. WALSH and TATSURU SASAGAWA

It has become evident in recent years that the vast majority of proteins are covalently modified during their cellular lifetime and that such modifications can influence their activity, localization, assembly or regulation.[1] These interpretations of genomic instructions involve modulation by proteolytic editing or by the covalent attachment of any of a wide variety of prosthetic groups to the side chains of specific amino acid residues. Such covalently modified amino acid residues are not necessarily recognized, even during analysis of complete amino acid sequences since the majority of the naturally occurring derivatives are acid-labile and thus undetected by amino acid analysis after conventional acid hydrolysis. Unless one is alerted by prior knowledge of cofactor involvement in function, or of unusual spectral characteristics, or by radiolabeling experiments, the presence and nature of a posttranslational modification of a protein structure can be missed. No generally applicable method for detection has been described, but several modifications have been found by continual attention to anomalous behavior of peptides after enzymatic fragmentation or of phenylthiohydantoins after Edman degradations. The flexibility and speed of HPLC methodology are playing an increasingly significant role in these identifications.

Identification as Phenylthiohydantoins

Edman degradation of proteins and their peptide fragments provides an opportunity to identify the amino acid residues sequentially.[2] Any chemically modified residue which survives repetitive exposure to the extremes of the degradation conditions (pH 9.0; anhydrous perfluoroorganic acids at 57°; 1 N HCl at 80°) should be identifiable as its phenylthiohydantoin, provided that the nature of the modification does not interfere with the degradation and that the derivative is extracted from the reaction mixture by conventional procedures.

In several instances the phenylthiohydantoins are not easily extracted.

---

[1] F. Wold, *Annu. Rev. Biochem.* **50,** 783 (1981).
[2] P. Edman, *Carlsberg Res. Commun.* **42,** 1 (1977).

METHODS IN ENZYMOLOGY, VOL. 106

For example, those of oligosaccharide-linked residues are insoluble in 1-chlorobutane, the organic solvent usually used as an extractant. The ^{32}P-labeled O-phosphoserine derivative β-eliminates to yield poorly extracted inorganic [^{32}P]phosphate.[3] The remainder of the serine is extracted as the phenylthiohydantoin of dehydroalanine, but this can also be derived from unmodified serine or cysteine. γ-Carboxyglutamyl residues slowly revert to glutamyl residues during the acidic phase of the degradation,[4] and residual γ-carboxyglutamyl derivatives require a more polar extractant such as 2-butanol.[5] Cross-links between residues within a fragment or between fragments complicate the release of the phenylthiohydantoin derivatives.

Modification of the α-amino group of a protein or of a fragment usually precludes Edman degradation. The two most common blocks of this nature are N^{α}-acetylation (which can be prevented during *in vitro* translation[6]) and pyrollidone carboxyl formation from amino-terminal glutamine (which can sometimes be removed enzymatically[7]). Occasionally N^{α}-methylation has been observed,[8] but in most cases this does not interfere with the Edman degradation.

After release and extraction of the product of Edman degradation, more than one method should be used to identify that product in order to optimize the probability of distinguishing a modified amino acid from a conventional one. The choice of a pair of methods will depend upon the experimental resources of a particular laboratory, but it is important that the methods differ from each other in principle as much as possible. Even with a highly resolving chromatographic system a modified aminoacyl phenylthiohydantoin may happen to coincide with and be mistaken for a conventional aminoacyl phenylthiohydantoin in any single system. This chance is greatly reduced if the alternate identification technique is based on a different property. Since HPLC methods are rapid, offer high resolution, and are nondestructive, they are the current methods of choice. Although not offering ideal complementary systems, some laboratories use two different HPLC analyses employing different solvents (e.g., methanol[9] vs acetonitrile[10]), different columns (e.g., octadecyl silica,

[3] C. G. Proud, D. B. Rylatt, S. J. Yeaman, and P. Cohen, *FEBS Lett.* **80**, 435 (1977).
[4] D. L. Enfield, L. H. Ericsson, K. Fujikawa, K. A. Walsh, H. Neurath, and K. Titani, *Biochemistry* **19**, 659 (1980).
[5] P. Fernlund and J. Stenflo, *J. Biol. Chem.* **257**, 12170 (1982).
[6] R. D. Palmiter, *J. Biol. Chem.* **252**, 8781 (1977).
[7] R. F. Doolittle, this series, Vol. 25, p. 231.
[8] M. A. Hermodson, K. C. S. Chen, and T. M. Buchanan, *Biochemistry* **17**, 442 (1978).
[9] P. J. Bridgen, G. A. M. Cross, and J. Bridgen, *Nature (London)* **263**, 613 (1976).
[10] C. L. Zimmerman, E. Appella, and J. J. Pisano, *Anal. Biochem.* **77**, 569 (1977).

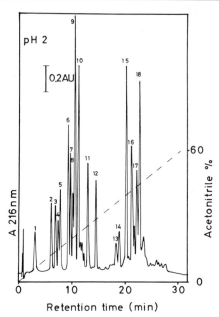

Retention time (min)

FIG. 1. Example of an experiment to calibrate an HPLC column. A tryptic digest of sperm whale myoglobin (30 nmol) was applied to a Hamilton PRP-1 column (4 × 150 mm) and eluted with the indicated gradient at 2 ml/min (the mobile phase is 0.1% aqueous trifluoroacetic acid). The various numbered fractions were collected, identified by composition and/or sequence analysis, and their mobilities recorded (Fig. 2) to calibrate this (or another) column. Peaks were identified as 1, YK; 2, HLK; 3, DIAAK + FK; 4, ASEDLKK; 5, TEAEMK + FDR; 6, SHPETLEK; 7, HKIPIK; 8, LFK; 9, ELGYOG + GHHWEAELKPLAQSHATK; 10, HPGNFGADAQGAMNK; 11, VEADVAGHGQDILIR; 12, ALELFR; 13, HGVTVLTALGAILKK; 14, HGVTVLTALGAILK; 15, VLSE-GEWQLVLHVWAK; 16, YLEFISEAIIHVLHSR; 17, not identified; 18, VLSE-GEWQLVLHVWAKVEADVAGHGQDILIR.

phenylalkylsilica,[11] or cyanopropyl silica[12]) or different programs.[13–16] Back hydrolysis of the phenylthiohydantoin to an amino acid,[17] although not an ideal primary identification technique because a modifying group may be lost, is a satisfactory approach to confirming the chromatographic

[11] L. E. Henderson, T. D. Copeland, and S. Oroszlan, *Anal. Biochem.* **102**, 1 (1980).
[12] M. W. Hunkapiller and L. E. Hood, *Science* **219**, 650 (1983).
[13] J. E. Harris, D. Robinson, and A. J. Johnson, *Anal. Biochem.* **105**, 239 (1980).
[14] S. D. Black and M. J. Coon, *Anal. Biochem.* **120**, 281 (1982).
[15] D. Hawke, P. Yuan, and J. E. Shively, *Anal. Biochem.* **120**, 302 (1982).
[16] M. N. Margolies and A. Brauer, *J. Chromatogr.* **148**, 429 (1978).
[17] J. Bridgen, A. P. Graffeo, B. L. Karger, and M. D. Waterfield, *in* "Instrumentation in Amino Acid Sequence Analysis" (R. N. Perham, ed.), p. 111. Academic Press, New York, 1975.

identification of a phenylthiohydantoin. Methods of thin-layer[18] or gas-liquid chromatography[19] can also be confirmatory and are convenient for large-scale applications.

Some modified amino acids are more easily identified after taking advantage of intrinsic chemical properties to generate derivatives that are more readily recognized, e.g., by converting O-phosphoserine into β-methylaminoalanine[20] prior to Edman degradation. The ultimate method for both confirmation and identification of an anomalous amino acid residue appears to be mass spectrometry of the phenylthiohydantoin after collection from HPLC,[18,21] or of small peptides derived from the parent protein.[22]

Anomalous Peptide Mobility on HPLC May Signal a Modification

Reversed-phase HPLC now serves as a powerful method of isolating peptides from enzymatic digests of proteins.[23] If a peptide contains less than 40 amino acid residues, it is possible to predict its retention time from its amino acid composition.[24-26] These estimations are based on the relative hydrophobicity of the different amino acids, each expressed as a retention constant, which in summation leads to a retention time for the peptide on a calibrated column (Fig. 2). A routine search for putative chemical modifications looks for discrepancies between observed retention times and those predicted either by the sequence or by the amino acid composition after acid hydrolysis.

Table I lists 12 cases in which chemical modification of a single amino acid residue in a short peptide has been reported to cause a change in HPLC retention time. In each case the mobility change is in a direction which is in qualitative accord with the difference in hydrophobicity associated with the modification.[27-29] In several cases, *in vivo* modifications

[18] B. Wittman-Liebold, A. W. Geissler, and E. Marzinzig, *J. Supramol. Struct.* **3,** 426 (1975).

[19] J. J. Pisano and T. J. Bronzert, *Anal. Biochem.* **45,** 43 (1972).

[20] W. D. Annan, W. Manson, and J. A. Nimmo, *Anal. Biochem.* **121,** 62 (1982).

[21] T. Fairwell and H. B. Brewer, Jr., *Anal. Biochem.* **107,** 140 (1980).

[22] S. A. Carr and K. Biemann, this volume [5].

[23] K. A. Walsh, L. H. Ericsson, D. C. Parmelee, and K. Titani, *Annu. Rev. Biochem.* **50,** 261 (1981).

[24] T. Sasagawa, T. Okuyama, and D. C. Teller, *J. Chromatogr.* **240,** 329 (1982).

[25] C. A. Browne, H. P. J. Bennett, and S. Solomon, *Anal. Biochem.* **124,** 201 (1982).

[26] J. L. Meek and Z. L. Rossetti, *J. Chromatogr.* **211,** 15 (1981).

[27] Recently we have induced such mobility changes purposefully *after a first chromatography* to specifically alter the hydrophobicity and mobility of peptides in a residue-specific manner for the isolation of methionine[28] and tryptophan[29] containing peptides.

[28] T. Sasagawa, K. Titani, and K. A. Walsh, *Anal. Biochem.* **128,** 371 (1983).

[29] T. Sasagawa, K. Titani, and K. A. Walsh, *Anal. Biochem.* **134,** 224 (1983).

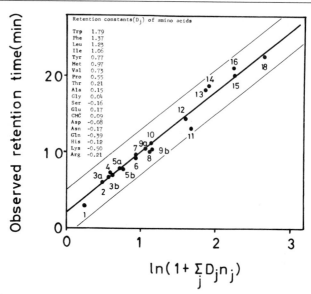

FIG. 2. Calibration of an HPLC column in order to compare observed peptide retention times (in this case, from Fig. 1) with those predicted from the product of the retention constant (D_j) and the number (n_j) of each amino acid residue (j) in a peptide by the relationship: predicted retention time $= A \ln(1 + \Sigma D_j \cdot n_j) + C$. Although for a given column and gradient system the constants A and C can be determined from the slope and intercept, the empirical graphical relationship is sufficient for a comparison of the mobility of an unknown peptide with that predicted from its amino acid composition. Most of the modified peptides in Table II would not fall on the line for conventional peptides. The area between the two lines shows the 70% confidence interval for these data.

have first been suspected on the basis of such altered mobility. Striking examples were seen in the catalytic subunit of bovine protein kinase,[30] in calcineurin,[31] and in a murine retrovirus protein,[32] where amino-terminal myristylation markedly increased the hydrophobicity of derived peptides (Table II). Glycosylated peptides are also recognized[25] by their anomalous mobility (Table II). Phosphorylated or carboxylated peptides can be recognized, especially if the modifying group carries a charge at the selected pH (Table II). On the other hand, the differences introduced by trimethylating a lysyl residue or hydroxylating an aspartyl residue are marginal as signals of modifications.

[30] S. A. Carr, K. Biemann, S. Shoji, D. C. Parmelee, and K. Titani, *Proc. Natl. Acad. Sci. U.S.A.* **79**, 6128 (1982).
[31] A. Aitken, P. Cohen, S. Santikarn, D. H. Williams, A. G. Calder, A. Smith, and C. B. Klee, *FEBS Lett.* **150**, 314 (1982).
[32] L. E. Henderson, H. C. Krutzsch, and S. Oroszlan, *Proc. Natl. Acad. Sci. U.S.A.* **80**, 339 (1983).

TABLE I
THE EFFECT OF SPECIFIC CHEMICAL MODIFICATIONS ON MOBILITY OF PEPTIDES ON REVERSED PHASE HPLC

Peptide	Sequence[a]	Nature of modified residue	Peptide retention time (min)	
			Modified	Unmodified
"CLIP"[a,b]	RPVKVYPNVAENESAEAFPLEF	Ser-14 is phosphorylated	34	38
α-MSH[c]	Acetyl-SYSMEHFRWGKPV-amide	Ser-1 is O-acetylated	11	7.5
Deamidated dermorphin[d]	YAFGYPS	Ser-7 is amidated	14	9
Deamidated dermorphin[d]	YAFGYPS	Pro-6 is hydroxylated	8	9
Oxytocin[e]	CYIQNCPLG-amide	Tyr-2 is monoiodinated	11.4	4.8
Oxytocin[e]	CYIQNCPLG-amide	Tyr-2 is diiodinated	15.6	4.8
Enkephalin[f]	YGGFM	Met-5 is S-methylated	12.5	16.0
Kemptide[g]	LRRASLG	Ser-5 is phosphorylated	7.6	11.5
Kemptide[g]	LRRASLG	Ser-5 is dehydroalanine	13.0	11.5
Kemptide[g]	LRRASLG	Ser-5 is cysteic acid	7.5	11.5
TRH[h]	PCA-His-Pro-amide	His-2 is 3-methyl-His	15	13
TRH[h]	Glu-His-Pro-amide	Glu-2 is PCA	63	13

[a] Underlined residue is modified.

[b] C. A. Browne, H. P. J. Bennett, and S. Solomon, *Biochemistry* **20**, 4538 (1981). Linear gradient elution with $CF_3(CF_2)_2COOH/CH_3CN$ system on Waters μBondapak C18 column (flow rate 1.5 ml/min). Gradient from 20 to 46% CH_3CN over 60 min.

[c] D. Rudman, R. K. Chawla, and B. M. Hollins, *J. Biol. Chem.* **254**, 10102 (1979). Isocratic elution with 0.01 M ammonium formate, pH 4/25% CH_3CN on a Waters μBondapak C18 column.

[d] L. Gozzini and P. C. Montecucchi, *J. Chromatogr.* **216**, 355 (1981). Isocratic elution with 0.01 M $CH_3COONa/50\%$ methanol on a Waters μBondapak C18 column (flow rate, 0.7 ml/min).

[e] T. Janáky, G. Tóth, B. Penke, K. Kovács, and F. A. László. *J. Liq. Chromatogr.* **5**, 1499 (1982). Isocratic elution with 0.01 M ammonium acetate, pH 4/20% CH_3CN on a Nucleosil 5 C18 column (flow rate, 2.0 ml/min). Note that the 2 Cys residues are connected by a disulfide bond.

[f] T. Sasagawa, K. Titani, and K. A. Walsh, *Anal. Biochem.* **128**, 371 (1983). Linear gradient elution with 0.1% trifluoroacetic acid/ acetonitrile on a Waters μBondapak C18 column. Gradient from 0 to 60% over 30 min (flow rate, 2.0 ml/min).

[g] T. R. Soderling and K. A. Walsh, *J. Chromatogr.* **253**, 243 (1982). Linear gradient elution with 0.1% trifluoroacetic acid/acetonitrile system on a Beckman Ultrasphere ODS column. Gradient from 15 to 30% over 20 min (flow rate, 1.0 ml/min).

[h] E. Spindel and R. J. Wurtman, *J. Chromatogr.* **175**, 198 (1979). Isocratic elution with 0.1% heptane sulfonic acid, 0.02 N acetic acid, 6.5% acetonitrile on a Waters μBondapak C18 column (flow rate, 2.0 ml/min).

TABLE II

COMPARISON OF OBSERVED AND PREDICTED HPLC MOBILITIES OF PEPTIDES FROM NATURALLY MODIFIED PROTEINS

Peptide sequence[a]	Nature of modification	Protein source	Retention time (min)		Mobile phase pH
			Observed	Predicted for unmodified	
GN	N-terminus is myristylated	Protein-kinase, catalytic subunit[b,c]	46.6	2	2[d]
CKDGLGEY	Asp is β-hydroxylated	Human coagulation factor x[e]	17.8	23 ± 4	2[d]
HHVNGSRPPCTGE	Asn is glycosylated	Cathepsin B[f]	14.5	21 ± 4	2[f]
VSIN	Ser is phosphorylated	Protein-kinase, catalytic subunit[b]	1.0	9 ± 2	8[g]
VSIN	Ser is phosphorylated	Protein-kinase, catalytic subunit[b]	10.5	9 ± 3	2[h]
ANSFLEEMK	Glu are γ-carboxylated	Human coagulation factor x[e]	1.0	11 ± 2	8[g]
HVMTNLGEKLTDEEVDEMIR	Lys is ε-trimethylated	Human brain calmodulin[i]	9.2	12 ± 2	8[g]

[a] Underlined residue is modified.

[b] S. Shoji, D. C. Parmelee, R. D. Wade, S. Kumar, L. H. Ericsson, K. A. Walsh, H. Neurath, G. L. Long, J. G. Demaille, E. H. Fischer, and K. Titani, Proc. Natl. Acad. Sci. U.S.A. **78**, 848 (1981).

[c] S. A. Carr, K. Biemann, S. Shoji, D. C. Parmelee, and K. Titani, Proc. Natl. Acad. Sci. U.S.A. **79**, 6128 (1982).

[d] Mobile phase and modifier were 0.1% trifluoroacetic acid and acetonitrile, respectively, in a Waters μBondapak C18 column. The gradient was 1%/min at a flow rate of 2 ml/min. T. Sasagawa, T. Okuyama, and D. C. Teller, J. Chromatogr. **240**, 329 (1982).

[e] B. A. McMullen, K. Fujikawa, W. Kisiel, T. Sasagawa, W. N. Howald, E. Y. Kwa, and B. Weinstein, Biochemistry **22**, 2875 (1983).

[f] K. Takio, T. Towatari, N. Katunuma, D. C. Teller, and K. Titani, Proc. Natl. Acad. Sci. U.S.A. **80**, 3666 (1983).

[g] T. Sasagawa, L. H. Ericsson, D. C. Teller, K. Titani, and K. A. Walsh, Abstr. 2nd. Int. Symp. HPLC Proteins, Peptides, Polynucleotides, 1982 p. 12. Mobile phase and modifier were 5 mM ammonium bicarbonate and acetonitrile, respectively, in a Hamilton PRP-1 polystyrene resin column. The gradient was 2%/min at a flow rate of 2 ml/min.

[h] Mobile phase and modifier were 0.1% trifluoroacetic acid and acetonitrile, respectively, in a Hamilton PRP-1 polystyrene resin column. The gradient was 2%/min at a flow rate of 2 ml/min.

[i] Sasagawa, L. H. Ericsson, K. A. Walsh, W. E. Schreiber, E. H. Fischer, and K. Titani, Biochemistry **21**, 2565 (1982).

The length of the subject peptide is an important consideration in the probability of detection.[24] The contribution of a modification to the overall hydrophobicity of a longer peptide may not be evident against a large background of hydrophobic contributions from the various conventional residues.

This volume serves to stress the ubiquity, variety, and importance of posttranslational modifications of protein structures. Current DNA sequencing procedures cannot locate these sites, and past protein sequencing methodology may well have missed important modifications, particularly those which are acid-labile. The importance of the modifications is deserving of more rigorous analyses aimed at their location and identification. Vigilant analyses of HPLC data at both the peptide and the phenylthiohydantoin levels offer simple approaches to detection of chemically altered residues.

Acknowledgments

The authors are grateful for discussions with Koiti Titani and Lowell Ericsson of many of the ideas considered herein, and for the continuing support of the NIH (GM 15731).

[4] Identification of Posttranslationally Modified Amino Acids in Proteins by Mass Spectrometry

By STEVEN A. CARR and KLAUS BIEMANN

I. Introduction

The identification and structure determination of modified amino acids in proteins are challenging analytical problems. Classical methods of amino acid analysis and protein sequence determination employ reaction conditions that may alter or destroy covalently modified residues. Furthermore, because identification is usually based on the known chromatographic mobilities of the 20 genetically coded for amino acids, modified residues may be lost or go unrecognized. Mass spectrometry (MS) is based on entirely different principles than the conventional approaches, and, therefore, is particularly well suited for the investigation of compounds whose structure is unknown and for which reference compounds do not exist. For these reasons MS has become the method of choice for the solution of certain problems to which the Edman degradation is not amenable, most notably the nature and structure of blocked NH_2-termini.[1,2] Mass spectrometry is currently undergoing a renaissance which should dramatically increase its role in the future of protein structure analysis. In the sections that follow we will try to acquaint the reader with these new approaches as well as successful mass spectrometric strategies that are currently in use.

This report is not meant to be exhaustive; rather our intent is to suggest strategies and methods. It should also be noted that examples where the modifying group has been removed prior to its structural characterization such as chromophores attached to proteins, e.g., refs. 3–5 will not be discussed here, even though mass spectrometry may have played a significant role in the process.

Recently, much attention has been focused on the determination of the primary structure of a protein by translation of the DNA sequence of the gene which is coded for the synthesis of the protein. However, it is important to note that although the nucleotide sequence can be used to determine the amino acid sequence of the end product of translation, it cannot, at least at this time, be used to identify the structure of the active protein because in many cases posttranslational modifications are necessary to generate full biological activity. From this perspective, the predictions concerning the imminent demise of direct protein sequencing have been greatly exaggerated.[6] An example of the mass spectrometric identification of a posttranslational modification at the NH_2-terminus of a large protein the structure of which had been otherwise correctly deduced from the DNA sequence will be discussed in Section III,A.

II. Strategies and Techniques for the Identification of Modified Amino Acids

The modern mass spectrometry laboratory has a veritable arsenal of effective techniques that can be brought to bear on problems of peptide

[1] K. Biemann, in "Biochemical Applications of Mass Spectrometry" (G. R. Waller, ed.), p. 405. Wiley (Interscience), New York, 1972.
[2] K. Biemann, in "Biochemical Applications of Mass Spectrometry, First Supplementary Volume" (G. R. Waller and O. C. Dermer, eds.), p. 469. Wiley (Interscience), New York, 1980.
[3] A. S. Brown, G. D. Offner, M. M. Ehrhardt, and R. F. Troxler, *J. Biol. Chem.* **254**, 7803 (1979).
[4] J. A. Duine, J. Frank, and P. E. J. Verwiel, *Eur. J. Biochem.* **108**, 187 (1980).
[5] O. D. Hensens, R. S. Dewey, J. M. Liesch, M. A. Napier, R. A. Reamer, J. L. Smith, G. Albers-Schönberg, and I. H. Goldberg, *Biochem. Biophys. Res. Commun.* **113**, 538 (1983).
[6] A. D. B. Malcolm, *Nature (London)* **275**, 90 (1978).

and protein structure. The particular MS method one employs is dictated by the level of information desired as well as the particular instrumentation available.

In principle, mass spectrometric protein sequencing strategies may be divided into two broad categories: those which employ conventional ionization methods, such as electron impact (EI) or chemical ionization (CI), and require that the amino acid or polypeptide be chemically derivatized, and those which are capable of directly analyzing underivatized large and/ or polar molecules—the so-called "soft ionization" methods (see below). For methods in the former category, the sample is vaporized prior to ionization which necessitates derivatization in order to impart sufficient volatility and thermal stability to the molecules. The extent of derivatization required strongly depends on how the sample is introduced to the ion source of the mass spectrometer, but generally involves esterification of the carboxyl and acylation of the amino terminus.

Strategies involving chemical derivatization may be further divided into those in which the derivative of a single amino acid or oligopeptide (or a very simple mixture thereof) is directly evaporated into the ion source of the mass spectrometer, and those involving derivatization of a complex mixture of amino acids or peptides which are subsequently separated and analyzed by a gas chromatograph coupled to a mass spectrometer–computer system.[1,2] In this section, the more commonly employed chemical derivatization-based strategies and "soft" ionization methods will be briefly discussed in terms of their applicability to posttranslationally modified proteins. The reader should be aware of the fact that these methods apply equally well to unmodified proteins.

Free Amino Acids. Mass spectrometry may be used for the identification of posttranslationally modified amino acids obtained from total acid, alkaline or enzymatic digestion of proteins or peptides (for a recent review, see Vetter[7]). This approach is most useful in cases where the protein structure has been completely defined except for the identity of the modified residue(s). In most cases, both molecular weight and detailed structural information may be obtained by electron impact or chemical ionization MS of suitable derivatives. The greatest volatility and most structurally informative mass spectra are obtained for the trimethylsilyl (TMS) and the N-perfluoroacyl alkyl ester derivatives of amino acids. These particular derivatives are sufficiently volatile for separation by gas chromatography (GC), and derivatives of all commonly occurring and many novel amino acids from animal, plant, fungal, and bacterial sources

[7] W. Vetter, *in* "Biochemical Applications of Mass Spectrometry, First Supplementary Volume" (G. R. Waller and O. C. Dermer, eds.), p. 439. Wiley (Interscience), New York, 1980.

have been structurally characterized by combined gas chromatography mass spectrometry (GCMS). These reference data are extremely valuable and aid in interpreting the mass spectra of unknowns (for lists of collections, see Vetter[7]). The use of GCMS has the advantage that amino acids may be identified typically at the nanomole level without prior separation from the total hydrolyzate. Of course, fractions isolated by high-performance liquid chromatography (HPLC) may be derivatized and introduced into the mass spectrometer either by GC or by direct insertion into the ion source. In addition, EI mass spectra of all derivatives commonly employed for chromatographic detection of amino acids such as dansyl, N-(2,4-dinitrophenyl), and fluorescamine have been obtained. Thus, MS may be used to resolve ambiguities arising from chromatographic identification. Unfortunately, the use of MS does not eliminate the principal disadvantages of analyzing total hydrolyzates, namely that the modified amino acid must be stable to the conditions used to release it from the protein, and to derivatize it to make it sufficiently volatile and chromatographically detectable.

Phenylthiohydantoin Derivatives. Mass spectrometry may also be used to identify the amino acids released in the stepwise Edman degradation of a protein.[1,2] Qualitative and quantitative identification of the phenylthiohydantoin (PTH) derivative or the product of back hydrolysis to the free amino acid is usually based on chromatographic retention on HPLC, which has proven to be an extremely reliable procedure. Covalently modified amino acids are likely to have different chromatographic mobilities than the 20 commonly occurring ones (or may not chromatograph at all using standard conditions), and, in such cases, MS must be employed for unequivocal structure identification. Mass spectra of the PTH derivatives of all commonly occurring amino acids have been obtained in both EI and CI ionization modes by direct introduction of the sample into the ion source. Most of these spectra exhibit abundant ions characteristic of molecular weight and structure, although certain of the more polar, less thermally stable amino acids such as arginine yield molecular ions in low abundance. Silylation or analysis of the PTH-amino acid derivative by a soft ionization technique (see below) can be used to obtain abundant molecular weight related ions even for the difficult amino acids (references to detailed procedures may be found in refs. 2,7). Again, these studies provide the reference spectra necessary for assignment of structures to unknowns.

In practice, several problems are associated with application of this strategy. It assumes that the Edman reaction conditions do not alter or destroy the modified side chain and that the NH_2-terminus of the protein is not blocked. The absence of a basic primary or secondary amino group while preventing the straightforward application of the Edman degradation actually makes the peptide more suitable for MS analysis, since the natural derivative is usually more volatile. Numerous applications of MS for the determination of blocked peptides have appeared, some of which are described in detail in Section III,A. The presence of sensitive functional groups is a difficult problem but may, in some cases, be recognized by altering the Edman reaction conditions, analysis of the intermediate anilinothiazolinone, or by the use of different derivatives. The two principal drawbacks of the Edman procedure as one penetrates further into the peptide chain are rising background and falling yield per cleavage cycle (for a detailed discussion, see Biemann[2]) which may obscure identification of the modified residues unless they are located near to the NH_2-terminus. As is the case in conventional sequence analysis, this necessitates the isolation of small (typically less than 50 aa) pure peptides which can only be obtained by multistep cleavage of the protein.

Amino Acids in Peptides. Mass spectrometry is capable of analyzing simple mixtures of components simultaneously, and two MS sequencing strategies that capitalize on this ability have been developed to the point where they are suitable for the analysis of peptide mixtures. These are the permethylation method, first suggested by Das *et al.*[8] and refined by many others,[9-12] and the gas chromatographic mass spectrometric (GCMS) sequencing strategy developed by Biemann and co-workers.[2,13-15] Both approaches provide unambiguous partial sequence data which in favorable circumstances may be reassembled into a complete primary structure.

The permethylation procedure (Scheme 1) involves reducing the polarity of an oligopeptide by acetylation of all amino groups followed by methylation of the newly formed amido groups, the –CO–NH– groups of the peptide bonds and of all carboxylic acid groups, namely the COOH-terminus and those present in side chains. The resulting derivative (**I**) is vaporized directly into the ion source of the mass spectrometer.[8,10] Sequence-determining peaks result from cleavage of the $CO-NCH_3$ bond and are often of significant intensity although diminishing at higher mass. Permethylation is suitable for individual oligopeptides of up to ten, in

[8] B. C. Das, S. D. Gero, and E. Lederer, *Biochem. Biophys. Res. Commun.* **29,** 211 (1967).

[9] P. A. Leclercq and D. M. Desiderio, Jr., *Anal. Lett.* **4,** 305 (1971).

[10] H. R. Morris, D. H. Williams, and R. P. Ambler, *Biochem. J.* **125,** 189 (1971).

[11] M. L. Polan, W. J. McMurray, S. R. Lipsky, and S. Lande, *J. Am. Chem. Soc.* **94,** 2847 (1972).

[12] H. R. Morris, *FEBS Lett.* **22,** 257 (1972).

[13] K. Biemann, F. Gapp, and J. Seibl, *J. Am. Chem. Soc.* **81,** 2274 (1959).

[14] S. A. Carr, W. C. Herlihy, and K. Biemann, *Biomed. Mass. Spectrom.* **8,** 51 (1981).

[15] W. C. Herlihy, R. J. Anderegg, and K. Biemann, *Biomed. Mass Spectrom.* **8,** 62 (1981).

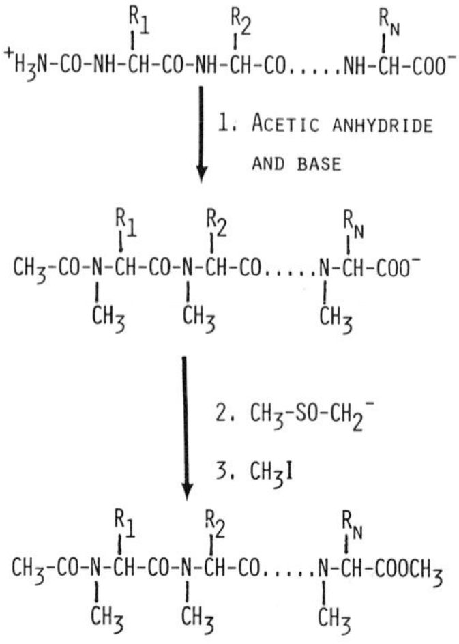

$$
\begin{array}{ccc}
R_1 & R_2 & R_N \\
| & | & | \\
{}^+H_3N-CO-NH-CH-CO-NH-CH-CO\ldots\ldots NH-CH-COO^-
\end{array}
$$

1. ACETIC ANHYDRIDE
AND BASE

$$
\begin{array}{ccc}
R_1 & R_2 & R_N \\
| & | & | \\
CH_3-CO-N-CH-CO-N-CH-CO\ldots\ldots N-CH-COO^- \\
| & | & | \\
CH_3 & CH_3 & CH_3
\end{array}
$$

2. $CH_3-SO-CH_2^-$

3. CH_3I

$$
\begin{array}{ccc}
R_1 & R_2 & R_N \\
| & | & | \\
CH_3-CO-N-CH-CO-N-CH-CO\ldots\ldots N-CH-COOCH_3 \\
| & | & | \\
CH_3 & CH_3 & CH_3
\end{array}
$$

I

SCHEME 1.

favorable cases even more amino acids, or very simple mixtures thereof if they can be thermally fractionated. The various permethylation procedures and their application have been reviewed[1,2] and detailed procedures have been published.[8,9,11,12] The best conditions involve brief exposure (1 min to 1 hr) to methylsulfinyl carbanion and methyl iodide to effect permethylation (Scheme 1).

In contrast, the GCMS sequencing strategy (Scheme 2) is applicable to very complex mixtures of oligopeptides such as acid or enzyme hydrolyzates of proteins. It involves conversion of the oligopeptides in a number of steps to the N-trifluoroethyl-O-trimethylsilyl polyamino alcohols (II).[1,2,13,14] The key step of this procedure is reduction of the amide backbone to a repeating ethylenediamine which drastically reduces the polarity of the peptides. The resulting complex mixture of polyamino alcohols is injected into the GCMS where the derivatives are separated and mass spectra of each fraction are obtained in the continuous scanning mode.[2,13,14] The mass spectra of these derivatives exhibit strong peaks due to cleavage of the bonds indicated (Scheme 2), generating a series of ions $A_1 \ldots A_n$ from which one can read the sequence from the NH_2-terminus and ions $Z_1 \ldots Z_n$ which indicate the sequence from the COOH-terminus. Modified amino acids are recognized by the difference in the incremental mass value compared to the commonly occurring amino acids. In addition to the simple and predictable mass spectra, the GC retention behavior of any O-TMS polyamino alcohol can be determined from the incremental contributions made by each of the amino acids present in the

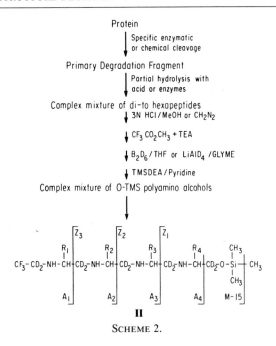

SCHEME 2.

peptide.[2] Thus, one can predict not only the major features of the mass spectrum of the O-TMS polyamino alcohol derived from any peptide, but also where in the gas chromatogram it will appear. This permits one to search the mass spectral data for the derivatives of specific peptides and to detect them even if present at very low concentration.

The principal drawback to the methods described above is that the chemical treatments employed to increase volatility may alter or destroy sensitive functional groups present in the molecule. In recent years a number of new ionization methods have been developed that allow MS analysis of peptides and other highly polar and thermally labile compounds without prior derivatization. The distinguishing feature of these methods is that ions are formed directly from the sample in the solid state.[16] Historically, field desorption (FD) MS[17,18] was the first of these techniques to be shown to have potential for the analysis of peptides, affording (predominantly) protonated and cationized molecular ions [i.e., $(M + H)^+$ and $(M + Na)^+$ or $(M + K)^+$]. However, FD of peptides is, in

[16] K. L. Busch and R. G. Cooks, *Science* **218**, 247 (1982).
[17] H. D. Beckey, "Principles of Field Ionization and Field Desorption Mass Spectrometry." Pergamon, Oxford, 1977.
[18] H.-R. Schulten, *Int. J. Mass Spectrom. Ion Phys.* **32**, 97 (1979).

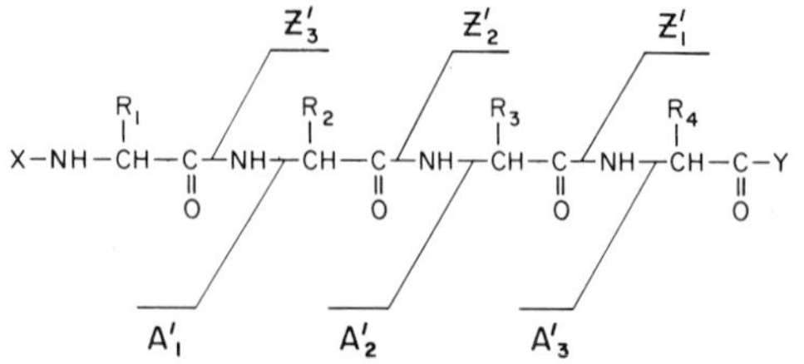

$$Z'_3 \qquad Z'_2 \qquad Z'_1$$

$$\begin{array}{ccccccccc} & R_1 & & R_2 & & R_3 & & R_4 & \\ & | & & | & & | & & | & \\ \text{X}-\text{NH}-\text{CH}-\text{C}-\text{NH}-\text{CH}-\text{C}-\text{NH}-\text{CH}-\text{C}-\text{NH}-\text{CH}-\text{C}-\text{Y} \\ & & \| & & \| & & \| & & \| \\ & & \text{O} & & \text{O} & & \text{O} & & \text{O} \end{array}$$

$$A'_1 \qquad A'_2 \qquad A'_3$$

SCHEME 3. Fragmentation observed in both ammonia–direct chemical ionization (NH$_3$-DCI) and fast atom bombardment (FAB) mass spectrometry. The most abundant ion series are X-NH-CHR$_m$-CO-NH$_3$ [(A$'_m$ + 2H)$^+$] and NH$_3$-CHR$_n$-CO-Y [Z$'_n$ + 2H)$^+$], which correspond to charge retention on the NH$_2$- and COOH-termini, respectively (X and Y denote the remainder of the peptide chain). Other ion series in FAB of peptides are described in ref. 20. These ions are also the dominant sequence ions in NH$_3$-DCI of peptides. Ammonium ion attachment is also common in NH$_3$-DCI resulting in ((A$'_m$ + H) + NH$_4$)$^+$ and ((Z$'_n$ + H) + NH$_4$)$^+$ ion series (see ref. 22).

general, technically difficult and relatively insensitive, and yields poorly reproducible mass spectra.[2]

Other, more recently developed techniques have been shown to be particularly well suited for the analysis of underivatized peptides. In fast atom bombardment [FAB, also known as liquid-SIMS (secondary ion mass spectrometry)] sample molecules dissolved in a viscous liquid such as glycerol are ionized and desorbed from a solid support by an impinging beam of atoms or ions.[16,19] Abundant protonated molecular ions (M + H)$^+$ are produced using microgram amounts of material or less. Protonated molecular ions have been obtained for peptides and small proteins with molecular weights approaching 10,000. For small peptides (generally less than 15 residues) sequence ions may also be obtained (see Scheme 3), although the presence of fragments is highly dependent on the composition and sequence of the particular polypeptide.[20]

Direct chemical ionization (DCI) MS, another recent addition to the mass spectroscopist's repertoire of soft ionization methods, is a sensitive (submicrogram) and reproducible means of obtaining molecular weight and sequence information on small (less than 10 residues) underivatized peptides. Samples are coated on a wire attached to an extended direct insertion probe which allows the sample to be rapidly heated within the charged reagent gas plasma.[21,22] Ammonia has been shown to be the most

[19] M. Barber, R. S. Bordoli, G. J. Elliot, D. Sedgwick, and A. N. Tyler, *Anal. Chem.* **54**, 645A (1982).

[20] D. H. Williams, C. V. Bradley, S. Santikarn, and G. Bojesen, *Biochem. J.* **201**, 105 (1982).

[21] M. A. Baldwin and F. W. McLafferty, *Org. Mass Spectrom.* **7**, 1353 (1973).

[22] S. A. Carr and V. N. Reinhold, *in* "Methods in Protein Sequence Analysis" (M. Elzinga, ed.), p. 263. Humana Press, Clifton, New Jersey, 1982.

suitable reagent gas for the analysis of underivatized peptides by DCI.[22] In general, the ions of highest mass correspond to $(M + H)^+$ and/or $(M + NH_4)^+$ and sequence ions arise by cleavage on either side of the amide nitrogen with hydrogen transfer and protonation (Scheme 3). These cleavages are also the dominant ones which are observed under FAB conditions.[20]

III. Applications

A. Amino- and Carboxy-Terminal Blocking Groups

Acylation of the free NH_2-terminus of a protein with a formyl or acetyl group or as a pyrrolidone ring (pyroglutamic acid, Pca) is one of the most common posttranslational modifications. Proteins with such blocked NH_2-termini cannot be sequenced by the Edman degradation since a free primary or secondary amine is required in the first step of the reaction sequence. Chemical methods have been devised for recognizing and distinguishing between acyl and Pca blocking groups, but they are quite laborious and require large amounts of material.[23] In contrast, blocking of the free NH_2-terminus is required for analysis by EI or CI MS, and, therefore, the mass spectrum of a naturally blocked peptide will readily establish the structure of the blocking group by the observed difference in mass, if the amino acid composition is known.

Pyroglutamic acid. Pyroglutamic acid is particularly difficult to characterize by chemical methods. Its presence is indirectly inferred by the absence of a free NH_2-terminus, negative test results for formate or acetate, and the ability to obtain a free-NH_2 terminus after treatment of the protein with a pyroglutamyl peptidase.[23] A more vexing problem is that one cannot be sure that the blocking group was present *in vivo* or was formed as a result of sample handling during isolation and purification. In acidic aqueous media, especially at elevated temperatures, Glu and Gln readily interconvert with the cyclized form. Thus, whenever possible, purification procedures using strong ion exchange resins should be avoided, and enzymatic digestion at physiological pH should be used to generate the peptide mixtures. Furthermore, Pca can also be formed thermally in the ion source of the mass spectrometer, and esterification and acylation conditions requiring acidic conditions or heating can also convert Gln and Glu to Pca.[12,24] Cyclization during derivatization may be minimized by using nonacidic reagents and conditions such as diazomethylation and base catalyzed acetylation.

[23] K. Narita, *in* "Protein Sequence Determination" (S. B. Needleman, ed.), p. 82. Springer-Verlag, Berlin and New York, 1970.
[24] J. Wietzerbin-Falszpan, B. C. Das, I. Gros, J.-F. Petit, and E. Lederer, *Eur. J. Biochem.* **32**, 525 (1973).

ACCURATE MASSES AND CORRESPONDING
ELEMENTAL COMPOSITIONS OF IONS FROM THE
N-TERMINUS OF BACTERIORHODOPSIN

Composition	Calculated mass	Observed mass
C_4H_6NO	84.0449	84.0445
$C_5H_6NO_2$	112.0398	112.0393
$C_7H_{11}N_2O_2$	155.0820	155.0819
$C_8H_{11}N_2O_3$	183.0770	183.0788

If a suitable short blocked peptide can be isolated and purified than minimal derivatization may be required prior to its analysis by MS. This was the case for bacteriorhodopsin where the blocked NH_2-terminal pentapeptide was diazomethylated then analyzed by high resolution EIMS (see the table) without any further chemical treatment.[25] High-resolution mass spectrometry (HRMS) allows the determination of the mass of an ion with an accuracy that defines its elemental composition.[26,27]

The data shown in the table list the accurate mass and elemental composition of abundant ions in the spectrum and their assignment. The same approach had been used by many others, both in the high-resolution and low-resolution modes. Needless to say, high-resolution data are more convincing, but strong signals at m/z 84 and 112, and the presence of the appropriate sequence ions for the amino acids that follow, support any such assignment from low-resolution data.

In the course of permethylation the pyrrolidone nitrogen is methylated which shifts Pca sequence ions to m/z 98 and 126, etc. This approach was used to identify, among others, the blocked NH_2-terminal sequence of bovine k-papacasein[28] and bovine plasma fibronectin.[29] The NH_2-terminal

[25] G. E. Gerber, R. J. Anderegg, W. C. Herlihy, C. P. Gray, K. Biemann, and H. G. Khorana, *Proc. Natl. Acad. Sci. U.S.A.* **76,** 227 (1979).

[26] K. Biemann, *Top. Org. Mass Spectrom.* **8,** 185 (1970).

[27] K. Biemann, "Mass Spectrometry: Organic Chemical Applications." McGraw-Hill, New York, 1962.

[28] G. Brignon, J. C. Mercier, B. Ripadeau-Dumas, and B. C. Das, *FEBS Lett.* **27,** 301 (1972).

[29] R. P. McDonagh, J. McDonagh, T. E. Petersen, H. C. Thogersen, K. Skorstengaard, L. Sottrup-Jensen, S. Magnusson, A. Dell, and H. R. Morris, *FEBS Lett.* **127,** 174 (1981).

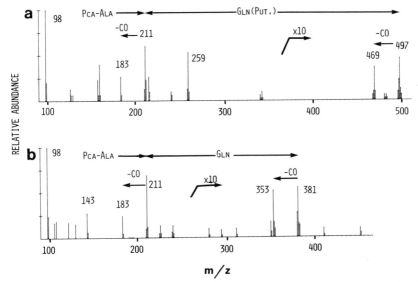

FIG. 1. Electron impact mass spectra obtained at 260° of (a) the deuteroacetylated permethylated tetrapeptide containing the putrescine (Put.) label, and (b) the unlabeled deuteroacetylated permethylated tetrapeptide (reproduced from ref. 29).

sequence of the latter was also shown by MS to contain the binding site for factor XIIIa which catalyzes the formation of (γ-glutamyl)lysyl amide bonds between chains of aggregated fibrin molecules. A pure blocked tetrapeptide was obtained by extensive proteolysis of fibronectin which had been labeled with radioactive putrescine. This primary amine was known to mimic the side chain of lysine and bind to the residue involved in cross-linking. The EI mass spectrum of the perdeuteroacetylated and permethylated derivative (Fig. 1a) clearly defines the sequence Pca-Ala-Glx(putrescine)--- by the NH_2-terminal ion series at m/z 98, 126, 183, 211,

469, and 497. The same peptide isolated from unlabeled fibronectin affords the mass spectrum of Pca-Ala-Gln--, confirming that the Glx of the binding site is a Gln (Fig. 1b). Neither of the peptides yielded a molecular ion by EIMS.

Acetylation of the peptide was required for derivatization of the primary amino group of putrescine which otherwise would have quaternized during permethylation. The acetylation step is, of course, always used for a peptide of unknown sequence, and the deuterated analog of acetic anhydride is often substituted so that native NH_2-acetylated proteins, if present, may be readily recognized (see below). As mentioned, the permethylation approach may be used for the analysis of simple mixtures. However, ambiguities may arise in the interpretation of mass spectra obtained from even simple mixtures of permethylated peptides because, for example, the mass of a Pca residue is equivalent to that of a number of other residues and commonly observed rearrangement products.[10,30] Comparison of the mass spectra obtained for peptides derivatized with deuteroacetyl and deuteromethyl reagents, with those obtained using unlabeled analogs helps, but does not eliminate such problems.[30]

The GCMS protein sequencing approach was used to define the NH_2-terminal sequence and identify the blocking group of a carboxypeptidase inhibitor from potatoes.[31] A mixture of small blocked polypeptides consisting of four and five residues, respectively, was obtained from the intact protein by selective chemical cleavage. Laborious isolation and purification of the individual peptides were avoided by subjecting the mixture directly to the reductive derivatization-GCMS procedure outlined in Section II. As mentioned before, the NH_2-terminus of a Pca-peptide will not undergo trifluoroacetylation but is simply converted to a dideuteropyrrolidone group.

Amino Terminal Acylation. Like Pca, NH_2-terminal acylated amino acids do not acylate in the appropriate step of the derivatization procedure. Therefore, a derivative with a series of NH_2-terminal sequence ions shifted in mass by the difference between the mass of the acyl group used during derivatization and that already present on the peptide is produced. Since acetyl is a common *in vivo* blocking group, one should use either

[30] A. Dell and H. R. Morris, *Biomed. Mass Spectrom.* **8**, 128 (1981).
[31] G. M. Haas, H. Nau, K. Biemann, D. T. Grahn, L. H. Ericsson, and H. Neurath, *Biochemistry* **14**, 1334 (1975).

deuterated or fluorinated acyl substituents to guard against coincidental identity of the two. For example, N-acetylserine was recognized as the NH_2-terminus of a *Neurospora* glutamase,[32] of trout hemoglobin,[33] and of an alcohol dehydrogenase[34] by the presence of an ion at m/z 158 rather than 161, indicating the presence of a CH_3CO group prior to acylation with $(CD_3CO)_2O$.

$$CH_3CO-NH-\overset{\overset{\displaystyle CH_2OCH_3}{|}}{CH}-CO^+ \qquad CD_3CO-NH-\overset{\overset{\displaystyle CH_2-OCH_3}{|}}{CH}-CO^+$$

$$m/z\ 158 \qquad\qquad m/z\ 161$$

If the presence of an N-acetyl group (or any other acyl group) is already suspected, the acylation step may be omitted and the sample permethylated directly. If the NH_2-terminus was indeed acetylated, the normal product will be obtained.[34,35] This method may also be used in conjunction with the polyamino alcohol technique which will then produce a derivative with a CH_3CD_2- instead of CF_3CD_2- at the NH_2-terminus.[36]

Another way of determining NH_2-terminal blocking groups uses an approach developed for confirming sequences of large proteins derived from the DNA sequence of the corresponding gene.[37,38] A tryptic digest of the protein is subjected to FAB MS which reveals the molecular weights of all tryptic peptides predicted for the hypothetical structure. The peptide corresponding to the modified NH_2-terminus will differ in mass from the predicted one and the mass difference reveals the modification. For methionyl-tRNA synthetase from yeast it was found that the amino terminal Met had been removed and the next amino acid (Ser) N-acetylated because instead of a peptide of molecular weight 1086, a peak was found at mass 998 corresponding to the protonated NH_2-terminal peptide. The mass difference points to removal of Met and addition of CH_2CO. This was further confirmed by accurate mass measurement (found 998.521, calculated 998.5199 for Ac-Ser-Phe-Leu-Ile-Ser-Phe-Asp-Lys).[39] Of

[32] H. R. Morris and A. Dell, *Biochem. J.* **149,** 754 (1975).
[33] S. Doonan, A. G. London, D. Bana, F. Bossa, and M. Brunori, *FEBS Lett.* **85,** 141 (1978).
[34] A. D. Aufret and D. H. Williams, *FEBS Lett.* **90,** 324 (1978).
[35] H. Jørnvall, H. Ohlsson, and L. Philipson, *Biochem. Biophys. Res. Commun.* **56,** 304 (1974).
[36] H. Nau, K. Lerch, and L. Witte, *FEBS Lett.* **79,** 203 (1977).
[37] K. Biemann, *Int. J. Mass Spectrom. Ion Phys.* **45,** 183 (1982).
[38] S. D. Putney, N. J. Royal, H. N. DeVegvar, W. C. Herlihy, K. Biemann, and P. Schimmel, *Science* **213,** 1497 (1981).
[39] B. W. Gibson, H. A. Scoble, and K. Biemann, *31st Annu. Conf. Mass Spectrom. Allied Top.* Paper ROB 4 (1983).

course, FAB MS can be used on isolated peptides as well. This was done in order to confirm the NH_2-terminal N-acetylserine on thymosin.[40]

Longer chain fatty acids are rarely found covalently attached to proteins.[41,42] This is at least partially the result of the hydrophobic nature of these peptidolipids which causes them to aggregate and irreversibly adsorb to columns, etc., thus making them very difficult to isolate. Until recently, the only well-documented example was that of the murein lipoprotein of the outer membrane of *Escherichia coli* the NH_2-terminus of which is a Cys-glycerol thioether. The two hydroxyls and the amino group of this modified amino acid are acylated with fatty acids.[43]

We have recently shown that the NH_2-terminus of the 349 amino acid catalytic subunit of bovine cardiac muscle cyclic AMP-dependent protein kinase is blocked by an *n*-tetradecanoyl (myristyl) group.[44] This blocking group prevented straightforward application of the Edman degradation on the intact protein. Initial evidence from carboxypeptidase digestion of small blocked peptides from the NH_2-terminus suggested that the sequence was X-Gly-Asn-Ala-Ala-Ala-Lys, where the blocking group, X, was thought to be far more hydrophobic than an acetyl group based on its unusually long retention on reversed phase HPLC.[45]

A combination of mass spectrometric techniques was employed to determine the structure of the blocking group. Evidence for the structure of X was obtained by FAB and DCI on the blocked tripeptide. The FAB mass spectrum exhibited abundant protonated $(M + H)^+$ and sodium-cationized $(M + Na)^+$ molecular ions at m/z 471 and 493, respectively. Although no sequence-related fragment ions were present in this mass spectrum, the calculated difference between the observed molecular weight and the mass of the peptide portion is 211, which corresponds to $C_{14}H_{27}O$. In contrast, the DCI mass spectrum obtained using NH_3 as the reagent gas (Fig. 2) is dominated by fragment ions arising from the blocked portion of the peptide. Ions at m/z 245 and 302 confirm the composition of the blocking group and its attachment to the NH_2-terminal Gly.[44] The detailed structure of the blocking group was determined by diazomethylation followed by GCMS on the extracts of total acid hydrolyzates of two blocked peptides (X-Gly-Asn-Ala and the intact NH_2-

[40] J. Michalewsky, T. F. Gabriel, D. P. Winter, R. Makofske, W. Danho, J. Shively, K. Biemann, and J. Meienhofer, *Int. J. Pept. Protein Res.* **21**, 93 (1983).

[41] M. J. Schlesinger, *Annu. Rev. Biochem.* **50**, 193 (1981).

[42] M. F. G. Schmidt, *Trends Biochem. Sci.* **7**, 322 (1982).

[43] K. Hantke and V. Braun, *Eur. J. Biochem.* **34**, 284 (1973).

[44] S. A. Carr, K. Biemann, S. Shoji, D. C. Parmelee, and K. Titani, *Proc. Natl. Acad. Sci. U.S.A.* **79**, 6128 (1982).

[45] S. Shoji, D. C. Parmelee, R. D. Wade, S. Kumar, L. H. Ericsson, K. A. Walsh, H. Neurath, G. L. Long, J. G. Demaille, E. H. Fischer, and K. Titani, *Proc. Natl. Acad. Sci. U.S.A.* **78**, 848 (1981).

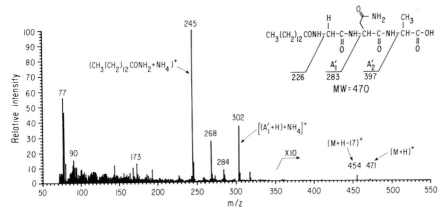

FIG. 2. Ammonia–direct chemical ionization mass spectrum of 0.3 μg of the underivatized, blocked peptide n(CH$_3$(CH$_2$)$_{12}$CO-NH-Gly-Asn-Ala-OH (reproduced from ref. 44).

terminal tryptic fragment). The product had a mass spectrum and GC retention behavior identical with that of authentic n-tetradecanoic acid methyl ester. Incidentally, the FAB mass spectrum of the alleged hexapeptide (Fig. 3) revealed that there was an additional alanine present in the NH$_2$-terminal tryptic peptide which had been previously overlooked.[44]

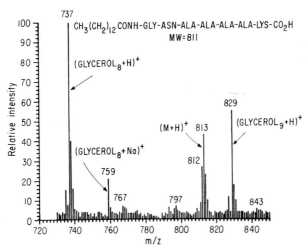

FIG. 3. Fast atom bombardment mass spectrum of the underivatized, blocked NH$_2$-terminal tryptic peptide from bovine heart protein kinase. Only the molecular ion region is shown; the enhanced contribution of m/z 813 is due to partial conversion of Asn to Asp (reproduced from ref. 44).

Tetradecanoic acid has recently also been identified as the NH_2-terminal blocking group of calcineurin, a calmodulin-dependent protein phosphatase[46] and of two murine retrovirus proteins.[47] The methods employed for structural characterization of the amino termini of these proteins were very similar to those described above. In each case, GCMS of organic extracts of total acid hydrolyzates esterified with either diazomethane[46] or bis(trimethylsilyl)trifluoroacetamide (BSTFA)[47] established the presence of n-tetradecanoic acid. Mass spectra and retention times were compared to authentic standards. The chromatographic mobility of blocked peptides obtained from enzymatic or cyanogen bromide digests gave further support to the hydrophobic character of the blocking group. As in the case of cyclic AMP-dependent protein kinase discussed above these blocked peptides eluted from reversed phase HPLC at high (40–60%) concentrations of acetonitrile, which belied the mobilities expected based solely on amino acid compositions.

In the case of calcineurin, the mass of the blocking group was also determined by FAB MS of blocked peptides of known composition obtained by cyanogen bromide and enzymatic cleavage of the intact protein.[46]

The NH_2-termini of the murine retrovirus proteins were sequenced by GCMS analysis of silylated dipeptides obtained by dipeptidyl carboxypeptidase[48] digestion of an HPLC purified thermolytic tripeptide X-Gly-Gln-Thr.[47] The two major peaks in the total ion chromatogram from this GCMS experiment gave mass spectra consistent with the silylated derivatives of Gln-Thr and n-tetradecanoyl-Gly. At present the physiological significance of this hydrophobic NH_2-terminal tail is unknown, but speculation has centered on possible roles in membrane translocation, binding, or subunit assembly. Specific acylation with tetradecanoic acid strongly suggests that this is an enzyme-mediated posttranslational event. The similarity of the NH_2-terminal sequences of these proteins, all of which begin with Gly, further supports the possibility of enzyme mediation.

Carboxy-Terminal Amidation. Posttranslational amidation of the carboxyl termini of proteins appears to be a far rarer event than NH_2-terminal acylation. The presence of a COOH-terminal blocking group is inferred by the lack of reactivity of a polypeptide or protein with carboxypeptidases, while the structure of the blocked terminus can be obtained only by Edman degradation of a sufficiently short carboxy-terminal peptide and comparison of the chromatographic mobility of the

[46] A. Aitken, P. Cohen, S. Santikarn, D. H. Williams, A. G. Calder, A. Smith, and C. B. Klee, *FEBS Lett.* **150**, 314 (1982).
[47] L. E. Henderson, H. C. Krutzsch, and S. Oroszlan, *Proc. Natl. Acad. Sci. U.S.A.* **80**, 339 (1983).
[48] H. C. Krutzsch, *Biochemistry* **19**, 5290 (1980).

blocked residue with authentic amidated amino acid standards. A simpler and more reliable approach is to analyze the blocked peptide either directly by FAB, DCI, or EI mass spectrometry of a suitable derivative. The latter approach has been used to identify the structure of the blocking group and COOH-terminal sequence (--Trp-Gly-Ser-NH$_2$) of a neurotoxic protein isolated from sea snake venom.[49]

Model peptides with blocked COOH-termini were analyzed by FAB and DCI mass spectrometry and the results clearly indicate the utility of these desorption methods for the analysis of blocked peptides. For example, mass spectra of the COOH-terminal octapeptide from substance P[22] and intact substance P[20] obtained using NH$_3$-DCI (Fig. 4) and FAB, respectively, exhibit protonated molecular ions and a series of NH$_2$- and COOH-terminal sequence ions (see Scheme 3) that are sufficient to establish the structures of the blocked peptides. The DCI spectrum clearly shows the protonated molecular ion at m/z 966 requiring all three carboxyl groups to be present as amides. The COOH-terminal amide is further supported by the mass of the Z_4–Z_7 ions (Fig. 4). The high sensitivity (microgram or less of material) and experimental simplicity of these techniques will undoubtedly result in their application to current structural problems of biochemical interest.

B. New Amino Acids Present in Proteins

γ-Carboxyglutamic Acid (Gla). Mass spectrometry, along with other spectroscopic techniques, has played an important role in elucidating the structure of covalently modified amino acids and determining of their sequence location within proteins.[2,7] Although not a known constituent of proteins, lysopine (**III**) was the first amino acid to have its structure

III   IV

(excluding stereochemistry) determined solely by mass spectrometry[50]; many other examples of structure elucidation of novel amino acids can be found in the literature, both by MS alone, or MS used in conjunction with other spectroscopic and chemical data (for recent reviews, see refs. 2,7).

A recent important example is γ-carboxyglutamic acid (**IV**), a new amino acid first found in the proteins responsible for blood coagula-

[49] N. Maeda and N. V. Tamiya, *Biochem. J.* **175,** 507 (1978).
[50] K. Biemann, C. Lioret, J. Asselinean, E. Lederer, and J. Polonsky, *Bull. Soc. Chim. Biol.* **42,** 979 (1960).

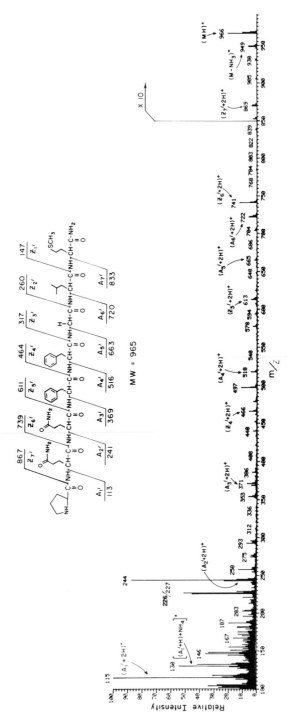

FIG. 4. Ammonia–direct chemical ionization mass spectrum of underivatized Pro-Gln-Gln-Phe-Phe-Gly-Leu-Met-NH₂ (reproduced from ref. 22).

tion.[51–53] γ-Carboxyglutamic acid is now known to be present in a diverse group of proteins all of which have in common the ability to bind calcium and the requirement that vitamin K and bicarbonate be present for the biosynthesis of their biologically active forms.[54]

The ease of decarboxylation of Gla led to its identification as glutamic acid (Glu) or its derivatives in the course of amino acid analyses or Edman degradations. The low yield in the latter and the electrophoretic mobility of some of the degradation peptides were not, however, compatible with the presence of Glu alone.[55,56]

The presence and positions of Gla (rather than Glu) in the N-terminal region of prothrombin were deduced from MS data simultaneously by Stenflo et al.,[57] Magnusson et al.,[58] and Nelsestuen et al.[59] Stenflo isolated a peptide Leu-Glx-Glx-Val from bovine prothrombin and found by electrophoresis that it was more negatively charged than could be accounted for if Glx was Glu. The mass spectrum of the N-acetylated permethyl derivative of this peptide gave mass increments for Glx-containing sequence ions 58 daltons ($-H + CO_2CH_3$) higher than expected, so the structure of this peptide was assigned as Leu-Gla-Gla-Val. Morris, in collaboration with Magnusson et al.,[58,60] showed that the assignment of this tetrapeptide was incorrect and determined that the sequence is Phe-Leu-Gla-Gla instead. Morris et al.[60] demonstrated with the use of isotopically labeled reagents and high-resolution MS that permethylation of Gla produces a mixture of up to four different derivatives (**V–VIII**). This leads

$$
\begin{array}{cc}
\text{COOCH}_3 & \text{COOCH}_3 \\
| & | \\
\text{CH}_3-\text{C}-\text{COOCH}_3 & \text{CHCH}_3 \\
| & | \\
\text{CH}_2 & \text{CH}_2 \\
| & | \\
\ldots-\text{N}-\text{CH}-\text{CO}\ldots & \ldots--\text{N}-\text{CH}-\text{CO}\ldots \\
| & | \\
\text{CH}_3 & \text{CH}_3 \\
\textbf{V} & \textbf{VI}
\end{array}
$$

[51] E. W. Davie and K. Fujikawa, *Annu. Rev. Biochem.* **44,** 799 (1975).
[52] J. Stenflo, *Adv. Enzymol.* **46,** 1 (1978).
[53] C. M. Jackson and Y. Nemerson, *Annu. Rev. Biochem.* **49,** 765 (1980).
[54] C. T. Esmon, J. A. Sadowski, and J. W. Suttie, *J. Biol. Chem.* **250,** 4744 (1975).
[55] P. Fernlund, J. Stenflo, P. Roepstorff, and J. Thomsen, *J. Biol. Chem.* **250,** 6125 (1975).
[56] D. Bucher, E. Nebelin, J. Thomsen, and J. Stenflo, *FEBS Lett.* **68,** 293 (1976).
[57] J. Stenflo, P. Fernlund, W. Egan, and P. Roepstorff, *Proc. Natl. Acad. Sci. U.S.A.* **71,** 2730 (1974).
[58] S. Magnusson, L. Sottrup-Jensen, T. E. Petersen, H. R. Morris, and A. Dell, *FEBS Lett.* **44,** 189 (1974).
[59] G. L. Nelsestuen, T. H. Zytokovicz, and J. B. Howard, *J. Biol. Chem.* **249,** 6347 (1974).
[60] H. R. Morris, A. Dell, T. E. Petersen, L. Sottrup-Jensen, and S. Magnusson, *Biochem. J.* **153,** 663 (1976).

VII   VIII

to rather complex mass spectra that are difficult to interpret. An additional complication is that peptides suitable for MS had to be obtained by enzymatic digestion to avoid decarboxylation, but Gla-rich proteins are unusually resistant to proteolysis.

We have developed a generally applicable strategy for the amino acid sequence determination of peptides and proteins containing Gla which overcomes most of these problems.[61] Briefly, the method relies on the incorporation of deuterium, a stable isotope label, during deliberate decarboxylation, rather than trying to generate a stable derivative of Gla itself. The intact peptide or protein is heated *in vacuo* following lyophilization from 0.05 *M* DCl (dry decarboxylation).[62,63] Under these conditions, Gla quantitatively decarboxylates and incorporates two atoms of deuterium per molecule, resulting in the formation of γ,γ-dideuteroglutamyl residues. The incorporation of deuterium occurs specifically and irreversibly at Gla residues. Following conversion to the polyamino alcohol derivatives (Scheme 2), peptide fragments in which Gla was present show sequence ions in their mass spectra corresponding to those of glutamic acid, but shifted upward by 2 mass units for each Gla residue.

This strategy has been used to determine the complete primary structure of the Gla-containing calcium-binding proteins (called osteocalcin) from chicken and monkey bone.[64-66] For example, in a mixture of *N*-trifluorodideuteroethyl, *O*-trimethylsilyl oligopeptide derivatives obtained by partial acid hydrolysis and subsequent derivatization of a decarboxylated 200 nmol sample of reduced and carboxymethylated chicken osteocalcin the derivatives of 52 peptides representing approximately 90% of the peptide bonds in the protein were identified.[64] An example of the mass spectrum of one of them is shown in Fig. 5. It can be interpreted as the derivative of Pro-xLeu-Gla-Ala-Glu (xLeu denotes leucine or isoleucine) and serves to illustrate the simplicity of the resulting

[61] S. A. Carr and K. Biemann, *Biomed. Mass Spectrom.* **7**, 172 (1980).
[62] P. V. Hauschka, *Biochemistry* **18**, 4992 (1979).
[63] J. W. Poser and P. A. Price, *J. Biol. Chem.* **254**, 431 (1979).
[64] S. A. Carr, P. V. Hauschka, and K. Biemann, *J. Biol.Chem.* **256**, 9944 (1981).
[65] P. V. Hauschka, S. A. Carr, and K. Biemann, *Biochemistry* **21**, 638 (1982).
[66] P. V. Hauschka and S. A. Carr, *Biochemistry* **21**, 2538 (1982).

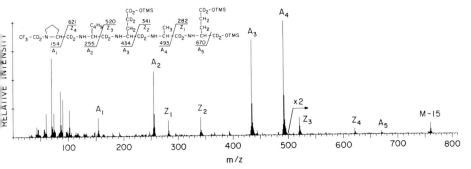

FIG. 5. Mass spectrum corresponding to the polyamino alcohol derivative of Pro-xLeu-Gla-Ala-Glu. The spectrum was produced during a GCMS experiment on a derivatized partial acid hydrolyzate of chicken osteocalcin. xLeu refers to either Leu or Ile; the distinction is made later in the analysis. OTMS, O-trimethysilyl (reproduced from ref. 64).

spectra and their interpretation.[2,64] All three Gla molecules present in the protein and their relative positions were determined in this way in a single experiment. The results are summarized in Fig. 6.

It has been suggested that posttranslational carboxylation of proteins may be a far more widespread phenomenon than previously thought.[67] The decarboxylation–GCMS approach could be used as a general procedure to screen previously sequenced or newly isolated proteins for other malonic acid derivatives such as β-carboxyaspartic acid.[68] This amino acid has indeed been identified in a total alkaline hydrolyzate of *E. coli* ribosomal protein.[69]

β-Hydroxyaspartic Acid (Bha). Recently, two groups using MS and proton NMR simultaneously and independently identified β-hydroxyaspartic acid (Bha) as a constituent of three Gla-containing vitamin K-dependent plasma proteins, human and bovine factor X, and protein C.[70,71] Although this posttranslationally derived amino acid (**IX**) had been found by mass spectrometry[27] in the antibiotic peptide duramycin[72] it had not been previously reported in proteins. It was apparently misidentified in

[67] P. M. Gallop and M. A. Paz, *Physiol. Rev.* **55**, 418 (1975).
[68] P. V. Hauschka, E. B. Henson, and P. M. Gallop, *Anal. Biochem.* **108**, 57 (1980).
[69] R. Christy, R. M. Barkley, T. H. Koch, J. J. Van Buskirk, and W. W. Kirsch, *J. Am. Chem. Soc.* **103**, 3935 (1981).
[70] T. Drakenberg, P. Ferlund, P. Roepstorff, and J. Stenflo, *Proc. Natl. Acad. Sci. U.S.A.* **80**, 1802 (1983).
[71] B. A. McMullen, K. Fujikawa, W. Kisiel, T. Sasagawa, W. N. Howald, E. Y. Kwa, and B. Weinstein, *Biochemistry* **22**, 2875 (1983).
[72] O. L. Shotwell, F. H. Stodola, W. R. Michael, L. A. Lindenfelser, R. G. Dworschack, and T. G. Pridham, *J. Am. Chem. Soc.* **80**, 3912 (1958).

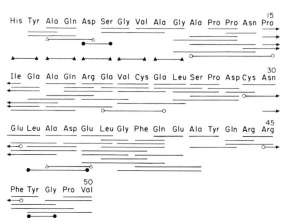

FIG. 6. Alignment of peptide sequences identified in GCMS experiments on chicken osteocalcin. ——, Partial acid hydrolyzate; ▲——▲, cathepsin C; ○——○, trypsin/thermolysin; ●——●, trypsin/proteinase K; △——△, pepsin. Partial sequences determined from GCMS experiments on enzymatic digests are shown only if they are not redundant with peptide derivatives identified in partial acid hydrolyzates of osteocalcin. Residues 1–16 were also identified by Edman degradation (reproduced from ref. 64).

the original sequence analyses of bovine factor X and protein C[73,74] due to the low extraction efficiency from the sequentator cup of the anilinothiazolinone derivative.

HOOC—CH—OH
|
H₂N—CH—COOH

IX

MW = 280

X

Using modified reaction conditions McMullen et al.[71] obtained the PTH derivative in good yield from human factor X. The methyl ester of the PTH derivative (**X**) was analyzed by chemical ionization MS using methane as the reagent gas. The spectra were identical to those of an

[73] D. L. Enfield, L. H. Ericsson, K. Fujikawa, K. A. Walsh, H. Neurath, and K. Titani, *Biochemistry* **19**, 659 (1980).
[74] P. Fernlund and J. Stenflo, *J. Biol. Chem.* **257**, 12170 (1982).

authentic sample of (**X**). The mass spectra exhibited abundant $(M + H)^+$ at m/z 281 and intense fragment ions at m/z 263, 231, and 193 corresponding, respectively, to the loss of H_2O, $H_2O + CH_3OH$, and the side chain (with concurrent H rearrangement, i.e., $+H$, $-CHOHCO_2CH_3$) from $(M + H)^+$.

Both Drakenberg et al.[70] and McMullen et al.[71] obtained mass spectra of acetylated and permethylated derivatives of peptides containing Bha. In the case of protein C, carboxymethyl-Cys-Ile-X-Gly-Leu-Gly-Gly was isolated and analyzed by direct insertion EIMS following conversion to the acetylated, permethylated derivative.[70] Although the interpretation of the mass spectra was complicated by the formation of at least four different products from the NH_2-terminal CMCys during permethylation the position of Bha could be established. A peptide of similar sequence containing the Bha residue was obtained from enzymatic digestion of human factor X: Cys-Lys-X-Gly-Leu-Gly-Glu-Tyr. Difficulties in interpretation of the mass spectra were avoided by subjecting the peptide to two cycles of manual Edman degradation prior to derivatization and mass spectrometry.[71] This is an example where FABMS on the underivatized isolated peptide fragments containing Bha would be a much more direct approach, thereby avoiding all of the problems associated with permethylation chemistry.

N-Methylphenylalanine. The identification of N-methylphenylalanine (**XI**) as the NH_2-terminal residue of pilin, a protein isolated from

XI

Pseudomonas aeruginosa, was less complex than that of Gla, but still presented problems for conventional sequence analysis methods.[75] This compound was not observed during amino acid analysis because the N-methylamine reacts slowly with ninhydrin. The presence of a modified amino acid was suspected as a result of the unusual chromatographic mobility of the PTH derivative obtained following one cycle of Edman degradation on the intact protein. The EI mass spectrum of this derivative exhibited peaks at m/z 296 $(M^{+\cdot})$, 205, 91, and 77, which confirmed the proposed structure.

[75] L. S. Frost, M. Carpenter, and W. Papanchych, *Nature (London)* **271**, 87 (1978).

C. Cross-Linking Components in Proteins

The biochemistry of intermolecular crosslinks in proteins has been the subject of intensive investigation over the past two decades. Cross-linking amino acids convey a range of properties including high tensile strength, rigidity, or elasticity to the proteins in which they are found. Such cross-links are essential for the structural and physiological integrity of the organism, and a number of inheritable diseases have been traced to defects in the genes coding for enzymes involved in this posttranslational covalent modification. Since this area has been extensively reviewed[67,76–79] only a brief attempt will be made here to acquaint the reader with the types of problems encountered, and of the mass spectrometric methods which have been used or are most applicable for their solution.

In collagen, lysine and hydroxylysine are enzymatically oxidized *in vivo* to the corresponding aldehydes which can then condense (intermolecularly) to form aldols or react with unoxidized amine side chains to form aldimine cross-links. Both the aldol and aldimine products are labile to heat and extremes of pH, precisely the conditions most commonly employed in efforts to isolate and characterize them. Reduction with borohydride has been frequently used to stabilize the *in vivo* products prior to their isolation by conversion to the corresponding alcohols and amines. However, neither the reagents nor the substrate are simple systems. Mixtures of products are often obtained, which has led to difficulties in interpretation of results and to misassignments of structure.

Mass spectrometry has provided key evidence to allow a number of structural controversies to be resolved. An early example involved the mistaken identification of α-amino aldehydes in collagen.[80,81] These workers showed by experiments in which specific activity was monitored that α-amino alcohols isolated from collagen reduced with tritiated sodium borohydride incorporated one rather than two atoms of tritium per molecule. On this basis they concluded that the precursor to the alcohol was an aldehyde carbonyl rather than an amide or ester carbonyl which would have incorporated two nonexchangeable tritium atoms upon reduction. However, mass spectrometric analysis of 2,4-dinitrophenyl-α-amino alcohols isolated from collagen reduced with $NaBD_4$ (labeled with tritium to

[76] M. L. Tanzer, *Science* **180**, 561 (1973).
[77] A. J. Bailey, S. P. Robins, and G. Balian, *Nature (London)* **251**, 105 (1974).
[78] D. R. Eyre, *Science* **207**, 1315 (1980).
[79] S. P. Robins, *Methods Biochem. Anal.* **28**, 329 (1982).
[80] P. M. Gallop, O. O. Blumenfeld, E. Hensen, and A. L. Schneider, *Biochem. J.* **7**, 2409 (1969).
[81] M. A. Paz, R. W. Lent, B. Farris, C. Franzblau, O. O. Blumenfeld, and P. M. Gallop, *Biochem. Biophys. Res. Commun.* **34**, 221 (1969).

facilitate isolation), clearly showed that the molecules incorporated two deuterium atoms at the carbonyl carbon.[82] This clearly eliminated the possibility of an aldehyde precursor and suggested that unanticipated peptide bond reduction had occurred.

The structure of the major reducible cross-link in collagen was also resolved using MS. The structures originally proposed (**XIIa** and **XIIb**) were suggested to have arisen by condensation of the oxidized (δ-semialdehyde) forms of lysine and hydroxylysine.[83,84] The correct structure of the isolated compound was later shown by MS to be dihydroxyly-sinonorleucine (**XIII**).[85–87] This is presumably formed by spontaneous *in*

vivo Amadori rearrangement of the Schiff base intermediate to the oxo compound which is subsequently reduced.[88,89] Fragments observed in the mass spectra of the acetyl permethyl, isobutyloxycarbonyl permethyl, and trifluoroacyl methyl ester were all consistent with this structure.

Isotopically labeled derivatizing reagents are particularly useful for obtaining additional structural information from low-resolution mass

[82] M. A. Paz, E. Hensen, R. Rombauer, L. Abrash, O. O. Blumenfeld, and P. M. Gallop, *Biochemistry* **9**, 2123 (1970).
[83] A. J. Bailey, L. J. Fowler, and C. M. Peach, *Biochem. Biophys. Res. Commun.* **35**, 663 (1969).
[84] A. J. Bailey, C. M. Peach, and L. J. Fowler, *Biochem. J.* **117**, 819 (1970).
[85] G. Mechanic and M. L. Tanzer, *Biochem. Biophys. Res. Commun.* **41**, 1597 (1970).
[86] G. Mechanic, P. M. Gallop, and M. L. Tanzer, *Biochem. Biophys. Res. Commun.* **45**, 644 (1971).
[87] N. R. Davis and A. J. Bailey, *Biochem. Biophys. Res. Commun.* **45**, 1416 (1971).
[88] S. P. Robins and A. J. Bailey, *FEBS Lett.* **38**, 334 (1974).
[89] D. R. Eyre and M. J. Glimcher, *Biochem. Biophys. Res. Commun.* **52**, 663 (1973).

spectra since the mass shifts observed can be used to determine the number and type of reactive groups present in the molecule.[27,90] Use of this "mass shift" technique has been central to structural studies on the cross-linking agents in collagen and elastin. These include the demonstration of the presence of isomers of desmosine (**XIV**) and isodesmosine (**XV**) as well as related compounds in lower oxidation states), in materials isolated from elastin,[91,92] and the identification of aldol-histidine (**XVI**) and histidinohydroxymerodesmosine (**XVII**) as potential cross-links in collagen.[90,93,94]

XIV   XV

XVI

The more subtle problems with the chemical methods used for stabilizing and isolating intermolecular bonds in connective tissues are the possibility that new structural elements may be generated that were not present in the native proteins, and that labile groups will be lost.[79] A more innoc-

[90] E. Hunt and H. R. Morris, *Biochem. J.* **135**, 833 (1973).
[91] M. A. Paz, B. Pereyra, P. M. Gallop, and S. Seifter, *J. Mechanochem. Cell Motil.* **2**, 231 (1974).
[92] M. A. Paz, P. M. Gallop, O. O. Blumenfeld, E. Hensen, and S. Seifter, *Biochem. Biophys. Res. Commun.* **43**, 289 (1971).
[93] M. L. Tanzer, R. Fairweather, and P. M. Gallop, *Biochim. Biophys. Acta* **310**, 130 (1973).
[94] T. Housley, M. L. Tanzer, E. Hensen, and P. M. Gallop, *Biochem. Biophys. Res. Commun.* **67**, 824 (1975).

$$
\begin{array}{c}
NH_2-CH-COOH \\
| \\
(CH_2)_3 \quad\quad HC{=\!=\!=}C^{CH_2-CH-COOH} \\
|\quad\quad\quad | \quad\quad\quad\quad | \\
COOH \quad\quad\quad\quad\quad\quad CH\text{———}N_{\diagdown C\diagup}N \quad\quad NH_2 \\
| \quad\quad\quad\quad\quad\quad\quad\quad\quad\quad | \\
CH-(CH_2)_2-CHOH-CH_2-NH-CH_2-CH \quad\quad H \\
| \quad\quad\quad\quad\quad\quad\quad\quad\quad\quad\quad\quad (CH_2)_2 \\
NH_2 \quad\quad\quad\quad\quad\quad\quad\quad\quad\quad\quad | \\
NH_2-CH-COOH
\end{array}
$$

XVII

uous approach might involve extensive or complete enzymatic digestion of unreduced collagen followed by HPLC separation of components and direct MS analysis by a soft ionization technique (see Section I). Soft ionization methods are increasingly being used in the analysis of underivatized cross-links. Recently, field desorption MS, NMR, and chemical synthesis were used to identify two new amino acid derivatives, lysinoalanine (**XVIII**) and histidinoalanine (**XIX**), in hydrolyzates of calci-

$$
\begin{array}{c}
HOOC-CH-(CH_2)_4-NH-CH_2-CH-COOH \\
| \quad\quad\quad\quad\quad\quad\quad\quad\quad\quad | \\
NH_2 \quad\quad\quad\quad\quad\quad\quad\quad\quad\quad NH_2
\end{array}
$$

XVIII

$$
\begin{array}{c}
HC{=\!=\!=}C^{CH_2-CH-COOH} \\
\,^5\quad^4\,| \quad\quad\quad | \\
^{1'}CH_2\text{———}N^{1\,\,2}_{\diagdown C\diagup}{}^{3}N \quad\quad NH_2 \\
| \quad\quad\quad\quad\quad\quad H \\
^{2'}CH-NH_2 \\
| \\
COOH
\end{array}
$$

XIX

fied tissue collagen.[95,96] Molecular ions and fragments related to the loss of alanine were observed in the FD mass spectrum of the latter compound. Another example involves pyridinoline (**V**), a recently proposed trifunctional cross-link in collagen which is postulated to form from two hydroxylylsyl aldehydes and a hydroxylylsine residue.[97-100] Both low- and high-resolution FAB mass spectra of this compound consist solely of a

[95] D. Fujimoto, M. Hirama, and T. Iwashita, *Biochem. Biophys. Res. Commun.* **103,** 1378 (1981).

[96] D. Fujimoto, M. Hirama, and T. Iwashita, *Biochem. Biophys. Res. Commun.* **104,** 1102 (1982).

[97] D. Fujimoto, K. Akiba, and N. Nakamura, *Biochem. Biophys. Res. Commun.* **76,** 1124 (1977).

[98] D. Fujimoto, T. Moriguchi, T. Ishida, and H. Hayashi, *Biochem. Biophys. Res. Commun.* **84,** 52 (1978).

[99] D. R. Eyre and H. Oguchi, *Biochem. Biophys. Res. Commun.* **92,** 403 (1980).

[100] D. F. Elsden, N. D. Light, and A. J. Bailey, *Biochem. J.* **185,** 531 (1980).

protonated molecular ion, the accurate mass of which confirms the proposed structure.[101]

$$
\begin{array}{l}
\overset{+}{H_3N}\diagdown \overset{H}{\underset{|}{C}} \diagup COO^- \\
 CH_2 \\
^-O \diagdown (CH_2)_2 - C \overset{H}{\underset{NH_3^+}{\diagup}} COO^- \\
 N^+ \\
 | \\
 CH_2 \\
 | \\
 CHOH \\
 | \\
 (CH_2)_2 \\
 | \\
\overset{+}{H_3N} \diagup \underset{H}{C} \diagdown COO^-
\end{array}
$$

Calculated MW = 428.1907
Observed $(M + H)^+$ = 429.1995
Calculated $(M + H)^+$ = 429.1985
diff. = 1.0 mmu

D. Glycoproteins

The task of unraveling the complete structure of a glycoprotein is extraordinarily complex. In addition to determining the sequences of the protein and carbohydrate, and the linkage and configuration of saccharide monomers, the carbohydrate–protein link must also be ascertained. The MS methods that have been developed for the elucidation of the carbohydrate moiety are beyond the scope of the present report, but have been reviewed elsewhere,[102] including applications of the newest soft ionization techniques.[103] Elucidation of the protein–carbohydrate linkage is often quite difficult and usually involves extensive proteolytic digestion to obtain the intact carbohydrate attached to a small peptide, or preferably, a single amino acid.[104]

Attempts to analyze glycopeptides by MS have met with varying degrees of success. The EI mass spectrum of an N-acetylated, permethylated, LiAlH$_4$ reduced glycopeptide from human transferrin allowed independent confirmation of the structure of the asparagine-linked nonasaccharide. Unfortunately, no information was obtained from the peptide portion of the molecule.[105]

More successful was the characterization of the O-glycosidically

[101] S. A. Carr, D. R. Eyre, and K. Biemann, unpublished observations.
[102] T. Radford and D. C. DeJongh, in "Biochemical Applications of Mass Spectrometry, First Supplementary Volume" (G. R. Waller and O. C. Dermer, eds.). Wiley (Interscience), New York, 1980.
[103] V. N. Reinhold and S. A. Carr, Mass Spectrom. Rev. 2, 153 (1983).
[104] N. Sharon, "Complex Carbohydrates: Their Chemistry, Biosynthesis and Function." Addison-Wesley, Reading, Massachusetts, 1975.
[105] K.-A. Karlsson, I. Pascher, B. E. Samuelsson, J. Finne, T. Krusius, and H. Ravala, FEBS Lett. 94, 413 (1978).

linked antifreeze glycopeptides isolated from the Antarctic fish *Trematomus borchgrevinki*.[106] Unlike the N-linked sugars, O-linked sugars are extremely labile under alkaline conditions.[104] In this case permethylation of the glycopeptide resulted in β-elimination of the carbohydrate side chain, an event which had several fortuitous consequences for the mass spectrometric analyses. Residues to which carbohydrate was formerly attached were converted to the corresponding dehydroamino acids. Although the permethyl derivative of dehydro-Thr in the peptide linkage has the same incremental mass as a Pro, the two were easily distinguished in a separate experiment utilizing CD_3I instead of CH_3I in the methylation step, which shifts the mass of the dehydro-Thr residues upward by 3 daltons but does not affect Pro residues. Thus the attachment sites of carbohydrate and peptide sequence were unambiguously established in two experiments. In addition, the eliminated sugar portion, which also permethylated during derivatization, turned out to be a homogeneous disaccharide. Owing to its small size this carbohydrate vaporized from the direct insertion probe of the mass spectrometer at a temperature well below that at which any peptide derivatives vaporize. Thus a "clean" spectrum of the permethylated disaccharide was also obtained without prior separation of this component from the mixture.

IV. Conclusions and Outlook

It is clear that mass spectrometry is a technique well suited for the detection of modified amino acids and for the determination of their structure. The more recently developed soft ionization methods eliminate the need for prior chemical derivatization which may destroy sensitive functional groups, requires more material and is often tedious.

Fast atom bombardment mass spectrometry has opened the way to tackle much larger molecules such as peptides up to 10,000 daltons. As outlined in Section III, if the "normal" amino acid composition is known one can deduce the mass of the modifying moiety from the molecular weight of the peptide as determined by FAB MS.

However, if the amino acid composition is not known molecular weight information alone is not of much use. In that case it is necessary to deduce as much sequence information as possible from the FAB mass spectrum. In most cases this will require enhancement of the abundance of fragment ions. This can be achieved by selecting the protonated molecular ion of the peptide using the first stage of a mass spectrometer, inducing this ion to fragment by collision with a gas (collision induced decom-

---

[106] H. R. Morris, M. R. Thompson, D. T. Osuga, A. I. Ahmed, S. M. Chan, J. R. Vandenheede, and R. E. Feeney, *J. Biol. Chem.* **253**, 5155 (1978).

position, CID) and then determining the mass of the decomposition products in second stage of analysis. These techniques, collectively referred to as mass spectrometry/mass spectrometry, have been demonstrated to yield sequence information for peptides of known structure[107–110] and to solve a sequence question on a cyclic peptide.[111] Using the same approach it should be possible to deduce the position and size, if not precise structure, of a modified amino acid present in peptides having a mass of many thousand daltons.

[107] J. J. Amster, M. A. Baldwin, M. T. Chang, C. J. Proctor, and F. W. McLafferty, *J. Am. Chem. Soc.* **105**, 1654 (1983).

[108] W. Heerma, J. P. Kamerling, A. J. Slotboom, G. J. M. van Scharrenburg, B. N. Green, and I. A. S. Lewis, *Biomed. Mass Spectrom.* **10**, 13 (1983).

[109] D. F. Hunt, W. M. Bone, J. Shabanowitz, J. Rhodes, and J. M. Ballard, *Anal. Chem.* **53**, 1704 (1981).

[110] D. Desiderio and I. Katakuse, *Anal. Biochem.* **129**, 425 (1983).

[111] M. L. Gross, D. A. McCrery, F. W. Crow, K. B. Tomer, M. R. Pope, L. M. Ciuffetti, H. W. Knoche, J. M. Daly, and L. D. Dunkle, *Tetrahedron Lett.* **23**, 5381 (1982).

[5] Complete Enzymatic Hydrolysis of Wool and Its Morphological Components

By Karin Röper, J. Föhles, and H. Klostermeyer

Enzymatic hydrolysis of proteins is of interest for a number of reasons, covering scientific objectives from the assessment of the nutritional value of proteins in food science and the gentle and limited alteration of cells and cellular organelles in cell biology, to the chemical analysis of protein structure in biochemistry. One of the major reasons for the efforts to achieve complete enzymatic hydrolysis of proteins is that it allows quantitative recovery of amino acids that are unstable in acid or alkali, both primary amino acids (e.g., asparagine, glutamine, tryptophan) and many of the co- and posttranslationally modified secondary amino acids which are the subjects of this volume. In this chapter the enzymatic hydrolysis of wool and its morphological components is described as an illustration of this latter objective.

The literature on the solubilization with the aid of proteases of native keratins, such as wool and human hair, is vast and has been reviewed comprehensively by Bradbury.[1] Since these proteins are rich in disulfide

[1] J. A. Bradbury, *Adv. Protein Chem.* **27**, 111 (1973).

bonds, and it is clearly established that the extent of protease attack is limited by the degree of cross-linking,[2] reduction of disulfide bonds is a prerequisite for extensive hydrolysis of keratins. Complete enzymatic hydrolysis of keratin fibers was described independently by Milligan *et al.*[3,4] and Cole *et al.*[5] both using a system introduced by Hill.[6] The method was improved by Schmitz *et al.* in 1975[7] and the following values were found for the content of unstable amino acids in wool: Gln, 466 μmol/g; Asn, 230 μmol/g; N^{ε}-(γ-glutamyl)lysine (Glu Lys), 11–13 μmol/g. More recently Hubbach[8] tested the known procedures for complete enzymatic hydrolysis for application to human hair and its morphological subunits. Because of inconsistent results she varied the system described by Schmitz[7] and recommended the following procedure: reduction and digestion with pronase is carried out simultaneously without carboxymethylation of cystine. After 24 hr, aminopeptidase and prolidase are added, and the digestion is allowed to proceed for an additional 24 hr.

The present paper deals with the enzymatic hydrolysis of wool, wool cuticle and its morphological subunits with respect to the distribution of diagnostic components in the fiber. As wool is subjected to drastic chemical treatments during industrial processing, in some cases specific degradation of definite morphological components, for example, exocuticle[9–11] and nonkeratinous proteins,[12] has been observed. The results of complete enzymatic hydrolysis are expected to support the findings made by means of electron microscopy.

Materials

For the experiments a wool top, which was chemically untreated (merino quality, ϕ 21.9 μm), was extracted with dry methylene chloride for

[2] J. A. Maclaren and B. Milligan, "Wool Science, The Chemical Reactivity of the Wool Fibre." Science Press, Marrickville, New South Wales, Australia, 1981.
[3] B. Milligan, L. A. Holt, and J. B. Caldwell, *Appl. Polym. Symp.* **18**, 113 (1971).
[4] L. A. Holt, B. Milligan, and C. M. Roxburgh, *Aust. J. Biol. Sci.* **24**, 509 (1971).
[5] M. Cole, J. C. Fletcher, K. L. Gardner, and M. C. Corfield, *Appl. Polym. Symp.* **18**, 147 (1971).
[6] R. L. Hill and W. R. Schmidt, *J. Biol. Chem.* **237**, 389 (1962).
[7] I. Schmitz, H. Baumann, and H. Zahn, *Proc. Int. Wool Text. Res. Conf., 5th, 1975* p. 313 (1976).
[8] M. Hubbuch, Thesis, RWTH Aachen (1981) [parts are prepared for publication in *J. Soc. Cosmet. Chem.* in press].
[9] G. Blankenburg and P. Kassenbeck, *Melliand Textilber./Int. Text. Rep. (Ger. Ed.)* **62**, 478 (1981).
[10] W. de Fries, U. Altenhofen, J. Föhles, and H. Zahn, *J. Soc. Dyers Colour.* **98**, 13 (1983).
[11] R. Hagege and M. Bauters, *Bull. Sci. Inst. Text. Fr.* **10**, 115 (1981).
[12] H. Baumann, *Forschungsber. Landes Nordrhein-Westfalen* **2465** (1975).

4 hr. The chemicals were of analytical grade: sodium dodecyl sulfate (Merck), dithioerythritol (DTE) (Serva), Triton X-100 (Serva), tris(hydroxymethyl)aminomethane (Merck), Pronase (Boehringer), aminopeptidase M (Merck), and prolidase (proline dipeptidase) (Sigma). The equipment used included a Shaker bath with thermostat (Julabo VL), centrifuge (Beckman J 21 C with Rotor JS-13), amino acid analyzer (Biotronic LC 6000 E) and a Starmix mixer.

Methods and Results

Isolation of Wool Cuticle

The cuticle cells were liberated by mechanical treatment of the fibers in a sodium dodecyl sulfate solution according to the method described by Ley et al.[13,14] With this method sufficient amounts of highly purified cuticle cell fragments for analytical purposes can be isolated in a short time.

Three grams of fiber material (cut to a length of about 3 mm) was mechanically treated in a mixer (consisting of a 1.5-liter vessel with a rotating knife on the bottom) with 150 ml of a 1% sodium dodecyl sulfate (SDS) solution. The total time of treatment was 10 min at 0.5- to 1-minute intervals with intermittent cooling on ice to keep the temperature below 20°. The vessel was chilled in the refrigerator overnight, effecting the collapse of the foam. The resulting suspension was filtered through a 100-μm steel screen and centrifuged (12,000 rpm, 30 min, 20°). The supernatant was collected and lyophilized for later examination of the SDS-soluble proteins. Residual SDS in the precipitate was removed by repeated washing with water/ethanol, increasing the content of ethanol in each succeeding washing step. The cell fragments were suspended in pure ethanol (10 mg/100 ml) and purified by sieving through steel screens with decreasing porosity (40, 28, 20 μm). Purity was controlled by light and electron microscopy. In the case of chemically untreated fibers a short mechanical pretreatment in water was necessary for the separation of skin flakes.

The composition of the small amount of SDS-soluble protein was determined by amino acid analysis after acid hydrolysis. It was rich in aspartic acid and glycine and contained a smaller amount of proline and cystine than typical wool proteins. The main fraction of the soluble proteins could be separated from the large excess of SDS by precipitating with acetone and a saturated solution of ammonium sulfate. In this fraction Glu Lys could be detected after enzymatic hydrolysis, but the soluble

[13] K. F. Ley and W. G. Crewther, Proc. Int. Wool Text. Res. Conf., 6th, 1980 p. 13 (1980).
[14] R. C. Marshall, personal communications.

protein was not found to originate from a definite morphological component.

Isolation of Endocuticle and Exocuticle Plus Cell Membrane Complex

"Keratins" and "nonkeratinous proteins" have been defined on the basis of relative resistance to digestion by proteolytic enzymes and results from microscopic studies. Nonkeratinous proteins are poor in cystine cross-links and belong to the morphological subunits endocuticle and cell membrane complex as well as to the intermacrofibrillar material and the nuclear remnants.[15-18] Recent findings based on protein fractionation have shown that the keratinous part of the wool fiber is not actually attacked by Pronase. Former doubts regarding the integrity of Pronase-treated keratins were found to be groundless.[19] In the case of wool cuticle, the keratins from the exocuticle and the proteins from the cell membrane complex are resistant to pronase digestion, while the nonkeratins from the endocuticle are dissolved.[20]

In our own experiments the separation of nonkeratinous material from the endocuticle and keratinous material from the exocuticle including the resistant membrane fractions was achieved by means of Pronase digestion following strictly the procedure described by Hubbuch.[8]

One hundred milligrams of isolated cuticle cells was incubated with 2 mg Pronase in 20 ml Tris buffer (0.05 M, pH 8.25) by shaking with a frequency of 140 rpm at 39° for 24 hr. The cell suspension was centrifuged (11,000 rpm, 30 min, 20°) and the supernatant was transferred to a 50-ml calibrated flask. The precipitate was washed twice with 5 ml of buffer and the wash liquors were added to the supernatant. The flask containing the Pronase-soluble proteins (endocuticle) was filled up to the mark and aliquots were subjected to further enzymatic hydrolysis or acid hydrolysis. The Pronase-resistant residue (exocuticle and cell membrane complex) was washed three times with 10 ml of water and was finally dried under reduced pressure over P_2O_5.

Complete Enzymatic Hydrolysis of Wool, Cuticle, and Its
*   Morphological Subunits*

The enzymatic hydrolysis of keratin components was carried out according to the method of Hubbuch.[8] For wool components the extent of

[15] P. H. Springell, *Aust. J. Biol. Sci.* **16,** 727 (1963).
[16] E. H. Mercer, *Text. Res. J.* **23,** 388 (1953).
[17] M. S. C. Birbeck and E. H. Mercer, *J. Biophys. Biochem. Cytol.* **3,** 203 (1957).
[18] M. S. C. Birbeck and E. H. Mercer, *J. Biophys. Biochem. Cytol.* **3,** 215 (1957).
[19] R. Greven, M. L. Klotz, and H. Zahn, *Text. Res. J.,* in press.
[20] J. H. Bradbury and K. F. Ley, *Aust. J. Biol. Sci.* **25,** 1235 (1972).

enzymatic hydrolysis was found to be poorer than that described for hair components. Therefore, the simultaneous treatment with Pronase and DTE was repeated and the total time of digestion was extended 72 hr. This modification led to acceptable rates of hydrolysis for wool and endo-cuticle as substrates. For the more resistant components, cuticle and exocuticle, the activity of some additional enzymes was studied. The use of proteinase K,[21] proteinase S,[22] thermolysin,[23] and collagenase[24] did not result in an increase in the rate and extent of hydrolysis. The cell membrane complex had been found to be particularly resistant to the attack of proteases. Hubbuch succeeded in obtaining a slight improvement of digestibility by addition of the nonionic detergent Triton X-100,[8] and the influence of detergents was investigated in this work. While the ionic detergents sodium dodecyl sulfate and N-dodecyl-N,N-dimethylamino-3-propane-1-sulfonic acid caused a decrease in the rate of enzymatic hydrolysis, addition of 0.1% Triton X-100 (in the buffer solution) resulted in a 12% increase on average for cuticle digestion from chemically untreated and treated wool. These observations led to the following procedure.

Five milligrams of substrate was incubated with 0.1 mg Pronase and 2.3 mg dithioerythritol in 1 ml Tris buffer (0.05 M, pH 8.25; in the case of cuticle and exocuticle the buffer contained 0.1% Triton X-100) at 39° for 24 hr, shaking with a frequency of 140 rpm. This procedure was repeated for another 24 hr. After 48 hr 0.1 mg aminopeptidase M and 0.25 mg prolidase were added, and the incubation was continued under the same conditions for 24 hr. The digestion was finally stopped by acidifying with crystals of citric acid, and the hydrolyzates were immediately diluted for automatic chromatographic separation of the amino acids. In the case of the endocuticle, an aliquot of the solution from the Pronase digestion containing approximately 5 mg of substrate was incubated with 0.1 mg aminopeptidase M and 0.25 mg prolidase for 24 hr.

Extent of Enzymatic Hydrolysis and the Distribution of Diagnostic Components

Results for unstable amino acids are listed in Table I. The cell membrane complex examined was isolated by means of papain/DTE incubation.[25]

[21] W. Ebeling, N. Hennrich, M. Klockow, H. Metz, H. D. Orth, and H. Lang, *Eur. J. Biochem.* **47**, 91 (1974).
[22] G. R. Drapeau, Y. Boily, and J. Houmard, *J. Biol. Chem.* **247**, 6720 (1972).
[23] M. B. Lees and D. S. Chan, *Neurochemistry* **25**, 595 (1975).
[24] S. Seifter and E. Harper, this series, Vol. 19, p. 613.
[25] A. Schwan, Thesis, RWTH Aachen (1981) [parts are prepared for publication in *Text. Res. J.* (1983)].

TABLE I
EXTENT OF ENZYMATIC HYDROLYSIS AND CONTENT OF DIAGNOSTIC
COMPONENTS OF THE INDIVIDUAL MORPHOLOGICAL SUBUNITS

Morphological component	Extent of enzymatic hydrolysis (%)	Mol%			
		Glu Lys	Cit	Asn	Gln
Wool	84	0.18	0.23	3.36	3.60
Cuticle	67	0.56	0.88	2.02	3.33
Endocuticle	84	0.76	1.15	3.15	2.72
Exocuticle + CMC	65	0.50	0.49	0.88	3.31
Cell membrane complex	35	0.27	0.15	3.98	0

The components citrulline, asparagine and N^{ε}-(γ-glutamyl)lysine, which have a lability comparable to the action of acids, were found to be particularly rich in the endocuticle. For glutamine a very low value was detected, which must be attributed to the known instability of the free species. Glu Lys bonds are thought to have the function of strengthening extracellular proteins and rendering them insoluble.[26,27] It has been proposed that these structures are preferably formed between membrane proteins,[28] and since nonproliferating cell membranes were found to contain more isopeptide bonds than those of actively dividing cells,[29] Glu Lys cross-links are thought to play a unique structural role in cell membranes.[30] The relatively small amount of Glu Lys bonds detected in the cell membrane fraction is not consistent with this hypothesis. Experiments aiming at an exhaustive enzymatic proteolysis by means of a method introduced by Rice and Green[31] gave no evidence for large amounts of isopeptide cross-links in the cell membrane complex. Addition of phospholipase to remove the protein/lipid interaction[32] did not result in an increase in the rate of enzymatic hydrolysis.[8]

The extent of enzymatic hydrolysis obtained for the different wool components is illustrated in Fig. 1. The values found after enzymatic

[26] A. G. Loewy, S. S. Matacic, and M. Showe, Fed. Proc., Fed. Am. Soc. Exp. Biol. 30, 1275 (1971).
[27] S. S. Matacic and A. G. Loewy, Biochim. Biophys. Acta 576, 263 (1979).
[28] P. J. Birckbichler, R. M. Dowben, S. S. Metacic, and A. G. Loewy, Biochim. Biophys. Acta 291, 149 (1973).
[29] P. J. Birckbichler, H. A. Carter, J. R. Orr, E. Conway, and M. K. Patterson, Biochem. Biophys. Res. Commun. 84, 232 (1978).
[30] R. B. Haughland, T. I. Lin, R. M. Dowben, and P. J. Birckbichler, Biophys. J. 37, 191 (1982).
[31] R. H. Rice and H. Green, Cell 11, 417 (1977).
[32] W. W. Christie, "Lipid Analysis." Pergamon, Oxford, 1973.

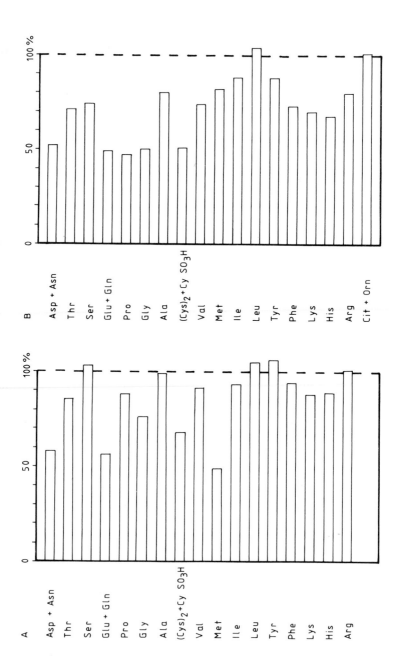

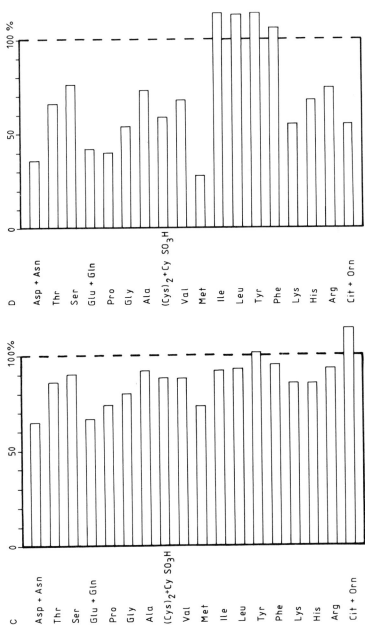

FIG. 1. Extent of enzymatic hydrolysis of wool and its morphological components. (A) Wool; (B) cuticle; (C) endocuticle; (D) exocuticle.

hydrolysis were corrected for the amounts of amino acids formed by autolysis of the enzyme mixture. Calculations of the extent of enzymatic hydrolysis are based upon the amounts released by acid hydrolysis.

Particularly poor results were obtained for the total amount of Asp + Asn and Glu + Gln, which can be attributed to the high resistance of the acidic amino acids Asp and Glu and to the sensitivity of Gln. (In the enzymatic hydrolysate of a substrate containing Asn; but no Asp, 94% of the theoretical amount of Asn was detected.) The amino acids Gly, Pro, Met and Cys were also released in a lower than average yield. On the other hand the amino acids Ile, Leu, Tyr, and Phe seemed to be completely released even to the point of exceeding 100% yield. This finding may be due to degradation or incomplete release during acid hydrolysis. The striking fact that the tendency to exceed 100% yields increases with

TABLE II

AMINO ACID COMPOSITION OF WOOL, CUTICLE, ENDOCUTICLE AND EXOCUTICLE +
CELL MEMBRANE COMPLEX OBTAINED BY COMPLETE ENZYMATIC HYDROLYSIS[a]

Amino acid	Wool	Cuticle	Endocuticle	Exocuticle + CMC
CySO$_3$H	0.43 ± 0.02	1.66 ± 0.01	1.02 ± 0.02	1.98 ± 0.08
Asp	1.35 ± 0.02	0.86 ± 0.02	1.96 ± 0.03	0.25 ± 0.04
Thr	6.54 ± 0.01	5.35 ± 0.02	6.06 ± 0.12	4.95 ± 0.03
Ser	12.53 ± 0.3	17.53 ± 0.17	12.89 ± 0.14	21.52 ± 0.42
Glu	4.63 ± 0.07	3.37 ± 0.01	4.96 ± 0.02	2.27 ± 0.12
Pro	6.61 ± 0.18	7.11 ± 0.09	6.16 ± 0.49	8.13 ± 0.33
Gly	8.18 ± 0.26	6.71 ± 0.01	7.16 ± 0.11	7.42 ± 0.18
Ala	6.30 ± 0.15	7.62 ± 0.05	7.03 ± 0.06	7.03 ± 0.07
(Cys)$_2$	4.33 ± 0.24	5.51 ± 0.01	1.49 ± 0.03	8.72 ± 0.19
Val	6.65 ± 0.02	8.22 ± 0.22	7.22 ± 0.05	7.62 ± 0.27
Met	0.30 ± 0.01	0.48 ± 0.03	1.01 ± 0.06	0.04 ± 0.01
Ile	3.74 ± 0.05	3.30 ± 0.02	4.34 ± 0.06	2.59 ± 0.03
Leu	9.82 ± 0.07	9.95 ± 0	10.40 ± 0.18	8.32 ± 0.11
Tyr	5.17 ± 0.06	3.81 ± 0.14	5.16 ± 0.13	3.70 ± 0.22
Phe	3.34 ± 0.10	2.00 ± 0.00	3.63 ± 0.1	1.65 ± 0.07
Lys	3.02 ± 0.08	2.75 ± 0	3.95 ± 0.1	1.64 ± 0.04
His	0.89 ± 0.03	0.92 ± 0.01	1.45 ± 0.03	0.67 ± 0.02
Arg	8.90 ± 0.06	6.03 ± 0.10	6.32 ± 0.18	5.75 ± 0.16
Gln	3.60 ± 0.11	3.33 ± 0.01	2.72 ± 0.06	3.31 ± 0.05
Cit	0.23	0.88 ± 0	1.15 ± 0.02	0.49 ± 0.02
Glu Lys	0.18 ± 0.03	0.56 ± 0	0.76 ± 0.02	0.50 ± 0.02
Asn	3.36 ± 0.16	2.02 ± 0.12	3.15 ± 0.19	0.88 ± 0.15

[a] Mol% ± q (95%).

TABLE III
EXTENT OF ENZYMATIC HYDROLYSIS AND CONTENT OF GLY
AND PRO IN THE INDIVIDUAL MORPHOLOGICAL COMPONENTS

Morphological component	Extent of enzymatic hydrolysis (%)	Gly (mol%)	Pro (mol%)
Wool	84	9.04	6.36
Cuticle	67	9.07	10.08
Endocuticle	84	7.55	7.00
Exocuticle + CMC	65	9.02	13.25
Cell membrane complex	35	9.8	4.8

decreasing extent of enzymatic hydrolysis may also suggest that peptides resistant to enzymatic hydrolysis (like Gly–Gly) interfere with the quantitation of the amino acids named above during automatic amino acid analysis.

In Table II the compositions of the amino acids after enzymatic hydrolysis are given for wool, cuticle, and its morphological subunits. The values are corrected for the amounts of amino acids formed by autolysis of the enzyme mixture.

Inhibitors for the Complete Enzymatic Hydrolysis Originating from the Substrate Preparations

Hubbuch[8] postulated a relationship between the extent of enzymatic hydrolysis obtained for the individual morphological components and their content of glycine. Comparing the values for glycine in the exocuticle of wool and hair, respectively, this postulate predicts that the exocuticle from wool should be degraded more extensively by proteolytic enzymes. Our experiments did not confirm this prediction, nor do the data in Table III. While for the cuticle and its morphological subunits this relationship seems to be possible, the relatively large amount of glycine in the wool fiber and the high extent of enzymatic hydrolysis are not consistent with the prediction. A relationship between the extent of enzymatic hydrolysis and the content of proline can equally well be proposed from the data in Table III, but again the cell membrane fraction represents a notable exception. All our attempts to find a correlation between the extent of enzymatic hydrolysis and the specific composition of the substrates by means of building a quotient according to the method of Zahn[33] were unsuccessful.

From the results in Table III it is noticeable that the cell membrane complex is particularly resistant to enzymatic proteolysis. While the whole fiber and the endocuticle are easily degradable the digestibility of the cuticle and exocuticle, which contain the cell membrane fraction, is reduced. These findings suggest a relationship between the extent of enzymatic hydrolysis and the content of lipids, which could also explain the different extent of enzymatic hydrolysis obtained for hair[8] and wool components. Hubbuch isolated the cuticle of hair fibers using the method introduced by Swift.[34] In this case the cuticle structure is fractured in the endocuticle. Using the SDS treatment for isolating cuticle cells, the fracture was found to occur in the cell membrane complex.[14] Considering these facts it is reasonable to assume that our preparation of cuticle from wool had a higher content of membrane lipids than did cuticle from hair. In order to test this assumption a cuticle preparation isolated by conventional SDS treatment was extracted with chloroform/methanol[25] to separate the membrane lipids. The residue was hydrolyzed by acidic and enzymatic methods. The increase found for the extent of enzymatic hydrolysis was not appreciable. Nevertheless, the results of these experiments do not exclude a possible dependence of digestibility on the content of lipids, as the solvent treatment may adversely affect the digestibility of the protein.

Conclusion

The procedure for complete enzymatic hydrolysis of keratins was improved for application to wool keratins. With this method the results of amino acid analysis after enzymatic hydrolysis were in good agreement with those obtained after acidic hydrolysis. Diagnostic components could be quantitated, permitting a correlation between morphological and chemical changes in the fiber to be made. With regard to the expected indicators of morphological changes caused by chemical treatments, the quantitation of asparagine, which is set free by proteolytic enzymes very easily, is of great promise. Similarly the behavior of the isopeptide crosslink Glu Lys during industrial processing is of special interest. Dependencies were sought for the digestibility and the amount of special amino acids and the content of lipids in the substrate. Final evidence on these

[33] H. Zhan, J. Föholes, P. Kusch, and A. Nadzeyka, *Schriftenr. Dtsch. Wollforschungsinst.* (*Tech. Hochsch. Aachen*), *Sonderb.* **1**, 90 (1979).

[34] J. A. Swift and B. Bews, *J. Soc. Cosmet. Chem.* **25**, 13 (1974).

points can be expected from sequencing the individual keratins and investigating synthetic model peptides.

Acknowledgments

Thanks are due to Forschungskuratorium Gesamttextil (AIF/No. 5127), the Australian Wool Corporation (AWC/No. TT/M/1202), and the International Wool Secretariat, Ilkley, London and Düsseldorf for financial support. Furthermore, the authors wish to thank Ms. D. Peters for careful technical assistance.

[6] Enzymatic Hydrolysis of Native, Soluble Proteins and Glycoproteins

By CHRISTOPHER C. Q. CHIN and FINN WOLD

During the last couple of decades a good deal of work has been devoted to the search for a combination of proteases that will permit complete hydrolysis of all the peptide bonds of both simple and extensively derivatized proteins. The main goal of these efforts is to make possible the complete analysis of all constituent amino acids in proteins, including those that are destroyed by routine acid (or alkaline) hydrolysis conditions. Some of this work has been reviewed,[1-3] and while there has been some notable success in the digestion of individual proteins and in isolating and quantifying specific, acid-labile amino acid derivatives in individual cases, it appears safe to conclude that at this stage there is no universal protocol for the combination of proteases and the digestion conditions that is likely to give complete digestion of any and all proteins.

The present chapter is included to make and document some practical points that probably have general validity whenever a new protein substrate is subjected to enzymatic hydrolysis for the purpose of achieving as complete a release of the constituent amino acids as possible. These points are well established in the literature, but nevertheless may be restated: (1) Protein substrates should be denatured to achieve optimal rate (and perhaps also extent) of hydrolysis. For native, globular proteins stable at neutral pH, we found that an initial digestion with an acid protease at pH 2–3 greatly facilitates the subsequent digestion with the more

---

[1] R. L. Hill, *Adv. Protein Chem.* **23,** 63 (1965).
[2] G. P. Royer, W. E. Schwartz, and F. A. Liberatore, this series, Vol. 47, p. 40.
[3] K. Röper, J. Föhles, and H. Klostermeyer, this volume [5].

METHODS IN ENZYMOLOGY, VOL. 106

commonly used neutral proteases. (2) Cross-links, notably disulfide bridges, significantly interfere with the digestion of the peptide bonds near the involved Cys residues,[3] and, as illustrated by the data below, proteins rich in disulfide bonds give incomplete release of amino acids under conditions where disulfide-free proteins appear to be completely hydrolyzed. (3) The co- and posttranslational amino acid derivatives may themselves seriously interfere with enzyme-catalyzed peptide bond cleavage. The refractory disulfide-containing peptides above represent only one example of this phenomenon; others are the protease resistance of the γ-carboxyglutamate-containing sequences in prothrombin[4] and of highly glycosylated glycopeptides[5] which will be illustrated below. Thus if the purpose of the enzymatic digestion of a protein is to liberate and characterize a given derivatized amino acid, the possibility must be considered that that derivative will not be released by the proteases used.

Methods and Results

The study to be presented was designed specifically to assess the digestion of proteins without any prior chemical treatment (oxidation, reduction, heat denaturation, etc.). The proteases used were all attached to an insoluble matrix (either aminopropyl glass beads or activated agarose) by standard methods.[6] This permitted the use of very high enzyme/substrate ratios (1/25–1/2) during the digestion and also made it easy to recover and reuse the various enzyme preparations. The following enzymes were used: crude acid aspergillopeptidase ("Molsin," Seishin Pharmaceutical Corp., Japan), Pronase (Sigma Chem. Co.), kidney aminopeptidase M (Henley & Co.), and prolidase (proline dipeptidase) (Miles Laboratories, Inc.). The routine digestion involved the initial digestion of 1–5 mg of protein substrate with aspergillopeptidase in 10^{-2} M HCl for 10–16 hr at 37°. The protease beads were removed by filtration and washed, and the combined filtrate and wash was lyophilized. The partial digest was then redissolved in 3–4 ml of 0.1 M sodium borate buffer, containing 10^{-3} M Ca^{2+}, pH 7.6, and the digestion was continued for an additional 16 hr with a simultaneous exposure to insolubilized Pronase and aminopeptidase M achieved by continuously pumping the substrate solution through two separate columns, each containing one of the two proteases. The digest was again collected by filtration and washing, Mn^{2+} was added to give a 10^{-2} M solution, and insolubilized prolidase was added. This final digestion step was allowed to proceed for 3–5 hr and

[4] G. L. Nelsestuen and J. W. Suttie, *Proc. Natl. Acad. Sci. U.S.A.* **70**, 3366 (1973).
[5] N. N. Aaronson and C. deDuve, *J. Biol. Chem.* **243**, 4564 (1968).
[6] C. C. Q. Chin and F. Wold, *Anal. Biochem.* **61**, 379 (1974).

after removing the protease by filtration the digest was subjected to analysis. The details of the digestion procedures have been described.[6]

Two groups of substrates were subjected to this digestion protocol, one consisting of common, well-characterized proteins with different disulfide content, pancreatic ribonuclease, egg white lysozyme, yeast enolase, and insulin, and the other consisting of glycoproteins with different carbohydrate types and content, ovalbumin, γ-globulin, and submaxillary mucin. For the first group the analysis of each digested protein consisted of direct amino acid analysis (Beckman Model 120 C) of the digest in comparison with parallel samples of the same protein sample subjected to acid hydrolysis (6 N HCl, 110°, 21 hr) with norleucine added as an internal standard. For the second group of substrates, the glycoproteins, the question to be explored was whether the glycosylated amino acids could be liberated from the polypeptide backbone as free derivatized amino acids, glycosyl-Asn from ovalbumin and IgG (immunoglobulin G) and glycosyl-Asn and glycosy-Ser (Thr) from the mucin. To this end, the glycosylated digestion products were separated from free amino acids (and small peptides) by gel filtration, and the purified carbohydrate-containing fractions were then analyzed for amino acid content after acid hydrolysis.

The results from the first group of substrates are given in the table as a percentage of each residue from the enzymatic digestion. They show that using this particular protocol, the proteolytic cleavage is quite complete for all residues except Cys. Enolase (from yeast), which has no disulfide bonds, and only one Cys residue per 45,000-dalton subunit (the Cys content is so low that the precision of the analysis is very poor) gave the most complete recovery of all the amino acids, while insulin with the highest disulfide content gave the most erratic results. However, the data in the table suggest that the main cause of the erratic results for insulin was the complete resistance of the disulfide loop peptide (Cys-Cys-Ala-Ser-Val-Cys) to the enzymes used, and that the remainder of the insulin structure in fact was hydrolyzed quite efficiently.

An additional point of caution in this type of work should be added to those stated above. In the analysis of protein digests by standard amino acid analysis as was done here, we generally assume that each elution position is unique to a given amino acid or amino acid derivative. The standard establishes the position of the known amino acids, and new derivatives can then be sought among the new and aberrant elution peaks. New derivatives coeluting with any of the known amino acids are thus likely to be lost in this type of single-assay analysis, especially since they are likely to be present in relatively small amounts. Another serious hazard in enzymatic hydrolysis of proteins is the high probability of encoun-

AMINO ACID RECOVERY FROM NATIVE PROTEINS AFTER
DIGESTION WITH PROTEASES[a]

Amino acid	RNase (95%)[b]	Lysozyme (102%)[b]	Enolase (100%)[b]	Insulin (103%)[b,c]
Asp	90	84	—[d]	
Thr	86	94	103	79
Ser	90	90	100	67 (100)
Asn	93	92	—[d]	100
Gln	—[e]	—[e]	—[e]	—[e]
Glu	108	120	—[d]	100
Pro	80	135	96	100
Gly	133	99	89	89
Ala	103	102	100	67 (100)
Cys	76	77	45	42 (93)
Val	95	111	85	80 (100)
Met	100	90	86	
Ile	126	102	93	92
Leu	185	106	100	109
Tyr	87	111	102	103
Phe	80	110	93	104
Trp		103	101	
Lys	93	88	96	100
His	98	107	95	98
Arg	100	90	100	100

[a] Given as percentage of theory.
[b] Percentage of total protein digested by enzymes in comparison to an acid hydrolyzed parallel sample (norleucine as internal standard).
[c] Numbers in parentheses calculated assuming that the segment Cys-Cys-Ala-Ser-Val-Cys was completely resistant.
[d] Gln, Asn content was not known.
[e] No attempt was made to estimate glutamine because of the cyclization of this amino acid to the ninhydrin negative pyrrolidone carboxylate. Based on the combined yield of asparagine and glutamine, it was estimated that 20–50% of the glutamine cyclized in different experiments.

tering dipeptides in the final analysis, either in unique peaks or in coelution with known compounds. The results for ribonuclease in the table may illustrate this problem. The abnormally high Leu analysis was disconcerting, but on examination of the leucine elution peaks it was found that the ratio of A_{570}/A_{440} was much lower than that observed for the leucine standard, and it was consequently concluded that an unknown component, likely to be a dipeptide by virtue of the high A_{440}, coeluted with leucine in the amino acid analysis.

The results for the glycoproteins will only be briefly summarized here. The amino acid analysis of the ovalbumin oligosaccharide fraction indicated that about 85% of the material was the free derivatized amino acid, glycosyl-Asn, and that 15% was the derivatized dipeptide glycosyl-Asn-Leu. Only traces of tri- and tetrapeptide components were indicated. This is consistent with results from other laboratories in which the pure glycosyl-Asn has been prepared on a preparative scale by Pronase digestion.[7] The γ-globulin samples under the same conditions gave only a 30% yield of glycosyl-Asn; the remaining 70% was primarily the glycosylated tripeptide glycosyl-Asn-Ser-Thr. In the case of the mucin, the amino acid composition of the oligosaccharide-containing fraction was very similar to that of the starting material, indicating that little or no amino acid release had taken place. The digestion was repeated with free enzymes to ascertain that the poor results were not caused by the matrix associated with the insolubilized enzymes, but the same very limited digestion was observed. This is again consistent with reports of the resistance of mucins to proteolytic digestion.[5]

Although the protocol used in this study does not provide a universal procedure for complete hydrolysis of proteins by proteases, it nevertheless may be, and indeed has been, of some use in the search for unstable amino acid derivatives in proteins. In the case of the *in vitro* modification of chymotrypsin and elastase with butyl isocyanate, the anticipated *O*-(butylcarbamoyl)serine could be isolated in 82 and 62% yield respectively after subjecting the two inactivated enzymes to the above protease digestion procedure.[8] An even less stable derivative, *S*-(butylcarbamoyl)cysteine, was also isolated from butyl isocyanate-inactivated yeast alcohol dehydrogenase after protease digestion.[9] In the latter case the lability of the thiocarbamate (the half-life at pH 8 is about 1 and at pH 7 about 6 hr) required a modified digestion procedure: pepsin for 4 hr at pH 2.0 followed by a 1 hr digestion with pronase and aminopeptidase M at pH 7.0.

Judging from general practice in this area, it appears that the choice of proteases and hydrolysis conditions most frequently is empirically established for the individual problem at hand. Many of the chapters in these volumes on posttranslational protein modifications contain descriptions of or references to specific hydrolysis conditions used to recover acid-labile derivatives from modified proteins. Some examples are the determi-

[7] C. C. Huang, H. E. Meyer, and R. Montgomery, *Carbohydr. Res.* **13**, 127 (1970).
[8] W. E. Brown and F. Wold, *Biochemistry* **12**, 835 (1973).
[9] J.-S. Twu and F. Wold, *Biochemistry* **12**, 381 (1973).

nation of γ-glutamyllysine and citrulline in wool,[3] γ-glutamyllysine in other proteins,[10] O^5-methyl glutamate,[11] O^4-methyl asparate,[12] β-aspartate lysine,[13] O^5-(ADPribosyl)$_n$-glutamate,[14] iodotyrosine,[15] O^4-phosphotyrosine.[16]

A wealth of data on available proteases is presented in other volumes of this series (Vols. 19, 45, 80).

[10] A. G. Loewy, this series, Vol. 107 [13].
[11] J. Stock, S. Clarke, and D. E. Koshland, Jr., this volume [29].
[12] S. Clarke, P. N. McFadden, C. M. O'Connor, and L. L. Lou, this volume [31].
[13] H. Klostermeyer, this series, Vol. 107 [14].
[14] K. Ueda and O. Hayaishi, this volume [47].
[15] J. Nunez, this series, Vol. 107 [30].
[16] W. B. Huttner, this series, Vol. 107 [11].

Section II

Nonenzymatic Modifications

[7] Measurement of Nonenzymatic Protein Glycosylation

By RUDOLF FLÜCKIGER and PAUL M. GALLOP

The development of quantitative methodology for the measurement of nonenzymatic protein glycosylation has been carried out mainly on hemoglobin. The level of glycosylated hemoglobin is a clinically useful measure of the averaged long-term glycemic control in a diabetic patient. The first contribution on the determination of glycosylated hemoglobins in this series, has dealt with methods of clinical interest[1]; the more general methodology for the detection of nonenzymatic protein glycosylation is discussed here with added reference to proteins other than hemoglobin which are also enzymatically glycosylated.

In the nonenzymatic glycosylation reaction, glucose and other aldoses and ketoses react with unprotonated amino groups to initially form a Schiff's base (aldimine) which can undergo the Amadori rearrangement to the corresponding 1-amino-1-deoxy-2-keto compound (ketoamine). The resulting fructosylvaline and ε-fructosyllysine (from glycohemoglobin) are not hydrolyzed under the strongly acidic conditions of the color reactions for glycoproteins, which renders the detection of nonenzymatic protein glycosylation difficult.[2,3]

Figure 1 summarizes the chemistry of useful alternative approaches for the detection of nonenzymatic protein glycosylation.

The charge-dependent separations employing ion exchange chromatography, isoelectric focusing, and agar gel electrophoresis rely on the decrease in positive charge caused by glycosylation of the N-terminal amino groups of the β chains of hemoglobin.[4] Glycosylation at all other sites does not alter the charge sufficiently to allow resolution of the adducts from unglycosylated hemoglobin. On ion exchange chromatography, the hemoglobins glycosylated at the N-terminus of the α chains and at some ε-amino groups elute in the leading edge of the main HbA peak.[5]

[1] K. H. Winterhalter, this series, Vol. 76, p. 732.
[2] H. S. Isabell and H. L. Frush, J. Am. Chem. Soc. 72, 1043 (1950).
[3] W. R. Holmquist and W. A. Schroder, Biochemistry 5, 2489 (1966).
[4] The β-N-terminally glycosylated hemoglobins which elute prior to HbA from a cation exchange resin are designated in order of their elution HbA_{a1}, HbA_{1a2}, HbA_{1b}, and HbA_{1c}. The two HbA_{1a} components are the fructose 1,6-bisphosphate and glucose 6-phosphate adduct, respectively; HbA_{1c} that of glucose; and in HbA_{1b}, a deamidation is thought to have occurred in the β chains of HbA_{1c}. For references, see ref. 1.
[5] H. F. Bunn, R. Shapiro, M. McManus, L. Garrick, M. J. McDonald, P. M. Gallop, and K. H. Gabbay, J. Biol. Chem. 254, 3892 (1979).

METHODS IN ENZYMOLOGY, VOL. 106

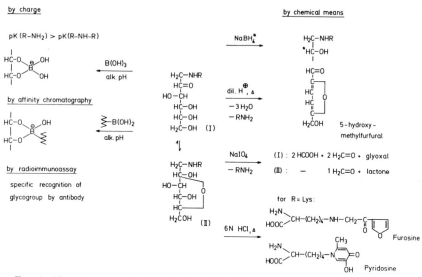

FIG. 1. Alternate approaches for the detection of nonenzymatic protein glycosylation.

Chromatographic resolution of these hemoglobins is possible in the presence of borate, which increases the negative charge by complex formation with the carbohydrate residues.[6] Apart from the resolution of glycosylated human serum albumin,[7] charge-dependent techniques have not been useful in detecting nonenzymatic glycosylation in other proteins. Total hemoglobin glycosylation may be determined chromatographically with boronate affinity chromatography or by one of several chemical techniques. Under certain conditions, the nonenzymatic glycosylation may even be determined in proteins which are also enzymatically glycosylated with these techniques (Table I).

Charge-Dependent Separations

Charge-dependent separations of the β-N-terminally glycosylated hemoglobins, HbA$_{1a-c}$, are widely used in clinical laboratories. It is essential to perform such determinations on fresh hemolysates, as ill-defined changes in charge occur during storage. These techniques yield reliable results with most samples. Erroneous estimates of hemoglobin glycosylation may, however, result in the presence of hemoglobin with a charge differing from that of HbA (Table II).

[6] R. Shapiro, M. J. McManus, C. Zalut, and H. F. Bunn, J. Biol. Chem. 255, 3120 (1980).
[7] J. F. Day, S. R. Thorpe, and J. W. Baynes, J. Biol. Chem. 254, 595 (1979).

TABLE I

COMPARISON OF SPECIFICITIES AND LIMITATIONS OF THE VARIOUS METHODS FOR THE
DETERMINATION OF NONENZYMATIC PROTEIN GLYCOSYLATION

Method	Limitations	Preventive Measures
Boronate affinity chromatography	Binding of enzymatically glyco-sylated proteins	Other eluting conditions, degly-cosylation
Borohydride reduction	Peptide bond reduction	Isolation of reduced amino acids
	Exchangeable protons	Extensive dialysis
Thiobarbituric acid method	Free carbohydrates	TCA precipitation
	Enzymatic glycosylation	Borohydride reduction
	Other chromogens	
Periodate oxidation	Enzymatic glycosylation	Borohydride reduction
Furosine method	Only lysine glycosylation measured	

Most often, it is desirable to determine stable HbA_{1c} only and to re-move labile HbA_{1c} either by incubation of erythrocytes with 0.9% saline or by hemolysis at pH 5.[8,9] With isoelectric focusing, the concentration of stable and labile HbA_{1c} can be determined in the same run.[10,11] Fetal hemoglobin does not interfere in the determination of HbA_{1c} by isoelectric focusing as HbF focuses cathodal to HbA and only the rather small pro-portion of acetylated HbF bands with HbA_{1c}.[12] However, on ion ex-change chromatography and agar gel electrophoresis HbF and HbA_{1c} are not resolved.[13,14] The glycosylated hemoglobins of the more positively charged hemoglobin variants HbS, C, or D do not separate on ion ex-change chromatography from the main hemoglobin. However, on isoelec-tric focusing and agar gel electrophoresis, bands representing the glycosy-lated hemoglobin variant species have been observed.[12,14] Reliable estimates of hemoglobin glycosylation may be obtained if their presence is realized and proper adjustments are performed.[15] All known hemoglobins

[8] D. E. Goldstein, S. B. Peth, J. D. England, R. L. Hess, and J. DaCosta, *Diabetes* **29**, 623 (1980).

[9] E. Bisse, W. Berger, and R. Flückiger, *Diabetes* **31**, 630 (1982).

[10] H. B. Mortensen, *Sci. Tools* **27**, 21 (1980).

[11] M. H. Stickland, C. M. Perkins, and J. K. Wales, *Diabetolgia* **22**, 315 (1982).

[12] M. Simon and J. Cuan, *Clin. Chem. (Winston-Salem, N.C.)* **28**, 9 (1982).

[13] E. C. Abraham, N. D. Cope, N. N. Braziel, and T. H. J. Huisman, *Biochim. Biophys. Acta* **577**, 159 (1979).

[14] L. Menard, M. E. Dempsy, L. A. Blankstein, H. Aleyassine, M. Wacks, and J. S. Soeldner, *Clin. Chem. (Winston-Salem, N.C.)* **26**, 1598 (1980).

[15] J. M. Sosenko, R. Flückiger, O. S. Platt, and K. H. Gabbay, *Diabetes Care* **3**, 590 (1980).

TABLE II
LIMITATIONS OF CHARGE-DEPENDENT TECHNIQUES

Method	Ion-exchange chromatography	Isoelectric focusing	Agar gel electrophoresis
Labile HbA$_{1c}$	Coelutes with stable HbA$_{1c}$	Resolved from stable HbA$_{1c}$	Comigrates with stable HbA$_{1c}$
HbF	Coelutes with HbA$_{1c}$	No interference	Chromatographs with HbA$_{1c}$
Variant hemoglobins	Adjustment possible	Resolved	Resolved
Other modified hemoglobins	Falsely increased HbA$_1$	Falsely increased HbA$_1$	Falsely increased HbA$_1$

modified by reactive molecules other than glucose are also more negatively charged than HbA and have chromatographic and electrophoretic
properties similar to those of the glycosylated HbA$_1$ components. These
posttranslational modifications comprise acetylation,[16] carbamylation,[17]
penicilloylation[18] of hemoglobin, and modification by acetaldehyde,[19] and
its condensation product with dihydroxyacetone phosphate, 5-deoxyxylu-
lose 1-phosphate.[20] Iodacetate adduct formation is an artifact caused by
use of iodoacetate-containing Vacutainers.[21]

Ion Exchange Chromatography

The resolution of the glycosylated hemoglobin components is better
on cation- than on anion-exchange resins. The weakly acidic cation ex-
changer Bio-Rex 70, minus 400 mesh (pK = 6.1), the resin corresponding
to Amberlite IRC-50 used in the early separations of glycosylated hemoglobins[22] is now commonly used for these separations. The glycosylated
hemoglobins are eluted with a developer (tabulated in ref. 23), and HbA
with a high salt concentration. Preparative[24] and analytical open column

[16] K. R. Bridges, G. J. Schmidt, A. Cerami, and H. F. Bunn, *J. Clin. Invest.* **56**, 201 (1975).
[17] R. Flückiger, W. Harmon, S. Loo, and K. H. Gabbay, *N. Engl. J. Med.* **304**, 823 (1981).
[18] R. Flückiger and W. Mathews, unpublished observation.
[19] V. J. Stevens, W. J. Fantl, C. B. Newman, R. V. Sims, A. Cerami, and C. M. Peterson, *J. Clin. Invest.* **67**, 361 (1981).
[20] H. D. Hoberman, *Biochem. Biophys. Res. Commun.* **90**, 757 (1979).
[21] S. Schoos-Barbette, J. Dodinval-Versie, H. Husquinet, and C. Lambotte, *Clin. Chim. Acta* **116**, 245 (1981).
[22] D. W. Allen, W. A. Schroeder, and J. Ballog, *J. Am. Chem. Soc.* **80**, 1628 (1957).
[23] R. A. Cole, J. S. Soeldner, P. J. Dunn, and H. F. Bunn, *Metab., Clin. Exp.* **27**, 289 (1978).
[24] M. J. McDonald, R. Shapiro, M. Bleichman, J. Solway, and H. F. Bunn, *J. Biol. Chem.* **253**, 2327 (1978).

systems,[25,26] as well as HPLC systems,[23] are in use, some of which have been optimized for the resolution of animal hemoglobins.[27,28] These separations are critically affected by variations in pH, ionic strength, and temperature.[29]

Specific Procedure. In our laboratory, HPLC is performed on a 4.5 × 125-mm column, maintained at 28°, and operated with a flow rate of 0.7 ml/min. The low-phosphate buffer, pH 6.68, contains 0.055 M NaH_2PO_4, 0.014 M Na_2HPO_4, and 0.015 M KCN; the high-phosphate buffer, pH 6.4, 0.104 M NaH_2PO_4, and 0.049 M Na_2HPO_4. The effluent is monitored at 460 nm, 10 μg of hemoglobin is injected, and elution of HbA with the high phosphate buffer is started after elution of HbA_{1c} at approximately 6 min. The equilibration time between runs is approximately 10 min.[30] Consistent near-baseline resolution between HbA_{1a+b} and HbA_{1c} is achieved if the low phosphate buffer is diluted 5:1 with water.

Isoelectric Focusing

Isoelectric focusing is performed on thin-layer polyacrylamide gels (1 mm), using a pH gradient of 6–8.[12,31,32] On such gels HbA_{1c} is separated from HbA by about 2 mm and HbA_{1a+b} is not detectable. Ready-to-use narrow-range (pH 5.0–6.5) Ampholine polyacrylamide gels are manufactured by LKB (Bromma, Sweden).

Specific Procedure.[12] Gels are prepared by dissolving 6.79 g of acrylamide and 0.21 g of N,N'-methylenebisacrylamide in 90 ml of distilled water to which 10 ml of ampholyte solution (4 g/10 ml water), pH 6–8, and 4 g of β-alanine are added. After degassing 25 ml of the mixture for 10 min, 5 μl of N,N,N',N'-tetramethylenediamine is added, followed by 100 μl of freshly prepared 50 g/liter ammonium persulfate solution. The solution is mixed, degassed, transferred to the molds, and left to polymerize at 4° for 1–2 hr. The gel is mounted on a cooling plate (10°) with a filter paper strip utilizing 0.1 N sodium hydroxide as catholyte and 1 M phosphoric acid as anolyte. The gel is prefocused at a constant voltage of 0.5 kV for

[25] L. A. Trivelli, H. M. Ranney, and H. T. Lai, *N. Engl. J. Med.* **824**, 353 (1971).
[26] G. T. Hammons, K. Junger, J. M. McDonald, and J. H. Ladenson, *Clin. Chem. (Winston-Salem, N.C.)* **28**, 1775 (1982).
[27] M. H. Blanc, F. H. Rhie, P. J. Dunn, and J. S. Soeldner, *Metab., Clin. Exp.* **30**, 317 (1981).
[28] P. J. Higgins, R. L. Garlick, and H. F. Bunn, *Diabetes* **31**, 743 (1982).
[29] M. Simon and J. Eissler, *Diabetes* **29**, 467 (1980).
[30] Equilibration time may be reduced by use of ethanol diluted buffers: R. S. Schifreen, J. M. Hickingbotham, and G. N. Bowers, *Clin. Chem. (Winston-Salem, N.C.)* **26**, 466 (1980).
[31] K. M. Spicer, R. C. Allen, and M. G. Buse, *Diabetes* **27**, 384 (1978).
[32] H. B. Mortensen, *J. Chromatogr.* **182**, 325 (1980).

30 min. Samples (2.5 μl hemolysate containing approximately 10 mg Hb and 1% KCN are applied to 0.5 × 0.5-cm squares of filter paper placed 0.5 cm apart in a straight line in the middle of the gel. The samples are focused at a constant voltage of 0.5 kV for 30 min, 1 kV for 1 hr, and finally at 1.2 kV until the bands are clearly resolved. The focused gel is fixed in 12.5% trichloroacetic acid (TCA) for 20 min, rinsed in a 50 ml/liter glycerol solution for 1 hr, and the hemoglobin concentration determined by microdensitometry. Quantitation by cutting off the bands of gels not fixed with TCA and eluting the hemoglobin has also been described.[11,32]

Agar Gel Electrophoresis

Separations on agar gel rely on the differential interaction of the negatively charged sulfate and pyruvate groups in the gel with β-N-terminally glycosylated and the residual hemoglobin. The migration of HbA is retarded under these conditions. Furthermore, because of the negative charge in the media, cations and their associated water molecules tend to migrate toward the cathode, creating an electroendosmotic flow, which seems to be the main cause for the cathodic movement of the hemoglobins rather than the direct field strength of electrophoresis. On agar gel electrophoresis, HbA$_1$ is resolved from HbA by approximately 10 mm, facilitating quantitation by densitometry. HbA$_{1c}$ is not resolved from HbA$_{1a+b}$ in this system.

Specific Procedure.[14] A mixture containing 20 g/liter agar, 40 g/liter sorbitol, 35 nmol of sodium citrate, 25 mmol of citric acid, and 0.9 mmol of disodium EDTA (pH 6.3) is prepared. The mixture is heated until the components dissolve. The gel is cast on a flexible polyester film after cooling of the heated mixture and sample wells are punched. The electrode solution is 0.1 M citrate buffer, pH 6.3. One microliter of hemolysate (30 μg Hb) is applied on the anodal side of the gel, and 60 V is applied for 40 min. The gel is heat fixed in an oven at 55° for 20 min and evaluated by scanning densitometry at 420 nm. Ready-to-use agar gel casettes are manufactured by Corning Glassworks (Sullivan Park, NY).

Boronate Affinity Chromatography

Under alkaline conditions, boronic acids form stable complexes with compounds containing coplanar cis-hydroxyl groups. These complexes may be dissociated by lowering the pH or with a competing polyol. Boronate affinity gels are prepared by coupling *m*-aminophenylboronic acid to a gel with an appropriate exclusion limit, the polyacrylamide gel BioGel P-6 (Affi-Gel, Bio-Rad) for work with amino acids and peptides and the cross-linked agarose Sepharose CL-6B (Glycogel B, Pierce and Matrix

Gel PBA, Amicon) for proteins.[33] With hemoglobin a low linear flow rate is essential for quantitative absorption of glycohemoglobin on the boronate-agarose.[34]

Specific Procedure. For hemoglobin: A column (14 × 130 mm) packed with Glycogel B and equilibrated with the wash buffer consisting of 0.25 M ammonium acetate and 0.05 M magnesium chloride, pH 8.5, with ammonium hydroxide, is operated with a flow rate of 0.5 ml/min preferably at 4°. Hemolysates (5–10 mg Hb) are applied and the unglycosylated hemoglobin is washed off the column (25 ml). The glycohemoglobin is eluted with 15 ml eluting buffer consisting of 0.1 M Tris, 0.2 M sorbitol, and 0.05 M EDTA, pH 8.5. The column is reactivated with 0.1 M HCl for further use. The oxidative degradation of the gel is minimized by storage in the dark either directly in the eluting buffer or in 0.01 N HCl.

For peptides and amino acids[35]: A column (1.5 × 14 cm) packed with Affi-Gel 601 is equilibrated with 0.025 M sodium phosphate, pH 9.0, and operated at a flow rate of 20 ml/hr. The pH of the sample is adjusted to 9.0 for application. The column is washed with 400 ml buffer to elute all nonabsorbed material and the column eluted with 0.025 M HCl. The effluent is monitored with ninhydrin.

Radioimmunological Assay and Phytic Acid Method

These two patented approaches, though elegant, are not generally useful.[36,37] The antibody against human HbA_{1c} is difficult to raise and not otherwise available.

The phytic acid assay relies on a rather small spectral change in the visible absorption spectrum of hemoglobin induced by the binding of organic polyphosphates such as phytic acid, the hexaphosphate of inositol, and hence is subject to error.

Chemical Procedures

Borohydride Reduction

The aldimine and ketoamine are reduced by either sodium borohydride or sodium cyanoborohydride. Sodium borohydride readily reduces

[33] A. K. Mallia, G. T. Hermanson, R. I. Krohn, E. K. Fujimoto, and P. K. Smith, *Anal. Lett.* **14**, 649 (1981).
[34] R. Flückiger, T. Woodtli, and W. Berger, submitted for publication.
[35] M. Brownlee, H. Vlassara, and A. Cerami, *Diabetes* **29**, 1044 (1980).
[36] J. Javid, P. K. Pettis, R. J. Koenig, and A. Cerami, *Br. J. Haematol.* **38**, 329 (1978).
[37] E. Moore, S. Stroupe, and L. Blecka, *Diabetes* **29**, 70A (1980).

imines to secondary amines in aqueous solution and neutral pH. Aldehydes and ketones are also reduced by sodium borohydride, the latter,
however, more slowly than aldehydes. The occurrence of a small level of
a peptide bond reductive side reaction leading to the formation of α-amino
alcohols has also been reported.[38] While carbonyl compounds or Schiff
bases are onefold reduced, i.e., one tritium/mol is incorpoated by use of
tritiated sodium borohydride, compounds resulting from the reduction of
peptide bonds incorporate two tritiums/mol of reduced group. Sodium
cyanoborohydride is a more selective reducing reagent which does not
reduce aldehydes and ketones under neutral conditions in aqueous solution. As the pH is lowered, the reduction becomes more rapid and at pH
3–4 aldehydes and ketones are readily reduced.[39] Reduction of nonenzymatically glycosylated proteins is usually performed with tritiated sodium
borohydride.[40] The mild reductive properties of sodium cyanoborohydride have been used to determine the kinetics of the Schiff base formation.[41] Glycosylated amino acids can be identified by amino acid analysis
of hydrolyzates of proteins previously reduced with radioactively labeled
borohydride. The glycosylated amino acids can be separated from unglycosylated material and nonspecifically incorporated radioactivity by boronate affinity chromatography.[35] With this approach, equal amounts of
glycosylated lysine and valine have been identified in hemoglobin.[35] On
amino acid analysis, glucitolvaline and glucitolaspartic acid elute early
from the column, while glucitollysine elutes in the region of tyrosine and
two peaks. The first one, glucitollysine, is usually larger than the second
one representing its dehydration product.[42] Hexosyl-hydroxylysine elutes
in front of glucitollysine.[43]

Specific Procedure. Reduction: Sodium borohydrate is dissolved in
ice-cold 0.01 N NaOH and a 100 molar excess added to the protein dissolved in 0.01 M phosphate buffer, pH 7.4. The mixture is incubated at
room temperature for 1 hr and the reaction stopped by adjusting the pH to
4 with acetic acid. The sample is exhaustively dialyzed against 0.05 M
acetic acid. The specific incorporation may be determined if $NaBH_4$ of a
known specific activity is used. The specific activity may be determined
by reduction of 4-p-nitrobenzamidobutyraldehyde and isolation of the reduced alcohol by thin-layer chromatography (TLC).[38]

[38] M. A. Paz, E. Henson, R., Rombauer, L. Abrash, O. O. Blumenfeld, and P. M. Gallop, *Biochemistry* **9**, 2123 (1970).
[39] R. F. Borch, M. D. Bernstein, and H. D. Durst, *J. Am. Chem. Soc.* **93**, 2897 (1971).
[40] R. M. Bookchin and P. M. Gallop, *Biochem. Biophys. Res. Commun.* **32**, 86 (1968).
[41] P. J. Higgins and H. F. Bunn, *J. Biol. Chem.* **256**, 5204 (1981).
[42] B. Trueb, G. J. Hughes, and K. H. Winterhalter, *Anal. Biochem.* **119**, 330 (1982).
[43] S. P. Robins and A. J. Bailey, *Biochem. Biophys. Res. Commun.* **48**, 76 (1972).

Thiobarbituric Acid Colorimetric Technique

In the thiobarbituric acid colorimetric technique (TBA method) 1-amino-2-deoxyketoses are dehydrated under defined acidic conditions and released as 5-hydroxymethylfurfural (5-HMF).[44] 5-HMF is quantitated by reaction with thiobarbituric acid. Under certain hydrolytic conditions, 5-HMF formation from free carbohydrates or N- and O-glycosides is small, allowing the specific determination of nonenzymatic protein glycosylation.[45] Because 5-HMF release is not quantitative and does not linearly depend on the protein concentration, the result is a relative measure of nonenzymatic protein glycosylation. With hemoglobin and albumin, incubation conditions were optimized for maximal color yield.[45–52] Applications of such conditions to the analysis of enzymatically glycosylated proteins requires a sodium borohydride reduced control.[53,54] Reduction converts the aminodeoxyketose to a 1-aminohexitol and hence prevents 5-HMF formation. The specificity of the 5-HMF detection may be increased by chromatographically determining its concentration. HPLC is performed on μBondapak C_{18}, with 1% acetic acid and UV monitoring at 280 or 254 nm.[55] Also, a fluorometric assay for 5-HMF yields a higher sensitivity than the detection with TBA.[56]

Specific Procedure. This modification of the TBA colorimetric technique has proven useful for the analysis of purified hemoglobin samples and for routine analysis of hemolysates[46] and its various reaction steps have been documented in detail.[47] Samples of hemolysates containing 10 mg hemoglobin are diluted to a volume of 1 ml with distilled water. After the addition of 0.5 ml 1.0 N oxalic acid, the tubes are mixed and capped

[44] A. Gottschalk, *Biochem. J.* **52**, 455 (1952).

[45] R. Flückiger and K. H. Winterhalter, *FEBS Lett.* **71**, 356 (1979).

[46] K. H. Gabbay, J. M. Sosenko, G. A. Banuchi, M. M. Mininsohn, and R. Flückiger, *Diabetes* **28**, 337 (1979).

[47] R. E. Pecoraro, R. J. Graf, J. B. Halter, H. Beiter, and D. Porte, *Diabetes* **28**, 1120 (1979).

[48] I. S. Ross and P. F. Gibson, *Clin. Chim. Acta* **98**, 53 (1979).

[49] R. W. Fischer, C. DeJong, E. Voigt, W. Berger, and K. H. Winterhalter, *Clin. Lab. Haematol.* **2**, 129 (1980).

[50] C. V. Subramaniam, B. Radhakrishnamurthy, and G. S. Berenson, *Clin. Chem.* (*Winston-Salem, N.C.*) **26**, 1683 (1980).

[51] K. M. Parker, J. D. England, J. DaCosta, R. L. Hess, and D. E. Goldstein, *Clin. Chem.* (*Winston-Salem, N.C.*) **27**, 669 (1981).

[52] K. A. Ney, K. J. Colley, and S. V. Pizzo, *Anal. Biochem.* **118**, 294 (1981).

[53] K. F. McFarland, E. W. Catalano, J. F. Day, S. R. Thorpe, and J. W. Baynes, *Diabetes* **28**, 1011 (1971).

[54] E. Elder and L. Kennedy, *Diabetologia* **24**, 70 (1983).

[55] O. H. Wieland, R. Dolhofer, and E. Schleicher, *Proc. Congr. Int. Diabetes Fed., 10th, 1979* p. 721 (1980).

[56] T. A. Walmsley and M. Lever, *Anal. Biochem.* **124**, 446 (1982).

with rubber stoppers vented with 26-gauge needles. The tubes are placed in a heating block at 100° and the needles are removed after a 10 min equilibration period. Heating is continued for a total of 5 hr. The tubes are removed and allowed to cool in an ice bath. Cold 40% trichloroacetic acid (0.5 ml) is added to each tube, the contents are mixed and centrifuged. A 1.5-ml aliquot of the supernatant is pipetted off and 0.5 ml of a 0.05 M aqueous 2-thiobarbituric acid solution is added with mixing. After a 15 min incubation at 37°, the tubes are allowed to stand at room temperature for 20 min before reading the samples at 443 nm.

Periodate Oxidation

Periodate oxidation of the 1-amino-1-deoxy-2-ketose in nonenzymatically glycosylated proteins yields 1 mol of formaldehyde and that of the 1-aminohexitol 2 mol of formaldehyde[57] (Fig. 1). The formaldehyde can be quantitated as the fluorescent 3,5-diacetyl-1,4-dihydrolutidine (DDL) formed from the condensation of formaldehyde with acetylacetone and ammonia. With hemoglobin this assay gives a direct measure of the number of glyco groups per hemoglobin molecule. It may also be used to determine the extent of nonenzymatic glycosylation of other proteins if the difference in the amount of formaldehyde released from periodate oxidation of $NaBH_4$ reduced and unreduced proteins is measured.

Specific Procedure. Globin prepared with acid tetrahydrofuran (THF) is required for this procedure as both hemin and acetone interfere in the fluorometric determination of formaldehyde. For globin preparation up to 1 ml of a hemolysate is added dropwise to 10 ml freshly prepared cold acid THF (1% 12 N HCl in THF) with vortexing. The globin is centrifuged (1000 g for 5 min) and is washed free of acid with at least two washes of 5 ml cold THF. If necessary, washing of the globin is continued until it has a light buff color. The THF is decanted and the protein dissolved in 1 to 2 ml H_2O. The concentration of globin is determined spectrophotometrically at 280 nm using an extinction coefficient of 8.5. For the assay, aliquots containing 2–4 mg protein are adjusted to 700 μl with H_2O and 20 μl of 1 N HCl is added. Oxidation is started by the addition of 100 μl 0.1 N $NaIO_4$ and is allowed to proceed at room temperature for 30 min.

The samples are cooled on ice, and 300 μl of ice-cold 10% zinc sulfate and 100 μl of 1.4 N NaOH are added. The samples are centrifuged (1000 g for 10 min) and the supernatant is collected. Freshly prepared formaldehyde detection reagent (1 ml of 2 M ammonium acetate and 0.02 M acetylacetone in water) are added and the fluorescence is allowed to develop for

[57] P. M. Gallop, R. Flückiger, A. Hanneken, M. M. Mininsohn, and K. H. Gabbay, *Anal. Biochem.* **117**, 427 (1981).

1 hr at 37°. The fluorescence is determined by excitation at 410 nm and with emission at 510 nm. A fructose standard curve containing 0 to 0.40 nmol is determined for calculation of results. One equivalent of formaldehyde is released from 1 mol of fructose and from 1 eq of a ketoamine glyco group.

Furosine Assay

Acid hydrolysis of proteins containing glycosylated lysine residues in 6 N HCL leads to the formation of the two dehydration products furosine and pyridosine with a yield of 30 and 10%, respectively.[58,59] These compounds can be separated by high-pressure liquid chromatography.[60] Similar degradation products arise from amino acids glycosylated at the α-amino group.[61] As all nonenzymatically glycosylated proteins are likely to contain some glycosylated ε-amino groups of lysine residues, and furosine does not arise from other sources, the measurement of furosine allows the specific estimation of generalized nonenzymatic glycosylation.

Specific Procedure. The protein is hydrolyzed for 18 hr in 6 N HCl at 95°. The HCl is evaporated under reduced pressure and the residue dissolved in a small volume of water, and aliquots thereof are applied to HPLC. High pressure liquid chromtography is performed on a 5-μm C_{18} column (4 × 200 mm), with isocratic elution with 5.6 mmol/liter H_3PO_4 at a flow rate of 1 ml/min. The effluent is monitored at 254 or 280 nm. Under these conditions, the elution time is approximately 3 min for furosine, 9 min for tyrosine, and 14 min for phenylalanine. A better resolution of furosine is obtained by use of an eluent consisting of 1.2 mmol/liter H_3PO_4 and 0.2 mmol/liter heptanesulfonic acid.[62] Calibration of results is based on furosine formation from synthetic fructosyllysine.[60]

Acknowledgment

This work was supported in part by NIH Grants AG00376 and HL20764.

[58] P. A. Finot, J. Bricout, R. Viani, and J. Mauron, *Experientia* 24, 1097 (1968).
[59] P. A. Finot, R. Viani, J. Bricout, and J. Mauron, *Experientia* 25, 134 (1969).
[60] E. Schleicher and O. H. Wieland, *J. Clin. Chem. Clin. Biochem.* 19, 81 (1981).
[61] K. Heyns, J. Heukeshoven, and K. H. Brose, *Angew. Chem.* 15, 627 (1968).
[62] B. W. Vogt, E. D. Schleicher, and O. H. Wieland, *Diabetes* 31, 1123 (1982).

[8] Nonenzymatic Glucosylation of Lysine Residues in Albumin

By JOHN W. BAYNES, SUZANNE R. THORPE, and
MARTHA H. MURTIASHAW

Nonenzymatic glucosylation is a common posttranslational modification of body protein, in which glucose reacts directly with primary amino groups on protein to yield stable covalent adducts.[1-4] As illustrated in Fig. 1, the sugar reacts primarily with the ε-amino groups of lysine residues in albumin[5] and other proteins, although reaction at the α-amino terminus also occurs, as in hemoglobin.[1,6,7] While the linear, aldehyde form of the sugar is considered to be the reactive species, the sugar in solution exists predominantly in the cyclic furanose and pyranose conformations. Similarly, during the early stages of the reaction, protein-bound glucose exists mainly as the cyclic glycosylamine (GA) and its Amadori rearrangement product, the N-substituted-1-amino-1-deoxyketopyranose (KP). Human serum albumin (HSA), as it is obtained from blood, contains both of these cyclic glucose adducts, although the labile GA derivative is probably completely discharged from the protein during most purification procedures. The purpose of this article is to describe the preparation and characterization of radiolabeled GA and KP derivatives of albumin, to describe procedures for determining kinetic constants involved in their formation and dissociation, and to describe a procedure for detecting glucose adducts to lysine residues in HSA and other proteins.

Reagents

In order to remove reactive contaminants,[8] radioactive glucose was purified within 2 weeks of each experiment by paper chromatography (Whatman 3MM) overnight in *n*-butanol:pyridine:water (9:5:4). The

---

[1] H. F. Bunn, K. H. Gabbay, and P. M. Gallop, *Science* **200,** 21 (1978).
[2] V. M. Monnier, V. J. Stevens, and A. Cerami, *Prog. Food Nutr. Sci.* **5,** 315 (1981).
[3] A. J. Bailey, *Horm. Metab. Res.* **11,** Suppl. 1, 90 (1981).
[4] S. R. Thorpe and J. W. Baynes, *in* "The Glycoconjugates" (M. I. Horowitz, ed.), Vol. 3, p. 113. Academic Press, New York, 1982.
[5] R. L. Garlick and J. S. Mazer, *J. Biol. Chem.* **258,** 6142 (1983).
[6] R. Shapiro, M. I. McManus, C. Zalut, and H. F. Bunn, *J. Biol. Chem.* **255,** 3120 (1980).
[7] P. J. Higgins and H. F. Bunn, *J. Biol. Chem.* **256,** 5204 (1981).
[8] B. Trüeb, C. G. Holenstein, R. W. Fischer, and K. H. Winterhalter, *J. Biol. Chem.* **255,** 6717 (1980).

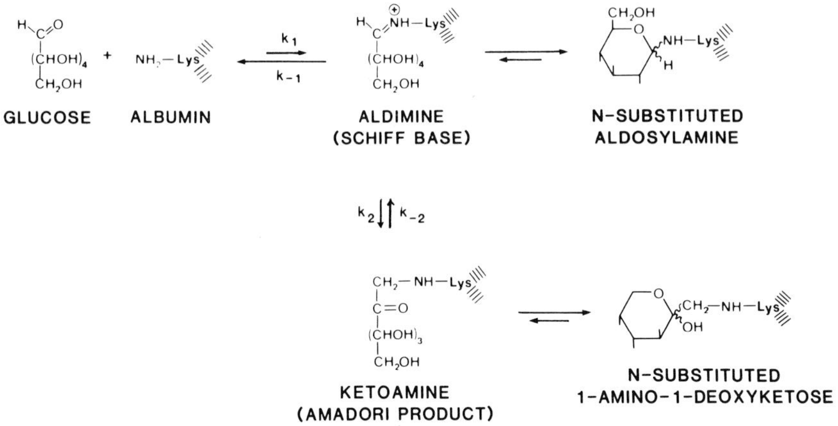

FIG. 1. Reaction scheme for formation of glucose adducts to albumin.

chromatogram was scanned for radioactivity (Packard Model 7201 Radiochromatogram Scanner), and the radioactive glucose region eluted with 1–2 ml distilled water. The glucose solution was concentrated on a rotary evaporator and stored frozen at $-20°$. Purification of the glucose to remove radioactive contaminants has, in our experience, been essential for careful studies on nonenzymatic glycosylation of any protein. An alternative procedure involving preincubation of the glucose with albumin and $NaCNBH_3$[7] has also proven satisfactory. Albumin was purified from fresh serum by affinity chromatography on Affi-Gel Blue (Pierce), as described by Travis and Pannel.[9,10] N^{α}-Formyl-N^{ε}-fructoselysine (FFL) was prepared from N^{α}-formyllysine[11] and glucose, essentially as described by Finot and Mauron.[12] Hexitollysine was prepared by reduction of FFL with a 100-fold molar excess of $NaBH_4$ in 0.05 N NaOH for 4 hr at room temperature, followed by acid hydrolysis (20 min at 100° in 2 N HCl) to remove the formyl group.

Procedures

Glucosylation of Albumin in Vitro. HSA concentration is determined spectrophotometrically at 280 nm using $E^{1\%} = 5.3$. Protein is incubated

[9] J. Travis and R. Pannell, *Clin. Chim. Acta* **49**, 49 (1973).
[10] K. F. McFarland, E. C. Catalano, J. F. Day, S. R. Thorpe, and J. W. Baynes, *Diabetes* **28**, 1011 (1979).
[11] K. Hofmann, E. Stutz, G. Spuhler, H. Yajima, and E. T. Schwartz, *J. Am. Chem. Soc.* **82**, 3727 (1960).
[12] P. A. Finot and J. Mauron, *Helv. Chim. Acta* **52**, 1488 (1969).

with radioactive glucose (diluted to desired specific activity) in 0.1 M Na phosphate buffer, pH 7.4. Incubation mixtures are sterilized by ultrafiltration through Gelman 0.2-μm Acrodisc filters. High protein concentrations (0.5–1 mM, 35–70 mg/ml) are used to increase the absolute yield (cpm) of radioactive protein in reaction aliquots. Reactions are carried out in a 37° incubator in sealed 1.5-ml plastic centrifuge tubes. For kinetic studies, aliquots are withdrawn and assayed immediately, or frozen at −20° until time of analysis.

Measurement of Glucose Incorporation. Aliquots of incubation mixtures are diluted to a final volume of 0.5 ml with ice cold Dulbecco's phosphate buffered saline (PBS,13), and applied to a Sephadex G-25 column (1 × 27 cm) equilibrated with PBS at 4°. Fractions (50 × 0.5 ml at 0.5 ml/min) are collected, and the entire fraction counted for radioactivity in 4 ml scintillation cocktail (Budget-Solv, Research Products International). The percentage of total glucose incorporated into protein is calculated by comparing radioactivity at the column void volume with total radioactivity recovered in column fractions. Mole glucose/mole protein is calculated from the amounts of protein and glucose radioactivity in the void volume fractions, and the specific activity of the glucose.

For determination of acid-stable (KP) adducts, GA derivatives are discharged by incubation in mild acid.[14] Reaction aliquots are adjusted to 0.5 ml with 0.2 M Na acetate, pH 4.5, incubated in a water bath at 37° for 1 hr, and then analyzed by chromatography, as above. The amount of acid-labile (GA) adduct is computed from the difference in the glucose content of the protein before and after acid treatment.

Determination of Kinetics of Glucosylation. Figure 2 shows the results of a typical experiment to determine the kinetics of incorporation of glucose into total, acid-stable and acid-labile adducts to HSA. The "total" curve displays biphasic kinetics, resulting from rapid attainment of the glycosylamine equilibrium (Fig. 1), followed by slower rearrangement to the Amadori product. Thus, the slopes of the "total" and "Amadori" progress curves become identical within 1–2 hr after starting the reaction. The difference between these curves, shown as the "glycosylamine" line, indicates that, at 5.5 M glucose and 1.1 mM HSA concentrations, 0.4% of the glucose in the incubation mixture is bound to HSA as the GA adduct. Therefore, under physiological conditions (5.5 mM Glc, pH 7.4, 37°), ~2% of HSA molecules contain glucose bound as a glycosylamine adduct. The equilibrium constant can then be calculated, as follows:

$$Glc + HSA \rightleftharpoons G{=}HSA \tag{1}$$

[13] R. Dulbecco and M. J. Vogt, *J. Exp. Med.* **98,** 169 (1954).
[14] E. Bissé, W. Berger, and R. Flückiger, *Diabetes* **31,** 630 (1982).

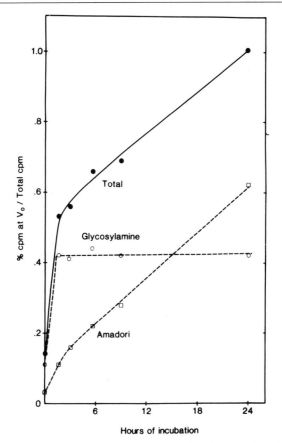

FIG. 2. Kinetics of glucosylation of human serum albumin. Reaction mixtures (500 μl) contained HSA (40 mg, 1.1 mM) and glucose (5.5 mM, 25 × 10^6 cpm). Samples (60 μl) were removed at various times for analysis by chromatography on Sephadex G-25. Incorporation of glucose into total (●---●) and acid-stable (□---□) adducts was determined on 25-μl aliquots, either directly or after acid treatment, as described under Procedures. Glyco-sylamine-bound glucose (○---○) was determined from the difference between total and acid-stable radioactivity.

$$K_{eq} = \frac{[\text{G=HSA}]}{[\text{Glc}][\text{HSA}]} = \frac{(0.004 \times 5.5 \times 10^{-3} \, M)}{(5.5 \times 10^{-3} \, M)(1.1 \times 10^{-3} \, M)} = 3.6 \, M^{-1} \quad (2)$$

That this is a true equilibrium constant is confirmed in the experiment shown in Fig. 3. Thus, over a range of 3.5–35 mM glucose (60–600 mg%), the acid-labile adduct to HSA represents a constant fraction of the total glucose in the reaction mixture. The average value for the equilibrium constant, obtained from a series of similar experiments conducted over

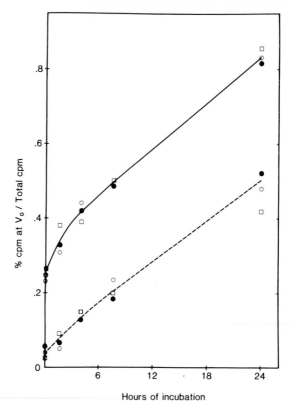

FIG. 3. Effect of glucose concentration on the glycosylamine content of albumin. HSA was incubated with 3.5 (●), 17.5 (○), or 35 (□) mM glucose. Reaction mixtures (300 μl) contained albumin (15 mg, 0.75 mM) and a constant amount of [6-^{3}H]glucose (2 × 10^6 cpm). Samples (50 μl) were withdrawn at various times and 20-μl aliquots analyzed on Sephadex G-25.

several months with different preparations of glucose and HSA, is 3.9 M^{-1} (SD = 0.3, n = 5). As shown in Fig. 4, the amounts of both total and acid-stable adduct formed during a fixed time period are also directly dependent on protein concentration.

Based on the progress curve in Fig. 2, the apparent rate of formation of the Amadori product in 5.5 mM glucose is ∼1.3% of the protein per day (0.25%/day/mM glucose). The apparent second order rate constant (k_2') may be calculated from the slope of this curve, as follows:

$$d\text{KP}/dt = k_2'[\text{HSA}][\text{Glc}] \tag{3}$$

$$k_2' = 0.1 \ \text{hr}^{-1} \ M^{-1} \tag{4}$$

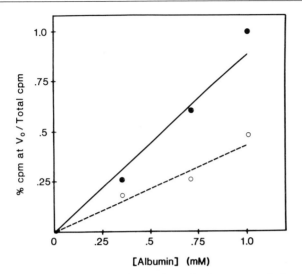

FIG. 4. Effect of protein concentration on glycosylamine content of albumin. Reaction mixtures (250 μl) contained HSA (20–70 mg/ml, 0.3–1 mM) and 5.5 mM glucose (4 × 10⁶ cpm [6-³H]glucose). Samples (200 μl) were withdrawn at 24 hr, and 85-μl aliquots analyzed on Sephadex G-25 for total (●---●) and acid-stable (○---○) adducts.

Similarly the actual first order rate constant, based on the equilibrium concentration of glycosylamine, may be estimated as

$$d\text{KP}/dt = k_2[G\text{=HSA}] = k_2 K_{\text{eq}_1}[\text{Glc}][\text{HSA}] \tag{5}$$

$$k_2 = 2.6 \times 10^{-3} \text{ hr}^{-1} \tag{6}$$

Measurement of Kinetics of Dissociation of Glucose Adducts. Because the rate of hydrolysis of glycosylamines is relatively slow at neutral pH and low temperature, the kinetics of dissociation of GA adducts can be measured by Sephadex G-25 chromatography. As shown in Fig. 5, the glycosylamine adduct to albumin has more than a 1 day half-life in PBS at 4°, while at 37°, dissociation is more rapid, with about a 1.8 hr half-life. The total radioactivity released after several half-lives at pH 7.4 agrees closely with the amount released during a short-term incubation at pH 4.5 (Fig. 5). From these data and the equilibrium constant, the kinetic constants k_1 and k_{-1} can be estimated, as follows:

$$k_{-1} = 0.693/t_{1/2} = 0.693/1.8 \text{ hr} = 0.39 \text{ hr}^{-1} \tag{7}$$

$$K_{\text{eq}_1} = k_1/k_{-1} \tag{8}$$

$$k_1 = K_{\text{eq}}k_{-1} = 3.9 \ M^{-1} \times 0.39 \text{ hr}^{-1} = 1.5 \text{ hr}^{-1} \ M^{-1} \tag{9}$$

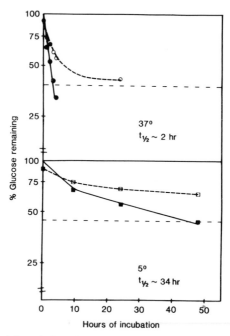

FIG. 5. Kinetics of dissociation of glycosylamine adducts from albumin. HSA (40 mg/0.5 ml, 1.1 mM) was glucosylated by incubation in 5.5 mM glucose (3.5 × 10⁶ cpm) for 12 hr, then isolated by chromatography on Sephadex G-25. The glucosylated protein (5100 cpm/ mg, 13 mg/ml) was then incubated at pH 7.4 in the absence of glucose at either 37° (●, ○) or 5° (■, □). Aliquots (350 μl) were withdrawn at various times and analyzed on Sephadex G-25. Data are expressed as either percentage of total glucose remaining on the protein (open symbols) or as percentage of labile glucose remaining bound to protein (closed symbols). For each preparation, labile glucose content (- - - -) was determined separately following acid treatment.

The Amadori adduct to the protein (after acid discharge of GA) was stable to extended incubation of the protein at 37° (data not shown), and had an estimated minimum half-life of about 3 weeks. Whether the Amadori rearrangement is actually reversible is uncertain, however, since the radioactive material released from the protein has not yet been identified. For HSA, reversal of the rearrangement is probably not physiologically relevant since the biological half-life of this protein is also about 3 weeks.

Summary of Kinetic Data for HSA; Comparison to Hemoglobin. The table compares kinetic constants obtained in this study for the glucosylation of HSA, with those obtained previously by Bunn and co-workers[6,7] for human hemoglobin. The kinetic and equilibrium constants for forma-

COMPARATIVE KINETICS OF GLUCOSYLATION OF
ALBUMIN AND HEMOGLOBIN[a]

Parameter	Albumin	Hemoglobin[b]
k_1 (M^{-1} hr^{-1})	1.5	0.9
k_{-1} (hr^{-1})	0.39	0.33
K_{eq1} (M^{-1})	3.9	2.7
k_2 (hr^{-1})	0.026	0.0075
k_2' (M^{-1} hr^{-1})	0.10	0.02

[a] Reaction scheme:

$$\text{Glc} + \text{Prot} \underset{k_{-1}}{\overset{k_1}{\rightleftarrows}} \text{G=Prot} \overset{k_2}{\underset{O}{\longrightarrow}} \text{GlcProt.}$$

[b] Data obtained or calculated from refs. 6 and 7.

tion of the GA adduct to the two proteins are remarkably similar, and approximately proportional to their relative lysine content (56 Lys/mol HSA, 44 Lys/mol Hb). In contrast, the rate of formation of the Amadori adduct appears to be about 4–5 times as rapid with HSA, compared to hemoglobin.

Detection and Quantitation of Glucose Adducts to Native Albumin. Recovery of glucosylated lysine residues from protein requires hydrolysis of the polypeptide chain. Although both the GA and KP adducts are labile to acid and alkaline hydrolysis, they can be stabilized by reduction to hexitollysine. The reduction is accomplished by mixing the protein with a 100-fold molar excess of $NaBH_4$ at pH 10–12, and incubating for 4 hr at room temperature (see legend to Fig. 6). Two products are obtained, the C-2 epimers, mannitol- and glucitol-lysine. These can be resolved on an amino acid analyzer (Fig. 6A), where they normally elute between the aromatic and basic amino acids. On treatment with acid under conditions used for protein hydrolysis, some degradation occurs (Fig. 6B), yielding a putative anhydroalditol derivative(s) of lysine.[15] A standard amino acid chromatogram of HSA (Fig. 6C) is flat in the hexitol-lysine region, consistent with the limited glucosylation of native HSA (<1 mol glucose/mol protein). However, when excess sample is applied (Fig. 6D), peaks for the hexitol-lysine adducts are clearly visible, but they cannot be accurately quantitated because of poor resolution from other unidentified trace compounds in the chromatogram. To avoid these problems it is convenient to measure radioactivity in the glucose adducts following reduction and labeling with [³H]NaBH₄. Typical radioactivity profiles obtained on hydrol-

[15] S. P. Robins and A. J. Bailey, *Biochem. Biophys. Res. Commun.* **48**, 76 (1972).

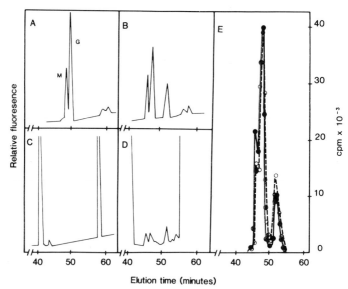

Elution time (minutes)

FIG. 6. Detection of glucose adducts to albumin by amino acid analysis. HSA (2 mg, 30 nmol, in 100 μl PBS) was reduced by adding 3 μmol [^{3}H]NaBH$_4$ (348 Ci/mmol) in 200 μl 0.1 N NaOH then incubating for 4 hr at room temperature. Excess NaBH$_4$ was discharged, and protein precipitated, by addition of 3 ml ice-cold tetrahydrofuran containing 1% concentrated HCl and centrifugation for 5 min at 2000 g. The protein pellet was dissolved in 2 ml 6 N HCl and hydrolyzed for 18 hr at 95°. Amino acid analyses were conducted by ion exchange chromatography on a Waters HPLC system using a Na citrate buffer gradient and postcolumn reaction with o-phthalaldehyde. (A) Hexitol-lysine, resolved into mannitol (M)- and glucitol (G)-lysine epimers. (B) Hexitollysine after hydrolysis in 6 N HCl for 24 hr at 110°, showing appearance of anhydroalditol degradation product. (C) Phenylalanine-lysine section from typical amino acid analysis of reduced HSA (50 μg), showing flat baseline in hexitol-lysine region. (D) Same section for analysis of 1 mg HSA. (E) Radioactivity profiles obtained from N^α-formyl-N^ε-fructoselysine (5 nmol, ○----○) and HSA (1 mg, ●——●), both reduced with [^{3}H]NaBH$_4$.

ysis and amino acid analysis of [^{3}H]NaBH$_4$-reduced fructoselysine and native HSA are shown in Fig. 6E. The yield of degradation product (anhydroalditol) is comparable for the model compound and the protein. The results shown in Fig. 6E, obtained with pooled human serum albumin, indicate that there are approximately 5 nmol glucose bound per mg HSA, or 0.29 mol Glc/mol HSA. This estimate is in good agreement with that calculated using the kinetic parameters obtained above (see the table). Thus, at steady state,

$$d\mathrm{GlcHSA}/dt = 0 = k_2'[\mathrm{Glc}][\mathrm{HSA}] - k_3[\mathrm{GlcHSA}] \qquad (10)$$

where k_3 is the fractional catabolic rate for HSA. Using a biological half-life of approximately 18 days for HSA ($k_3 = 0.039$/day),[16] and assuming an average plasma glucose concentration of 5.5×10^{-3} M (100 mg/dl):

$$[\text{GlcHSA}]/[\text{HSA}] = (k_2'/k_3) \times [5.5 \times 10^{-3}\ M]$$

$$= (0.24\ M^{-1}/\text{day})/(0.039/\text{day}) \times 5.5 \times 10^{-3}\ M \qquad (11)$$

$$[\text{GlcHSA}]/[\text{HSA}] = 0.34 \qquad (12)$$

These calculations suggest a normal steady state level of 0.25 mol glucose bound per mole HSA *in vivo*.

Other chemical procedures for measuring glucosylation of native albumin have also been described. These methods include (1) measurement of 5-hydroxymethyl-2-furfural (HMF) released on mild acid hydrolysis of glucosylated HSA, either directly by HPLC,[17] or by a colorimetric assay following reaction with 2-thiobarbituric acid (TBA)[4,10,18–21]; and (2) measurement of furosine released on strong acid hydrolysis of the protein, by HPLC.[22,23] The yield of HMF is indicative of total glucosylation, i.e., glucose bound both to intrachain lysines and amino terminal amino acids, while the furosine assay is specific for modified lysine residues. Limitations in each of these methods should be noted. Thus, the yield of HMF from Amadori products is variable,[20] and a substantial amount of the HMF released may also be destroyed under the hydrolysis conditions used.[21] Similarly, the quantitation of KP adducts by measuring furosine formation assumes that the yield of furosine from KP adducts to protein is identical to that obtained with a standard preparation of fructoselysine. Results with the thiobarbituric acid assay in various laboratories suggest a range of 8–16% glucosylation of HSA isolated from normal serum.[4] The estimate obtained here by a radiochemical procedure (29%), is more consistent with predictions from kinetic measurements, and is in excellent agreement with results of the furosine assay[23] which suggests about 28% glucosylation of native HSA. Despite limitations in accuracy of the TBA

[16] H. E. Schultze and J. F. Heremans, "Molecular Biology of Human Proteins" p. 450. Am. Elsevier, New York, 1966.

[17] O. H. Wieland, R. Dolhofer, and E. Schleicher, *Proc. Congr. Int. Diabetes Fed., 10th, 1979* p. 721 (1980).

[18] M. Keeney and R. Bassette, *J. Dairy Sci.* **42,** 945 (1959).

[19] R. Flückiger and K. H. Winterhalter, *FEBS Lett.* **71,** 356 (1976).

[20] R. Dolhofer and O. H. Wieland, *Diabetes* **29,** 417 (1980).

[21] K. A. Ney, K. J. Colley, and S. V. Pizzo, *Anal. Biochem.* **118,** 294 (1981).

[22] E. Schleicher, L. Scheller, and O. H. Wieland, *Biochem. Biophys. Res. Commun.* **99,** 1011 (1981).

[23] E. Schleicher and O. H. Wieland, *J. Clin. Chem. Clin. Biochem.* **19,** 81 (1981).

assay, good precision is obtainable, and measurements of glucosylation of albumin and total serum proteins by this method have proven useful for evaluating glycemic control in diabetes.[10] Gallop et al.[24] have recently introduced a chemical procedure for measuring nonenzymatic glucosylation of hemoglobin, based on quantitation of formaldehyde released by periodate treatment of the protein. One mole of HCHO is formed from each mole of GA or KP adduct bound to native protein, and 2 mol per mol of hexitol derivative on reduced proteins.[24] This method appears to provide an accurate measure of hemoglobin glucosylation, and should be adaptable for assays on albumin.

Acknowledgments

This work was supported by USPHS Research Grants AM-19971 and 25373, a Research Career Development Award to J.W.B. (AM-00931), and a Research Service Award to M.H.M. (AM-06712).

[24] P. M. Gallop, R. Flückiger, A. Hannekin, M. M. Mininsohn, and K. H. Gabbay, *Anal. Biochem.* **117**, 427 (1981).

[9] *In Vivo* Racemization in Mammalian Proteins

By JEFFREY L. BADA

Only L-amino acids are usually present in mammalian proteins. This system is thermodynamically unstable, however, since under conditions of chemical equilibrium the D and L enantiomers are present in equal abundances.

The amino acid racemization reaction (shown in Scheme 1) will, after a period of time which is characteristic of each amino acid, eventually convert optically active amino acids into a racemic mixture.[1] It has been known for nearly a hundred years that amino acids undergo racemization in acidic and basic solutions heated at elevated temperatures. Within the last 10–12 years it has also become established that racemization takes place at neutral pH, and the rates are comparable with those observed in dilute acid and base. Moreover, racemization has been detected in fossils and this reaction has been used to estimate the age of various fossil materials.[1–3]

[1] J. L. Bada, *ISR, Interdiscip. Sci. Rev.* **7**, 30 (1982).
[2] J. L. Bada and E. H. Man, *Earth Sci. Rev.* **16**, 21 (1980).
[3] J. L. Bada, *Earth Planet. Sci. Lett.* **55**, 292 (1981).

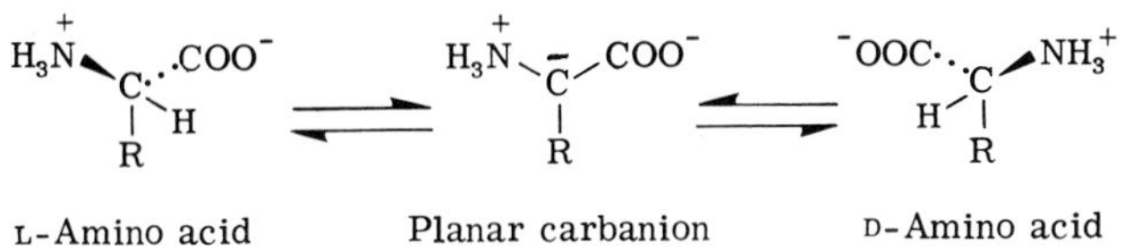

L-Amino acid   Planar carbanion   D-Amino acid

SCHEME 1. The mechanism of amino acid racemization showing the formation of the carbanion intermediate. Base abstracts the α-proton; this is the rate-limiting step in the reaction. Readdition of the proton occurs by the reaction of water with the carbanion. This mechanism is applicable to both free and peptide-bound amino acids.

Racemization has also been detected in the metabolically stable proteins of living mammals.[1,4–6] *In vivo* racemization represents a unique posttranslational stereochemical modification. As a consequence of this reaction, protein structure–function relationships may be altered.[1] In the following discussion *in vivo* racemization in various mammalian protein systems is considered as are the implications and consequences of this reaction.

Analytical Methods Used to Detect *in Vivo* Racemization

In order to detect *in vivo* racemization in living organisms very precise analytical methods are required since in general the enantiomeric ratios which are measured are exceedingly small. Although a variety of analytical methods are available for separating amino acid enantiomers many do not provide highly reproducible enantiomeric ratios.

In this laboratory most investigations of *in vivo* racemization have concentrated on aspartic acid because this amino acid is one of the most prone to racemization of the various amino acids.[1] In a typical analysis, the protein system under investigation is isolated, hydrolyzed in 6 M HCl for 6 hr at 100° and then desalted by cation exchange chromatography. A 6 hr hydrolysis is used in order to reduce the extent of acid-catalyzed racemization.[4,5] In order to carry out precise aspartic acid enantiomeric measurements it is first necessary to separate aspartic acid from the other amino acids by anion exchange chromatography.[7] The isolated aspartic acid is reacted with L-leucine-N-carboxy anhydride (L-Leu-NCA obtained from Miles Laboratory) in order to synthesize the diastereomeric dipeptides L-leucyl-L-aspartic acid and L-leucyl-D-aspartic acid.[8] These

[4] P. M. Helfman and J. L. Bada, *Nature* (*London*) **262**, 279 (1976).
[5] P. M. Masters, J. L. Bada, and J. S. Zigler, *Nature* (*London*) **268**, 71 (1977).
[6] E. H. Man, M. E. Sandhouse, J. Burg, and G. H. Fisher, *Science* **220**, 1407 (1983).
[7] C. H. W. Hirs, S. Moore, and W. H. Stein, *J. Am. Chem. Soc.* **76**, 6063 (1954).
[8] J. M. Manning and S. Moore, *J. Biol. Chem.* **243**, 5591 (1968).

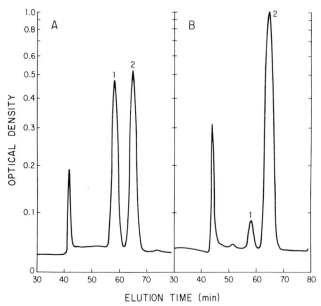

FIG. 1. Separation of the diastereomeric dipeptides L-leucyl-D-aspartic acid (peak 1) and L-leucyl-L-aspartic acid (peak 2). A 56 × 0.9 cm column filled with Beckman-Spinco AA-15 resin was used for the separation. The column was eluted with pH 3.24 buffer for 1 min; then the buffer was switched to pH 4.25 for the remainder of the run. (A) A derivatized solution of DL-aspartic acid. The ninhydrin color yield is lower for L-Leu-D-Asp than for L-Leu-L-Asp. In order to obtain the actual aspartic acid D/L ratio in a sample, the measured enantiomeric ratio must be multiplied by a conversion factor in order to correct for the different color yields. The factor determined for our analyzer is 1.20, when DMSO is used as the ninhydrin solvent. The peak which appears before the L-Leu-D-Asp dipeptide is underivatized aspartic acid, which in the elution scheme shown here, elutes at the buffer change. (B) The diastereomeric Leu-Asp dipeptides for a male narwhal (*Mondon monoceros*) tooth. The identity of the peaks is the same as in (A). The measured D/L aspartic acid ratio in this sample was 0.058.

diastereomeric dipeptides are then separated on an automatic amino acid analyzer.[9,10] Representative chromatograms are shown in Fig. 1.

The amino acid analyzer (Beckman Model 119C) is interfaced with a DEC PDP-11 computer which is used for data acquisition and analysis. The computer is used to integrate the peaks corresponding to the respective diastereomeric dipeptides and the D/L ratio is thus calculated. During

[9] J. L. Bada and R. Protsch, *Proc. Natl. Acad. Sci. U.S.A.* **70,** 1331 (1973).
[10] J. L. Bada, P. M. Masters, E. Hoopes, and D. Darling, *in* "Radiocarbon Dating" (R. Berger and H. Suess, eds.), p. 740. Univ. of California Press, Los Angeles and Berkeley, California, 1979.

a series of analyses, standards (i.e., derivatized solutions with known D/L aspartic acid ratios) are run repeatedly along with the samples under investigation. The reproducibility and precision of a D/L aspartic acid ratio is generally on the order of ±5–10%.

Another amino acid which also might be prone to *in vivo* racemization is serine.[1] Serine racemization measurements are complicated by the inherent instability of this amino acid. The derivatization procedures used in gas chromatographic analyses often result in extensive serine decomposition which make the measurement of accurate D/L serine ratios difficult. We have found that, as was the case for aspartic acid, in order to obtain accurate enantiomeric measurements, serine first must be isolated from the sample under investigation. This is accomplished by collecting fractions eluting from the buffered column of the amino acid analyzer. The serine-containing fractions are identified by high-pressure liquid chromatography (HPLC) using the *o*-phthaldialdehyde (OPT) technique.[11] The fractions must only contain serine since other amino acids interfere with serine racemization analyses. The diastereomeric dipeptides L-leucyl-L-serine and L-leucyl-D-serine are then synthesized using L-Leu-NCA using the standard procedure.[8–10] Leucine derived from the hydrolysis of L-Leu-NCA interferes with the analyses of the leucyl–serine dipeptides. However, leucine can be separated from the dipeptides by chromatography on Cu^{2+}-saturated Chelex 100 resin. The purified dipeptides are analyzed by either HPLC (see Fig. 2) or on the amino acid analyzer. Serine measurements carried out by this procedure probably have a precision on the order ±15–20%.

Kinetics and Mechanism of Amino Acid Racemization

In order to interpret the results obtained in racemization studies of various mammalian proteins, it is necessary to consider the kinetics and mechanism of the reaction. Racemization is a reversible first-order reaction and the kinetic equation for this reaction is[1]

$$\ln\left\{\frac{1 + \text{D/L}}{1 - K \cdot \text{D/L}}\right\} - \ln\left\{\frac{1 + \text{D/L}}{1 - K \cdot \text{D/L}}\right\}_{t=0} = (1 + K)k_i t \tag{1}$$

where D/L is the amino acid enantiomeric ratio at a particular time (t), k_i is the first order rate constant for the interconversion of amino acid enantiomers, and $K = 1/(\text{D/L})_{eq}$. For amino acids with one center of asymmetry, $(\text{D/L})_{eq} = 1.0$, while for the diastereomeric amino acids (i.e., isoleucine, threonine, and hydroxyproline) $(\text{D/L})_{eq}$ is different from unity.[1] The $t = 0$

---

[11] P. Lindroth and K. Mopper, *Anal. Chem.* **51,** 1667 (1979).

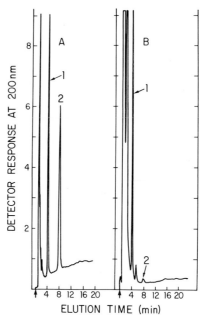

ELUTION TIME (min)

FIG. 2. Separation of the diastereomeric dipeptides L-leucyl-L-serine and L-leucyl-D-serine using reversed phase high-pressure liquid chromatography (HPLC). The separations were carried out on a Sperisorb 5-S 5-μm C$_{18}$ column. The mobile phase was 99.5% 0.2 M KH$_2$PO$_4$ titrated to pH 3 with H$_3$PO$_4$. The flow rate was 1.0 ml/min. The effluent was monitored at 200 nm using a Hitachi Model 100-10 UV-visible spectrophotometer. The time of sample injection is indicated by the arrow. (A) A derivatized DL-serine solution. (B) The diastereomeric Leu-Ser dipeptides in the dentin fraction of a 42-year-old human tooth. The estimated serine D/L ratio is ~0.03.

term in Eq. (1) is needed to account for the fact that the initial D/L ratio in the system under investigation is not exactly 0 mainly because some slight racemization takes place during the protein hydrolysis step.[4,5,12] When the extent of racemization in the system is small, i.e., D/L < 0.15, racemization can be considered as an irreversible first order reaction and Eq. (1) can thus be simplified to

$$\ln[1 + \text{D/L}] - \ln[1 + \text{D/L}]_{t=0} = k_i t \qquad (2)$$

Investigations of the racemization of free amino acids in buffered aqueous solutions indicate that the reaction obeys the expected reversible first order rate law.[13,14] However, for amino acids in some peptides and

[12] R. Liardon and R. Jost, Int. J. Pept. Protein Res. 18, 500 (1981).
[13] J. L. Bada and R. A. Schroeder, Naturwissenschaften 62, 71 (1975).
[14] G. G. Smith and T. Sivakua, J. Org. Chem. 48, 627 (1983).

TABLE I
RACEMIZATION HALF-LIVES[a] AT pH 7.6 OF SEVERAL
FREE AMINO ACIDS AT VARIOUS TEMPERATURES

Amino acid	0° (years)	37° (years)	100° (days)
Serine[b]	—	~50(?)	~4(?)
Aspartic acid	430,000	460	35
Lysine[c]	—	—	40
Histidine[c]	—	—	40
Alanine	1.4×10^6	1500	120
Isoleucine	$\sim 6 \times 10^6$	~6500	~300

[a] Taken in part from Bada.[1]
[b] Estimated from measurements [G. G. Smith and B. S. de Sol, *Science* **207**, 765 (1980)] at 122° and assuming that the ratio of the rates of racemization of serine and aspartic acid is the same at all temperatures.
[c] Calculated from measurements reported by H. M. Engel and P. E. Hare [Year Book—*Carnegie Inst. Washington* **81**, 422 (1982)].

proteins and in certain types of fossils (notably carbonates) the racemization kinetics are more complex.[1,15,16]

The rate of racemization (i.e., $2k_i$) depends upon the particular amino acid. Rates are also greatly influenced by temperature, pH, ionic strength, metal ion chelation, etc. Listed in Table I are the racemization half-lives (i.e., the time required to reach a D/L ratio of 0.33) for several free amino acids at pH 7.6 and 0, 37, and 100°. The racemization half-lives range from a few days for serine at 100° to several million years for isoleucine at 0°.

The mechanism of amino acid racemization (as first proposed by Neuberger[17]) involves the removal of the α-hydrogen of the amino acid resulting in the formation of a planar carbanion (see Scheme 1). This mechanism is supported by isotopic exchange studies.[13,14] The stability of the carbanion intermediate for any particular amino acid depends upon electron withdrawing and resonance stabilizing capacities of the various substituents on the carbanion.[1,18] The more effective a substituent is in stabilizing the carbanion intermediate the more rapid the rate of racemization. Investigations[1,18] of the rate of racemization of various ionic species

[15] G. G. Smith and R. C. Evans, *in* "Biogeochemistry of Amino Acids" (P. E. Hare, T. C. Hoering, and K. King, eds.), p. 257. Wiley, New York, 1980.
[16] M. Friedman and P. M. Masters, *J. Food Sci.* **47**, 760 (1982).
[17] A. Neuberger, *Adv. Protein Chem.* **4**, 298 (1948).
[18] J. L. Bada, *in* "Chemistry and Biochemistry of the Amino Acids" (G. C. Barrett, ed.). Chapman & Hall, London (in press).

of amino acids indicate that the ionic form AA^{0+} [$RCH(NH_3^+)COOH$] racemizes much more rapidly than AA^{+-} [$RCH(NH_3^+)COO^-$]. This result is consistent with the carbanion mechanism since the protonated carboxyl group greatly enhances the stability of the carbanion intermediate by both electron withdrawal and resonance effects.[1,18]

The abundance of a particular ionic species plays a critical role in determining the rate of racemization of free amino acids. Even though a particular ionic species may be present in only trace quantities at a certain pH the racemization rate of that particular ionic species may be much greater than that of the more abundant species. As a result, the observed racemization only occurs in the more reactive species. For example, racemization of the ionic form AA^{0-} [$RCH(NH_2)COO^-$] is observed[14,18] only in fairly concentrated basic solutions (1–2 M NaOH) even though this is the predominant ionic species at pH values greater than ~9. Since $-NH_3^+$ has a much greater electron withdrawing capacity relative to $-NH_2$, the racemization rate of the protonated amino group species is several orders of magnitude greater than that of the unprotonated species.[1,18] Only in highly basic solutions where the concentration of the protonated amino species is very low does the racemization of the unprotonated amino group species become the major reaction.

Besides the effects which ionic species have on the stability of the carbanion, the R substituents of the various amino acids are important in determining the relative order of racemization rates.[1,14,18] The more strongly electron withdrawing a particular R group, the more stable the carbanion intermediate and thus the faster the rate of racemization. The electron withdrawing ability of a particular substituent is measured by the quantity σ^*, the polar substituent, or Taft constant, which is an indicator of how well a substituent can stabilize the negatively charged carbon atom. Plots of log k_i against σ^* for the ionic species AA^{0+} and AA^{+-} indicate that amino acids which have R groups that are highly electron withdrawing, e.g., serine, have the fastest racemization rates at neutral pH while those which have substituents that are electron donating, e.g., the leucines, have the slowest racemization rates[18] (see Table I).

In living organisms, the amino acids are contained mainly in proteinaceous material. The formation of the peptide bond can greatly affect the rate of amino acid racemization. For the amino acid residues in peptides, the carboxyl group has a form roughly analogous to $-COOH$ of free amino acids.[18] From studies with free amino acids,[1,18] it has been found that $k_i^{COOH}/k_i^{COO^-} = ~10^4$. Thus, at neutral pH peptide-bound amino acids might be expected to racemize much faster than the corresponding free amino acids. On the other hand, the amino group of peptide bound amino acids can no longer be protonated and this should greatly retard racemization. However, it is difficult to theoretically predict the relative magnitude of these two effects. In experimental studies using the tripeptide glycyl-leucylglycine heated at 131° and pH 7.6, the internally bound leucine was found to racemize at a rate which is 2–4 times faster than that of free leucine.[18,19] In studies with a variety of peptides and proteins[15,16] it has been found that in the neutral pH region the racemization rates are 2–4 times those of the corresponding free amino acids.

Amino acids at terminal positions in peptides would be expected to have racemization rates which differ from those of internally bound amino acids. For N-terminal amino acids the amino group can be protonated while the carboxyl group is effectively equivalent to that of the interior residues. Since the $-NH_3^+$ group greatly enhances racemization one would predict that N-terminal racemization rates should be greater than those for internally bound residues at neutral pH. In contrast, amino acids at the C-terminal position might be expected to racemize more slowly than internally bound amino acids at neutral pH since the negatively charged carboxyl group should greatly destabilize the carbanion intermediate. In studies of dipeptides at 131° and neutral pH it has been found that N-terminal amino acid residues racemize ~15–100 times faster than those at the C-terminal positions.[18,19] Investigations[18,19] using tripeptides indicate that at neutral pH the ratio of N-terminal to internal racemization rates is on the order of 5–8. Investigations of isoleucylglycine at 131° and pH 7 indicate that N-terminal isoleucine epimerizes ~20 times faster than free isoleucine under the same conditions.[18–20] These studies demonstrate that for amino acids bound at either the N-terminal or internal positions of a protein, the racemization rates are faster than those shown in Table I for free amino acids.

As was the case for free amino acids, the electron withdrawing capacities of the various R substituents appear to be the principal factor which determines the relative racemization rates for peptide bound amino acids.[18] For a number of proteins heated at elevated temperatures at neutral and basic pH, the relative order of racemization rates are Asp > Ala = Glu > Leu, which is consistent with the σ^* values of the various substituents. There are some exceptions to this pattern which have been attributed to the intermolecular interactions of neighboring amino acid residues. For example, Engel et al.[21] showed that in the peptide hormone β_p-MSH (melanocyte-stimulating hormone), aspartic acid in only slowly

[19] S. Steinberg, Ph.D. Thesis, Scripps Institution of Oceanography, University of California, San Diego (1982).
[20] S. Steinberg and J. L. Bada, Science 213, 544 (1981).
[21] M. H. Engel, T. K. Sawyer, M. E. Hadley, and V. J. Hruby, Anal. Biochem. 116, 303 (1981).

racemized in 0.1 M NaOH at 97°. In this peptide hormone, the aspartic acid racemization rate was comparable to that of the hydrophobic amino acids. Engel *et al.* suggested that this might be due to neighboring group effects which somehow retard the racemization of aspartic acid. However, in the peptide hormone that was investigated by Engel and co-workers, aspartic acid was at the N- and C-terminal positions, and thus the peptide did not contain any internally bound aspartic acid residues. In 0.1 M NaOH the amino and carboxyl groups for terminal amino acid residues would both be unprotonated which would have a great inhibitory effect upon the racemization rate in comparison to internally bound aspartic acid residues. Thus the apparent slow racemization rate of aspartic acid observed in β_p-MSH in 0.1 M NaOH was simply due to the retarding effects of the $-NH_2$ and $-COO^-$ substituents and is not the result of intermolecular neighboring group interactions.

The effects which a neighboring residue may have on racemization rates of peptide bound amino acids is not well established at the present time. Comparisons[22] of the aspartic acid racemization rate in different metabolically stable mammalian proteins indicate that the rate may vary by about a factor of 20. The factors which contribute to these differences are not known.

It is easy to envision that certain residues in peptides and proteins may be more prone to racemization than others. One factor which could enhance the racemization rates of glutamyl or aspartyl residues is a change in the pK_a the β-carboxyl group. At physiological pH, the unprotonated β-carboxyl group is normally the prevalent ionic species but in some biological systems this can be altered. For example, in bovine trypsin, the pK_a of the aspartyl 102 residue is such that at neutral pH the protonated carboxyl group species is the dominant form.[23] This change in ionic character should have an enhancing effect on the racemization rates of that particular residue. If a larger proportion of aspartyl or glutamyl residues in a particular protein were more highly protonated than in others, these residues might be more susceptible to racemization.

This brief summary of amino acid racemization in various systems demonstrates that factors such as position of an amino acid in a peptide chain, the R substituent of the amino acid, and alteration of the protein microstructure due to changes in the degree of protonation of acidic residues all greatly affect amino acid racemization rates.

[22] P. M. Masters, *J. Am. Geriatr. Soc.* **31**, 426 (1983).
[23] R. H. Koeppe and R. M. Stroud, *Biochemistry* **15**, 3450 (1976).

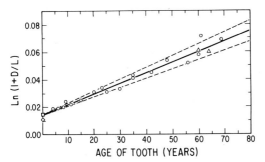

FIG. 3. A plot of aspartic acid racemization in dentin isolated from human teeth of various ages. Since the measured extent of racemization is small, the results are plotted in the form of Eq. (2). The slope (i.e., 7.87×10^{-4} yr^{-1}) is the k_{Asp} value in human dentin. The dashed lines are the changes in the k_{Asp} value calculated for a $\pm0.7°$ temperature difference, the range of the body temperature of humans. (O) Individuals of known age; (△) exact ages of individuals not known but were estimated by dentists; (□) albumin; and (◇ bovine tendon analyzed as controls. Taken from Helfman and Bada.[4]

In Vivo Racemization in Mammalian Proteins

Based on amino acid geochemical studies, it was estimated that the racemization rate of aspartyl residues at 37° should be sufficiently rapid that detectable *in vivo* racemization should take place in the metabolically stable proteins in long-lived mammals.[24] In proteins which are synthesized early in the developmental stage of a mammal's lifetime and which have remained essentially unaltered throughout the animal's life, detectable racemization of aspartic acid should occur since these proteins have been incubated at 37° for many decades. In order to test this, a protein system which could be obtained in pure form from known age mammals was required. The most suitable proteins appeared to be those present in teeth, i.e., enamel and dentin, since they satisfied the criterion of being metabolically stable. Moreover, appropriate known-age human teeth could be easily obtained from dentists. Known age human teeth were collected, the enamel and dentin isolated, and aspartic acid racemization measurements carried out. A detectable increase in the extent of aspartic acid racemization in both enamel and dentin with increasing age of the tooth was observed.[4,25] The results of the dentin analyses are shown in Fig. 3.

[24] J. L. Bada and P. M. Helfman, *World Arch.* **7**, 160 (1975).
[25] P. M. Helfman and J. L. Bada, *Proc. Natl. Acad. Sci. U.S.A.* **74**, 2891 (1975).

Besides enamel and dentin constituents of human teeth, racemization investigations have also been carried out on some of the other protein components. Masters[26] has shown that in the highly phosphorylated non-collagenous protein component present in human teeth the rate of aspartic acid racemization is substantially greater than that observed in unfractionated human dentin, and this in turn is greater than that observed in the collagenous component. A 20 times rate enhancement was observed in the phosphoprotein component in comparison to the collagenous fraction. It is possible that in the highly acidic, hydrophilic phosphoprotein the pK_a of the β-carboxyl group of the aspartyl residues is different for that in the collagenous fraction. As was discussed earlier, a change in the pK_a such that the predominant species at physiological pH would be the protonated β-carboxyl group would greatly enhance the apparent aspartic acid racemization rate. The highly phosphorylated protein fraction in human teeth may represent the maximum rate enhancement that might arise from a particular amino acid sequence. Other sequences which contain aspartyl residues but which are not nearly as acidic or hydrophilic may have racemization rates less than those measured in the phosphoprotein component in human teeth.

The racemization studies with human teeth suggested that in any protein system sequestered from active metabolic processes the aspartyl residues would be prone to racemization. Another protein system which fulfills the criterion of metabolic stability are the various crystallins contained in the ocular lens nucleus. Since these proteins are some of the most metabolically inert in mammalian tissues,[27] they provide a good system for studying racemization in a soft, rather than a calcified, tissue. A series of known age human lenses were analyzed for their extent of aspartic acid racemization and an increase in the extent of aspartic acid racemization was observed with increasing age.[5] No significant increase in the extent of aspartic acid racemization was found in the lens outer cortex[28] which is consistent with labeling studies which have shown that active protein synthesis takes place in this section of the lens. In addition to these studies, investigations of the extent of aspartic acid racemization in various protein components isolated from the ocular lens nucleus have also been conducted.[28] Water-insoluble (WI) proteins were found to contain much more highly racemized aspartyl residues than the water-soluble (WS) component. In the WI component, the extent of racemization increased with increasing age, whereas in the WS material no systematic

[26] P. M. Masters, *Calcif. Tissue Int.* **35,** 43 (1983).

[27] J. S. Zigler and J. Goosey, *Trends Biochem. Sci.* **6,** 133 (1981).

[28] P. M. Masters, J. L. Bada, and J. S. Zigler, *Proc. Natl. Acad. Sci. U.S.A.* **75,** 1204 (1978).

increase in the extent of aspartic acid racemization was observed with age.

Racemization of aspartic acid has also been observed in the white, myelin-rich, inner portion of the aging human brain.[6] The rate of racemization in this tissue is similar to that observed in the lens nucleus. The brain racemization results differ from those of the lens and teeth, however, in that no detectable racemization was observed during the first ~20 years of the human lifetime. Only after about 20 years of age is a detectable increase in the extent of aspartic acid racemization observed. It is tempting to speculate that during the first 20 years of life there are active metabolic processes involving aspartic acid which take place in the brain's white matter. If the aspartic acid turnover rate was substantially faster than the aspartic acid racemization rate, racemization would not be detectable. These active metabolic processes apparently cease after about 20 years, however, and as a result D-aspartyl residues then begin to accumulate because of racemization.

So far this discussion of *in vivo* racemization of mammalian proteins has been concerned with investigations of metabolically stable protein systems. The extent of aspartic acid racemization which has been detected in these systems is quite small, amounting to an increase in the D-aspartic acid content of the tissue of a few tenths of a percent per year. However, these results indicate that even in proteins which are actively turned-over, racemization would also take place, but the extent of racemization would be exceedingly small. In Table II are listed the D/L aspartic

TABLE II

EXTENT OF ASPARTIC ACID RACEMIZATION (ASP
D/L RATIOS) PREDICTED IN PROTEINS WITH
VARIOUS LIFETIMES IN MAMMALS[a]

Protein lifetime at 37°	Predicted aspartic acid D/L ratio
1 hr	2.5×10^{-7}
1 day	3.6×10^{-6}
30 days	0.0001
6 months	0.0006
2 years	0.003
10 years	0.015

[a] The enantiomeric ratios were calculated from Eq. (2) using $k_{Asp} = 1.3 \times 10^{-3}$ yr^{-1}, the rate constant observed in the human ocular lens nucleus.[5] The (D/L)$_{t=0}$ value in Eq. (2) was assumed to be zero.

acid ratios that would be expected for mammalian proteins with various lifetimes. In the procedures which are generally used to detect *in vivo* aspartic acid racemization, the protein is hydrolyzed for 6 hr at 100° in 6 *M* HCl. This procedure itself results in some racemization and generates a aspartic acid D/L ratio of 0.01 to 0.05, depending on the tissue.[4,5] This hydrolysis-induced racemization would completely obscure the minute racemization that might take place in rapidly turned over proteins. In order to detect racemization in short-lived proteins modified analytical procedures are required.

Recently Clarke and co-workers[29] (see also Clark, McFadden, O'Connor, and Lou, this volume [31]) have demonstrated that the enzyme protein carboxyl methyltransferase (PCM) specifically methylates only the D-aspartyl residues in a particular protein. This enzyme can be used to assay for minute levels of aspartic acid racemization that would be undetectable using the analytical procedures that involve a hydrolysis step. Using the enzymatic assay, Clarke and co-workers have shown that detectable racemization of aspartyl residues could be observed in aging human erythrocyte membrane proteins.[30] Since the erythrocytes have a finite lifetime of ~120 days, it is apparent that even in short-lived metabolically active proteins, the aspartyl residues are susceptible to racemization.

In Vivo Racemization of Amino Acids Other than Aspartic Acid

The σ^* values of the various amino acid R groups indicate that with the exception of serine, aspartic acid should have a racemization rate which greatly exceeds that of the other amino acids. As a result it would be expected that no detectable racemization of amino acids other than aspartic acid and possibly serine should take place in metabolically stable proteins. Indeed, investigations of the extent of racemization of alanine, glutamic acid, phenylalanine, and leucine in enamel[25] and the ocular lens nucleus[5] have shown that these amino acids are not susceptible to *in vivo* racemization.

The σ^* value for serine suggests that this amino acid could have a racemization rate that exceeds that of aspartic acid.[1,18] In fact, measurements carried out at 122° and pH 7.6 indicate that the racemization rate of free serine is about 10 times that of aspartic acid (see Table I). Also, for several proteins heated in basic solutions the racemization rate of seryl residues is greater than that of aspartyl residues.[21,31,32] It might thus be

[29] P. N. McFadden and S. Clarke, *Proc. Natl. Acad. Sci. U.S.A.* **79**, 2460 (1982).
[30] J. R. Barber and S. Clarke, *J. Biol. Chem.* **258**, 1189 (1983).
[31] J. R. Whitaker, *ACS Symp. Ser.* **123**, 154 (1980).
[32] R. Liardon and R. F. Hurrell, *J. Agric. Food Chem.* **31**, 432 (1983).

TABLE III
In Vivo RACEMIZATION OF SERINE AND
ASPARTIC ACID IN DENTIN AND THE OCULAR
LENS NUCLEUS IN HUMANS

Source	Age (years)	D/L ratios Ser	Asp
Dentin	8	0.06	0.020
	30	<0.01	0.033
	42	0.03	0.045
	60	<0.03	0.060
Lens nucleus	16	0.07	0.079
	41	0.08	0.109

expected that in metabolically stable proteins serine would show extensive racemization and the rate of accumulation of D-seryl residues would exceed that of D-aspartyl residues. To test this a seres of known age human teeth and ocular lens nucleus samples were analyzed[33] for the extent of serine racemization. These results are shown in Table III. As can be seen, no systematic increase in the extent of serine racemization was observed with increasing age. Moreover, the serine D/L ratios which were measured were similar to the aspartic acid D/L values, which indicates that in metabolically stable proteins serine racemizes at a rate which is either comparable to or less than that of the aspartic acid. This is inconsistent with the observations made for the free amino acids heated at elevated temperatures at neutral pH.

One possible explanation for the observed slow racemization of serine is that at physiological pH aspartic acid racemizes faster than serine in proteins. Although measurements of the rate of serine racemization in various proteins have been made in basic solutions,[21,32] no investigations have been carried out in the neutral pH range. We were thus interested in establishing the relative serine-aspartic acid racemization rates in proteins at neutral pH. A series of elevated temperature experiments were conducted[33] at various pH values using free serine and aspartic acid and poly(L-serine) and poly(L-aspartic) acid. From these experiments it was possible to determine the ratio of racemization rates of both free and peptide-bound serine and aspartic acid in the neutral pH region. These results are summarized in Table IV. At neutral pH peptide-bound serine and aspartic acid have essentially identical racemization rates. Thus our

[33] S. Steinberg, P. M. Masters, and J. L. Bada, unpublished results.

TABLE IV
RATIO OF RACEMIZATION RATES OF FREE AND
PEPTIDE-BOUND SERINE AND ASPARTIC ACID AT
SEVERAL pH VALUES AT 100°

pH	Free k_{Ser}/k_{Asp}	Peptide-bound[a] k_{Ser}/k_{Asp}
3.60	0.28	0.002
4.66	1.63	0.05
5.80	—	0.19
6.65	1.73	1.17
7.65	9.36	2.13

[a] Poly(L-serine) and poly(L-aspartic) acid were used as model systems for these investigations.

failure to observe extensive serine racemization in the metabolically stable proteins is consistent with the relative serine–aspartic acid racemization rate determined at elevated temperatures for peptide-bound residues.

These various results suggest that only aspartyl residues would be prone to *in vivo* racemization. Other amino acids have racemization rates at physiological pH which are less than that of aspartic acid. It should be emphasized, however, that so far only total aspartic acid (asparaginyl plus aspartyl residues) racemization measurements have been carried out. Based on the σ^* values it would be predicted that asparaginyl residues should racemize much more rapidly than the aspartyl residues.[16] Thus it is conceivable that if racemization measurements of asparaginyl residues in metabolically stable proteins could be carried out the extent of racemization of this amino acid could exceed that observed for the total aspartyl residues.

Also, as was discussed in the section on kinetics and mechanism, N-terminal amino acids would be expected to have racemization rates which are greater than those of internally bound residues. Therefore, amino acid residues at the N-terminal position could be much more prone to racemization than internally bound residues. If a particular amino acid is contained only at the N-terminal position in a particular protein its apparent racemization rate might thus exceed that which would be observed for aspartic acid in the total protein. At the present time, however, no investigations of the extent of *in vivo* racemization of N-terminal amino acids in mammalian proteins have been carried out.

Applications and Implications of *in Vivo* Racemization

The observation of *in vivo* racemization in metabolically stable proteins in long-lived mammals provides the basis for a biochronological tool useful for assessing the ages of mammals whose ages are difficult or impossible to determine by other methods. Investigations of the extent of racemization in the teeth from various known-age mammalian species indicate the racemization rate in other mammals is similar to that in humans.[34] The rate constant which has been determined for aspartic acid racemization in human teeth can thus be used to calculate the ages of other living mammals. One of the more interesting applications of this method has been the use of the extent of aspartic acid racemization in teeth to determine the ages of various marine mammals. For example, studies of the teeth from narwhals (*Mondon monoceros*) indicate that the aspartic acid racemization method can be used to determine accurate ages for these Arctic cetaceans.[35] In the case of narwhals, the racemization method is particularly useful for determining the ages of mature females which have been previously difficult or impossible to accurately age.

Another potentially important biochronological application of the amino acid method would be to use this technique to test the authenticity of the extremely old ages claimed by individuals in certain human populations. People living in the Georgian Province of Russia, in Hunza, Pakistan, in Vilcabamba, Ecuador often claim to have ages well in excess of 100 years. There has been a major controversy as to whether these people are really as old as they claim or whether they are simply misrepresenting their ages for the prestige of being noted as the oldest living humans on earth. We have suggested[4,25] that the racemization method could be used to determine whether these people are really as old as they claim, provided a suitable tooth could be obtained. Some preliminary tests have already been carried out in collaboration with some Russian gerontologists. The racemization method was used to determine the age of a tooth (submitted without prior knowledge of the documented age) from a 96-year-old woman; the amino acid age was determined to be 99 years. Further racemization tests should help establish whether certain human populations have individuals with extraordinary longevities.

The extent of aspartic acid racemization in enamel and dentin can also be used to estimate the age at death of humans from ancient burials. The

[34] J. L. Bada and S. E. Brown, *Trends Biochem. Sci.* **5,** III–V (1980).
[35] J. L. Bada, E. Mitchell, and B. Kemper, *Nature* (London) **303,** 418 (1983).

main requirement in this application is that the burial should be fairly recent or that the burial environmental temperature should be relatively cool so that the extent of postmortem racemization is small compared to that which occurred during the lifetime of the individual. This application of the racemization method has been used to calculate the age at death of several mummified Alaskan Eskimos[36,37] and also to assess the ages at death of individuals in a medieval cemetery in Czechoslovakia.[38] The racemization ages were found to be in good agreement with those determined independently by anthropologists. The racemization method is particularly useful in determining the ages of older individuals, whose age at death is often difficult to estimate from anatomical evidence.

In vivo racemization in mammalian proteins may have important implications concerning the aging process and perhaps even protein turnover.[39] Racemization in proteins may induce structural changes which in turn could affect the functionality of the protein. Thus racemization may play some role in the complex aging process observed in long-lived mammals. A good example of the possible relationship between racemization and aging is provided by the studies carried out on the extent of aspartic acid racemization in the ocular lens nucleus. Several structural changes take place in the lens nucleus proteins during the normal aging process. For example, the proportion of water-insoluble protein increases relative to water-soluble protein and there is an increase in the extent of cross-linking.[27] These observations indicate that changes in native conformations of the lens proteins are occurring during aging. Fractional analysis of the protein present in the lens nucleus from various age individuals indicate that racemization of aspartyl residues may lead to the formation of heavy molecular weight aggregates which are then converted to water-insoluble proteins by various reactions.[22,28] In the lens nucleus it therefore appears that racemization may contribute to some of the age-related changes observed in the proteins in this system. Additional studies are required, however, before any definitive conclusions can be reached about what part, if any, of the aging process in mammals can be attributed to racemization.

It is possible that even in short-lived proteins, racemization could play some role in protein turnover, provided very small amounts of racemization have deleterious effects on protein function. It has recently been suggested by Clarke and co-workers[29] that the main function of the protein carboxyl methyltransferase enzyme in various mammalian tissues is

---

[36] P. M. Masters and M. R. Zimmerman, *Science* **201,** 811 (1978).
[37] P. M. Masters, *Arct. Anthrop.* (in press).
[38] P. M. Masters and J. L. Bada, *Adv. Chem. Ser.* **171,** 117 (1978).
[39] P. M. Helfman, J. L. Bada, and M.-Y. Shou, *Gerontology* **23,** 419 (1977).

to invert the D-aspartyl residues produced from racemization back into the L-enantiomer. This enzyme may thus have the function of eliminating or repairing racemized aspartyl residues. That such a repair system should exist strongly suggests that *in vivo* racemization in peptides and proteins has detrimental biological consequences.

It is of interest to consider the possibility that *in vivo* racemization also takes place in ectothermic organisms. At temperatures less than 37°, the aspartyl racemization rate would be less than that in mammalian proteins. If an organism lived in an environment where the average temperature was 25°, only minute amounts of aspartic acid racemization would occur even over time periods as long as several centuries. Some recent measurements[40] of the extent of aspartic acid racemization in the ocular lens nucleus of Hawaiian sea turtles, which live at temperatures between 22 and 28°, indicate that little, if any, detectable *in vivo* racemization occurs during the life of these animals. Thus, for organisms which reside at temperatures less than 37°, the aspartyl residues in their proteins would be much less prone to racemization than in mammalian systems.

On the other hand, the aspartyl residues in organisms that live at temperatures greater than 37° should be more susceptible to racemization than those in mammals. In birds which have body temperatures on the order of 40°, racemization should occur in metabolically stable proteins at a rate which should allow for detection of the accumulation of D-aspartyl residues on time scales as short as a few years. One group of organisms in which racemization rates might be extraordinarily rapid are thermophilic bacteria, which live in geothermal waters where temperatures are ~80–90°. At these temperatures, the half-life for aspartyl racemization could be on the order of a few days or so. Thus, the aspartyl residues in the proteins in these organisms might be highly susceptible to racemization. In fact, racemization of aspartyl residues may represent a major post-translational modification which is important in regulating the life-span of these bacteria. No investigations of aspartic acid racemization in thermophilic bacteria have been conducted.

Acknowledgments

I thank P. M. Masters and S. Steinberg for helpful discussions and assistance with some of the studies summarized here. This work was supported by Grant PHS AG00638 from the National Institute on Aging.

---

[40] J. L. Bada, unpublished results.

[10] Pyridoxal Phosphate-Binding Site in Enzymes; Reduction and Comparison of Sequences

By Yoshimasa Morino and Fujio Nagashima

In pyridoxal-P (pyridoxal phosphate)-dependent enzymes, the formyl group at position 4 of pyridoxal-P forms an aldimine bond with the ε-amino group of a specific lysyl residue.[1–3] There are a number of lysyl residues in many proteins and any lysyl residue is in principle capable of forming an aldimine linkage to pyridoxal-P. However, involvement of a single particular lysyl residue indicates that there is a specific domain structure favorable for stabilizing the aldimine bond which, otherwise, is readily hydrolyzed. X-Ray analyses of both cytosolic[4] and mitochondrial[5] isoenzymes of aspartate aminotransferase, representative pyridoxal-P-dependent enzymes, provided details of interaction between pyridoxal-P and the enzyme side chains in the coenzyme-accommodating active site as follows: (1) the aldimine double bond and the coenzyme pyridine ring are roughly coplanar; (2) the aldimine linkage is directed toward the 3 hydroxyl of the coenzyme, i.e., cisoid conformation; (3) an aspartyl carboxyl is hydrogen-bonded to the protonated nitrogen of the coenzyme pyridine ring; (4) a tyrosyl hydroxyl group forms a hydrogen bond to the ionized 3 hydroxyl of pyridoxal-P; and (5) a positively charged arginyl guanidinium group forms an ion pair with the 5'-phosphate group of the coenzyme. Such multiple interactions entail a strictly specific mode of combination of the coenzyme with apoenzyme. Thus only the *Re* side of the aldimine bond is open to solvent, the *Si* face being shielded from solvent.[6,7] This stereospecific status of the bound coenzyme has been remarkably identical among a variety of pyridoxal-P-dependent enzymes

[1] E. H. Fischer, A. B. Kent, E. R. Synder, and E. G. Krebs, *J. Am. Chem. Soc.* **80**, 2906 (1958).
[2] R. C. Hughes, W. T. Jenkins, and E. H. Fischer, *Proc. Natl. Acad. Sci. U.S.A.* **48**, 1615 (1962).
[3] O. L. Polyanovskii and B. A. Keil, *Biokhimiya* **28**, 372 (1963).
[4] A. Arnone, P. D. Briley, P. H. Rogers, C. C. Hyde, C. M. Metzler, and D. M. Metzler, *in* "Molecular Structure and Biological Activity" (J. F. Griffen and W. L. Duax, eds.), p. 57. Elsevier/North-Holland, New York, 1982.
[5] G. C. Ford, G. Eichele, and J. N. Jansonius, *Proc. Natl. Acad. Sci. U.S.A.* **7**, 2559 (1980).
[6] E. Austermuhle-Bertola, Dissertation, No. 5009, Eidgenossischen Technischen Hochschule, Zurich (1973).
[7] S. W. Zito and M. Martinez-Carrion, *J. Biol. Chem.* **255**, 8645 (1980).

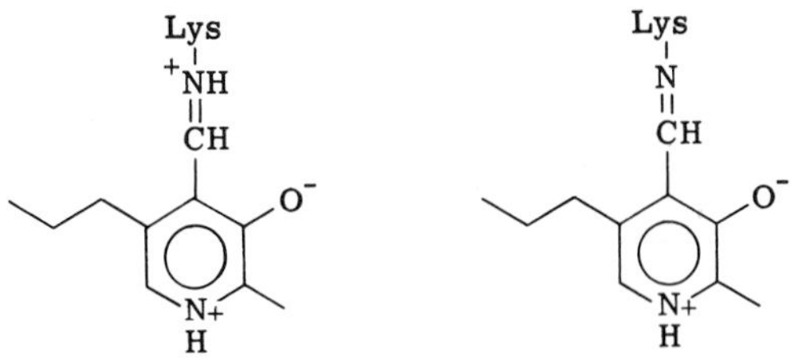

Lys
$^+$NH
CH

Absorbs at ~ 420 nm

[I]

Lys
N
CH

~ 350 nm

[II]

SCHEME 1.

so far studied.[8–10] The striking identity in the stereospecificity of borohydride reduction, in turn, strongly suggests a degree of similarity in the mode of interaction of the coenzyme molecule with apoenzyme side chains in the active site.

Spectral Properties of Pyridoxal-P Dependent Enzymes

At neutral pH, pyridoxal-P absorbs at 390 nm, which shifts more or less to a longer wavelength (410–430 nm) upon the formation of an aldimine linkage with amino compounds or apoenzymes. There are two spectral species which have been ascribed to a protonated I and unprotonated II forms of the coenzyme aldimine (Scheme 1) (see Vol. 18 [77][11] and Vol. 62 [79][12] for details of interpretation of the spectral characteristics of pyridoxal-P aldimine).

The spectral characteristics of many pyridoxal-P enzymes are listed in Table I. Spectral changes accompanying the protonation and deprotonation are observed with only a few enzymes such as aspartate aminotransferases, alanine aminotransferase, tryptophanase, and glutamate decarboxylase. Most other pyridoxal-P enzymes do not exhibit such pH-dependent spectral changes. Many pyridoxal-P enzymes show a spectral feature reflecting structure I, while aspartate β-decarboxylase exhibits a 360-nm species, II, over a wide range of pH. Phosphorylase is an exception. Pyridoxal-P in this enzyme absorbs at 330 nm. This is ascribed

---

[8] J. C. Vederas, E. Schleicher, M. D. Tsai, and H. G. Floss, *J. Biol. Chem.* **253,** 5350 (1978).

[9] J. C. Vedras, I. D. Reingold, and H. W. Sellers, *J. Biol. Chem.* **254,** 5053 (1979).

[10] J. C. Vedras and H. G. Floss, *Acc. Chem. Res.* **13,** 455 (1980).

[11] R. J. Johnson and D. E. Metzler, this series, Vol. 18, p. 433.

[12] B. I.-U. Yang and D. E. Metzler, this series, Vol. 62, p. 528.

TABLE I
SPECTRAL PROPERTIES OF PYRIDOXAL-P ENZYMES

Enzyme (source)	EC number	λ_{max}(pH) (nm)	ε_M (M^{-1} cm^{-1})	Circular dichroism	$\Delta\varepsilon/\varepsilon^a$ ($\times 10^{-4}$)	Number of subunit[b]	References[c]
Aminotransferase							
Aspartate (pig heart)							
Cytosolic	2.6.1.1	430 (5.5)	7370	Positive	33.7	2	1–3
		362 (8.0)	8400	Positive	19.2		
Mitochondrial		436 (5.5)	7380	Positive	32.7	2	1–3
Alanine (pig heart)	2.6.1.2	427 (5.5)	9060	Positive	26.0	2	4
Lysine (A. liquidum)	2.6.1.36	415 (7.4)				2	5
Ornithine (rat liver)	2.6.1.13	420 (7.5)				4	6
Leucine (pig heart)	2.6.1.6	414 (7.0)				1	7
Tyrosine (rat liver)	2.6.1.5	415 (6.4)		Positive		4	8
Histidinolphosphate (S. typhimurium)	2.6.1.9	430 (8.6)				1	9
γ-Aminobutyrate (pig brain)	2.6.1.19	415 (7.4)				2	10
β-Alanine (Pseudomonas sp.)	2.6.1.18	390 (7.4)				1	11
D-Alanine (B. subtilis)	2.6.1.21	415 (9.2)		Negative		1	12
Pyridoxamine (Pseudomonas MA-1)	2.6.1.30	410 (7.0)				2	13
Racemase							
Amino acid (P. striata)	5.1.1.10	420 (7.0)	9300	Negative	20	2	14, 15
Alanine (P. putida)	5.1.1.1	420 (7.4)				1	16
Arginine (P. gravelens)	5.1.1.9	420 (7.2)				4	17
Elimination-catalyzing enzyme							
Tryptophanase (E. coli)	4.1.99.1	420 (6.1)	6880	Positive		4	18, 19
		337 (9.2)	6020				

Enzyme	EC number	Absorption max (pH)		Reactivity		Subunits	Reference
Tryptophan synthase β_2 (E. coli)	4.2.1.20	407 (7.5)				2	20
D-Serine dehydratase (E. coli)	4.2.1.14	415 (6.0)				1	21
Threonine dehydratase, biosynthetic (S. typhimurium)	4.2.1.16	410 (7.4)		Positive		4	22
Threonine dehydratase, degradative							
AMP-dependent (E. coli)		415 (6.8)		Positive	15	4	23
ADP-dependent (C. tetranomorphum)		415 (7.0)				8	24
β-Tyrosinase (E. intermedia)		430 (7.0)				2	25
O-Acetylserine (thiol)-lyase (S. typhimurium)	4.2.99.8	412 (7.6)				2	26
Cystathionine γ-lyase (rat liver)	4.4.1.1	427 (7.4)	6360	Positive	13.5	4	27, 28
Cystathionine γ-synthase (S. typhimurium)	4.2.99.9	422 (7.3)	9750			4	29
Methionine γ-lyase (P. ovalis)	4.4.1.11	420 (7.2)				4	30
Kynureninase (P. marginalis)	3.7.1.3	430 (7.2)				1	31
Decarboxylase							
Glutamate (E. coli)	4.1.1.15	415 (4.6)		Positive	8.9	6	32, 33
		340 (6.3)		Positive	0.9		
Lysine (B. cadaveris)	4.1.1.18	425 (6.2)				10	34
Arginine (E. coli)	4.1.1.19	420 (6.6)				10	35
Ornithine (E. coli)	4.1.1.17	420 (6.0)				2	36
Dialkylamino acid (P. cepacia)	4.1.1.64	405 (7.5)				4	37
Aspartate,β (A. faecalis)	4.1.1.12	360 (6.8)				12	38
Threonine aldolase (C. humicala)	4.1.2.5	420 (6.4)				6	39

(continued)

TABLE I (continued)

Enzyme (source)	EC number	λ_{max}(pH) (nm)	ε_M (M^{-1} cm^{-1})	Circular dichroism	$\Delta\varepsilon/\varepsilon^a$ ($\times 10^{-4}$)	Number of subunit[b]	References[c]
Serine hydroxymethyltransferase (rabbit liver)	2.1.2.1	428 (7.3)				6	40
Phosphorylase (rabbit muscle)	2.4.1.1	333		Positive	11	2	41

[a] Anisotropic factor.

[b] Correspond to the number of pyridoxal-P bound per mole of enzyme.

[c] Key to references:

1. M. Martinez-Carrion, D. C. Tiemier, and D. L. Peterson, *Biochemistry* **9**, 2574 (1970).
2. A. E. Braunstein, *in* "The Enzymes" (P. D. Boyer, ed.), 3rd ed., Vol. 9, p. 379. Academic Press, New York, 1973.
3. S. Tanase and Y. Morino, unpublished.
4. M. H. Saier and W. T. Jenkins, *J. Biol. Chem.* **242**, 91 (1967); H. Kojima, S. Tanase, and Y. Morino, unpublished.
5. H. Misono and K. Soda, *J. Biochem. (Tokyo)* **82**, 535 (1977).
6. Y. Sanada, T. Shiotani, E. Okuno, and N. Katunuma, *Eur. J. Biochem.* **69**, 507 (1976).
7. R. T. Taylor and W. T. Jenkins, *J. Biol. Chem.* **241**, 4396 (1966).
8. F. Valeriote, F. Auricchio, D. Riley, and G. Tomkins, *J. Biol. Chem.* **244**, 3618 (1969); M. Inoue and Y. Morino, unpublished.
9. R. G. Martin, *Arch. Biochem. Biophys.* **138**, 239 (1970).
10. T. Beeler and J. E. Churchich, *Eur. J. Biochem.* **85**, 365 (1978).
11. K. Yonaha, S. Toyama, M. Yasuda, and K. Soda, *FEBS Lett.* **71**, 21 (1976).
12. M. Martinez-Carrion and W. T. Jenkins, *J. Biol. Chem.* **240**, 3538 (1965).
13. H. Wada and E. E. Snell, *J. Biol. Chem.* **237**, 133 (1962).
14. K. Soda and T. Osumi, *Biochem. Biophys. Res. Commun.* **35**, 363 (1969).
15. S. Tanase, Y. Morino, T. Osumi, and K. Soda. *Seikagaku* **45**, 443 (1973).
16. G. Rosso, K. Takashima, and E. Adams, *Biochem. Biophys. Res. Commun.* **34**, 134 (1969).
17. T. Yorifuji, K. Ogata, and K. Soda, *J. Biol. Chem.* **246**, 5058 (1971).

18. W. A. Newton, Y. Morino, and E. E. Snell, *J. Biol. Chem.* **240**, 1211 (1965).
19. Y. Morino and E. E. Snell, *J. Biol. Chem.* **242**, 5206 (1967).
20. G. M. Hathaway, S. Kida, and D. M. Crawford, *Biochemistry* **8**, 989 (1969).
21. D. Dupourque, W. A. Newton, and E. E. Snell, *J. Biol. Chem.* **241**, 1233 (1966).
22. R. O. Burns and M. H. Zarlengo, *J. Biol. Chem.* **243**, 178 (1968).
23. T. Tanabe, Y. Shizuta, K. Inoue, A. Kurosawa, and O. Hayaishi, *J. Biol. Chem.* **249**, 873 (1974).
24. A. Nakazawa and O. Hayaishi, *J. Biol. Chem.* **242**, 1146 (1967).
25. H. Kumagai, H. Yamada, H. Matsui, H. Ohkishi, and K. Ogata, *J. Biol. Chem.* **245**, 1773 (1970).
26. M. A. Becker, N. M. Kredich, and G. M. Tomkins, *J. Biol. Chem.* **244**, 2418 (1969).
27. Y. Matsuo and D. M. Greenberg, *J. Biol. Chem.* **230**, 545 (1958).
28. H. Sakamoto and Y. Morino, unpublished.
29. M. M. Kaplan and M. Flavin, *J. Biol. Chem.* **241**, 5781 (1966).
30. H. Tanaka, N. Esaki, and K. Soda, *Biochemistry* **16**, 100 (1977).
31. M. Moriguchi, T. Yamamoto, and K. Soda, *Biochem. Biophys. Res. Commun.* **26**, 109 (1967).
32. T. E. Huntly and D. E. Metzler, *Biochem. Biophys. Res. Commun.* **12**, 2969 (1973).
33. P. H. Strausbach and E. H. Fischer, *Biochemistry* **9**, 226 (1970).
34. S. S. Tate and A. Meister, *Biochemistry* **9**, 2626 (1970).
35. S. L. Blethen, E. A. Boeker, and E. E. Snell, *J. Biol. Chem.* **243**, 1671 (1968).
36. D. Applebaum, D. L. Sabo, E. H. Fischer, and D. R. Morris, *Biochemistry* **14**, 3675 (1975).
37. G. B. Bailey and W. B. Dempsy, *Biochemistry* **6**, 1526 (1967).
38. E. M. Wilson and A. Meister, *Biochemistry* **5**, 1166 (1966).
39. H. Kumagai, T. Nagata, H. Yoshida, and H. Yamada, *Biochim. Biophys. Acta* **258**, 779 (1972).
40. L. Schirch and M. Mason, *J. Biol. Chem.* **237**, 2578 (1962).
41. G. F. Johnson and D. J. Graves, *Biochemistry* **5**, 2906 (1966).

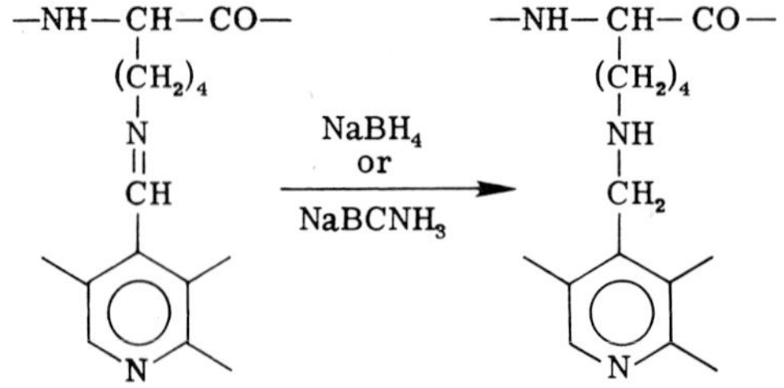

FIG. 1. Reduction of the aldimine bond between pyridoxal-P and ε-amino group of a lysyl residue.

to the neutral tautomer of structure **II**, which could exist in a strongly hydrophobic environment.[13] In contrast to a majority of pyridoxal-P enzymes in which the coenzyme–aldimine bond is crucial to their catalytic function, the 5′-phosphate group is catalytically important in the phosphorylase-catalyzed reaction.[14,15] This difference in the mode of action of pyridoxal-P is reflected by that in the binding mode between phosphorylase and other pyridoxal-P enzymes. The bound pyridoxal-P exhibits a variable degree of circular dichroism where it absorbs (Table I). Its magnitude may reflect the degree of structural asymmetry of the bound coenzyme.

Isolation and Structural Determination of Phosphopyridoxyl Peptide
    from Pyridoxyl-P-Combining Site

Principle

    Pyridoxal-P–aldimine bond is reduced by borohydride or borocyanohydride under mild conditions (Fig. 1).[1,2,16] The reduction renders the aldimine bond highly stable to hydrolysis. The resulting reduced enzyme is fragmented by either proteolytic or chemical cleavage, followed by isolation of a phosphopyridoxyl peptide and its structural determination.

Procedure (General)

    Reduction of Pyridoxal-P–Aldimine Bond. NaBH$_4$ decomposes rapidly in water. Hence an excess amount must be used to ensure a complete

[13] G. F. Johnson, J.-I. Tu, M. L. S. Bartlett, and D. J. Graves, *J. Biol. Chem.* **245,** 5560 (1970).
[14] K. Feldman and W. E. Hull, *Proc. Natl. Acad. Sci. U.S.A.* **74,** 856 (1977).
[15] S. G. Withers, N. B. Madsen, B. D. Sykes, M. Takagi, S. Shimomura, and T. Fukui, *J. Biol. Chem.* **256,** 10795 (1981).
[16] Y. Matuso and D. M. Greenberg, *J. Biol. Chem.* **234,** 507 (1959).

reduction of enzyme samples in solution. Reduction proceeds rapidly. Foaming can be prevented by an antifoaming agent such as 1-octanol. Solid $NaBH_4$ is often used with success. In some cases, particularly when NaB^3H_4 is used, it is convenient to use as a solution. For this purpose, a weighed amount of $NaBH_4$ is dissolved in a volume of freshly distilled anhydrous dimethylformamide to make a solution at a known concentration. This can be stored for at least 5 months under dry conditions at $-10°$. $NaCNBH_3$ is much weaker as reducing reagent than $NaBH_4$. However, it is sufficiently stable in H_2O. Most pyridoxal enzymes are yellow. The reduction can be judged by bleaching of enzyme solution. In our experiences, 1 mM aspartate aminotransferase in a solution at pH 6–7 was reduced by 2-fold molar excess of $NaCNBH_3$ within 30 min at 25°.

Isolation of a Peptide Fragment Containing Phosphopyridoxyllysyl Residue. In principle, any proteolytic and chemical methods can be used for fragmentation. To obtain a high yield of the peptide, the use of reagents with a high specificity of peptide bond cleavage (i.e., trypsin, cyanogen bromide, etc.) is recommended. Purification is achieved by combined uses of gel permeation, ion-exchange column chromatography, paper/thin-layer chromatography, and electrophoresis. All procedures are preferably carried out under protection from light since pyridoxyl compounds are labile under light. A few representative examples will be described below.

Detection of the Phosphopyridoxyl Peptides. Phosphopyridoxyl peptides are detected by fluorescence (at 390 nm) which is characteristic for a pyridoxyl derivative (excited at 325 nm at neutral pH) or by absorbance at 325 nm (which is less sensitive than fluorescence). An alternative method is the detection of ^3H-radioactivity incorporated into the pyridoxyl moiety upon reduction of the aldimine with NaB^3H_4.

Procedures for Isolation and Sequence Determination of a Phosphopyridoxyl Peptide from Alanine Aminotransferase[17]

Reduction, Carboxymethylation, and Tryptic Digestion of the Enzyme. The pyridoxal form of alanine aminotransferase[18–20] (100 mg in 7 ml of 0.1 M Tris–HCl buffer, pH 7.5) was reduced by adding 0.1 ml of NaB^3H_4 (4 mCi/mmol). The reducing reagent was prepared by diluting 10 mCi of NaB^3H_4 (6 Ci/mmol, New England Nuclear) with 100 mg of cold $NaBH_4$ in 1 ml of freshly distilled anhydrous dimethylformamide. After

[17] S. Tanase, H. Kojima, and Y. Morino, *Biochemistry* **18**, 3002 (1979).
[18] M. H. Saier and W. T. Jenkins, *J. Biol. Chem.* **242**, 91 (1967).
[19] T. Matsuzawa and H. L. Segal, *J. Biol. Chem.* **243**, 5929 (1968).
[20] H. Kojima, *Kumamoto Med. J.* **32**, 23 (1979).

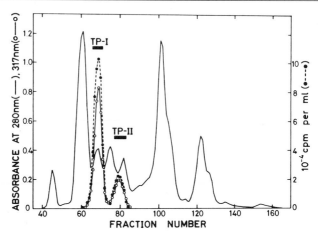

FIG. 2. Elution pattern of tryptic digests of NaB^3H_4-reduced alanine aminotransferase on BioGel P-10. Fractions (3.5 ml) were collected and monitored for absorbance and radioactivity: absorbance at 280 nm (——); absorbance at 317 nm (○); radioactivity (●). The bar indicates the fractions pooled. See the text for further details. From S. Tanase, H. Kojima, and Y. Morino, *Biochemistry* **18,** 3002 (1979).

incubation for 20 min at 25°, the reduced preparation was dialyzed against 50 mM Tris–HCl buffer (pH 8.0) for 20 hr at 5° with two changes of the buffer. The resulting preparation exhibited an absorption band at 325 nm, which is characteristic of a pyridoxyl derivative. The specific radioactivity was 1.54×10^6 cpm/μmol of monomeric unit of the enzyme. The reduced enzyme preparation was incubated for 2 hr at 37° under nitrogen in 5 ml of 6 M guanidine HCl containing 0.1 M Tris–HCl buffer (pH 8.0), 1 mM EDTA, and 2 mM dithiothreitol. This is to ensure extensive unfolding of the protein. Sulfhydryl residues in the protein were then carboxymethylated by adding sodium iodoacetate to a final concentration of 5 mM and incubating at 37° for 20 min. The carboxymethylated preparation was extensively dialyzed against distilled water. The resulting heavy precipitate was collected by centrifugation and suspended in 5 ml of 0.1 M NH_4HCO_3, pH 8.2. The suspension was then adjusted to a protein concentration of 15 mg/ml, and 1 mg of L-(1-tosylamino-2-phenylethyl)chloromethylketon (TPCK)-treated trypsin[21] was added. During incubation for 3 hr at 37° under occasional stirring, the solution became clear.

Purification of Phosphopyridoxyl Peptides. The tryptic digest was passed over a 2.6 × 140 cm column of BioGel P-10 (200–400 mesh) equilibrated with 0.1 M NH_4HCO_3 (pH 8.5). The elution profile revealed one major radioactive peak (TP-I) and one minor peak (TP-II) (Fig. 2). Materi-

[21] F. H. Carpenter, this series, Vol. 11, p. 237.

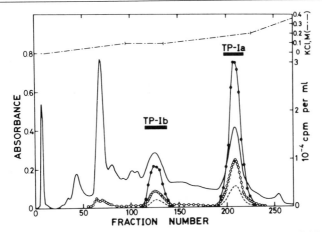

FIG. 3. Isolation of pyridoxyl-P peptides on a column of SP-Sephadex C-25. Fractions (1.1 ml) were collected and analyzed for absorbance and radioactivity: absorbance at 230 nm (——); absorbance at 290 nm (○); absorbance at 325 nm (---); radioactivity (●). The absorption maximum of a pyridoxyl derivative at an acidic pH is at 290 nm. The bars indicate the fractions pooled. Further details are given in the text. From S. Tanase, H. Kojima, and Y. Morino, *Biochemistry* **18**, 3002 (1979).

als in both peaks exhibited an intense blue fluorescence under an ultraviolet lamp (Mineralight 325 nm) and showed an absorption band at 317 nm which is attributed to the pyridoxyl derivative. Approximately 90% of the radioactivity in the original digest was recovered in both peak fractions. Each peak fraction was separately lyophilized. Lyophilized TP-I was dissolved in 0.1 M formic acid and applied to a column of SP-Sephadex C-25 (1 × 15 cm) equilibrated with 30 mM formic acid, pH 2.7. A gradient elution yielded two peak fractions containing pyridoxyl peptides as monitored by fluorescence and radioactivity (Fig. 3). Both fractions (TP-Ia and TP-Ib) were separately lyophilized and desalted on BioGel P-6 (200–400 mesh) column equilibrated with 0.1 M NH_4HCO_3. Fraction TP-II was further purified by SP-Sephadex column chromatography, followed by two-dimensional paper chromatography and electrophoresis.[22] Overall yields of these purified peptides were 25, 8, and 3%, with TP-Ia, TP-Ib, and TP-II, respectively.

Sequence Analysis of Phosphopyridoxyl Peptides. The result of amino-terminal analysis (the amino-terminus of TP-Ib was blocked), the chromatographic behavior and the amino acid composition, taken together, indicated that TP-Ib was formed from the parent peptide TP-Ia by cyclization of the amino-terminal glutaminyl into pyroglutamyl residue.

[22] J. C. Bennett, this series, Vol. 11, p. 330.

Sequence determination was performed by a manual Edman procedure[23] and the dansyl-monitored Edman method.[24] Phenylthiohydantoin hormone (PTH)amino acids were identified either on silica gel thin-layer glass plates (Kieselgel 60 F_{254}, Merck) or on polyamide sheets (Cheng Chin Trading Co.) as described. Dansyl amino acids were identified on polyamide sheets. The PTH derivative of N^6-(phosphopyridoxyl)lysine remained in the acidic aqueous phase. It was identified by its distinct blue fluorescence on the silica gel thin-layer chromatogram and confirmed by its characteristic absorption band at 325 nm at pH 7 and by its radioactivity which was incorporated upon reduction with NaB^3H_4. PTH derivatives of pyridoxyllysine and histidine appeared as red-orange and red-brown spots, respectively, upon staining the filter paper with Pauly's reagent. Carboxypeptidase Y digestion[25] was performed at 25°, pH 5.6 (50 mM sodium acetate buffer), at an enzyme to substrate molar ratio of 1 : 200. The enzymic hydrolysis was monitored for a period of 60 min, and the products were assayed in an amino acid analyzer. In a two-column system of amino acid analysis, phosphopyridoxyllysine was eluted just before leucine on the long column and pyridoxyllysine emerged from the short column between lysine and histidine.

The Edman degradation of the peptide TP-Ia (0.1 μmol) was carried out up to the nineteenth step. The sequence from the amino-terminus to the fourteenth residue was established by the identification of each residue as its PTH derivative. The remainder sequence was analyzed by the dansyl–Edman procedure. The sequence was further confirmed by the analysis of the structure of peptide fragments obtained by chymotryptic and thermolytic digestion of the parent peptide TP-Ia.

TP-Ib was found to be derived from TP-Ia as described above. TP-II was found to be a fragment formed from TP-Ia and TP-Ib by the removal of seven residues from the amino-terminal sequence. Thus it was concluded that all these peptides are derived from a single coenzyme-combining site of this enzyme. The sequence thus established is shown in Table II.

"Diagonal" or "Differential" Procedure for Purifying Pyridoxyl Peptides

This is based on the charge difference of a phosphopyridoxyl peptide before and after the hydrolytic cleavage of 5'-phosphate group by alkaline phosphatase. The resulting pyridoxyl peptide which is less negatively

[23] P. Edman and A. Henschen, in "Protein Sequence Determination" (S. B. Needleman, ed.), 2nd ed., p. 232. Springer-Verlag, Berlin and New York, 1975.
[24] W. R. Gray, this series, Vol. 25, p. 333.
[25] R. Hayashi, this series, Vol. 45, p. 568.

TABLE II

AMINO ACID SEQUENCE OF PYRIDOXYL PEPTIDES FROM VARIOUS PYRIDOXAL-P ENZYMES

Enzyme (source)	EC number	Sequence[a]	References[b]
Aminotransferases			
Aspartate	2.6.1.1		
Cytosolic			
(Pig heart)		Y F V S E G F E L F C A Q S F S K* N F G L Y N E R	1, 2
(Human heart)		Y F V S E G F E F F C A Q S F S K* N F G L Y N E R	3
(Sheep liver)		S K* N F	4
(Chicken heart)		S K* N F	5
Mitochondrial			
(Pig heart)		H F I E Q G I N V C L C Q S Y A K* N M G L Y G E R	6-8
(Human heart)		H F I E Q G I N V C L C Q S Y A K* N M G L Y G E R	3
(Sheep liver)		A K* N M G L Y	4
(Beef kidney)		Y A K* N M G	9
Alanine (pig heart)	2.6.1.2	Q E L A S F H S V S K* G F M G E C G F R	10
Pyridoxamine (Ps. MA-1)	2.6.1.30	A D I Y V T G P D K* C L (P_2, G_2, A_2, M)	11
Elimination-catalyzing enzymes			
D-Serine dehydratase (E. coli)	4.2.1.14	S K* G R I N K A T	12
Tryptophanase (E. coli)	4.1.99.1	Y A D M L A M S A K K* D A M V P M G G L L C M	13
Tryptophan synthase (E. coli)	4.2.1.20	R E D L L H G G A H K* T N Q V L G Q A L L A K	14
(Ps. putida)		L N H T G A H K* V N N C I G Q V L L	15
Cystathionine γ-lyase (rat liver)	4.4.1.1	C S A T K* Y M	16

(continued)

TABLE II (continued)

Enzyme (source)	EC number	Sequence[a]	References[b]
Decarboxylases			
Arginine (E. coli)	4.1.1.9	A T H S T H K* L L N A L S Q A S Y	17
Lysine (E. coli)	4.1.1.18	V I Y Q T E S T H K* L L A A F	18
Ornithine (E. coli)	4.1.1.17	V H K* Q Q A G Q	19
Glutamate (E. coli)	4.1.1.15	S I S A S G H K* F	20
Aromatic-L-amino-acid (pig kidney)	4.1.1.28	N F N P H K* W	21
Serine hydroxymethyl-transferase (rabbit liver)	2.1.2.1	V V T T T H K* T L	22
Glycogen phosphorylase	2.4.1.1		
(Rabbit muscle)		Q I S T A G T E A S G T G N M K* F M L N G A L T I T I G	23
(Yeast)		I S T A G T E A S G T S N M K* F V M	24
(Potato)		H I S T A G M E A S G T S N M K* F A M N G (C,E,T,G,L$_2$)L D G	25

[a] K*, lysyl residue involved in aldimine formation with pyridoxal-P.
[b] Key to references:
1. Yu. Ovchinnikov, C. A. Egorov, N. A. Aldanova, M. Yu. Feigina, V. M. Lipkin, N. G. Abdulaev, E. V. Grishin, A. P. Kiselev, N. N. Modyanov, A. E. Braunstein, O. L. Polyanovsky, and V. V. Nosikov, FEBS Lett. **29**, 31 (1973).
2. Y. Morino and M. Okamoto, Biochem. Biophys. Res. Commun. **47**, 1061 (1973).
3. H. Teranishi, H. Kagamiyama, K. Teranishi, H. Wada, T. Yamano, and Y. Morino, J. Biol. Chem. **253**, 8842 (1978).
4. M. Campos-Cavieres and C. P. Milstein, Biochem. J. **147**, 275 (1975).

5. Y. M. Torchinsky, V. M. Kochkina, and M. Sajgo, *Acta Biochim. Biophys. Acad. Sci. Hung.* **9**, 213 (1974).
6. H. Kagamiyama, R. Sakakibara, H. Wada, S. Tanase, and Y. Morino, *J. Biochem. (Tokyo)* **82**, 291 (1977).
7. H. Kagamiyama, R. Sakakibara, S. Tanase, Y. Morino, and H. Wada, *J. Biol. Chem.* **255**, 6153 (1980).
8. Y. Morino and S. Tanase, *J. Biol. Chem.* **253**, 252 (1978).
9. F. Bossa, G. Polidoro, D. Barra, A. Liverzani, and P. Fasella, *Int. J. Pept. Protein Res.* **8**, 499 (1976).
10. S. Tanase and Y. Morino, *Biochemistry* **18**, 3002 (1979).
11. J. Hodsdon, H. Kolb, E. E. Snell, and R. D. Cole, *Biochem. J.* **169**, 429 (1978).
12. Y. Z. Huang and E. E. Snell, *J. Biol. Chem.* **247**, 7358 (1972).
13. H. Kagamiyama, Y. Morino, and E. E. Snell, *J. Biol. Chem.* **245**, 2819 (1970).
14. R. Fluri, L. E. Jackson, W. E. Lee, and I. P. Crawford, *J. Biol. Chem.* **246**, 6620 (1971).
15. R. Maurer and I. P. Crawford, *J. Biol. Chem.* **246**, 6625 (1971).
16. C. W. Fearon, J. A. Rodky, and R. H. Abeles, *Biochemistry* **21**, 3790 (1982).
17. E. A. Boeker, E. H. Fischer, and E. E. Snell, *J. Biol. Chem.* **246**, 6776 (1971).
18. D. L. Sabo and E. H. Fischer, *Biochemistry* **13**, 670 (1974).
19. D. Appelbaum, D. L. Sabo, E. H. Fischer, and D. R. Morris, *Biochemistry* **14**, 3675 (1975).
20. P. H. Strausbauch and E. H. Fischer, *Biochemistry* **9**, 233 (1970).
21. F. Bossa, F. Martini, D. Barra, C. B. Voltattorni, A. Minelli, and C. Turano, *Biochem. Biophys. Res. Commun.* **78**, 177 (1977).
22. F. Bossa, D. Barra, F. Martini, L. V. Schirch, and P. Fasella, *Eur. J. Biochem.* **70**, 397 (1976).
23. K. Titani, A. Koide, J. Hermann, L. H. Ericsson, S. Kumar, R. D. Wade, K. A. Walsh, H. Neurath, and E. H. Fischer, *Proc. Natl. Acad. Sci. U.S.A.* **74**, 4762 (1977).
24. K. Lerch and E. H. Fischer, *Biochemistry* **14**, 2009 (1975).
25. K. Nakano, S. Wakabayashi, T. Hase, H. Matsubara, and T. Fukui, *J. Biochem. (Tokyo)* **83**, 1085 (1978).

charged should behave differently from the parent peptide on ion-exchange chromatography and electrophoresis.

Procedure for Isolation of Pyridoxyl Peptide from Glutamic Decarboxylase[26]

A crude peptide mixture obtained by trypsin–chymotrypsin digestion of borohydride-reduced glutamic decarboxylase from *Escherichia coli* was passed over a Dowex 1-X2 column. The phosphopyridoxyl peptide emerged when the pH of the eluting buffer was 5.9. The fluorescent fractions, containing the phosphopyridoxyl peptide as well as many contaminating peptides, were pooled and lyophilized. The residue was dissolved in 0.05 M NaHCO$_3$ (pH 8.2) and treated with pure *E. coli* alkaline phosphatase (66 μg/μmol of the peptide) for 2.5 hr at room temperature. This treatment caused the hydrolysis of the 5'-phosphate group of the bound coenzyme, thereby altering the charge of the substituted peptide. The phosphatase digest was lyophilized and the residue was dissolved in water and chromatographed on the same column of Dowex 1-X2 exactly as described above. The elution pattern was altered because two negative charges of the original phosphopyridoxyl peptide was removed; the pyridoxyl peptide now emerges at pH 7.2 while the contaminating peptides emerge, as before, at pH 5.9. This differential procedure was successfully utilized for the isolation of pyridoxyl peptides from the following enzymes; arginine decarboxylase,[27] D-serine dehydratase,[28] lysine decarboxylase,[29] aspartate aminotransferase (chicken),[30] ornithine decarboxylase,[31] phosphorylase (yeast),[32] serine hydroxymethyltransferase,[33] and phosphorylase (potato).[34] Sequences so far described are listed in Table II.

[26] P. H. Strausbauch and E. H. Fischer, *Biochemistry* **9**, 233 (1970).

[27] E. A. Boeker, E. H. Fischer, and E. E. Snell, *J. Biol. Chem.* **246**, 6776 (1971).

[28] Y. Z. Huang and E. E. Snell, *J. Biol. Chem.* **247**, 7358 (1972).

[29] D. L. Sabo and E. H. Fischer, *Biochemistry* **13**, 670 (1974).

[30] Y. M. Torchinsky, V. M. Kochkina, and M. Sajgo, *Acta Biochim. Biophys. Acad. Sci. Hung.* **9**, 213 (1974).

[31] D. Appelbaum, D. L. Sabo, E. H. Fischer, and D. R. Morris, *Biochemistry* **14**, 3675 (1975).

[32] K. Lerch and E. H. Fischer, *Biochemistry* **4**, 2009 (1975).

[33] F. Bossa, D. Barra, F. Martini, L. V. Schirch, and P. Fasella, *Eur. J. Biochem.* **70**, 392 (1976).

[34] K. Nakano, S. Wakabayashi, T. Hase, H. Matsubara, and T. Fukui, *J. Biochem. (Tokyo)* **83**, 1085 (1978).

Structural Analysis of Affinity-Labeled Site in Pyridoxal-P-Dependent Enzymes

The structural study of the pyridoxal-P binding site is important for the assessment of functional roles of the structural components. One of the experimental approaches to this problem is the affinity labeling technique using the mechanism-based irreversible inhibitors (see Vol. 46 [3][35]). With pyridoxal-P enzymes, there are some cases in which the affinity labeling results in the modification of the lysyl residue involved in combining coenzyme. Cytosolic[36–40] and mitochondrial[38–42] aspartate aminotransferases and γ-cystathionase (cystathionine γ-lyase)[43] were successfully examined from this viewpoint. The results confirm the proximity of a bound quasisubstrate with the coenzyme-combining lysyl ε-amino group and have driven one to speculate that the lysyl side chain might act as an important catalytic residue involved in mobilization of a proton from the bound substrate, a key catalytic process in the reactions catalyzed by this type of pyridoxal-P-dependent enzymes.[44–46]

Identification of the Active Site of Aspartate Aminotransferase Labeled with 3-Chloro-L-Alanine[36,42]

Both cytosolic and mitochondrial isoenzymes of aspartate aminotransferase undergo irreversible inactivation during α,β-elimination reaction with 3-chloro-L-alanine.[41,47] The finding that this process is greatly accelerated by the presence of formate ion facilitated the labeling of the active site of these isoenzymes by this quasisubstrate. Details will be described for the labeling procedure and the identification of the labeled site of the mitochondrial isoenzyme.[42]

[35] R. R. Rando, this series, Vol. 46, p. 28.
[36] Y. Morino and M. Okamoto, *Biochem. Biophys. Res. Commun.* **50**, 1061 (1973).
[37] M. Yamasaki, S. Tanase, and Y. Morino, *Biochem. Biophys. Res. Commun.* **65**, 652 (1975).
[38] S. Tanase and Y. Morino, *Biochem. Biophys. Res. Commun.* **68**, 1301 (1976).
[39] H. Gehring, R. R. Rando, and P. Christen, *Biochemistry* **16**, 4832 (1977).
[40] Y. Morino and M. Okamoto, *Biochem. Biophys. Res. Commun.* **47**, 498 (1972).
[41] Y. Morino, A. M. Osman, and M. Okamoto, *J. Biol. Chem.* **249**, 6684 (1974).
[42] Y. Morino and S. Tanase, *J. Biol. Chem.* **253**, 252 (1978).
[43] C. W. Fearon, J. A. Rodkey, and R. H. Abeles, *Biochemistry* **21**, 3790 (1982).
[44] D. E. Metzler, M. Ikawa, and E. E. Snell, *J. Am. Chem. Soc.* **76**, 648 (1954).
[45] A. E. Braunstein and M. M. Shemiyakin, *Biokhimiya* **18**, 393 (1953).
[46] L. Davis and D. E. Metzler, *in* "The Enzymes" (P. D. Boyer, ed.), 3rd ed., Vol. 7, p. 33. Academic Press, New York, 1972.
[47] Y. Morino and M. Okamoto, *Biochem. Biophys. Res. Commun.* **40**, 600 (1970).

Labeling Procedure. The pyridoxal form of mitochrondrial aspartate aminotransferase from pig heart (350 mg, 7.6 μmol of monomeric unit) was incubated for 10 min at 30° with 2.4 mmol of 3-chloro-L-[U-^{14}C]alanine (9.7 $\times$ 10^4 dpm/μmol) in a total volume of 20 ml of 0.1 M potassium pyrophosphate buffer (pH 7.5) containing 3 M potassium formate. pH was adjusted to 7.5–8.0 with NaOH during the reaction. The inactivation was 92% complete after 10 min. At this time, 30 mg of NaBH$_4$ and a few drops of 1-octanol were added to the reaction mixture. After standing for 30 min at 30°, the mixture was dialyzed overnight against distilled water at 5°. The dialyzed solution was lyophilized. The preparation contained an amount of radioactivity corresponding to 0.97 mol of the label per monomer of the enzyme.

Denaturation, Carboxymethylation, and Tryptic Digestion of the Labeled Enzyme. The above lyophilized preparation was dissolved in 10 ml of 6 M guanidine–HCl containing 0.1 M Tris–HCl buffer (pH 8.5), 0.5 mM dithiothreitol, and 10 mM EDTA. The mixture was incubated at 37° for an hour to ensure the unfolding of the protein. Then, sodium iodoacetate was added to a final concentration of 10 mM, followed by incubation for 15 min at 37°. The mixture was dialyzed overnight against distilled water at 5°. The resulting heavy precipitate was collected by centrifugation, suspended in 20 ml of 0.1 M NH$_4$HCO$_3$ and digested with 2% (w/w) TPCK-treated trypsin for 3 hr at 30° with occasional stirring.

Isolation of the Labeled-Site Peptide. The tryptic digest retained the radioactive label corresponding to 0.89 mol/monomer of the enzyme, indicating that the labeling was of a covalent nature. The digest was passed over a Sephadex G-50 (superfine) column (4 $\times$ 78 cm) previously equilibrated with 0.1 M NH$_4$HCO$_3$. The column was developed with the same solvent at a flow rate of 50 ml/hr. Fractions of 9 ml were collected. Earlier fractions exhibiting radioactivity as well as blue fluorescence were pooled and lyophilized (Fig. 4A). The peptides were dissolved in 2.5 ml of 20 mM Tris–HCl buffer (pH 7.2) and passed over a DE-32 column (1.2 $\times$ 30 cm) which was equilibrated with the same buffer and then developed with a gradient formed between 20 mM Tris–HCl buffer (pH 7.2) and the same buffer containing 0.2 M KCl (100 ml each). As can be seen in Fig. 4B, only fractions fluorescent and absorbing at 325 nm exhibited radioactivity. Both paper chromatography and electrophoresis of an aliquot from these pooled fractions revealed a single peptide remaining at the origin. Overall recovery of the labeled site peptide was approximately 50% as determined either from the absorbance at 325 nm or from the amount of radioactivity derived from the affinity label, 3-chloro-L-[U-^{14}C]alanine. The constant ratio (1 : 1) of the bound radioactivity to the pyridoxyl derivative observed in preparations from various steps of purification procedures indicated

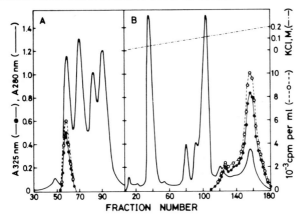

FIG. 4. Elution diagrams of the labeled site peptide from tryptic digests of mitochondrial aspartate aminotransferase affinity-labeled by 3-chloro-L-alanine. (A) On Sephadex G-50; (B) on DEAE-cellulose (DE-32). Absorbance at 280 nm (——); absorbance at 325 nm (●); radioactivity (○); salt gradient (- - -). From Y. Morino and S. Tanase, *J. Biol. Chem.* **253**, 252 (1978).

that equimolar amounts of the coenzyme and the 3-carbon moiety of the affinity label are attached to the labeled site peptide via covalent linkages.

Structure of the Labeled-Site Peptide. The amino acid composition of the purified radioactive peptide was the same as that of the phosphopyridoxyl peptide isolated from tryptic digests of the borohydride-reduced enzyme except for the absence of pyridoxyllysine residue in the labeled site peptide. Cellulose thin-layer (microcrystalline, Tokyo Chemical Industry Co. Ltd.) chromatography [solvent: 1-butanol/acetic acid/water (3:1:1)] of the acid hydrolyzate of the labeled site peptide revealed a single fluorescent spot at an R_f value of 0.1, which was also radioactive. On the other hand, a single fluorescent spot (R_f 0.47) on the chromatogram of the acid hydrolyzate of the phosphopyridoxyl peptide showed a mobility identical with that of the authentic N^6-(pyridoxyl)lysine.[6,48,49]

The covalent structure of the labeled site peptide was determined by sequential Edman degradation up to the sixteenth residue. The remaining part of the sequence was determined by analysis of the peptides obtained by further fragmentation of the parent peptide by chymotrypsin.

Relevance of the Labeling Process to the Catalytic Mechanism. In an analogous fashion to 3-chloro-L-alanine, L-serine O-sulfate inactivates the

[48] D. Heyl, S. A. Harris, and K. Folkers, *J. Am. Chem. Soc.* **70**, 3429 (1948).
[49] A. W. Forrey, R. B. Olsgaard, C. Nolan, and E. H. Fischer, *Biochimie* **53**, 269 (1971).

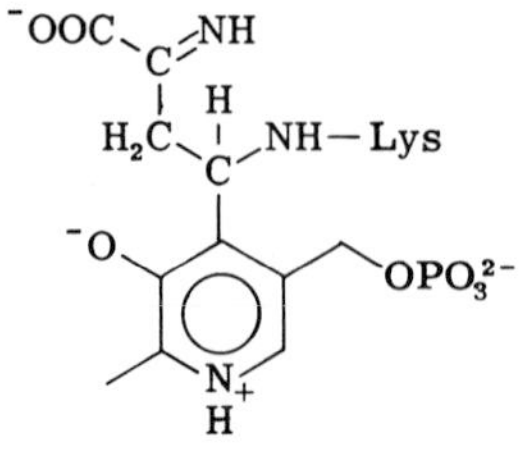

SCHEME 2.

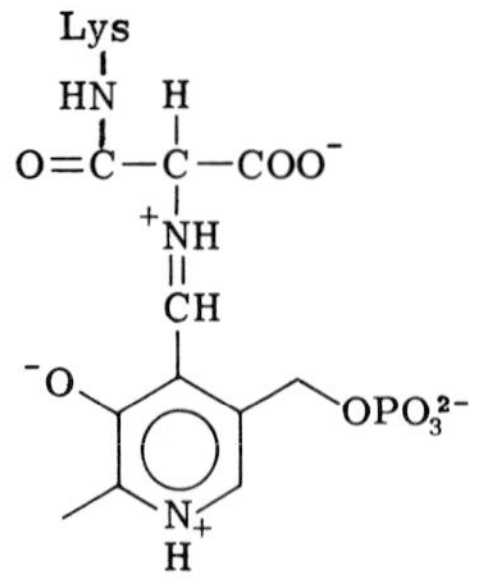

cytosolic aspartate aminotransferase.[50] Chemical analyses of the inactivation product suggested structure **III** in Scheme 2 for the enzyme-bound pyridoxal-P-affinity label adduct,[51] which would also apply to that formed upon reaction with chloroalanine, although a final decision awaits a direct identification of the structure of the adduct. On the other hand, structure **IV** was assigned to the adduct formed upon inactivation of γ-cystathionase by 3,3,3-trifluoro-L-alanine.[43] These results indicate that the ε-amino group of the lysyl residue involved in an aldimine formation with pyridoxal-P is sufficiently close to the α-carbon of these quasisubstrates. In transamination, withdrawal and addition of a proton must occur at the α-carbon of a substrate and the 4'-carbon of the coenzyme. In the γ-cystathionase-catalyzed α,γ-elimination reaction, protonation and deprotonation occur at the α-, β-, and γ-carbons of the substrate. Thus the results obtained from the affinity labeling study suggest that the lysyl side chain which forms an aldimine with the coenzyme participates in a general acid or base catalysis in the enzymic reactions. This suggestion is consistent with the observation of a large isotope effect for the proton transfer catalyzed by pyridoxamine–pyruvate transaminase[52,53] and cystathionine β-synthase.[54]

Pyridoxal-P Binding Site of Enzymes in Which Pyridoxal-P Acts as Inhibitor

Many enzymes have been reported to be inhibited by pyridoxal-P. In most cases, pyridoxal-P forms an aldimine linkage of a particular lysyl

[50] R. A. John and P. Fasella, *Biochemistry* **8**, 4477 (1969).
[51] H. Ueno, J. J. Likos, and D. E. Metzler, *Biochemistry* **21**, 4387 (1982).
[52] J. E. Ayling, H. C. Dunathan, and E. E. Snell, *Biochemistry* **7**, 4537 (1968).
[53] H. C. Dunathan, *Adv. Enzymol. Relat. Areas Mol. Biol.* **35**, 79 (1972).
[54] B. I. Posner and M. Flavin, *J. Biol. Chem.* **247**, 6412 (1972).

TABLE III

AMINO ACID SEQUENCE OF PYRIDOXAL-P-COMBINING SITE IN ENZYMES IN WHICH PYRIDOXAL-P ACTS AS INHIBITOR

Enzyme (source)	EC number	Sequence	(Position)	References[a]
Dehydrogenase				
Glyceraldehyde-phosphate (rabbit muscle)	1.2.1.12	I I P A S T G A A K* A V G K	(K 212)	1
Alcohol (horse liver)	1.1.1.1	I I G V D I N K D K* F	(K 228)	2
Glutamate (bovine liver)	1.4.1.2	G G A K* A G V K	(K 126)	3, 4
		Q L T K* S N A P R V K	(K 333)	
(NADP) (N. crassa)	1.4.1.4	L S M G G G K* G G A D F D P K	(K 113)	5
(NAD) (N. crassa)	1.4.1.3	R L A K* A R		6
		Q R K* N K		
		K S R S K* E		
		K A T K* N T K		
		L E K* K		
		G G S K* G V I L		
Malate (bovine heart, mitochondrial)	1.1.1.37	K* P G M T R		7
Phosphogluconate (decarboxylating) (Candida utilis)	1.1.1.44	A G G K* G (E) T K		8
		T V S K* V D H F I		
Aldolase (rabbit muscle)	4.1.2.13	G G V V G I K* V D K		9
Phospholipase A₂ (snake venom)	3.1.1.4	F G N M I N K* M G Q S V F	(K 11)	10
		G G K G K* P I D A T D R C C F	(K 33)	
		D T K* W	(K 58)	
		G H S S S K* C T G T E Q C	(K 111)	
Aspartate carbamoyltransferase (E. coli)	2.1.3.2	T S L G K K* G Q T L A N T I S -	(K 80)	11

(continued)

TABLE III (continued)

Enzyme (source)	EC number	Sequence	(Position)	References[a]
Ribonuclease A (bovine pancreas)	3.1.27.5	D R C K* P V N T F K E T A A A K* F	(K 41) (K 7)	12, 13
Ribulosebisphosphate carboxylase/oxygenase (Rhodospirillum rubrum)	4.1.1.39	- V G T I I K* P K -		14

[a] Key to references:

1. B. G. Foreina, G. Ferri, M. C. Zapponi, and S. Ronchi, *Eur. J. Biochem.* **20**, 535 (1971).
2. D. C. Sogin and B. V. Plapp, *J. Biol. Chem.* **250**, 205 (1975).
3. K. Moon, D. Piszkiewicz, and E. L. Smith, *Proc. Natl. Acad. Sci. U.S.A.* **69**, 1380 (1972).
4. J. C. Talbot, C. Gross, M. P. Cosson, and D. Pantaloni, *Biochim. Biophys. Acta* **494**, 19 (1977).
5. J. C. Woitton, G. K. Chambers, A. A. Holder, A. J. Baron, J. G. Taylor, J. R. S. Finchau, K. M. Blumenthal, K. Moon, and E. L. Smith, *Proc. Natl. Acad. Sci. U.S.A.* **71**, 4361 (1974).
6. B. M. Austen, M. E. Haberland, J. F. Nyc, and E. L. Smith, *J. Biol. Chem.* **252**, 8142 (1977).
7. M. J. Wimmaer and J. H. Harrison, *J. Biol. Chem.* **250**, 8768 (1975).
8. L. Minchiotti, S. Ronchi, and M. Rippa, *Biochim. Biophys. Acta* **657**, 232 (1981).
9. M. Anai, C. Y. Lai, and B. L. Horecker, *Arch. Biochem. Biophys.* **156**, 712 (1973).
10. C. C. Viljoen, L. Visser, and D. B. Potes, *Biochim. Biophys. Acta* **483**, 107 (1977).
11. T. D. Kempe and R. K. Stark, *J. Biol. Chem.* **250**, 6861 (1975).
12. C. H. R. Raetz and D. S. Auld, *Biochemistry* **11**, 2229 (1972).
13. S. S. Chen and P. C. Engel, *Biochem. J.* **147**, 351 (1975).
14. C. S. Herndon, I. L. Norton, and F. C. Hartman, *Biochemistry* **21**, 1380 (1982).

residue. Reduction with NaBH$_4$ renders the inhibition irreversible. The structure of the modified site can be studied by the method used for the pyridoxal-P-dependent enzymes as described above. The results obtained with several enzymes are summarized in Table III. With hemoglobin in which pyridoxal-P acts like an allosteric effector,[55] the modified site was located to the α-amino group of the amino-terminal residue, Val, of both α- and β-chains,[56] indicating a high reactivity of the amino-terminus and a nature of its environment favorable for binding pyridoxal-P.

[55] R. E. Benesch, R. Benesch, R. D. Renthal, and N. Maeda, *Biochemistry* **11,** 3576 (1972).
[56] R. Benesch, R. E. Benesch, S. Kwong, A. S. Acharya, and J. M. Manning, *J. Biol. Chem.* **257,** 1320 (1982).

Section III

Modifications at the Level of Aminoacyl-tRNA

[11] Properties and Specificity of Methionyl-tRNA$_f^{Met}$ Formyltransferase from *Escherichia coli*

By Sylvain Blanquet, Philippe Dessen, and Daniel Kahn

Formyl-L-methionyl-tRNA$_f^{Met}$ appeared to be a requirement for the initiation of protein synthesis,[1-3] although prokaryotic cells have been reported to be capable of growing without formylation of methionyl-tRNA$_f^{Met}$.[4,5] More recently, it was suggested that formylation of initiator tRNA might control the relative expression of the distal-to-proximal cistrons in polycistronic messenger RNAs, through the coupling between transcription termination and initiation of translation.[6,7]

The N-transformylation of initiator methionyl-tRNA$_f^{Met}$ is catalyzed by methionyl-tRNA$_f^{Met}$ formyltransferase[8,9] [10-formyltetrahydrofolate:L-methionyl-tRNA N-formyltransferase (E.C. 2.1.2.9); methionyl-tRNA$_f^{Met}$ transformylase]:

$$10\text{-f-}H_4\text{folate} + \text{Met-tRNA}_f^{Met} \rightleftharpoons \text{fMet-tRNA}_f^{Met} + H_4\text{folate}$$

This chapter describes a large-scale procedure for the purification of *Escherichia coli* methionyl-tRNA$_f^{Met}$ formyltransferase to homogeneity. The procedure represents a distinct improvement if compared to previous attempts to obtain this enzyme.[10,11] Large amounts of pure methionyl-tRNA$_f^{Met}$ formyltransferase became available to study the properties of this enzyme and to measure its ligand binding stoichiometries. Studies with unformylatable tRNAs are described as well.

[1] B. F. C. Clark and K. Marcker, *J. Mol. Biol.* **17**, 394 (1966).
[2] M. R. Cappechi, *Proc. Natl. Acad. Sci. U.S.A.* **55**, 1517 (1966).
[3] J. Eisenstadt and P. Lengyel, *Science* **25**, 527 (1966).
[4] C. E. Samuel, L. D'Ari, and J. C. Rabinowitz, *J. Biol. Chem.* **245**, 5115 (1970).
[5] B. N. White and S. T. Bailey, *Biochim. Biophys. Acta* **272**, 583 (1972).
[6] A. Danchin, *FEBS Lett.* **34**, 327 (1973).
[7] A. Danchin and A. Ullmann, *Trends Biochem. Sci.* **2**, 51 (1980).
[8] K. Marcker, *J. Mol. Biol.* **14**, 63 (1965).
[9] H. W. Dickermann, E. Steers, B. Redfield, and H. Weissbach, *J. Biol. Chem.* **242**, 1522 (1967).
[10] H. W. Dickermann and H. Weissbach, this series, Vol. 12, Part B, p. 681.
[11] H. W. Dickermann, this series, Vol. 20, Part C, p. 182.

Assay Method

Principle

Enzymatic activity of formyltransferase can be followed by formylating L-methionyl-tRNA$_f^{Met}$ with 10-[^{14}C]formyltetrahydrofolate and measuring the incorporation of radioactivity into material insoluble in trichloroacetic acid.[10] More conveniently, it can be followed by measuring the resistance conferred by formylation against deacylation of L-[^{14}C]methionyl-tRNA$_f^{Met}$.[12]

Reagents

5-Formyltetrahydrofolate (citrovorum factor) was obtained from Serva.

Unfractionated *E. coli* tRNA (80 pmol of methionine acceptance per A_{260} unit) was obtained from the pilot facilities of the Institut des Substances Naturelles (Centre National de la Recherche Scientifique, Gif/Yvette, France) or from Boehringer (Mannheim).

tRNA$_f^{Met}$ (1300–1500 pmol/A_{260} unit) was prepared by the procedure of Seno *et al.*[13] and stored at $-20°$ in the dark in 1 mM potassium acetate (pH 5.5) containing 0.1 mM MgCl$_2$.

L-[^{14}C]Methionine (55 Ci/mol) was from the Commissariat à l'Energie Atomique (Saclay, France).

Methionyl-tRNA synthetase was purified at homogeneity from *E. coli* strain EM20031 carrying the F32 episome, and stored (20–40 mg/ml) at $-20°$ in 20 mM imidazole–HCl (pH 7.6) containing 50% glycerol and 10 mM 2-mercaptoethanol.[14]

Bovine serum albumin (crystallized and lyophilized) was from Sigma.

Procedure

The final volume of the assay mixture was 100 μl.

Step 1. Preparation of 5,10-Methenyltetrahydrofolate.[15] 5-Formyltetrahydrofolate [7 mg in 1.5 ml distilled water containing 1% (v/v) 2-mercaptoethanol] was converted to 5,10-methenyltetrahydrofolate by lowering the pH to 1.9 with dropwise addition of 0.1 N HCl. The solution

[12] D. Kahn, M. Fromant, G. Fayat, P. Dessen, and S. Blanquet, *Eur. J. Biochem.* **105,** 489 (1980).
[13] T. Seno, M. Kobayashi, and S. Nishimura, *Biochim. Biophys. Acta* **169,** 80 (1968).
[14] Cassio and J. P. Waller, *Eur. J. Biochem.* **20,** 283 (1971).
[15] J. C. Rabinowitz, in "The Enzymes" (P. B. Boyer, ed.), 2nd rev. ed., Vol. 2, p. 185. Academic Press, New York, 1960.

was then completed to 2.2 ml with distilled water and left to stand for 4 hr at room temperature during which time the increase in absorbance at 355 nm, characteristic of 5,10-methenyltetrahydrofolate formation was followed [~0.17 OD$_{355}$ after 1/500 (v/v) dilution in H$_2$O]. The compound obtained (6 mM 5,10-methenyltetrahydrofolate in 0.1 M 2-mercaptoethanol) was stable when stored at $-15°$.

Step 2. Conversion of 5,10-Methenyltetrahydrofolate into 10-Formyltetrahydrofolate. Ten minutes before adding transformylase activity, 5 μl of the above solution containing 5,10-methenyltetrahydrofolate was added to 68 μl of a solution containing standard buffer (20 mM imidazole–HCl, pH 7.6, 0.1 mM EDTA, 10 mM 2-mercaptoethanol, 150 mM KCl, 7 mM MgCl$_2$), 160 μM unfractionated *E. coli* tRNA or 4 μM tRNA$_f^{Met}$, 50 μM L-[^{14}C]methionine and 300 μM ATP. The dilution of 5,10-methenyltetrahydrofolate in a pH 7.6 buffered solution ensured full conversion into 10-formyltetrahydrofolate.

Step 3. Aminoacylation of tRNA$_f^{Met}$. One minute before adding transformylase activity, 2 μl homogeneous *E. coli* methionyl-tRNA synthetase (0.4 μM in standard buffer containing 200 μg/ml bovine serum albumin) was added to the above 73 μl solution containing tRNA, L-[^{14}C]methionine, ATP, and 10-formyltetrahydrofolate.

Step 4. Formylation Reaction. Twenty-five microliters of catalytic amount of transformylase in standard buffer containing 200 μg/ml bovine serum albumin was added to 75 μl of the above [^{14}C]methionyl-tRNA solution containing 400 μM 10-formyl-DL-tetrahydrofolate. The reaction mixture was incubated at 25° for 10 min.

Step 5. Deacylation of Unformylated Methionyl-tRNA. One hundred microliters 0.36 M CuSO$_4$ in 1.1 M Tris (pH 7.3) was added to the above incubation mixture and the incubation continued for 5–10 min at 25°. This procedure hydrolyzed completely the ester bound of any unformylated Met-tRNA but left fMet-tRNA$_f^{Met}$ intact.[16] At the end of the deacylation by CuSO$_4$, tRNA was precipitated by the addition of 1–2 ml of 10% trichloroacetic acid (0°) [0.1 ml carrier RNA from yeast (2 mg/ml) was further added to precipitate tRNA when pure tRNA$_f^{Met}$ was used in the assay]. The precipitate was filtered on Whatman GF/C filters, washed three times by 20 ml 5% trichloroacetic acid containing 0.5% (w/v) DL-methionine. The radioactivity on each filter was measured in Bray's solution using a scintillator counter.

Unit. One unit of activity was defined as that amount of enzyme required to formylate 1 μmol of methionyl-tRNA$_f^{Met}$ per minute.

---

[16] P. Schofield, *Biochemistry* **9**, 1694 (1970).

Purification Procedure[12]

This method requires about 4–5 weeks for the isolation of ~20 mg homogeneous formyltransferase. The time for the preparation of the cells is not included.

Purification Steps

Step 1. Preparation of Cell Extract Supernatant. E. coli strain EM20031 (F32) was grown at 37° in 1.5 liters of a shaken culture in M63 medium containing in grams per liter: 10 g KH_2PO_4, 20 g K_2HPO_4, 5 g $(NH_4)_2SO_4$, 0.2 g $MgSO_4$, 15 g glucose, and 0.01 g thiamin. At 0.6 OD_{650}, the culture was used to inoculate a 30-liter culture in the same medium. The culture was grown overnight at 37° and two fractions of 15 liters each were used to inoculate two 200-liter cultures in CHEMAPEC (Männedorf, Switzerland) fermentors containing the same medium. Cells were harvested in mid-exponential phase (9–10 OD_{650}). Five kilogram wet weight cells were obtained and suspended in 5 liters 20 mM potassium phosphate (pH 8) containing 10 mM 2-mercaptoethanol at 0°. Cells were disrupted by two successive passages (40–50 liters/hr) through a refrigerated Menton-Gaulin apparatus (500 kg/cm^2). Then the extract was centrifuged for 3 hr in a refrigerated Sharples centrifuge (15000 g) and the pellet was discarded. All the following steps were carried out in the cold room at 4–8°.

Step 2. Streptomycin Fractionation. Nucleic acids and ribosomes were precipitated by addition of 6–10% (v/v) of a saturated streptomycin solution to the supernatant (6.7 liters) from step 1, which was then centrifuged for 2 hr (8000 g).

Step 3. Ammonium Sulfate Fractionation. The supernatant from step 2 was brought to 30% ammonium sulfate saturation, left to stand overnight and then centrifuged for 2 hr (8000 g). The residue was discarded. The ammonium sulfate concentration of the supernatant was raised to 50% saturation. The suspension was left to stand for 3 hr and centrifuged for 2 hr (8000 g). The supernatant was discarded and the sediment was dissolved in a minimal volume of 20 mM potassium phosphate (pH 8) containing 10 mM 2-mercaptoethanol, and dialyzed overnight against 130 liters of the same buffer with stirring.

Step 4. Chromatography on Sephadex G-200. The protein solution (3.5 liters) was passed through two Sephadex G-200 columns (50 × 25.4 cm, from Wright, Kenley, England) coupled in series. The same buffer as in step 3 was pumped upward at a flow rate of 0.8 liter/hr, and 0.8 liter fractions were collected and assayed for formyltransferase activity.

Step 5. Concentration by Ammonium Sulfate Precipitation. The active fractions from step 4 were pooled (15 liters), brought to 55% ammonium sulfate saturation, and then centrifuged for 2 hr (8000 *g*). The pellet was dissolved and dialyzed as in step 3.

Step 6. Second Chromatography on Sephadex G-200. The protein solution (2 liters) was chromatographed as in step 4. This step achieved a twofold further purification.

Step 7. Chromatography on DEAE-Sephadex. The active fractions from step 6 were pooled (10 liters) and applied on a column of DEAE-Sephadex A-50 (23 × 15 cm) equilibrated with 20 m*M* potassium phosphate (pH 8) and 10 m*M* 2-mercaptoethanol. The formyltransferase activity was eluted at 175 m*M* KCl using a 60-liter linear gradient (100–300 m*M*) in the same buffer. The flow rate was 1 liter/hr and 0.5-liter fractions were collected.

Step 8. Chromatography on Hydroxyapatite. The active fractions from step 7 were pooled (11.2 liters) and the pH lowered to 6.75 by dropwise addition of 1 *N* HCl. The protein solution was applied to a hydroxyapatite column (20 × 9 cm) equilibrated with 20 m*M* potassium phosphate (pH 6.75) and 10 m*M* 2-mercaptoethanol. The formyltransferase activity was eluted at 150 m*M* potassium phosphate with a 20-liter linear gradient (20–200 m*M*) at pH 6.75 containing 10 m*M* 2-mercaptoethanol. The flow rate was 840 ml/hr and 210 ml fractions were collected. Active fractions were pooled (3.2 liters) and concentrated to 0.2 liter by ultradialysis.

Step 9. Chromatography on Sepharose-Bound tRNA. After dialysis against 10 liters 20 m*M* imidazole–HCl (pH 7.6) containing 100 m*M* KCl and 10 m*M* 2-mercaptoethanol, the protein solution from step 8 was separated into two batches and applied on two columns (20 × 2.2 cm) of Sepharose 4B (Pharmacia) coupled to *E. coli* bulk tRNA (40 OD$_{260}$ units/ml packed gel)[17] equilibrated with the above dialysis buffer. Of the same buffer, 0.7 liters was passed through each column at a flow rate of 26 ml/hr. Thereafter, the active protein was eluted stepwise with the same buffer containing 200 m*M* KCl. Fractions (20 ml) were collected. Active fractions from the two affinity columns were pooled, concentrated to 5.5 ml, dialyzed 72 hr against 20 m*M* imidazole–HCl (pH 7.6) containing 50% glycerol and 150 m*M* KCl, and then stored at −15°. The enzyme thus obtained was homogeneous as evidenced by polyacrylamide gel electrophoresis analysis in the presence of sodium dodecyl sulfate.

---

[17] P. Rémy, C. Birmele, and J. P. Ebel, *FEBS Lett.* **27,** 134 (1972).

STEPS FOR PURIFICATION OF *E. coli* METHIONYL-tRNA
FORMYLTRANSFERASE[a]

Step	Total protein (mg)	Total activity (unit)	Specific activity (unit/mg)
1. Cell extract	500,000	1,740	0.0035
2. Streptomycin precipitation	480,000	1,680	0.0035
3. Ammonium sulfate fractionation	210,000	1,500	0.0072
4. Sephadex G-200	95,000	1,320	0.0139
5. Ammonium sulfate precipitation	46,000	1,260	0.0274
6. Sephadex G-200	22,000	1,140	0.0518
7. DEAE-Sephadex A-50	8,000	1,020	0.1275
8. Hydroxyapatite	400	900	2.25
9. Sepharose-tRNA	22	780	35.4

[a] D. Kahn, M. Fromant, G. Fayat, P. Dessen, and S. Blanquet, *Eur. J. Biochem.* **105,** 489 (1980).

Comment

The final steps of the purification were conducted in buffers containing a minimum 100 mM KCl concentration. Otherwise the purified enzyme adsorbed on glassware and the activity was lost (see the table).

Properties

Stability

The purified preparations of methionyl-tRNA$_f^{Met}$ formyltransferase (4 mg/ml) were stable for months when kept in 20 mM imidazole–HCl (pH 7.6), 0.1 mM EDTA, 10 mM 2-mercaptoethanol containing 50% glycerol, and 150 mM KCl at −15°. The activity of the homogeneous enzyme (1 μM) remained constant over 2 days at 25° in the same buffer without glycerol. In the absence of added ionic strength, the formyltransferase activity decreased within 10 min, reaching a plateau value depending on the initial concentration of the enzyme and the surface of glass in contact with the enzyme.

Optical Spectrum and Absorption Coefficient

The optical spectrum of methionyl-tRNA$_f^{Met}$ formyltransferase displayed a maximum at 280 nm with two shoulders at 275 and 288 nm. The absorption coefficient at 280 nm was 1.39 cm^2 mg^{-1} (1 OD$_{280}^{1 cm}$ unit =

0.72 mg/ml) (from refractive index increment measurements using bovine serum albumin as standard).

Molecular Ratio

Native methionyl-tRNA$_f^{Met}$ formyltransferase in buffer (pH 7.6) containing 150 mM KCl showed an M_r of 33,000 ± 5000, as measured by sedimentation equilibrium, molecular sieving, and small-angle neutron scattering experiments. It displayed a unique polypeptide chain of M_r 32,000 ± 5000, as evidenced by polyacrylamide gel electrophoresis analysis in the presence of sodium dodecyl sulfate.

Radius of Gyration

According to the slope of Guinier plots from small-angle neutron scattering measurements (Fig. 1), native monomeric methionyl-tRNA$_f^{Met}$

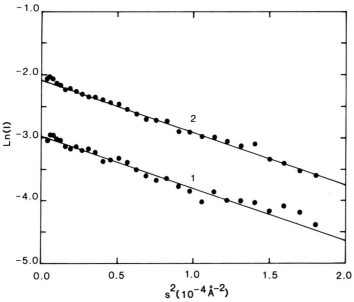

FIG. 1. Neutron-scattering curves obtained for methionyl-tRNA$_f^{Met}$ formyltransferase at 2.6 (1) and 6.0 mg/ml (2) in 0.15 ml of 20 mM imidazole–HCl, pH 7.6, 0.1 mM EDTA, 10 mM 2-mercaptoethanol, containing 150 mM KCl. Samples were exposed for 60 min on the D11 camera (Institut Laue Langevin, Grenoble) with a sample-to-detector distance of 2.55 m and a neutron wavelength of 10 Å. The straight lines were obtained by linear regression of the ln(I) values as a function of s^2 in the Guinier region (0.002 Å^{-1} < s < 0.008 Å^{-1}), s = 2sin(θ/λ), with 2θ, the observed angle from the incident neutron beam). An M_r value of 33 ± 5K and an R_G value of 2.55 ± 0.03 nm were obtained using the $I(0)$ and R_G^2 values calculated from the regressions.

formyltransferase showed an R_G of 2.55 ± 0.05 nm, reflecting a prolate ellipsoid with an axial ratio of 1:3 (10.2 × 3.2 × 3.2 nm).

Kinetic Parameters

Formyltransferase activity at 25° in 20 mM imidazole–HCl, pH 7.6, 0.1 mM EDTA, 10 mM 2-mercaptoethanol, 0.2 mg/ml bovine serum albumin, was maximum in the presence of 150 mM KCl and 7 mM MgCl$_2$. Under these conditions, Michaelian parameters were the following: K_m for 10-formyl-L-tetrahydrofolate (in the presence of 2 μM Met-tRNA$_f^{Met}$) was 13.5 ± 1.5 μM. K_m for Met-tRNA$_f^{Met}$ (in the presence of 150 μM 10-formyl-L-tetrahydrofolate) was 0.35 ± 0.05 μM. V was 20 ± 2 sec^{-1} (37.5 units/mg of formyltransferase) at saturation of both substrates.

Binding of Ligands

Binding to methionyl-tRNA$_f^{Met}$ formyltransferase of various ligands could be easily followed by fluorescence measurements.

Fluorescence Titration

The excitation light (295 nm) was provided by a stabilized HBO 450-W/2 Osram xenon source analyzed by a HRS2 monochromator (Jobin et Yvon, France). The emission of fluorescence was registered at 332 nm using another HRS2 monochromator followed by an EMI 6456 photomultiplier. Both monochromators were equipped with diffraction gratings blazed at 340 nm. The quartz cell (1 × 0.4 cm, Hellma) was filled with 0.8 ml of the enzyme solution (25°) in 20 mM imidazole–HCl, pH 7.6, 0.1 mM EDTA, 10 mM 2-mercaptoethanol containing variable KCl and/or MgCl$_2$ concentrations. Saturation curves were obtained by adding successive aliquots of 5 μl of the buffer containing the ligand under study. All saturation curves were corrected for dilution and inner-filter effects.

Reagents

Tween 80 was from Serlabo.

Methionyl-tRNA$_f^{Met}$: 1400 pmol methionine incorporated/A_{260} unit of tRNA.[18]

Formylmethionyl-tRNA$_f^{Met}$ was purified on a benzoylated DEAE-cellulose column[19] after enzymatic formylation.

[18] A. Sourgoutchov, S. Blanquet, G. Fayat, and J. P. Waller, *Eur. J. Biochem.* **46,** 431 (1974).

[19] C. E. Samuel and J. C. Rabinowitz, *Anal. Biochem.* **47,** 244 (1972).

10-L-*Formyltetrahydrofolate*

One mole monomeric enzyme bound 1.1 mol 10-L-formyltetrahydrofolate. The equilibrium constant was 0.7 μM^{-1} in 150 mM KCl and 1.4 μM^{-1} in 25 mM KCl plus 8 mM MgCl$_2$ (Fig. 2). At saturation of the substrate, the initial fluorescence of the enzyme was reduced by 33%. Under the same conditions, the nonsubstrate analog 5-formyltetrahydrofolate had no effect on the enzyme fluorescence.

tRNA$_f^{Met}$

In 5 mM KCl, in the presence of 0.1% (v/v) Tween 80, which reduced the adsorption of formyltransferase on the fluorescence quartz cell, the saturation curves indicated 10 formyltransferase molecules binding to the same tRNA molecule. With increasing ionic strength (KCl or/and MgCl$_2$) the stoichiometry tended toward a one-to-one stoichiometry. In 150 mM KCl, the stoichiometry was 0.65 mol tRNA$_f^{Met}$ bound/mol enzyme with

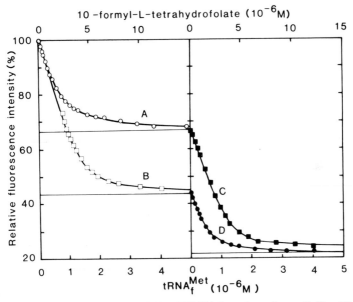

FIG. 2. Fluorescence titrations of Met-tRNA$_f^{Met}$ formyltransferase (1.85 μM) by 10-formyltetrahydrofolate (A and D), in the presence (D) or in the absence (A) of 4.1 μM tRNA$_f^{Met}$. Fluorescence titrations of Met-tRNA$_f^{Met}$ formyltransferase (1.85 μM) by tRNA$_f^{Met}$ (B and C) in the presence (C) or in the absence (B) of 14.6 μM 10-L-formyltetrahydrofolate (29.2 μM 10-DL-formyltetrahydrofolate). The curves through the experimental points were the results of the fit with binding parameters obtained by iterative nonlinear least-squares regression procedures.

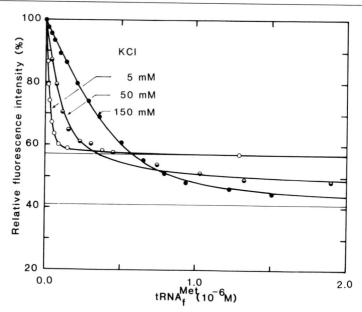

FIG. 3. Fluorescence titrations of methionyl-tRNA$_f^{Met}$ formyltransferase by tRNA$_f^{Met}$ in 20 mM imidazole–HCl, pH 7.6, 0.1 mM EDTA, 10 mM 2-mercaptoethanol, 0.1% Tween 80 containing 5, 50, and 150 mM KCl. Enzyme concentrations were 0.41, 0.97, and 0.79 μM, respectively.

60% quenching of the initial fluorescence of the formyltransferase. The stability constant was 13 μM^{-1} (Figs. 2 and 3).

The favored formation of a 1 : 1 enzyme : tRNA$_f^{Met}$ complex under high ionic strength conditions was further evidenced by tRNA$_f^{Met}$ labeling of methionyl-tRNA$_f^{Met}$ formyltransferase.[20] The method consisted in allowing a free amino group (probably the ε-amino group of a lysine) of the enzyme to form a Schiff's base with the 2′,3′-aldehyde groups created at the 3′-terminal ribose of tRNA by periodate oxidation. The reaction was monitored in the presence of cyanohydridoborate, a mild reducing agent which has the property to leave intact the dialdehyde, but converts the imino moiety of the Schiff's base specifically and continuously into a stable secondary amine.[21] Saturating amounts of periodate-treated tRNA$_f^{Met}$ with a ^{14}C-labeled 3′-terminal adenosine were left to react with the enzyme at 37° in 20 mM imidazole–HCl (pH 8.5) containing 150 mM KCl, 25% glycerol, 0.1% Tween 80, and 2 mM cyanohydridoborate. Both the residual formyltransferase activity and the concomitant amount of

[20] C. Hountondji, G. Fayat, and S. Blanquet, *Eur. J. Biochem.* **107**, 403 (1980).
[21] R. F. Borch, M. D. Bernstein, and H. D. Durst, *J. Am. Chem. Soc.* **93**, 2897 (1971).

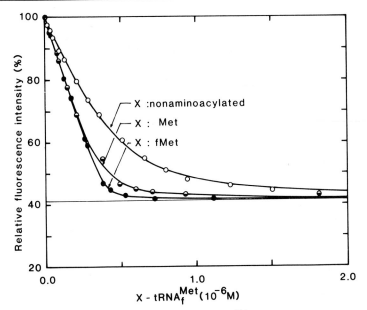

FIG. 4. Fluorescence titrations of methionyl-tRNA$_f^{Met}$ formyltransferase by tRNA$_f^{Met}$, methionyl-tRNA$_f^{Met}$, and formylmethionyl-tRNA$_f^{Met}$ in 20 mM imidazole–HCl, pH 7.6, 0.1 mM EDTA, 10 mM 2-mercaptoethanol, containing 150 mM KCl and 0.1% Tween 80. Formyltransferase concentrations were 0.79, 0.45, and 0.45 μM, respectively.

covalently incorporated tRNA were followed at various times during the incubation. Incorporated [^{14}C]adenosine was measured after ribonuclease treatment of the samples, in the material insoluble in trichloroacetic acid. One mole methionyl-tRNA$_f^{Met}$ transformylase was fully inactivated by the incorporation of 0.85 mol of the ^{14}C-labeled tRNA$_f^{Met}$ derivative.

Methionyl-tRNA$_f^{Met}$

Methionyl-tRNA$_f^{Met}$ showed enhanced binding to formyltransferase if compared to tRNA$_f^{Met}$ (Fig. 4). For example at 150 mM KCl, the binding constant increased to 49 μM^{-1} upon aminoacylation. Corresponding binding stoichiometry and fluorescence quenching were $n = 0.81$ and $\Delta\Phi = 59\%$, respectively. The relative stabilizing effect of the methionyl residue increased with increasing MgCl$_2$ concentration.

Formylmethionyl-tRNA$_f^{Met}$

The affinity of methionyl-tRNA$_f^{Met}$, the substrate of formyltransferase, was further improved by formylation (Fig. 4). At 150 mM KCl, the binding constant of formylmethionyl-tRNA$_f^{Met}$ was 350 μM^{-1}. Binding stoichiometry and fluorescence quenching remained unchanged ($n = 0.87$ and $\Delta\Phi = 59\%$). However, the reinforcing effect of the formyl moiety on the stability of the enzyme: methionyl-tRNA$_f^{Met}$ complex disappeared upon the addition of 11 mM MgCl$_2$ to the buffer containing 150 mM KCl.

Specificity

Mischarged *E. coli* tRNA$_f^{Met}$ molecules were formylatable,[22,23] while methionyl-tRNA$_m^{Met}$ was not,[8] even in the presence of 2 μM homogeneous formyltransferase in the assay. Nevertheless, both tRNA$_m^{Met}$ and methionyl-tRNA$_m^{Met}$ were capable of binding the enzyme, as shown by monitoring the intrinsic fluorescence of formyltransferase.[24] The binding of tRNA$_m^{Met}$ significantly differed, however, from that of tRNA$_f^{Met}$. In 150 mM KCl, tRNA$_m^{Met}$ bound formyltransferase with an affinity constant of 0.8 μM^{-1}, one order of magnitude smaller than that of tRNA$_m^{Met}$. The quenching of fluorescence at tRNA$_m^{Met}$ saturation was 37%, to be compared to 60% in the case of tRNA$_f^{Met}$. Moreover, upon aminoacylation, the behavior of tRNA$_m^{Met}$ radically differed from that of tRNA$_f^{Met}$: in 150 mM KCl, the affinity constant of methionyl-tRNA$_m^{Met}$ was smaller (2- to 3-fold) than that of tRNA$_m^{Met}$.

The broad specificity of methionyl-tRNA$_f^{Met}$ formyltransferase at the level of tRNA binding was further documented using *E. coli* bulk tRNA or pure tRNA^{Phe} from yeast. Each tRNA species bound the enzyme with concomitant quenching of its intrinsic fluorescence. However, both the stability constant and the extent of fluorescence quenching were smaller than those observed using pure *E. coli* tRNA$_f^{Met}$.

[22] R. Giégé, J. P. Ebel, and B. F. C. Clark, *FEBS Lett.* **30**, 291 (1973).
[23] J. M. Old and D. S. Jones, *FEBS Lett.* **66**, 264 (1976).

[12] Glu-tRNA^{Gln}: An Intermediate in Yeast Mitochondrial Protein Synthesis

By Nancy C. Martin and Murray Rabinowitz

Mitochondria contain their own DNA which codes for transfer RNA (tRNA), ribosomal RNA (rRNA), and messenger RNA (mRNA) used in organelle protein synthesis.[1,2] Mitochondrial DNA specifies only a limited number of products, however, and most structural and functional mitochondrial proteins are coded on nuclear genes.[1,2] Early estimates of the number of transfer RNA genes in mitochondrial DNA of all species were low and it did not appear that mitochondrial DNA could code for enough tRNAs to support mitochondrial protein synthesis if it proceeded according to the wobble rules proposed by Crick.[3] When DNA sequence analysis of several mitochondrial genomes,[4-6] and RNA sequence analysis of the tRNAs they encode,[7,8] unequivocally demonstrated that mitochondria indeed contain fewer tRNAs than necessary to translate mRNA following Crick's wobble rules, it became clear that unique mitochondrial wobble rules must function to allow all codons to be read by the limited number of mitochondrial encoded tRNAs. Considering the limited number of tRNA genes one would expect that few isoacceptor tRNAs coded by different genes are found in mitochondrial systems, and it was therefore surprising to find two isoaccepting tRNAs for glutamic acid,[9] as glutamic acid has only two codons. Subsequent experiments demonstrated that one of these two mitochondrial tRNAs accepts glutamic acid but responds to glutamine codons in ribosome binding studies.

The isoaccepting Glu-tRNAs were detected by charging isolated yeast mitochondrial tRNAs with [^{3}H]glutamic acid using synthetase preparations obtained from mitochondria and fractionating the resulting aminoacyl-tRNA by RPC-5 (reversed-phase chromatography) (Fig. 1). Two distinct peaks of tRNA which accept glutamic acid were separated. Both fractions were isolated and shown to be gene products of mitochondrial DNA by nucleic acid hybridization as shown in Fig. 2. The observation that the hybridization of the two tRNAs to mitochondrial DNA was additive (Fig. 2) indicated the two tRNAs were coded by two different genes.

[1] P. Borst and L. A. Grivell, *Cell* **15,** 705 (1978).

[2] G. Attardi, *TIBS* **6,** 86, 101 (1981).

[3] F. H. C. Crick, *J. Mol. Biol.* **19,** 548 (1966).

[4] S. G. Bonitz, R. Berlani, G. Coruzzi, M. Li, G. Macino, F. G. Nobrega, M. P. Nobrega, B. E. Thalenfeld, and A. Tzagoloff, *Proc. Natl. Acad. Sci. U.S.A.* **77,** 3167 (1980).

[5] B. G. Barrell, S. Anderson, A. T. Bankier, M. H. L. DeBruijn, E. Chen, A. R. Coulson, J. Drouin, I. C. Eperon, D. P. Nierlich, B. A. Roe, F. Sanger, P. H. Schreier, A. J. H. Smith, R. Staden, and I. G. Young, *Proc. Natl. Acad. Sci. U.S.A.* **77,** 3164 (1980).

[6] H. G. Kochel, C. M. Lazarus, N. Basak, and H. Kuntzel, *Cell* **23,** 625 (1981).

[7] J. E. Heckman, J. Sarnoff, B. Alzner-DeWeerd, S. Yin, and U. L. RajBhandary, *Proc. Natl. Acad. Sci. U.S.A.* **77,** 3159 (1980).

[8] B. A. Roe, J. F. Wong, E. Y. Chen, P. W. Armstrong, A. Stankiewicz, D. P. Ma, and J. McDonough, *in* "Mitochondrial Genes" (P. Slonimiski, P. Borst, and G. Attardi, eds.), p. 45. Cold Spring Harbor Lab., Cold Spring Harbor, New York, 1982.

[9] N. Martin, M. Rabinowitz, and H. Fukuhara, *J. Mol. Biol.* **101,** 285 (1976).

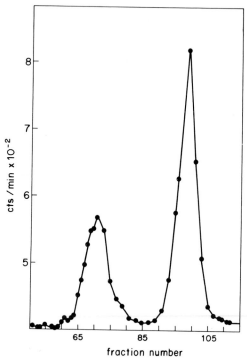

FIG. 1. RPC-5 chromatographic elution pattern of mitochondrial tRNA charged with [³H]glutamic acid by mitochondrial aminoacyl synthetase. The tRNAs were eluted with a linear NaCl gradient from 0.4 to 0.7 M in 10 mM ammonium acetate (pH 4.6), 10 mM MgCl$_2$, and 0.2 mM dithiothreitol.

Hybridization competition experiments and mapping experiments gave further confirmation that these tRNAs were different in primary sequence.[9]

Subsequent experiments (Table I) examined the codon responses of these two tRNAs in ribosome binding studies with synthetic oligonucleotides using the methods of Nirenberg and Leder.[10] The tRNA eluting first from the RPC-5 column (GluI-tRNA) responded, as expected, to oligonucleotides including the glutamic acid GAA and GAG codons.[11] The tRNA eluting at higher salt (GluII-tRNA) did not bind to ribosomes in response to glutamic acid codons but did bind when an oligonucleotide including the glutamine codon CAA was used. The results of this analysis suggested that yeast mitochondria did contain a tRNA that recognized

[10] M. Nirenberg and P. Leder, *Science* **145**, 1399 (1964).
[11] N. C. Martin, M. Rabinowitz, and H. Fukuhara, *Biochemistry* **16**, 4672 (1977).

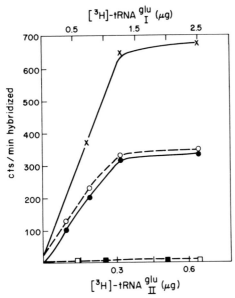

FIG. 2. Hybridization of mitochondrial GluI-tRNA and GluII-tRNA to yeast mitochondrial and nuclear DNA. Increasing amounts of [^{3}H]GluI-tRNA (●) and [^{3}H]GluII-tRNA (○) were hybridized individually or together (×) with filters containing 20 μg yeast mitochondrial DNA. Both tRNAs were also hybridized to nuclear DNA GluI-tRNA (■); GluII-tRNA (□).

glutamine codons even though neither we nor others[12] have been able to aminoacylate yeast mitochondrial tRNA with glutamine. In addition, the aminoacylation of mitochondrial tRNA with glutamic acid cannot be competed by adding excess glutamine (Table II). These two observations suggest that yeast mitochondria do not have the enzymatic machinery necessary to charge tRNA directly with glutamine. The glutaminyl tRNA is first aminoacylated with glutamic acid to form a glutamyl tRNA^{Gln} and then, presumably in a subsequent step, an amidotransferase converts the glutamic acid to glutamine to form a glutaminyl-tRNA^{Gln} prior to its use in protein synthesis. Dirheimer et al.[12] have reported that such an amidotransferase activity is present in yeast mitochondria.

We do not know how general this mechanism is to mitochondria. Lynch and Attardi[13] report they were unable to charge mitochondrial

---

[12] G. Dirheimer, G. Keith, A. P. Sibler, and R. P. Martin, In "Transfer RNA: Structure, Properties and Recognition" (P. R. Schimmel, D. Soll, and J. N. Abelson, eds.), p. 19. Cold Spring Harbor Lab., Cold Spring Harbor, New York, 1979.
[13] P. Lynch and G. Attardi, J. Mol. Biol. 102, 125 (1976).

TABLE I

CODON RECOGNITION OF YEAST GLU-tRNAs:
INCREASE IN AMINOACYL-tRNA BOUND TO RIBOSOMES
IN RESPONSE TO OLIGONUCLEOTIDES[a]

Oligonucleotides	tRNAs (cpm)		
	Cytoplasmic	mt GluI	mt GluII
None	63	38	32
U	96	28	40
GA	610	352	40
CA	109	38	227

[a] GA includes GAA and GAG glutamic acid codons. CA includes the CAA glutamine codon. Binding assays were carried out as described in Martin et al.[11] and followed the procedure of Nirenberg and Leder.[10]

TABLE II

EFFECT OF UNLABELED GLUTAMIC ACID AND GLUTAMINE ON THE
ACCEPTANCE OF RADIOACTIVELY LABELED AMINO ACID BY
MITOCHONDRIAL AND CYTOPLASMIC tRNAs[a]

tRNA	Labeled amino acid	Unlabeled amino acid	Acceptance of labeled amino acid (%)
Mitochondrial	32 ng [³H]Glu	—	100
		64 ng Glu	67
		124 ng Glu	35
		64 ng Gln	103
		2920 ng Gln	101
Cytoplasmic	620 ng [¹⁴C]Gln	—	100
		1400 ng Gln	72
		2900 ng Gln	36

[a] Aminoacylation reactions were carried out as described in Martin et al.[11]

tRNA from HeLa cells with glutamine but in the same experiments several other tRNAs now known to be present could not be charged with their cognate amino acids either. As far as we know, there have been no other reports on whether or not mitochondrial tRNAs from other organisms can be charged directly with glutamine.

The use of tRNA^{Gln} as a cofactor for the synthesis of glutamine destined for protein is not unique to yeast mitochondria. This novel enzymatic pathway was first discovered by Wilcox and Nirenberg in their studies of the protein synthetic apparatus of certain gram-positive bacteria.[14] The amidation step is catalyzed by a specific amidotransferase and requires divalent cations, ATP and L-glutamine or L-asparagine as the amide donor. Wilcox[15] has published a detailed account of the biochemistry of this reaction.

[14] M. Wilcox and M. Nirenberg, *Proc. Natl. Acad. Sci. U.S.A.* **61**, 229 (1968).
[15] M. Wilcox, *Eur. J. Biochem.* **11**, 405 (1969).

[13] Characterizing the Function of O^β-Phosphoseryl-tRNA

By THOMAS S. STEWART and STEPHEN SHARP

The characterization of unique species of O^β-phosphoseryl-tRNA (SerP-tRNA) from various tissue sources including chicken liver,[1] bovine mammary gland,[2] and bovine liver,[3,4] poses the question as to the *in vivo* function of such a tRNA. Can SerP-tRNA incorporate phosphoserine (SerP) at the translational level producing phosphorylated products? On the basis of their function, the phosphate moieties of phosphoproteins may be considered to be of two types. One type can be termed "regulatory" because the phosphorylation mechanism acts to regulate metabolic activity of certain enzymes and proteins. "Regulatory" phosphoproteins are equally controlled by protein kinase activities as they are by protein phosphatase activities. In the second type of phosphoprotein, the phosphate transfer reaction is essentially irreversible. Once incorporated into protein these phosphates are not readily removed but act as a "structural" phosphate, examples of which are found in casein[5] and phosvitin.[6] Studies of casein phosphorylation using completely dephosphorylated casein lead to the suggestion that some phosphate must be present in the

[1] P. H. Mäenpää and M. R. Bernfield, *Proc. Natl. Acad. Sci. U.S.A.* **67**, 688 (1970).
[2] S. J. Sharp and T. S. Stewart, *Nucleic Acids Res.* **4**, 2123 (1977).
[3] A. Diamond, B. Dudock, and D. Hatfield, *Cell* **25**, 497 (1981).
[4] D. Hatfield, A. Diamond, and B. Dudock, *Proc. Natl. Acad. Sci. U.S.A.* **79**, 6215 (1982).
[5] D. T. Davies and A. J. R. Law, *J. Dairy Res.* **46**, 447 (1977).
[6] M. R. Redshaw and B. K. Follet, *Biochem. J.* **124**, 759 (1971).

nascent casein to allow specific casein kinase recognition. Conceivably, as proposed in an earlier model,[2] the initial phosphate(s) of casein could be incorporated at the translational level by SerP-tRNA. For this to be achieved there must be unique codons within the casein messenger RNA (mRNA) that are recognized by this tRNA.

The approaches used to determine whether SerP-tRNA has a translational role in the synthesis of "structural" phosphoproteins include determining the structure of tRNA[SerP] and casein mRNAs by nucleic acid sequencing methods. The sequence data predict that SerP-tRNA does not incorporate SerP into casein as part of the mechanism for casein phosphorylation. These findings are discussed as are possible functions for SerP-tRNA.

tRNA[SerP]: Features of an Anomalous tRNA

The nucleotide sequences of two bovine liver seryl-tRNA species[3,4] and one from mouse liver,[7] which can be phosphorylated *in vitro* to form SerP-tRNA, reveals several unusual structural features. These tRNAs, although aminoacylated to form seryl tRNA, have a tryptophan anticodon (C_mCA^3 and $NCA^{4,7}$) but respond, *in vitro,* to the termination codon UGA.[3,4] They are the largest tRNAs sequenced to date containing 90 nucleotides. In the cloverleaf secondary structure, two extra nucleotides occur in the D stem and loop and there is one extra unpaired nucleotide in the TΨC stem. These structural features must contribute to an allowed $C_m = A$ pairing as indicated by the UGA rather than UGG codon–anticodon interaction.

A study of the involvement of SerP-tRNA in *in vitro* translation systems has demonstrated that this tRNA can suppress the UGA termination codon of β-globin mRNA in a reticulocyte lysate producing an identifiable globin readthrough product.[3,4] Although it is not possible to deduce from this experiment whether serine or Pserine was incorporated in this *in vitro* suppression assay, it was shown that SerP-tRNA was bound to ribosomes in response to the UGA triplet.[4] That is, the presence of a phosphoserine did not prevent ribosome binding.

SerineP Codon Usage in Casein mRNA (cDNA)

A direct approach to determining a role for SerP-tRNA in phosphoprotein synthesis is to examine the codon usage for SerP residues in such proteins. That is, to examine the mRNA nucleotide sequence (in practice,

[7] C. J. Hutchins and T. S. Stewart, unpublished results.

CODON USAGE FOR PHOSPHOSERINE RESIDUES IN
α_{S1}-CASEIN[8]

Codon	Residue number
AGU	46, 68, 75, 115
AGC	64
UCA	48
UCG	66

the cloned complementary DNA sequence) to determine if there are unique codons for any SerP residues. The cDNA sequences for part of bovine α_{S1}-casein,[8] α-,[9] β-,[10] and γ-[9]caseins as well as mouse ε-casein[11] are known. The cDNA sequences for bovine α_{S1}-casein confirmed the amino acid sequence determined earlier in which the positions of SerP residues were given.[12] For the rat and mouse caseins, the amino acid sequences were deduced from the nucleotide sequence and the positions of SerP residues inferred by comparison with bovine caseins or from the specificity of the casein kinase.[13] As shown in the table, all SerP residues in α_{S1}-casein use serine codons. This is also the case for the rat and mouse casein cDNA sequences where the potential sites for SerP have serine codons.[9-11] Therefore, given that tRNA[Pser] responds to the UGA codon *in vitro* suppression assays and ribosome binding, it appears that this tRNA is not involved in the incorporation of SerP residues into casein polypeptide by this mechanism.

Outlook

Given that this tRNA does respond to the UGA termination codon, it is most unlikely that this codon could appear in the correct reading frame within the message without causing premature termination. For a direct involvement of SerP-tRNA in translation, a mechanism in which the UGA could be recognized "out of phase" would be essential.

In the cDNA sequence for α_{S1}-casein the AGU codon is used for four phosphorylated serine residues. Of these, two occur in the sequence

[8] I. M. Willis, A. F. Stewart, A. Caputo, A. R. Thompson, and A. G. Mackinlay, *DNA* **1**, 375 (1982).

[9] A. A. Hobbs and J. M. Rosen, *Nucleic Acids Res.* **10**, 8079 (1982).

[10] D. E. Blackburn, A. A. Hobbs, and J. M. Rosen, *Nucleic Acids Res.* **10**, 2295 (1982).

[11] L. G. Hennighausen, A. Steudle, and A. E. Sippel, *Eur. J. Biochem.* **126**, 569 (1982).

[12] J.-C. Mercier, F. Grosclaude, and B. Ribadeau-Dumas, *Eur. J. Biochem.* **23**, 41 (1971).

[13] J.-C. Mercier, *Biochimie* **63**, 1 (1981).

AGUGA. The proportion is similar in rat and mouse caseins. The extra nucleotide in the TΨC stem of yeast mitochondrial tRNA^{Phe},[14] has been suggested as a structural feature that allows the tRNA^{Phe} to function, *in vivo*, as a frameshift suppressor.[15] An intriguing possibility therefore, is that the SerP-tRNA, which has extra nucleotides in the TΨC stem and the D-loop, could act by a "frameshift"-type mechanism to read the UGA of the AGUGA sequence and incorporate SerP at these positions.

We are presently analyzing the *in vivo* function of tRNA^{SerP} using a *Xenopus* oocyte translation system[16] by coinjecting purified tRNA and appropriate mRNAs into live oocytes. *Xenopus* liver contains a seryl tRNA with the properties of SerP-tRNA^4 which may indicate the appropriate phosphotransferase for SerP-tRNA formation are present. It should be possible to determine (e.g., by electrophoretic mobility) if the translation products have SerP incorporated. Using casein mRNAs it will be necessary to distinguish any readthrough products (all casein mRNAs so far sequenced have UGA termination codons) from those in which SerP may be incorporated internally.

A direct involvement of tRNA^{SerP} in the phosphorylation of "structural" phosphoproteins seems unlikely by consideration of the codon usage in the caseins. Also, the widespread occurrence of tRNA^{SerP} argues against a cellular function of tRNA^{SerP} specifically for phosphorylation of "structural" phosphoproteins. For completeness of discussion, other possible functions for tRNA^{SerP} should be considered.

The anomalous structure of tRNA^{SerP} raises the question of whether this RNA actually functions in translation as a transfer RNA. Roles for tRNAs other than in the translation elongation process have been identified[17] and furthermore several cytoplasmic RNAs having tRNA-like structures have been identified.[18] These RNAs include the structures found on the 3' end of plant single-strand RNA virus genomes.[19] Some of the tRNA-like structures of viral genomes are able to be aminoacylated. Whether this aminoacylation is a consequence of a coincidental similarity to tRNA structure or is involved in the mechanism of viral growth, is not known. Also, the amino acid involved usually is not the one correspond-

[14] R. P. Martin, A. P. Sibler, J. M. Schneller, G. Keither, A. J. C. Stahl, and G. Dirheimer, *Nucleic Acids. Res.* **5,** 4579 (1978).

[15] T. D. Fox and B. Weiss-Brummer, *Nature* (*London*) **288,** 60 (1980).

[16] M. Bienz, E. Kubli, J. Kohli, S. de Henau, and H. Grosjean, *Nucleic Acids. Res.* **8,** 5169 (1980).

[17] R. LaRossa and D. Söll, *in* "Transfer RNA" (S. Altman, ed.), p. 136. MIT Press, Cambridge, Massachusetts, 1978.

[18] G. Zieve and S. Penman, *Cell* **8,** 19 (1976).

[19] A. H. Haenni, S. Joshi, and F. Chapeville, *Prog. Nucleic Acids Res.* **27,** 85 (1982).

ing to the anticodon of the tRNA-like structure. This is of interest since as already noted tRNA^{SerP} actually has a tryptophan anticodon. Thus, is the aminoacylation reaction of tRNA^{SerP} required for its function? The possibility of tRNA^{SerP} having a function in translation other than for the transfer of an amino acid to the nascent polypeptide, should also be considered since this is a possible function for many small cytoplasmic RNA species.[20] There are two tRNA-like small RNAs encoded by adenovirus, VAI and VAII. The VAI RNA is required for efficient translation of viral mRNAs at late times after infection and appears to have a role during initiation of translation.[21] It is conceivable therefore that tRNA^{SerP} may not have a function of directing amino acid transfer but may have a function in translational control.

[20] W. McCormick and S. Penman, *J. Mol. Biol.* **39**, 315 (1969).
[21] J. S. Logan and T. Shenk, *Microbiol. Rev.* **46**, 377 (1982).

Section IV

Covalent Modifications of the α-Amino and α-Carboxyl Groups of Proteins

[14] Amino-Terminal Acetylation of Proteins: An Overview[1]

By Susumu Tsunasawa and Fumio Sakiyama

Since the discovery of an acetyl group at the amino-terminus of tobacco mosaic virus coat protein,[2] a number of N^α-acetylated proteins have been found in animals, plants, and their viruses, and also, although more rarely, in bacteria and fungi. N^α-Acetylation is therefore considered one of the typical modification of proteins in living organisms. Moreover, in some eukaryotic cells, it has been suggested that more than half of the intracellular soluble proteins are N^α-acetylated: the extent of acetylation is 90, 80, and >50% for the soluble proteins of mouse L cells, Ehrlich ascites cells, and *Neurospora crassa,* respectively.[3,4] It would appear that N^α-acetylation is characteristic of intracellular proteins in eukaryotes.

Experiments with ovalbumin,[5] α-crystallin,[6] and histone[7] have shown that N^α-acetylation is a cotranslational event. In the biosynthesis of ovalbumin, a secretory protein, the amino-terminal methionine was removed when the nascent peptide chain had extended about 20 residues, while the N^α-acetylation of the new amino-terminal glycine took place after the peptide chain had been elongated up to about 40 residues. A similar phenomenon was also observed in the processing of α-crystallin, a nonsecretory protein. These results suggest that for both secretory and nonsecretory proteins, N^α-acetylation occurs at a stage when the amino-terminal portion of the growing chain has protruded from the ribosome. This is consistent with the facts that the majority of the N^α-acetylated proteins so far isolated are intracellular and that the structural genes for the secretory proteins that are N^α-acetylated generally lack the code for an amino-terminal signal sequence.[8–11]

[1] We dedicate this article to the late Professor Kozo Narita.
[2] K. Narita, *Biochim. Biophys. Acta* **28,** 184 (1958).
[3] J. L. Brown, *J. Biol. Chem.* **254,** 1447 (1979).
[4] J. L. Brown and W. K. Roberts, *J. Biol. Chem.* **251,** 1009 (1976).
[5] R. D. Palmiter, J. Gagnon, and K. A. Walsh, *Proc. Natl. Acad. Sci. U.S.A.* **75,** 94 (1978).
[6] G. J. A. M. Straus, A. J. M. Berns, and H. Bloemendal, *Biochem. Biophys. Res. Commun.* **58,** 876 (1974).
[7] E. Kecskes, I. Sures, and D. Gallwitz, *Biochemistry* **15,** 2541 (1976).
[8] L. McReynolds, B. W. O'Malley, A. D. Nisbet, J. E. Fothergill, D. Givol, S. Fields, M. Robertson, and G. G. Brownlee, *Nature (London)* **273,** 723 (1978).
[9] I. Schechter, Y. Burnstein, R. Zammell, E. Ziv, F. Kantor, and D. Papermaster, *Proc. Natl. Acad. Sci. U.S.A.* **76,** 2654 (1979).

METHODS IN ENZYMOLOGY, VOL. 106

An alternative mechanism has been proposed for the N^α-acetylation of β-endorphin,[12] α-melanotropin,[12] and α- or β-thymosin,[13] which all are synthesized as precursors and of mature proteins such as *Escherichia coli* ribosomal L12/S5/S18 proteins[14] and *Dictyostelium discoideum* actin.[15]

N^α-Acetylation is an enzyme-catalyzed reaction in which the protein accepts the acetyl group from acetyl-CoA. The enzyme N^α-acetyltransferase has been found in various cells and tissues such as rabbit reticulocytes,[16,17] rat liver,[18,19] calf lens,[20] rat pituitary,[21,21–24] hen oviduct,[25] and wheat germ.[13] The purification and enzymatic property of this enzyme from several sources are described in other chapters of this volume. According to the mode of action in catalysis, the transferases are classified in two major groups. One group includes the enzyme that catalyzes N^α-acetylation of the nascent peptide chain growing on ribosomes. The enzymes in this group are probably ribosome bound and/or membrane bound. The other group of enzymes includes those that are associated with the processing of bioactive peptides and mature proteins. In this case, it seems that each peptide and protein substrate may be N^α-acetylated by a unique and highly specific enzyme.

It is found that glycine, alanine, serine, methionine, and aspartic acid are dominant as the amino-termini of the N^α-acetylated proteins (see the table). This fact suggests that the amino-terminal residue is a primary recognition site for N^α-acetyltransferase. However, proteins having these amino acids at their amino-termini are not always acetylated, indicating that the enzyme recognizes some structural characteristics of the amino-

---

[10] S. Matsuura, M. Arpin, C. Hannum, E. Margoliash, D. D. Sabatini, and T. Morimoto, *Proc. Natl. Acad. Sci. U.S.A.* **78**, 4368 (1981).

[11] W. A. Braell and H. F. Lodish, *Cell* **28**, 23 (1982).

[12] C. C. Glembotski, *J. Biol. Chem.* **257**, 10493 (1982).

[13] H. Kido, A. Vita, E. Hannappel, and B. L. Horecker, *Arch. Biochem. Biophys.* **208**, 95 (1981).

[14] A. G. Cumberlidge and K. Isono, *J. Mol. Biol.* **131**, 169 (1979).

[15] R. Rubenstein, P. Smith, J. Deuchler, and K. Redman, *J. Biol. Chem.* **256**, 8149 (1981).

[16] J. Traugh and S. B. Sharp, *J. Biol. Chem.* **252**, 3738 (1977).

[17] C. C. Liew and C. C. Yip, *Proc. Natl. Acad. Sci. U.S.A.* **71**, 2988 (1974).

[18] A. Pestana and H. C. Pitot, *Biochemistry* **14**, 1397 (1975).

[19] R. M. Green, J. S. Elce, and R. Kisilevsky, *Can. J. Biochem.* **56**, 1075 (1978).

[20] M. Granger, G. I. Tesser, W. W. DeJong, and H. Bloemendal, *Proc. Natl. Acad. Sci. U.S.A.* **73**, 3010 (1976).

[21] T. A. Woodford and J. E. Dixon, *J. Biol. Chem.* **254**, 4993 (1979).

[22] K. A. Pease and J. E. Dixon, *Arch. Biochem. Biophys.* **212**, 177 (1981).

[23] C. C. Gelmbotski, *J. Biol. Chem.* **257**, 10501 (1982).

[24] T. L. O'Donohue, *J. Biol. Chem.* **258**, 2163 (1983).

[25] S. Tsunasawa, K. Kamitani, and K. Narita, *J. Biochem.* (*Tokyo*) **87**, 645 (1980).

N-Terminal Acetylated Proteins[a]

N-Terminal amino acid	Protein
Ac-Ala-	Adenovirus 2 hexon protein, arginine kinase (lobster muscle), calmodulin (eukaryotes), carbonic anhydrase [human (B), monkey (I), equine (D)], cytochrome b_5 (rat liver), cytochrome c (plants), cytochrome oxidase VI (bovine heart), cucumber virus 4 coat protein, enolase (rabbit liver), fructose-1,6-bisphosphatase (rabbit liver), glucosephosphate isomerase (rabbit muscle), hemoglobin α chain [frog, tadpole (III)], hemoglobin β chain (human variant: Raleigh, crocodile), histone H1 (mouse testis), keratin [sheep (SCMK-B, SCMKB-III-B), silber gull feathers], lactate dehydrogenase (mammals), leghemoglobin b (soybean), myelin basic protein (vertebrates), myoglobin (yellow tuna fish), parvalbumin [rabbit, pike (III), cod, carp (3)], pyruvate kinase (yeast), ribosomal protein S5, S18, (*E. coli*), southern bean mosaic virus coat protein, superoxide dismutase (human, bovine red cell, horse liver), tobacco mosaic virus coat protein (cowpea strain), troponin I (rabbit cardiac muscle), tymosin β_8/β_9 (calf thymus)
Ac-Asn-	Cytochrome c 550 (*Paracoccus denitrificans*)
Ac-Asp-	Actin (rabbit skeletal muscle, bovine brain, *D. discoideum*), troponin C (human, rabbit skeletal muscle)
Ac-Gly-	Cytochrome c (vertebrates, starfish, *Euglena gracilis*), egg albumin (hen), hemoglobin (human fetal), leghemoglobin d (soybean)
Ac-Met-	Adenovirus fiber protein, adenylate kinase (human, pig skeletal muscle), avian oncovirus *gag* precurssor polyprotein (Pr 76 *gag*), core protein (Semliki Forest virus, Sindbis virus), α-crystallin [mammals (A_2, B_2 chain)], Ca^{2+}-dependent ATPase (rabbit skeletal muscle), cucumber mosaic virus protein, fatty acid binding protein (rat liver), metallothionein [horse kidney (1B), human liver (II), mouse liver (I, II)], phosphatidylcholine-transfer protein (bovine liver), protein phosphatase inhibitor-1 (rabbit skeletal muscle), red cell membrane band-3 protein (human), rhodopsin (bovine), Simian virus 40 large and small tumor antigen, α-tropomyosin (rabbit skeletal, cardiac muscle), troponin C (bovine, rabbit cardiac muscle, rabbit skeletal), turnip yellow mosaic virus coat protein, Z-protein (rat liver)
Ac-Ser-	Acylphosphatase (horse muscle), adenovirus component IX protein, alfalfa mosaic virus coat protein (S, 425 strains), alcohol dehydrogenase (horse, rat liver, yeast, *Drosophila melanogasta*), apoferritin (human, horse spleen), ATP/ADP carrier protein (beef heart mitochondria), brome mosaic virus coat protein, Ca^{2+}-binding protein (pig intestinal), carbonic anhydrase C (human, bovine, sheep red cell), coagulation

(*continued*)

N-Terminal Acetylated Proteins (continued)

N-Terminal amino acid	Protein
	factor XIII (human, bovine), cowpea chlorotic mottle virus coat protein, cyclic AMP-dependent protein kinase subunit type II (bovine heart muscle), cytochrome c (green alga), elongation factor Tu (E. coli), globin (aplysia), glyceraldehyde-3-phosphate dehydrogenase (lobster), glycogen phosphorylase b (rabbit muscle), hemoglobin α chain [carp, catstomid fish, trout (I, IV), shark], hemoglobin β chain (felines), histone [H4, H2A, H1 (rabbit thymus, sea urchin)], β-keratin (emu feathers), α-melanotropin (mammals), mitochondria structure protein (bovine heart), myosin (rabbit skeletal muscle), NADP-dependent glutamate dehydrogenase (Neurospora crassa), parvalbumin [rat skeletal muscle, frog, pike (II), thornback-ray], phosphoglycerate kinase (human, horse, yeast), P2 protein (rabbit, bovine, sciatic nerve myelin), ribosomal protein L7 (E. coli), tobacco mosaic virus coat protein (Hr, vulgare strains etc.), troponin I [chicken skeletal muscle (fast)], troponin T (rabbit skeletal muscle), thymosin α_1/β_4 (calf thymus), tyrosinase (Neurospora crassa)
Ac-Tyr-	β-Endorphin (mammals, salmon)

[a] Based on literature reports to 1982.

terminal portion in addition to a particular amino acid at the terminus. In fact, the β chain of human hemoglobin which has the unacetylated amino-terminal valine is N^α-acetylated when the valine is replaced by alanine.[26] By substitution of valine with serine the β chain of a variant of cat hemoglobin also becomes susceptible to N^α-acetylation.[27] Similar phenomena are observed in frameshift mutants of Neurospora NADP-specific glutamate dehydrogenase[28] and yeast iso-1-cytochrome c.[29]

The biological significance of N^α-acetylation of proteins is still an open question. It has been proposed that this posttranslational modification protects intracellular proteins from proteolysis.[30] However, the results so

[26] W. F. Moo-Penn, K. C. Bechtel, R. M. Schmidt, M. H. Johnson, D. L. Jue, D. E. Schmidt, Jr., W. M. Dunlap, S. J. Opella, J. Bonaventura, and C. Bonaventura, Biochemistry 16, 4872 (1977).

[27] J. K. Jolly and F. Taketa, Arch. Biochem. Biophys. 192, 336 (1979).

[28] M. A. M. Siddig, J. A. Kinsey, J. R. S. Fincham, and M. Keighern, J. Mol. Biol. 137, 125 (1980).

[29] F. Sherman, J. W. Stewart, and A. M. Schweingruber, Cell 20, 215 (1980).

[30] H. Jörnvall, J. Theor. Biol. 55, 1 (1975).

far reported do not all support this proposal. In the case of actin from slime mold, proteolytic degradation becomes slower when the protein is N^α-acetylated,[31] but cat hemoglobin turns over at the same rate irrespective of N^α-acetylation.[32] In addition, the acetylated intracellular protein in mouse L cells shows no special stability against the proteolysis.[33]

The recent result from DNA sequencing has shown that in structural genes for the secretory proteins that are N^α-acetylated the codon for the acetylated amino-terminal residue is directly preceded by the initiation codon without the insertion of additional codons for amino acids. To understand the meaning of this fact, further study is necessary. Little effort has been made to understand the relationship between N^α-acetylation and the transport of secretory proteins across biological membranes.

The functional role of the N^α-acetyl group in β-endorphin and α-melanotropin is more obvious in directly affecting biological activity: in contrast to N^α-unacetylated native β-endorphin, the acetylated peptide is nonanalgesic and cannot be bound to opiate receptors *in vitro*.[34] Conversely, native α-melanocyte-stimulating hormone (α-MSH) is more potent than the N^α-deacetylated peptide in causing melanocyte dispersion.[35]

To really understand the function of N^α-acetylation, it will be important to identify the N^α-acetylated amino acid in proteins and peptides on a microanalytical scale. For this purpose, removal of the N^α-acetyl group or the N^α-acetyl amino acid must be efficiently achieved. No practical method has yet been devised for the preferential elimination of the N^α-acetyl group from N^α-acetylated protein, but N^α-acylamino acid-releasing enzymes have been successfully used.[36,37] One such enzyme which was purified from animal liver can liberate the N^α-acetylamino acid from rather short peptides derived from N^α-acetylated proteins. Although substrate specificity is narrow for the amino-terminal residue, this enzyme is now the first choice for the release of the N^α-acetylated amino-terminal residue from N^α-acetylated peptides after proteolysis of the parent protein. This in turn permits the sequencing of the amino-terminal sequence, starting with the second residue. To identify the N^α-acetylamino acid release, several techniques such as mass spectrometry,[38] reverse-phase

[31] P. Rubenstein and J. Deuchler, *J. Biol. Chem.* **254**, 1142 (1979).

[32] M. R. Mank, G. R. Putz, and F. Taketa, *Biochem. Biophys. Res. Commun.* **71**, 768 (1976).

[33] J. L. Brown, *J. Biol. Chem.* **254**, 1447 (1979).

[34] D. G. Smyth, D. E. Massey, S. Zakarian, and M. D. A. Finnie, *Nature (London)* **279**, 252 (1979).

[35] J. P. Waller and H. B. F. Dixon, *Biochem. J.* **75**, 320 (1960).

[36] S. Tsunasawa and K. Narita, this series, Vol. 45, p. 552.

[37] J. Gagnon, R. D. Palmiter, and K. A. Walsh, *J. Biol. Chem.* **253**, 7464 (1978).

[38] D. H. Williams, G. Bojesen, A. D. Auffret, and L. C. E. Tayer, *FEBS Lett.* **128**, 37 (1981).

HPLC,[39] and high-resolution proton NMR spectroscopy[40] have been recently reported. The former two methods are approximately two orders of magnitude more sensitive than the classical methods. At present, HPLC is a practical technique for rapid quantitation of N^α-acetylamino acids with high sensitivity. The details of other methods for the identification of N^α-acetylamino acids are described elsewhere by Narita,[41] Doolittle,[42] and Allen.[43]

[39] S. Tsunasawa and K. Narita, this series, Vol. 91, p. 84.
[40] J. P. Van Eerd, J. P. Capony, C. Ferraz, and J. F. Pechere, *Eur. J. Biochem.* **91,** 321 (1978).
[41] K. Narita, *in* "Protein Sequence Determination" (S. B. Needleman, ed.), p. 95. Springer-Verlag, Berlin and New York, 1975.
[42] R. F. Doolittle, *in* "Advanced Methods in Protein Sequence Determination" (S. B. Needleman, ed.), p. 51. Springer-Verlag, Berlin and New York, 1977.
[43] G. Allen, "Sequence of Proteins and Peptides," p. 245. Am. Elsevier, New York, 1981.

[15] Rat Pituitary N^α-Acetyltransferase

By JACK E. DIXON and TERRY A. WOODFORD

Several laboratories have demonstrated that proteins such as ovalbumin,[1] α-crystalline,[2] and peptides obtained from rat liver polysomes[3,4] are N^α-acetylated following addition of 20–50 amino acids to a nascent polypeptide. Acetyltransferases which are associated with ribosomal fractions of rabbit reticulocytes have also been described and partially purified.[5] Although there are numerous examples of N^α-acetylated proteins,[6] the functional significance of modifying nascent peptides is not understood. Possible functions for N^α-acetylation include altered rates of protein deg-

[1] R. D. Palmiter, J. Gagnon, and K. A. Walsh, *Proc. Natl. Acad. Sci. U.S.A.* **75,** 94 (1978).
[2] H. Bloemendal and G. J. Strous, *in* "Lipmann Symposium: Energy, Regulation and Biosynthesis in Molecular Biology" (D. Richter, ed.), p. 89. deGruyter, Berlin, 1979.
[3] A. Pestana and H. C. Pitot, *Biochemistry* **14,** 1397 (1975).
[4] A. Pestana and H. C. Pitot, *Biochemistry* **14,** 1404 (1975).
[5] J. A. Traugh and S. B. Sharp, *J. Biol. Chem.* **252,** 3738 (1977).
[6] M. Dayhoff, ed., "Atlas of Protein Sequences and Structure," Vol. 5, p. 172. Natl. Biomed. Res. Found., Washington, D.C., 1975; also, Suppl. 1, 2, to Vol. 5 (1976).

radation[7,8]; however, Brown[9] did not find evidence to support this suggestion in L cell proteins. The possibility that N^{α}-acetylation plays a role in protein secretion has also been suggested.[1]

N^{α}-Acetylation is not restricted to nascent polypeptides. The pituitary hormones, α-melanotropin (α-MSH) and β-endorphin are synthesized initially in the form of a "polyprotein,"[10-13] which has been referred to as pro-opiomelanocortin. The polyprotein is in turn cleaved via proteolysis and modified by several posttranslational events.[14] One of these posttranslational modifications includes the N^{α}-acetylation of these hormones. In the case of α-MSH an amino terminal serine is modified while β-endorphin is acetylated on the amino terminal tyrosine. Although the functional significance of N^{α}-acetylation of proteins is poorly understood, the presence of the acetyl group on α-MSH and β-endorphin has a profound effect upon hormone activity. In the case of α-MSH, acetylation distinguishes the biological properties of α-melanotropin from those of corticotropin (ACTH).[15] Acetylation of β-endorphin leads to a loss in biological activity.[16]

In order to form α-MSH from ACTH several modifications must occur: shortening of the polypeptide chain by proteolysis and amidation of the valine at position 13, forming a carboxyl terminal valine amide. In addition, the amino-terminal serine must be acetylated. These transformations are outlined in Fig. 1. Woodford and Dixon[17] first described the properties of an enzyme present in rat pituitary which catalyzed the N^{α}-acetylation of the serine residue of ACTH and ACTH fragments. This paper will review some of the properties of the rat pituitary N^{α}-acetyltransferase and will focus primarily on results obtained in the authors' laboratory. In addition, Glembotski[18-20] as well as Eipper and Mains[21]

[7] H. Jornvall, *J. Theor. Biol.* **55,** 1 (1975).
[8] J. L. Brown and W. K. Roberts, *J. Biol. Chem.* **251,** 1009 (1976).
[9] J. L. Brown, *J. Biol. Chem.* **254,** 1447 (1979).
[10] R. E. Mains, B. A. Eipper, and N. Ling, *Proc. Natl. Acad. Sci. U.S.A.* **74,** 3014 (1977).
[11] J. Roberts and E. Herbert, *Proc. Natl. Acad. Sci. U.S.A.* **74,** 5300 (1977).
[12] S. Nakanishi, A. Inoue, T. Kita, M. Nakamura, A. Chang, S. Cohen, and S. Numa, *Nature (London)* **278,** 423 (1979).
[13] E. Herbert and M. Uhler, *Cell* **30,** 1 (1982), and references therein.
[14] B. A. Eipper and R. E. Mains, *Endocr. Rev.* **1,** 1 (1980).
[15] P. T. Lowry, R. E. Silman, J. Hope, and A. P. Scott, *Ann. N. Y. Acad. Sci.* **297,** 49 (1977).
[16] D. Smyth, D. Massey, S. Zakanian, and M. Finnie, *Nature (London)* **279,** 252 (1979).
[17] T. A. Woodford and J. E. Dixon, *J. Biol. Chem.* **254,** 4993 (1979).
[18] C. C. Glembotski, *J. Biol. Chem.* **256,** 7433 (1981).
[19] C. C. Glembotski, *J. Biol. Chem.* **257,** 10493 (1982).
[20] C. C. Glembotski, *J. Biol. Chem.* **257,** 10501 (1982).
[21] B. A. Eipper and R. E. Mains, *J. Biol. Chem.* **256,** 5689 (1981).

```
          1                 5                      10                                    39        O
ACTH    Ser · Tyr · Ser · Met · Glu · His · Phe · Arg · Trp · Gly · Lys · Pro · Val · · · · · · · · · ·  Phe—C
                                                                                                           OH
```

```
          O       1              5                      10                    O
          ‖                                                                    ‖
α-MSH   CH₃C—N—SER · Tyr · Ser · Met · Glu · His · Phe · Arg · Trp · Gly · Lys · Pro · Val · C—NH₂
              |
              H
```

$$\alpha\text{-MSH}\quad CH_3C\overset{O}{\overset{\|}{}}-\underset{H}{N}-\underset{1}{SER}\cdot Tyr\cdot Ser\cdot \underset{5}{Met}\cdot Glu\cdot His\cdot Phe\cdot Arg\cdot \underset{10}{Trp}\cdot Gly\cdot Lys\cdot Pro\cdot Val\cdot \overset{O}{\overset{\|}{C}}-NH_2$$

FIG. 1. The molecular transformations necessary to convert ACTH to α-MSH require cleavage of the polypeptide, amidation of the valine residue 13, and NH₂-terminal acetylation.

have recently described a series of elegant studies on the rat and bovine N^α-acetyltransferase and have also characterized a number of the biosynthetic intermediates formed during posttranslational processing. The reader is also referred to a recent study by O'Donohue[22] describing the properties of an enzyme which acetylates β-endorphin.

[22] T. L. O'Donohue, *J. Biol. Chem.* **258**, 2163 (1983).

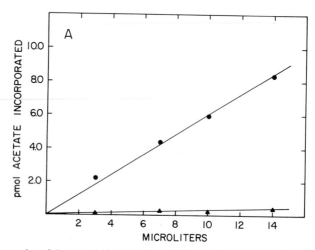

FIG. 2. Properties of the acetylation reaction. (A) Increasing concentrations of the pituitary fraction (13.5 μg of protein/μl) were incubated for 3 min with [³H]acetyl-CoA and ACTH(1–39). The picomoles of [³H]acetate incorporated in the presence (●) and absence (▲) of substrate was determined. (B) Time course for acetylation. Two hundred microliters of the soluble pituitary extract (11.04 μg of protein/μl; 8.45 μg of protein/μl in the inset) was incubated with [³H]acetyl-CoA and ACTH(1–39) (total volume 400 μl). At designated time intervals, 35-μl aliquots were removed from the incubation mixture and the picomoles of acetate incorporated was determined (picomoles per 35 μl). (C) The soluble pituitary extract was incubated for 3 min with [³H]acetyl-CoA and ACTH(1–39) in 50 mM Tris–HCl, 300 mM sucrose, 50 mM KCl adjusted to various pH values.

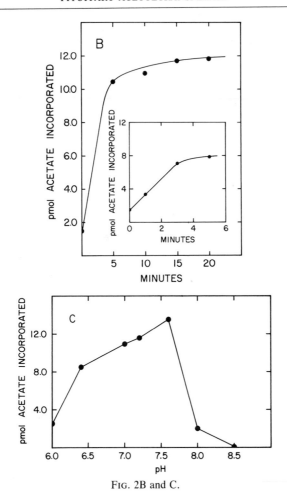

FIG. 2B and C.

Properties of the Rat N$^\alpha$-Acetyltransferase. The assay employed by numerous investigators is similar to or identical to that originally described by Woodford and Dixon.[17] Several characteristics of the rat pituitary acetyltransferase are given in Fig. 2. The incorporation of [³H]acetate arising from radiolabeled acetyl-CoA into ACTH(1–39) is dependent upon the enzyme concentration, the time of incubation and the pH of the assay mixture as shown in Fig. 2A, B, and C, respectively. The broad pH versus rate profile which extends to pH 6.0 and below has also been noted by Chappell *et al.*[23] At pH values above 7.6, there was a pronounced decline in reaction rate.

[23] M. Chappell, Y. Loh, and T. O'Donahue, *Peptides* **3**, 405 (1982).

TABLE I
RELATIVE EFFICIENCY OF ACETYLATION OF
ACTH FRAGMENTS

Substrate	Activity[a]
ACTH 1–24	100 ± 2
1–10	84 ± 18
2–10	7 ± 1
3–10	12 ± 3
4–10	48 ± 8
1–8	6 ± 2

[a] Incorporation of [^{3}H]acetate into ACTH(1–24) expressed in pmoles incorporated/mg protein was set at 100 and the values for the other peptides computed accordingly. The data represent the average of at least two trials performed in duplicate, using 1.0 mM substrate.

The location of the ^{3}H[acetyl] group within ACTH and ACTH analogs was established by analysis of the reaction products obtained following exhaustive proteolytic digestion of the radiolabeled peptide(s). The radioactive material from the enzymatic digest comigrated with N-acetyl-L-serine. Further evidence that the acetylation occurs on the NH$_2$-terminal serine (as opposed to ε-amino groups of lysine) was obtained by demonstrating that the radioactive material in the acidic fraction of the enzymic digest cocrystallized with N-acetylserine.

The substrate specificity of acetyltransferase was explored with regard to the amino terminal residue and the chain length that was required for N^α-acetylation.[17,24] ACTH analogs were used at saturating concentrations. An additional control using heat-inactivated enzyme was used to determine nonenzymatic acetylation. This accounted for less than 2% of the total acetylated products. The values listed in Table I are corrected for acetylation of endogenous material and are expressed as percentages based on the acetylation of ACTH(1–24). The peptides, ACTH(1–24) and ACTH(1–10), at saturating concentrations are comparable substrates for the N^α-acetyltransferase. Trypsin cleavage of ACTH(1–10) resulted in an octapeptide, ACTH(1–8) and Trp-Gly. Acetylation of the octapeptide was drastically reduced compared to the decapeptide. Removing the amino terminal serine of ACTH(1–10) essentially abolished enzymatic acetylation of the peptide. The removal of tyrosine from ACTH(2–10) did not

[24] K. A. Pease and J. E. Dixon, *Arch. Biochem. Biophys.* **212,** 177 (1981).

TABLE II
SUBCELLULAR LOCALIZATION OF N^α-ACETYLTRANSFERASE[a]

Cell fraction	Volume (ml)	Protein (mg)	cpm (10^{-4})	Specific activity (pmol acetate incorporated/mg protein)	Total activity (pmol [^{3}H] acetate incorporated)
Crude extract	2.0	4.2	25.8	33	139
10,000 g pellet	1.0	1.9	3.94	11	21
27,000 g supernatant	1.6	1.6	20.4	69	110
27,000 g pellet	0.5	0.2	9.95	270	54
Percentage recovery		88.1			133%

[a] Subcellular fractionation and the assay for N^α-acetyltransferase activity were carried out as described in the text using ACTH(1–10) as the substrate and [^{3}H]acetyl-CoA as the acetate donor.

substantially improve the peptide's performance as a substrate even though the amino terminus of ACTH(3–10) is again serine. Surprisingly, ACTH(4–10), which has methionine at the amino terminus, is a substrate for the acetyltransferase. It was acetylated approximately 50% as well as ACTH(1–24). These results clearly establish the importance of the amino-terminus in the N^α-acetylation reaction and also suggest that additional information within the ACTH fragments is necessary for enzyme specificity.

The rat pituitary acetyltransferase is not a stable enzyme in our hands. Glycerol has been used to extend the half-life of the enzyme, but even this reagent did not provide the necessary stability required for purification of the enzyme.

Subcellular Localization of the N^α-Acetyltransferase. The subcellular distribution of the N^α-acetyltransferase activity is shown in Table II. The majority of the activity is observed in the 27,000 g supernatant and pellet fractions. The total enzyme activity in the 10,000 g pellet, 27,000 g supernatant, and 27,000 pellet was consistently greater than that observed in the crude extract (ranging from 133 to 160%). This may be indicative of the presence of an inhibitor or result from hydrolysis of acetyl-CoA in the crude homogenate. The recovery of protein in the various subcellular

TABLE III
DISTRIBUTION OF $N^{\alpha\gamma}$-ACETYLTRANSFERASE ACTIVITY IN
VARIOUS ORGANS[a]

Organ	[³H]Acetate incorporated (pmol)		μg of protein/μl of extract	Specific activity[b]
	+ACTH	−ACTH		
Pituitary	17.9	4.6	10.2	93.0
Lens	31.7	3.3	120.0	16.9
Heart	19.9	19.0	6.3	10.2
Lung	12.7	7.2	17.8	22.1
Kidney	8.1	5.7	12.1	14.2
Liver	13.4	9.5	16.8	16.6
Muscle	10.4	7.8	6.4	29.0
Whole brain (minus pituitary)	4.5	0.9	6.8	37.8
Serum	7.1	5.9	1.2	1.0

[a] [³H]Acetyl-CoA(1.2 μM) was incubated for 3 min with 14 μl of tissue extract using 0.28 mM ACTH(1–10) as the substrate. Acetylation of endogenous proteins in the various tissues was determined by substituting the reaction buffer for the substrate.

[b] Picomole [³H]acetate incorporated/mg protein (minus the acetate incorporated into endogenous proteins).

fractions was 88% of that present in the crude homogenate. The specific activities in the particulate fractions were generally 4 to 10 times greater than those noted in the homogenate or high speed supernatant.[24] These results suggest that there could be two acetyltransferase activities (although other explanations cannot be ruled out), one enzyme which is soluble and another which is particulate. It is anticipated that the maturation of ACTH to α-MSH would occur within granules and that an acetyltransferase contained within these granules would most likely account for the maturation of this hormone *in vivo*.[18–20,22]

Distribution of N^{α}-Acetyltransferase Activity in Various Organs. Incorporation of [³H]acetate into ACTH(1–10) and endogenous rates of acetylation were determined with various soluble rat tissue extracts.[17] It is important to note that by using ACTH(1–10), the possibility of ε-amino acetylation of lysine residues is excluded. The picomoles of acetate incorporated with each tissue extract is shown in Table III. The presence of endogenous acetyl-CoA or CoA in the various tissue homogenates was

TABLE IV
N^α-ACETYLTRANSFERASE ACTIVITY IN RAT PITUITARY LOBES[a]

Lobe	[^{3}H]Acetate incorporated (pmol)	Protein (mg)	Specific activity (pmol/mg)
Anterior	10.2	0.156	65.4
	7.55	0.180	41.9
	8.28	0.192	43.1
Posterior-intermediate	7.12	0.186	38.3
	5.37	0.174	30.9
	7.10	0.156	45.5

[a] Results of three experiments.

not taken into account. Available data on the levels of these cofactors in rat liver and heart would suggest that they are probably insignificant in comparison to those used in the assay.[25] The extracts from lung, muscle, and brain, were all capable of using ACTH(1–10) as a substrate although none was as efficient as the pituitary, which showed the highest specific activity of all tissues examined.

Acetyltransferase Activity in Anterior and Posterior Intermediate Pituitary Lobes. Since ACTH is present largely in the anterior pituitary and α-MSH in the intermediate lobe of the rat,[26] it was of interest to determine the distribution of the acetyltransferase in these two lobes of the pituitary. The intermediate lobe is very small and difficult to separate from the posterior lobe. The pituitary was dissected into the anterior and the posterior–intermediate lobes as described.[27] The lobes from groups of 10 male rats were prepared and the enzyme activity in the extracts was assayed. The results of three determinations are shown in Table IV. The specific activities of the acetyltransferase in the two tissues are very similar. An effort was also made to separate the intermediate lobe cells from the posterior lobe using the method of Crine *et al.*[27] Acetyltransferase activity in cells in the intermediate lobe and posterior lobe was consistently observed; however, quantitative determinations were not possible due to the low levels of acetate incorporation. More recent evidence has established that the rat intermediate pituitary does indeed harbor an acetyltransferase(s) which appears to be responsible for N-acetylation of

[25] J. R. Williamson and B. Corky, this series, Vol. 13, p. 434.
[26] P. J. Lowry and A. P. Scott, *Gen. Comp. Endocrinol.* **26**, 16 (1975).
[27] P. Crine, C. Gianoulakis, N. Seidah, F. Gossard, P. Pezalla, M. Lis, and M. Chrétien, *Proc. Natl. Acad. Sci. U.S.A.* **75**, 4719 (1978).

TABLE V
KINETIC CONSTANTS FOR ACETYLATION OF
ACTH FRAGMENTS

Substrate	K_m (μM)
ACTH(1–24)	4.2
ACTH(1–10)	96
ACTH(4–10)	37
AcCoA	2.2

both ACTH and β-endorphin. Glembotski[28] has demonstrated that an N^α-acetyltransferase activity (as measured by acetylation of β-endorphin) present in the rat neurointermediate lobe can be competitively inhibited by α-MSH analogs as well as β-endorphin analogs. These experiments would suggest that a single enzyme within these cells is responsible for acetylation of both ACTH and β-endorphin. This observation raises the important point that regulation of the acetyltransferase would control the relative cellular concentrations of α-MSH/ACTH and N-acetyl-β-endorphin/β-endorphin.

Kinetic Constants of the N^α-Acetyltransferase. The kinetics of the rat pituitary N^α-acetyltransferase have also been explored.[20,24] A summary of the kinetic constants obtained by Pease and Dixon[24] is given in Table V. Determining kinetic constants with the particulate enzyme was difficult and a number of control experiments were carried out to establish the validity of these determinations. A brief summary of the experimental conditions employed in obtaining these values is outlined below. The apparent Michaelis constants for [³H]AcCoA and ACTH(1–24) were determined. The K_m for AcCoA was 2.2 μM when 0.27 mM ACTH(1–24) was employed as the substrate and the [³H]AcCoA concentration varied from 4.2 μM to 0.42 mM. The K_m for ACTH(1–24) was 4.2 μM in the presence of 4.2 μM [³H]AcCoA when ACTH(1–24) concentrations were varied from 14 to 1.4 μM. The concentration of [³H]AcCoA used to determine the K_m for ACTH(1–24) was only twofold higher than the K_m. Raising the [³H]AcCoA concentration increased the acetylation of endogenous material obscuring interpretation of results at higher concentrations. There was concern over the validity of K_m determinations, since as much as 20% of the [³H]AcCoA was consumed during some of the reactions. A time course of the reaction with 4.2 μM [³H]AcCoA and 14 μM

[28] C. Glembotski, *Am. Pept. Symp. 8th, 1983* (in press).

ACTH(1–24) (the high concentrations used in the determination of the K_m) was still linear at 3 min. Thus, the kinetic constants should be valid under the assay conditions employed. The apparent K_m values for ACTH(1–10) and ACTH(4–10) were also determined in a similar manner. Again the [³H]AcCoA concentration was maintained at 4.2 μM, and the peptide concentrations were varied from 10 to 100 μM. Under these conditions, the K_m for ACTH(1–10) was 96 μM, while that for ACTH(4–10) was 37 μM.

Summary

The rat pituitary contains an enzyme which will acetylate certain ACTH fragments using acetyl coenzyme A. This acetyltransferase activity was found in all three lobes of the rat pituitary as well as in all other tissues examined. The rat pituitary enzyme appears to be largely particulate in nature. The enzyme sedimenting at 27,000 and 100,000 g had specific activities 4–10 times greater than the soluble fraction. The acetyltransferase activity was dependent on substrate concentration (ACTH) and pH, was linear with time, and was inactivated at 55°. The enzyme would acetylate ACTH(1–24), (1–10), and ACTH(4–10), but would not use ACTH(2–10), (3–10), or ACTH(1–8) as substrates. The apparent K_m values for the substrates were as follows: AcCoA, 2.2 μM, ACTH(1–24), 4.2 μM; ACTH(1–10), 96 μM; and ACTH(4–10), 37 μM.

Acknowledgment

This work was supported in part by a grant from the National Institutes of Health, AM 18024.

[16] Actin Amino-Terminal Acetylation and Processing in a Rabbit Reticulocyte Lysate

By Kent L. Redman and Peter A. Rubenstein

As many as 80% of the proteins in a mammalian cell may be amino-terminally acetylated.[1] The properties of the enzyme(s) carrying out this acetylation reaction or the functional significance of the modification have

[1] J. L. Brown and W. K. Roberts, *J. Biol. Chem.* **251**, 1009 (1976).

not been described. It is known, though, that the rabbit reticulocyte lysate will acetylate those proteins normally NH_2-terminally acetylated *in vivo*.[2]

Previous attempts to characterize the protein acetyltransferases in a reticulocyte lysate utilized proteins such as casein and histones as substrates.[3] In those studies, the acetylation observed was probably on the ε-amino groups of lysine residues in the proteins and not at the amino-terminus. The reason for this statement is that histones and most other proteins are NH_2-terminally acetylated early in translation in an irreversible fashion. Moreover, experiments were not performed in the studies cited above to localize the exact site of incorporation of labeled acetyl groups.

We decided to study protein NH_2-terminal acetylation by synthesizing as a substrate a nonacetylated form of a protein that normally carries an NH_2-terminal acetyl group. With such a substrate, we could then examine directly in a reticulocyte lysate the NH_2-terminal acetylation of a single protein species. For this purpose, we chose *Dictyostelium discoideum* actin. This protein is made in high abundance in an mRNA-dependent reticulocyte lysate primed with *Dictyostelium* mRNA.[4] The acetylated and nonacetylated forms of actin are easily separable by two-dimensional polyacrylamide gel electrophoresis according to the method of O'Farrell.[5] *D. discoideum* mRNA can be obtained in large quantities, and the *Dictyostelium* actin gene and polypeptide have been sequenced.[6,7]

Mature *Dictyostelium* actin possesses an acetyl-Asp-Gly amino-terminus. However, the gene codes for a Met-Asp-Gly amino-terminal sequence. In eukaryotic cells removal of the initiator methionine from nascent polypeptide chains occurs *in vivo* for a number of proteins and has been demonstrated in a reticulocyte lysate as well. For proteins previously examined, the methionine is removed early in translation, after polymerization of about 40 amino acids, only if the second amino acid is small and uncharged.[8] *Dictyostelium* actin, as well as yeast and mammalian nonmuscle actins, would seem to be an anomaly since the second amino acid, Asp or sometimes Glu, is acidic.

In investigating this problem, we discovered that actin amino terminal processing proceeds *in vivo* and *in vitro* by a novel posttranslational pathway involving removal of acetylmethionine instead of free methionine as

[2] H. Bloemendal, *Science* **197,** 127 (1977).
[3] J. A. Traugh and S. B. Sharp, this series, Vol. 60, p. 534.
[4] P. A. Rubenstein, P. Smith, J. Deuchler, and K. Redman, *J. Biol. Chem.* **256,** 8149 (1981).
[5] P. H. O'Farrell, *J. Biol. Chem.* **250,** 4007 (1975).
[6] J. Vandekerckhove and K. Weber, *Nature (London)* **284,** 475 (1980).
[7] R. A. Firtel, R. Timm, A. R. Kimmel, and M. McKeown, *Proc. Natl. Acad. Sci. U.S.A.* **76,** 6206 (1979).
[8] Y. Burstein and I. Schechter, *Biochemistry* **17,** 2392 (1978).

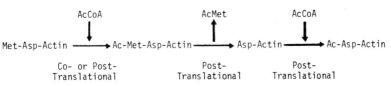

FIG. 1. Pathway for NH$_2$-terminal processing of *D. discoideum* actin.

occurs for other proteins.[4,9,10] This pathway is shown in Fig. 1. The methionine probably is acetylated early in translation. The completed 43,000-dalton actin polypeptide containing an acetylmethionine amino terminus is released from the ribosome. Acetylmethionine then is removed, and the newly exposed Asp is acetylated. Inhibition of the methionine acetylation prevents methionine removal.

In this chapter, we will describe a method for inhibiting the acetylation of proteins in the reticulocyte lysate. Using this system, we then will describe a procedure for assaying the acetylation of the actin NH$_2$-terminal methionine residue and the subsequent removal of acetylmethionine from the completed actin polypeptide chain. We will not deal specifically with the final acetylation of the NH$_2$-terminal aspartic acid residue.

I. Inhibition of Protein Acetylation

In 1977, Palmiter[11] reported a method for inhibiting protein acetylation *in vitro* by converting endogenous acetyl-CoA to citrate with citrate synthase and oxaloacetate. Using this method, however, we never could achieve more than 50% inhibition of acetylation. To improve the degree of acetylation inhibition achievable we synthesized a nonreactive analog of acetyl-CoA, S-acetonyl-CoA, to serve as a potential competitive inhibitor of the protein acetyltransferase.[12] If we first treated the reticulocyte lysate according to Palmiter and then added S-acetonyl-CoA, 85–95% inhibition of acetylation could be achieved.[4]

A. Synthesis of S-Acetonyl CoA

As a reagent, monobromoacetone was synthesized according to Catch *et al.*[13] The product was collected by distillation at 50 mm between 63.5 and 64° as a water-clear liquid and stored in the dark at −20°.

[9] K. Redman and P. A. Rubenstein, *J. Biol. Chem.* **256**, 13226 (1981).
[10] P. A. Rubenstein and D. Martin, *J. Biol. Chem.* **258**, 3961 (1983).
[11] R. D. Palmiter, *J. Biol. Chem.* **252**, 8781 (1977).
[12] P. A. Rubenstein and R. D. Dryer, *J. Biol. Chem.* **255**, 7858 (1980).
[13] J. R. Catch, D. F. Elliott, D. H. Hey, and E. R. Jones, *J. Chem. Soc.* p. 272 (1948).

Coenzyme A (200 μmol) was dissolved in 15 ml of freshly degassed ice-cold water with magnetic stirring. Dithiothreitol (20 μmol) was added to ensure complete reduction of the coenzyme. The pH was brought to 8.0 to 8.2 with NaOH. Alkylation was performed by addition of monobromoacetone (250 μmol) dissolved in 5 ml of 95% ethanol just prior to use. Disappearance of –SH groups was monitored spectrophotometrically at 412 nm after adding 5 μl of the reaction mixture to 1.00 ml of 0.1 mM DTNB [5,5'-dithiobis(2-nitrobenzoic acid)] adjusted to pH 8.5 with 50 mM Tris buffer. The reaction with each aliquot of bromoacetone was usually complete in 1 min or less. Solvent and excess bromoacetone were removed by lyophilization, and the product (S-acetonyl-CoA) was dissolved in 1.5 ml of water and chromatographed on a column (1.5 $\times$ 45 cm) of Sephadex G-15 equilibrated with water. Eluant fractions containing the highest concentration of material absorbing at 260 nm were pooled. The pooled fractions were lyophilized to dryness and stored at $-20\cdot$. The material is also stable as a frozen aqueous solution. Acetonyl-CoA has a E_{mM}^{260} of 15.4.

B. Preparation of Nonacetylated Actin with a NH$_2$-Terminal Methionine

Reagents

mRNA-dependent rabbit reticulocyte lysate system with [^{35}S]methionine as the labeled amino acid.[14]

Oxaloacetic acid, 10 mM, made fresh daily by dissolving the free acid in 50 mM imidazole $\cdot$ HCl, pH 7.5, and adjusting the final pH to 7.5 with NaOH.

Porcine citrate synthase, 750 units/ml. Citrate synthase (Sigma) in 2.2 M (NH$_4$)$_2$SO$_4$ is pelleted by centrifugation at 59000 g for 5 min in a Beckman Airfuge. The pellet is then dissolved in distilled water to produce the desired concentration.

S-Acetonyl-CoA, 2 mM, in water.

Whole cell $D.$ $discoideum$ RNA, 6 mg/ml in water, containing translatable actin mRNA. The RNA is prepared from late log phase cells according to Ullrich et $al.$[15]

Procedure. The basic reaction mixture used for synthesis of $D.$ $discoideum$ actin contains 35 μl of mRNA-dependent rabbit reticulocyte lysate supplemented with 19 unlabeled amino acids excluding methionine,

[14] H. R. B. Pelham and R. J. Jackson, *Eur. J. Biochem.* **67,** 247 (1976).
[15] A. Ullrich, J. Shine, J. Chirgwin, R. Pictet, E. Tischer, W. Rutter, and H. M. Goodman, *Science* **196,** 1313 (1977).

30 μCi of L-[^{35}S]methionine (1000 Ci/mmol), and the requisite energy mix.[14] To this solution is added 1 mM oxaloacetate and 35 units/ml of citrate synthase. Following incubation of the reaction mixture for 7 min at 25°,[11] 75 μM acetonyl-CoA is added.[4] The translation is initiated by adding 40 μg of whole cell *D. discoideum* RNA giving a final volume of 70 μl. After the desired length of time, the translation products are analyzed by two-dimensional gel electrophoresis according to O'Farrell[5] and autoradiography. Alternatively, the gel can be prepared for fluorography as follows. The fixed and stained gel is washed 30 min with 10 volumes of distilled water and is then placed in 8–10 volumes of an aqueous solution of 1 M sodium salicylate for 30 min.[16] The gel is dried *in vacuo* and is exposed to X-ray film at $-100°$. Nonacetylated actin migrates at a more basic isoelectric point than does acetylated actin. Results of a typical experiment are shown in Fig. 2 (see especially 2d). The presence of an acetylated NH$_2$-terminus or acetylated lysine residue on any protein can in theory be determined by this procedure as has been shown with rat tyrosine aminotransferase.[17] This procedure should also be useful in unblocking NH$_2$-terminally acetylated proteins for sequencing studies.

II. Posttranslational NH$_2$-Terminal Acetylation of Actin in a Reticulocyte Lysate

For a number of proteins so far examined, it has been demonstrated that NH$_2$-terminal acetylation can occur as a cotranslational process after the first 30–40 amino acids of the nascent polypeptide chain have been polymerized.[18,19] Using our acetylation inhibition system, we were able to make actin as a completed 43,000-dalton polypeptide with a free NH$_2$-terminal methionine. We could then demonstrate that acetylation of the NH$_2$-terminal methionine can occur in a fully posttranslational acetyl-CoA-dependent reaction in the reticulocyte lysate.[4]

Procedure

[^{35}S]Methionine-labeled nonacetylated actin is made as described in Section I. An aliquot of the translation mixture containing newly synthesized nonacetylated actin is diluted 3-fold with fresh lysate and then made 2 mM in acetyl-CoA. At various times aliquots are withdrawn and ana-

[16] J. P. Chamberlain, *Anal. Biochem.* **98,** 132 (1979).
[17] J. L. Hargrove and D. K. Granner, *J. Biol. Chem.* **256,** 8012 (1981).
[18] G. J. Strous, A. J. Bearns, and H. Bloemendal, *Biochem. Biophys. Res. Commun.* **58,** 876 (1974).
[19] A. Pestana and H. Pitot, *Biochemistry* **14,** 1397 (1975).

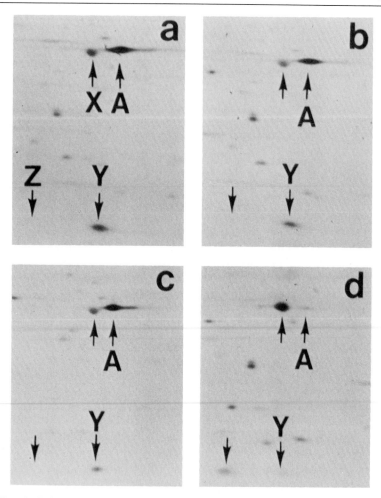

Fig. 2. Enhanced inhibition of rabbit reticulocyte lysate protein acetylation with S-acetonyl-CoA. Proteins coded for by $D.$ $discoideum$ mRNA were synthesized in a reticulocyte lysate under four different conditions. The proteins were then analyzed by two-dimensional gel electrophoresis. The actin regions of autoradiograms of the gels are shown here. A, Acetylated actin; X, nonacetylated actin; Y, a 25,000-dalton acetylated protein; Z, position of nonacetylated form of this protein. (a) Endogenous acetylation was uninhibited; (b) oxaloacetate (1 mM) and citrate synthase (35 units/ml) were added to the lysate at 25° for 7 min prior to RNA addition; (c) S-acetonyl-CoA (100 μM) was added to the lysate with the RNA; (d) the lysate was treated with citrate synthase and oxaloacetate was described in b. Then 75 μM S-acetonyl-CoA was added with the RNA.

lyzed by two-dimensional gel electrophoresis and fluorography as described above. Acetylation is gauged by the acidic shift of actin from its nonacetylated to its normal position on the gel. Confirmation that acetylation is occurring at the NH_2-terminus is obtained by a peptide-mapping procedure described in Section III,B.

III. Removal of Acetylmethionine from the Amino-Terminus of Newly Synthesized Actin

As discussed previously, when *D. discoideum* actin is synthesized in a rabbit reticulocyte lysate, approximately 85–95% of the actin possesses an NH_2-terminal acetylmethionine if endogenous acetylation is not inhibited. Subsequently the acetylmethionine is slowly removed and the newly exposed actin NH_2-terminus acetylated to form mature actin.[9]

Removal of acetylmethionine is assayed in our laboratory either by two-dimensional gel electrophoresis involving isoelectric focusing and sodium dodecyl sulfate–polyacrylamide gel electrophoresis (SDS–PAGE) or by analysis of the actin amino-terminal tryptic peptide. Two-dimensional gel electrophoresis, as discussed previously, separates the different actin species on the basis of charge. Actin, from which the acetylmethionine has been cleaved, will have a single net positive charge more than will actin still retaining the acetylmethionine. The difference in mass between these protein species is too small to make an appreciable difference in the migration of the actins during SDS–PAGE in the second dimension. The second dimension is required, however, since the total cell RNA used in the lysate codes for other proteins having the same isoelectric point as actin.

This gel electrophoresis method of analysis for the actin biosynthetic system can be misleading if the molecular structures of the various intermediates are not known. In this system, mature actin with an acetyl-Asp amino-terminus will comigrate with a precursor actin possessing an acetyl-Met-Asp amino-terminus. Likewise, NH_2-Met-Asp-actin will comigrate with NH_2-Asp-actin.

To establish the nature of these intermediates, it is necessary to analyze the amino-terminal actin tryptic peptide for the presence or absence of the initiating methionine. This method is based on a thin-layer electrophoresis procedure developed by Vandekerckhove and Weber[20] and is made possible by three characteristics of the actin NH_2-terminal tryptic peptide. First, as seen in Fig. 3, the performic acid-oxidized actin tryptic peptide has a high negative charge/mass ratio and as a result migrates at

---

[20] J. Vandekerckhove and K. Weber, *J. Mol. Biol.* **126**, 783 (1978).

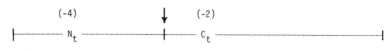

A   Ac-ASP-GLY-GLU-ASP-VAL-GLN-ALA-LEU-VAL-ILE-ASP-ASN-GLY-SER-GLY-*MET-CYS-LYS

B   Ac-*MET-ASP-GLY-GLU-ASP-VAL-GLN-ALA . . . .

FIG. 3. NH₂-terminal tryptic peptide of *D. discoideum* actin. The arrow denotes the thermolysin cleavage site, and the N_t and C_t peptides produced by this enzyme. The numbers in parentheses denote the net negative charge present on each peptide at pH 6.3 following thermolysin cleavage of the performic acid oxidized tryptic peptide. (A) The acetylated NH₂-terminal tryptic peptide of mature actin; (B) the acetylated NH₂-terminal tryptic peptide of the major actin precursor made in the reticulocyte lysate for 1 hr. The * denotes the presence of [³⁵S]methionine if actin synthesis was carried out in a medium containing this labeled amino acid.

pH 6.3 far ahead of other methionine containing peptides. Second, there is a major thermolysin cleavage site between Ala_7 and Ile_8 which splits the tryptic peptide into a highly acidic N_t portion and a less acidic C_t portion. Third, the C_t portion contains a methionine residue. If the protein is labeled with [³⁵S]methionine, the C_t peptide will be labeled. The N_t peptide will be labeled only if the initiator methionine is still retained.

Conventional procedures such as Edman degradation can be employed to determine if the amino-terminus of the actin tryptic peptide is blocked or free. Experiments of this nature have shown that when *D. discoideum* actin is synthesized for 1 hr in a reticulocyte lysate under conditions where acetylation is allowed, 85–95% of the acetylated actin retains the initiator methionine.

In the actin-processing assay, no matter which method is employed, acetylated actin labeled with [³⁵S]methionine is made in a reticulocyte lysate for 45 min. To the lysate is then added 1 mM unlabeled methionine plus cycloheximide, 100 μg/ml, to inhibit further incorporation of labeled methionine, and the acetylation inhibition system is added to prevent acetylation of the processed actin. The processing is initiated by the addition of nine volumes of a solution containing the acetylmethionine cleaving enzyme. The source of this solution can be either a cell extract, enzyme fraction or fresh reticulocyte lysate. This operation effectively dilutes out endogenous acetylmethionine cleaving activity present in the original translation mixture. For the two-dimensional gel procedure, a shift of labeled actin to a more basic isoelectric point is measured as a function of time. For the peptide map procedure, relative amounts of radioactivity present in the NH₂-terminal tryptic peptides of the different actin species are measured. In the latter procedure, since the peptides will have either 1 or 2 methionines, normalization based on methionine content is required. Since the two-dimensional gel procedure depends on the total label in actin which contains 16 methionines, the addition of the initiator methionine in the precursor does not make enough of a difference in the quantification to require this correction.

Materials

Translation mixture containing [^{35}S]methionine-labeled acetylated actin (see previous section).

Cycloheximide, 1 mg/ml, and unlabeled methionine, 25 mM, in water.

Citrate synthase, oxaloacetate, and acetonyl CoA (see previous section).

Source of processing enzyme.

Cellulose thin-layer plates.

Electrophoresis tank with pH 6.3 buffer—pyridine : HOAc : H$_2$O (25 : 1 : 225).

Formic acid, 97%, and performic acid (9.5 ml of 97% formic acid + 0.5 ml 30% hydrogen peroxide—2 hr at room temperature).

TPCK-treated trypsin (1 mg/ml) in 0.1 N NH$_4$HCO$_3$.

Thermolysin (0.3 mg/ml) in 0.1 N NaHCO$_3$.

CaCl$_2$, 6 mM.

DNase I agarose (Worthington).

Procedure

Method A—Two-Dimensional Gel Assay. Actin, as described previously, is synthesized for 45–60 min in a mRNA-dependent rabbit reticulocyte lysate in the presence of [^{35}S]methionine under conditions where endogenous acetylation is permitted. At this point, 85–95% of the actin possesses an acetyl methionine amino-terminus.

Translation is stopped by the addition of cycloheximide, 100 μg/ml, and unlabeled methionine, 1 mM. Citrate synthase and oxaloacetate followed by S-acetonyl-CoA are added as previously described to inhibit acetylation of processed material. Aliquots consisting of 5–20 μl of this translation mixture are diluted 10-fold with a pH 7.5 buffered protein solution containing putative actin processing activity. This solution also should contain cycloheximide, unlabeled methionine, and the acetylation inhibition system. Phosphate, imidazole, and Tris [tris(hydroxymethylaminomethane] buffers work equally well in this system. Following incubation of the reaction mixture at 25° for the desired time, an aliquot is removed and analyzed by two-dimensional gel electrophoresis,[5] and fluorography. The reaction is monitored by following the basic shift in the

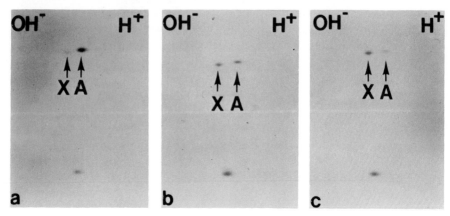

FIG. 4. Removal of NH$_2$-terminal acetylmethionine from *D. discoideum* actin synthesized in a reticulocyte lysate (Method A). Following synthesis of the actin (a), an aliquot of fresh lysate was added as described in the text, and samples were withdrawn 60 min (b) and 135 min (c) following addition of the lysate. Autoradiograms of the actin regions of the two dimensional gels of these samples are shown here. A denotes the position of unprocessed actin. X denotes the position of processed action. OH$^-$ and H$^+$ show the acidic and basic ends of the pH gradient in the first dimension.

actin polypeptide due to processing, and the percentage conversion at each time point is quantitated using scanning densitometry. The actin regions of fluorograms from a typical experiment are shown in Fig. 4. A time course of actin processing using reticulocyte lysate as a source of processing enzyme is shown in Fig. 5.

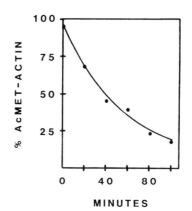

MINUTES

FIG. 5. Time course of acetylmethionine removal. An experiment like that described in Fig. 4 was performed. The processed and unprocessed actin regions at each time point were quantitated by densitomeric measurements of the autoradiograms using a Transidyne General RFT II scanning densitometer at 550 nm. The sum of the two species at each time point was taken as 100%.

Method B—Actin NH₂-Terminal Peptide Mapping. Processing of actin is initiated as described in Method A. At the desired time, an aliquot, 0.1–1 ml, is removed and passed over a 1.0 ml DNase I actin affinity column previously equilibrated with 10 mM Tris–HCl, pH 8.0 containing 2.0 mM CaCl₂, 0.2 mM ATP, and 10% (v/v) formamide.[21] The column is washed with 0.75 M guanidine · HCl in the same buffer and the actin is eluted with 3 M guanidine · HCl containing 0.5 M sodium acetate.[22] Following exhaustive dialysis, the actin is lyophilized to dryness and dissolved in 0.1 ml of 97% formic acid. To this mixture is added 0.1 ml of performic acid, and oxidation is allowed to proceed for 30 min at 4°. The reaction solution is diluted 10-fold with water and lyophilized to dryness.

The oxidized actin is taken up in 0.5 ml of 0.1 M NH₄HCO₃ and 15 μl of stock trypsin solution is added. Following trypsin digestion for 2 hr at room temperature, the NH₄HCO₃ is removed by lyophilization, the residue is dissolved in water, and the resulting solution again lyophilized to dryness.

The tryptic peptides are dissolved in 20 μl of the pH 6.3 electrophoresis buffer and subjected to thin-layer electrophoresis at pH 6.3 at 400 V using Orange G as a marker dye. An autoradiogram of the electrophoresis plate is then made. The labeled NH₂-terminal actin tryptic peptide with an acetylmethionine migrates just behind the Orange G. Removal of the acetylmethionine results in a significant retardation in the rate of migration of the NH₂-terminal tryptic peptide. A typical electrophoretogram is shown in Fig. 6.

The analysis described so far is sufficient if the structures of the NH₂-terminal tryptic peptides are known. If one wishes to determine whether a particular tryptic peptide is derived from the actin amino-terminus and still contains the initiator methionine, the following procedure is used. The tryptic peptide in question is eluted from the cellulose layer with water and lyophilized to dryness. The residue is taken up in 20 μl of 10 mM NaHCO₃ containing 2 mM CaCl₂ and digested with 1.5 μg of thermolysin for 2.5 hr at 30°. The digest is applied directly to a thin layer of cellulose and subjected to electrophoresis at pH 6.3. When applying the solution to the thin-layer plate, a drying fan should not be used since this often results in the appearance of doublets on the electrophoretogram. If the original tryptic peptide was derived from actin, a labeled C_t peptide (Fig. 3) will be observed at about 0.35 times the distance traveled by the Orange G. This peptide is diagnostic for nonmuscle actins in general.[20] If the initiator methionine still remains at the NH₂-terminus of the actin amino-terminal tryptic peptide, a labeled N_t peptide (Fig. 3) will also be

[21] K. Zechel, *Eur. J. Biochem.* **110**, 343 (1980).
[22] E. Lazarides and U. Lindberg, *Proc. Natl. Acad. Sci. U.S.A.* **71**, 4742 (1975).

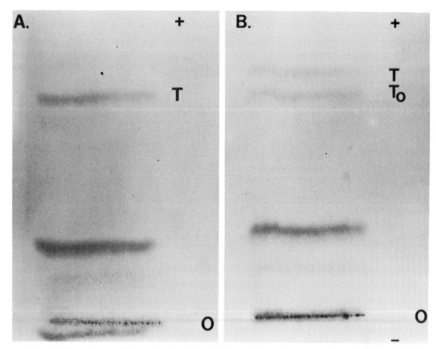

FIG. 6. Tryptic peptide analysis of the NH$_2$-terminus of processed actin uniformly labeled with [^{35}S]methionine. Acetylmethionyl actin labeled as indicated was processed with fresh lysate for 1.5 hr and subjected to tryptic digestion following performic acid oxidation as described in the text. The tryptic peptides were subjected to thin-layer electrophoresis at 6.3, and an autoradiogram was made of the gel. T denotes the position of the NH$_2$-terminal tryptic peptide of Ac-Met-Asp-actin, and T$_0$, that of actin from which Ac-Met has been removed. The samples were run on different thin-layer chromatography plates, which accounts for the mobility differences in the two panels. Other bands represent non-amino-terminal-labeled actin tryptic peptides. The double origin in A resulted from excessive application of buffer to the electrophoretogram. A, Unprocessed actin; B, processed actin. O denotes the origin.

observed at a distance approximately 1.3 times that traveled by Orange G. Scanning of the autoradiogram provides a means of quantitation for these peptides. An autoradiogram from such an experiment is shown in Fig. 7.

Additional Comments

The methods outlined above for assaying actin NH$_2$-terminal acetylation and removal of the actin NH$_2$-terminal acetylmethionine moiety are not rapid. If pure actin mRNA could be obtained in large quantities, these two-dimensional gel assays could be carried out using one-dimensional isoelectric focusing. Likewise, purification of the native form of the actin

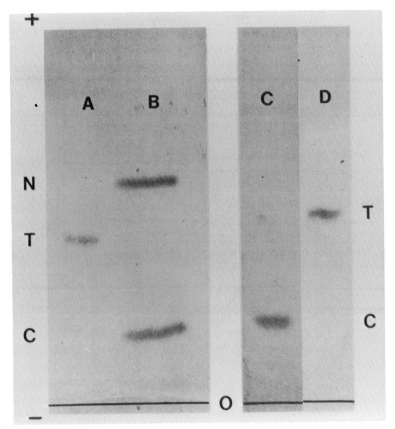

FIG. 7. Thermolysin digestion of [^{35}S]methionine-labeled NH$_2$-terminal tryptic peptides of acetylated and nonacetylated *D. discoideum* actin synthesized in a reticulocyte lysate. Experimental details are given in the text. Electrophoretograms were run at pH 6.3. Left: A, NH$_2$-terminal tryptic peptide of actin synthesized *in vitro* and labeled uniformly with [^{35}S]methionine; B, thermolysin digest of the *in vitro* NH$_2$-terminal tryptic peptide; C, thermolysin digest of the NH$_2$-terminal tryptic peptide of actin synthesized *in vivo* and uniformly labeled with [^{35}S]methionine; D, NH$_2$-terminal tryptic peptide of *D. discoideum* actin synthesized *in vivo* and labeled with [^{35}S]methionine; T, position of the tryptic peptide; C$_t$, carboxyl-terminal thermolysin peptide; N$_t$, amino-terminal thermolysin peptide. An autoradiogram is shown.

made *in vitro* for use as a substrate for subsequent processing experiments[23] might make the assays simpler. However, such a purification procedure would probably involve significant losses of labeled actin, thus making this approach counterproductive. A third possibility for an im-

[23] D. G. Uyemura, S. S. Brown, and J. A. Spudich, *J. Biol. Chem.* **253**, 9088 (1978).

proved assay involves the use of small chemically synthesized peptides as artificial substrates coupled with a rapid chromatographic or electrophoretic analysis. If, however, an acetylmethionine processing activity is detected with such a substrate, the enzyme catalyzing the reaction also must be shown to be responsible for actin processing.

The processing sequence outlined above applies only to those actins containing an initiator methionine directly preceding the aspartic or glutamic acid residue that will be the final NH_2-terminus. Examples of these types of actin are *D. discoideum*,[7] yeast,[24] and β-nonmuscle actins in avian and mammalian cells.[25] There is, however, a second class of actins in which *two* amino acids must be removed to produce the final NH_2-terminal aspartate or glutamate. The genes for these actins code for the amino terminal sequence Met-X-Asp(Glu), where X is usually Cys but can be also Gly or Ala. Proven examples of this class of actins are cardiac actin,[26] skeletal muscle actin,[27] *Acanthamoeba castellanii* actin,[28] sea urchin actin,[27] *Drosophila melanogaster* actin,[29] and soybean actin.[30] The mode of processing of these actins is under active investigation in our lab. Preliminary experiments suggest that for these actins acetylmethionine cleavage is not involved.

Finally, while the acetylmethionine cleavage enzyme may be involved with processing of actin and possibly a small number of other proteins,[9] the methods outlined for inhibition of acetylation and for studying NH_2-terminal protein acetylation should be applicable to many systems.

Acknowledgments

This work was supported by grants to P.A.R. from the National Institutes of Health (GM 24702) and the American Heart Association and its Iowa Affiliate (81-868). P.A.R. is an Established Investigator of the American Heart Assoc. We thank Dr. Arthur Strauch for his suggestions during preparation of this manuscript.

[24] R. Ng and J. Abelson, *Proc. Natl. Acad. Sci. U.S.A.* **77,** 3912 (1980).

[25] R. Zakut, M. Shani, D. Givol, S. Neuman, D. Yaffe, and U. Nudel, *Nature (London)* **298,** 857 (1982).

[26] H. Hamada, M. Petrino, and T. Kakunaga, *Proc. Natl. Acad. Sci. U.S.A.* **79,** 5901 (1982).

[27] A. D. Cooper and W. R. Crain, Jr., *Nucleic Acids Res.* **10,** 4081 (1982).

[28] W. Nellen and D. Gallwitz, *J. Mol. Biol.* **159,** 1 (1983).

[29] E. A. Fyrberg, B. J. Bond, N. D. Hershey, K. A. Mixter, and N. Davidson, *Cell* **24,** 107 (1981).

[30] D. M. Shah, R. Hightower, and R. Meagher, *Proc. Natl. Acad. Sci. U.S.A.* **79,** 1022 (1982).

[17] Amino-Terminal Protein Transacetylase from Wheat Germ

By Hiroshi Kido, Alberto Vita, and B. L. Horecker

Many proteins, including both enzymes and structural proteins, are acetylated at the amino-terminus[1] and transacetylase (acetyltransferase) is widely distributed in animal tissues.[1-6] It has also been found in wheat germ,[7,8] and the formation of acetylated products *in vitro* by the wheat germ translation system has been demonstrated.[7,9]

Alternative mechanisms may account for the biosynthesis of NH_2-terminal acetylated peptide hormones from larger precursor polypeptides. One is the processing of completed peptide chains by specific proteases, followed by acetylation of the newly formed NH_2-terminal amino acid residues. The other is acetylation of nascent chains on ribosomes followed by proteolytic processing as the chains reach an appropriate length. The second mechanism may operate in the case of larger polypeptide chains. The wheat germ transacetylase, which catalyzes NH_2-terminal acetylation of peptides, may provide important information on the mechanism of biosynthesis and acetylation of peptide hormones. It is also useful for the acetylation of peptide hormones that have been synthesized in the deacetyl form by recombinant DNA techniques[10,11] which may require acetylation of the NH_2-termini. NH_2-terminal acetylation has been shown to have profound effects on the biological activities of peptide hormones such as α-MSH,[12,13] β-endorphin,[14] and leucine-enkephalin.[15] The acetyla-

---

[1] H. Bloemendal, *Science* **197**, 127 (1977).

[2] J. Traugh and S. B. Sharp, *J. Biol. Chem.* **252**, 3738 (1977).

[3] R. D. Palmiter, J. Gagnon, and K. A. Walsh, *Proc. Natl. Acad. Sci. U.S.A.* **75**, 94 (1978).

[4] A. Pestana and H. C. Pitot, *Biochemistry* **14**, 1397 (1975).

[5] S. Tsunasawa, K. Kamitani, and K. Narita, *J. Biochem. (Tokyo)* **87**, 645 (1980).

[6] C. C. Glembotski, *J. Biol. Chem.* **257**, 10493 (1982).

[7] D. S. Shin and P. Kaesberg, *Proc. Natl. Acad. Sci. U.S.A.* **70**, 1799 (1973).

[8] H. Kido, A. Vita, E. Hannappel, and B. L. Horecker, *Arch. Biochem. Biophys.* **208**, 95 (1981).

[9] J. I. Garrels and T. Hunter, *Biochim. Biophys. Acta* **564**, 517 (1979).

[10] J. F. Morrow, this series, Vol. 68, p. 3.

[11] R. Wetzel, H. L. Heyneker, D. V. Goeddel, P. Thurani, J. Shapiro, R. Crea, T. L. K. Low, J. E. McClure, and A. L. Goldstein, *Biochemistry* **19**, 6096 (1980).

[12] D. Rudman, R. K. Chawla, and B. M. Hollins, *J. Biol. Chem.* **254**, 10102 (1979).

[13] The following abbreviations are used in this article. DTT, dithiothreitol; HPLC, high-performance liquid chromatography; MSH, melanocyte-stimulating hormone.

tion of peptide hormones by heterologous or homologous enzyme systems *in vitro* has been reported from a number of laboratories,[3,5,16,17] but the yields reported for the conversion of the deacetyl precursors were less than 1%. With the conditions described here, the wheat germ ribosomal preparation efficiently converts at least 15% of added deacetyl thymosin α_1[18] to thymosin α_1.[8,19]

This article describes a method for assay of wheat germ transacetylase using as substrate deacetyl thymosin α_1 and some properties of the enzyme.

Assay Methods[8]

Transacetylase was assayed by monitoring the incorporation of label from radiolabeled acetyl-CoA or acetate into various peptides and histones. In this work, deacetyl thymosin α_1 was used as substrate.

Reagents

Buffer I: 50 mM Tris–HCl, pH 7.5, containing 3 mM $MgCl_2$ and 1 mM DTT.

Buffer II: 1 M formic acid and 0.2 M pyridine, pH 2.8. These reagents were distilled twice over ninhydrin before use.

Wheat germ extract: An extract of wheat germ (Niblack Raw, Niblack Foods, Rochester, NY) containing 13 mg of protein/ml was prepared freshly before use.

Substrate and acetyl-CoA: Solutions of 30 μl of [^{3}H]acetyl-CoA (2.1 nmol, 1.4 Ci/mmol) (New England Nuclear) and 1.2 μg (400 pmol, 100 μg/ml) of deacetyl thymosin α_1 (a gift from Dr. R. B. Merrifield of the Rockefeller University) were lyophilized in micropolypropylene test tubes (Walter Sarstedt, Inc.).

Thymosin α_1 (Synthetic, kindly provided by Dr. J. Meienhofer of Hoffmann-La Roche, Inc.).

0.2 M $NaHCO_3$, pH 8.5

10 N NaOH and 12 N HCl

[14] D. G. Smith, D. E. Massey, S. Zakarian, and M. Finnie *Nature (London)* **299**, 352 (1979).

[15] B. R. Seizinger, V. Höllt, and A. Herz, *Biochem. Biophys. Res. Commun.* **101**, 289 (1981).

[16] M. Granger, G. I. Tesser, W. W. De Jong, and H. Bloemendal, *Proc. Natl. Acad. Sci. U.S.A.* **73**, 3010 (1976).

[17] T. A. Woodford and J. E. Dixon, *J. Biol. Chem.* **254**, 4993 (1979).

[18] T. W. Wong and R. D. Merrifield, *Biochemistry* **19**, 3233 (1980).

[19] A. L. Goldstein, T. L. K. Low, M. McAdoo, J. McClure, G. B. Thurman, J. Rossio, C. Y. Lai, D. Chang, S. S. Wang, C. Harvey, A. H. Ramel, and J. Meienhofer, *Proc. Natl. Acad. Sci. U.S.A.* **74**, 725 (1977).

1-Propanol

70% Formic acid

Trypsin solution: A stock solution (1 mg/ml) was prepared by dissolving TPCK-treated trypsin (Millipore Corporation) in 1 mM HCl,
and was stored in small aliquots at $-20°$.

Preparation of Wheat Germ Extracts

The 30,000 g supernatant (S30) of wheat germ was prepared by modifications of the procedure of Roberts and Paterson.[20] Wheat germ (6 g) was
ground for 1 min in a chilled mortar with an equal weight of sand and 20 ml
of buffer I. The homogenate was centrifuged at 30,000 g for 20 min at 0–
2°, and the supernatant (S30) was removed, without the surface layer of
fat. The S30 fraction was passed through a column (60 × 2 cm) of Sephadex G-25 (coarse), equilibrated with buffer I, at a flow rate of 1.4 ml/min.
The turbid fractions (6 ml) were pooled and centrifuged at 180,000 g for 2
hr at 2°. The precipitate was suspended in 0.75 ml of buffer I, and is
referred to as the wheat germ ribosome preparation (approximately 13
mg/ml). It was freshly prepared each day, because the acetylation activity
was found to decrease during storage at 2 or $-20°$. The recovery of
acetylation activity in the ribosome fraction was only 30% of the total
activity in the S30 fraction, but this step increased the specific activity
3.5-fold.

Acetylation of Deacetyl Thymosin α_1

Solutions containing [³H]acetyl-CoA (2.1 nmol) and deacetyl thymosin
α_1 (400 pmol) were lyophilized in micropolypropylene test tubes. Reactions were started by addition of 10 μl of wheat germ ribosome preparation (13 μg of protein) in buffer I. After incubation at 35° for 20 min, 40 μl
of solution containing 80 μg (24 nmol) of thymosin α_1 was added as carrier. Then 10 N NaOH solution was added to give a final concentration of
1 N NaOH and the mixture was incubated at 35° for 15 min to destroy
excess acetyl-CoA. This solution was then neutralized with concentrated
HCl. After neutralization, the solution was treated with 200 μl of buffer
II, centrifuged at 18,000 g for 30 min, and analyzed by high-performance
liquid chromatography (HPLC) to detect the radioactive peak that coeluted with thymosin α_1 carrier.

High-Performance Reversed-Phase Liquid Chromatography

This was carried out on a reversed-phase column (Ultrasphere-Octyl,
5 μm, or Ultrasphere-ODS, 5 μm, 4.5 × 250 mm, Altex Scientific, Inc.) as

[20] B. E. Roberts and B. M. Paterson, *Proc. Natl. Acad. Sci. U.S.A.* **70**, 2330 (1973).

described by Rubinstein.[21] Material was eluted in 2 hr with a linear gradient of 0–40% (v/v) 1-propanol in buffer II at a flow rate of 0.33 ml/min and fractions were collected every 2.5 min. At 20-sec intervals, 5-μl samples were diverted to a fluorescence-detection system.[22] Aliquots (50 μl) of each fraction were taken for measurement of the radioactive peak that coeluted with carrier thymosin α_1. This peak was not observed with control reaction mixture from which deacetyl thymosin α_1 was omitted or with a mixture in which thymosin α_1 was used in place of deacetyl thymosin α_1 as substrate.[8] In the latter case, radioactivity was recovered only in peaks in positions corresponding to acetic acid and acetyl-CoA and there was thus no evidence of acetylation of lysyl ε-NH$_2$ groups with the experimental conditions employed.

Characterization of the Labeled Product Formed from Deacetyl Thymosin α_1

Confirmation that the labeled product was thymosin α_1 was provided by analysis of the peptides formed on digestion with TPCK-trypsin and analysis of the tryptic digest by HPLC. Radioactivity was recovered only in the fractions containing the NH$_2$-terminal tryptic tetradecapeptide.[8]

Extent of Conversion

From the results of amino acid analysis of the tryptic peptide and the radioactivity in this fraction, it was calculated that 59.6 pmol of radioactive thymosin α_1 was formed from 400 pmol of added deacetyl thymosin α_1, representing a conversion of 15%.[8]

Properties of Wheat Germ Transacetylase

The recovery of activity in the 180,000 g pellet suggests that the enzyme is associated with ribosomes of the rough endoplasmic reticulum of wheat germ. Transacetylases from rabbit reticulocytes,[2] rat liver,[4] and hen oviduct[5] were also associated with ribosomes of the rough endoplasmic reticulum. High concentrations of KCl-enhanced NH$_2$-terminal acetylation (Fig. 1A), but the addition of MgCl$_2$ (1–3 mM) had no effect (data not shown). The reaction was also dependent on the concentration of substrate (Fig. 1B) and the quantity of enzyme added.[8] It was abolished by heating the enzyme preparation at 100° for 10 min. The optimum pH was 6.5 to 7.5; above this pH the extent of NH$_2$-terminal acetylation was greatly decreased, and below pH 6.5, a mixture of radioactive products was formed from added deacetyl thymosin α_1 that was eluted later from

[21] M. Rubinstein, *Anal. Biochem.* **98**, 1 (1979).
[22] P. Böhlen, S. Stein, J. Stone, and S. Udenfriend, *Anal. Biochem.* **67**, 438 (1975).

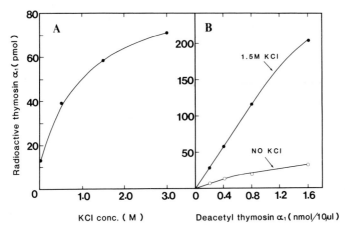

FIG. 1. Effects of KCl and concentration of deacetyl thymosin α_1 on the formation of thymosin. Reactions were carried out with 400 pmol of deacetyl thymosin α_1, 13 μg of wheat germ polysomes suspended in 10 μl of buffer I, and the concentrations of KCl indicated in A. In the experiment shown in B, the amounts of deacetyl thymosin α_1 and KCl were varied as shown. After 20 min at 35°, the reaction mixtures were processed and analyzed by HPLC. Values were corrected for the recovery of carrier thymosin α_1.

the Ultrasphere-octyl column employed for reversed-phase HPLC; these products were not identified but they may reflect acetylation of ε-amino groups of lysine. The addition of KCl reduced this nonspecific acetylation.

Comments

Some acetylation of endogenous wheat germ proteins was also observed but this was small compared with that of deacetyl thymosin α_1.

Losses of carrier thymosin α_1 and the NH_2-terminal tryptic peptides occurred during preparation of the samples for HPLC. The recoveries can be greatly improved by desalting the samples using Sep-Pak cartridges as described by Hannappel *et al.*[23] This procedure has been shown to result in overall recoveries of 80% when employed for the isolation of peptides from crude tissue extracts.

Acknowledgments

The authors are indebted to Dr. J. Meienhofer of the Roche Research Center, Hoffmann-La Roche, Inc., and to Dr. R. B. Merrifield of the Rockefeller University for generous supplies of synthetic thymosin α_1 and deacetyl thymosin α_1. They also thank Dr. N. Katunuma for helpful discussions.

[23] E. Hannappel, G.-J. Xu, J. Morgan, J. Hempstead, and B. L. Horecker, *Proc. Natl. Acad. Sci. U.S.A.* **79**, 2172 (1982).

[18] Aminoacyl-tRNA : Protein Transferases

By CHARLES E. DEUTCH

Aminoacyl-tRNA : protein transferases are soluble enzymes that catalyze the transfer of specific amino acids from tRNAs to the NH_2-terminus of suitable protein or peptide substrates. The amino acids are incorporated through a normal peptide bond at the α-amino group. Two transfer activities have been described. The enzyme L-leucyl-tRNA : protein transferase (EC 2.3.2.6, leucyltransferase) catalyzes the addition of leucine, phenylalanine, or methionine to substrates possessing arginine or lysine as the initial NH_2-terminal residue. The enzyme L-arginyl-tRNA : protein transferase (EC 2.3.2.8, arginyltransferase) is specific for arginine and requires aspartate or glutamate at the acceptor site. The two enzymes are similar in their insensitivity to antibiotics that inhibit protein synthesis at the level of the ribosome. The leucine transfer activity has been demonstrated in a number of gram-negative bacteria. The arginine transfer activity has been found in all eukaryotes tested, including yeast, higher plants, and mammalian tissues.

Assay Method[1,2]

Principle. Transfer activities are measured *in vitro* by the conversion of a radioactive amino acid to a hot acid-insoluble form. The radioactivity can be added to the reactions either as a ^{14}C-labeled aminoacyl-tRNA or as a ^{14}C-labeled amino acid together with ATP, tRNA, and a partially purified aminoacyl-tRNA synthetase. The former method is preferable for the demonstration of transfer activity and for studies on substrate specificity. The latter method is convenient for stoichiometric aminoacylation of particular proteins. In addition, the reactions require a sulfhydryl-reducing agent, relatively high concentrations of a monovalent cation, and a protein having a suitable NH_2-terminus. For assays involving crude extracts, an antibiotic such as chloramphenicol, puromycin, or cycloheximide is added to inhibit ribosome-dependent protein synthesis.

[1] M. J. Leibowitz and R. L. Soffer, *Biochem. Biophys. Res. Commun.* **36,** 47 (1969).
[2] R. L. Soffer and H. Horinishi, *J. Mol. Biol.* **43,** 163 (1969).

Reagents

2.0 *M* Tris–HCl at pH 8.2 for L-leucyl-tRNA: protein transferase or at pH 9.0 for L-arginyl-tRNA: protein transferase

2.0 *M* KCl

2.0 *M* 2-mercaptoethanol

25 mg/ml α-casein for L-leucyl-tRNA: protein transferase or 25 mg/ml bovine serum albumin for L-arginyl-tRNA: protein transferase

[^{14}C]Phenylalanyl-tRNA, [^{14}C]leucyl-tRNA, or [^{14}C]arginyl-tRNA, approximately 20 μM in amino acid (250 μCi/μmol)

Procedure. The reagents are combined to give a concentrated reaction mixture, 30 μl of which is then added to each sample to be assayed to give a total volume of 50 μl. The final concentrations of reagents in the reactions are 50 m*M* Tris–HCl, 150 m*M* KCl, 50 m*M* 2-mercaptoethanol, 1 mg/ml α-casein or 5 mg/ml bovine serum albumin, and 2 μM aminoacyl-tRNA. The reactions are incubated at 37° for 10–30 min. A portion of each reaction is then spotted on a 2.5-cm Whatmann 3MM filter paper disk, which is dropped into cold 10% trichloroacetic acid (TCA) to terminate the assay. When all the disks are collected, the TCA is decanted and replaced with hot 5% TCA. After heating for 10 min at 90° to hydrolyze the radioactive aminoacyl-tRNA, the disks are washed three times with room temperature 5% TCA, twice with a 1:1 mixture of ethanol and ether, and twice with ether. After drying at room temperature, the disks are counted in a toluene-based scintillation fluid in a liquid scintillation counter. One unit of activity is defined as the amount of enzyme catalyzing the transfer of 1 nmol of amino acid per minute at 37°.

Comments. ^{14}C-labeled aminoacyl-tRNAs are prepared from deacylated tRNA using a partially purified aminoacyl-tRNA synthetase (see Vol. 5 [96] and Vol. 59 [19] in this series). Since the aminoacyl-tRNAs are unstable at the alkaline pHs that are optimal for transfer activity, the incorporation of amino acids is linear for only short times. For assays using a coupled aminoacyl-tRNA generating system in place of the ^{14}C-labeled aminoacyl-tRNA, the following are added to each reaction: 10 m*M* magnesium acetate, 3 m*M* ATP, 5 mg/ml tRNA, 20 μM ^{14}C-labeled amino acid (100 μCi/μmol), and approximately 0.01 units/ml aminoacyl-tRNA synthetase. For crude extracts 150 μg/ml chloramphenicol (leucine transfer) or 0.2 m*M* puromycin (arginine transfer) is added. A number of proteins other than α-casein or bovine serum albumin can be used; α-lactalbumin has been reported to give higher rates of arginine transfer activity.[3]

---

[3] H. Horinishi, M. Kato, and T. Takahashi, *Anal. Biochem.* **75,** 22 (1976).

TABLE I
PURIFICATION OF L-LEUCYL-tRNA : PROTEIN TRANSFERASE FROM *Escherichia coli* B[a]

Fraction	Vol (ml)	Protein (mg)	Specific activity (units/mg)	
			Phe transfer	Leu transfer
105,000 g extract	380	11,200	0.046	0.047
First PC eluate	730	330	0.95	0.78
Ammonium sulfate	2.0	53.6	4.6	4.4
Sephadex G-100	24	11.5	7.9	7.6
Second PC eluate	1.0	0.29	61	63

[a] R. L. Soffer, *J. Biol. Chem.* **248,** 8424 (1973).

Purification of Enzymes

L-Leucyl-tRNA : protein transferase and L-arginyl-tRNA : protein transferase have been extensively purified from *Escherichia coli* B and rabbit liver, respectively. A characteristic feature of both purifications is that the elution behavior of the transfer activity from a column of a cation exchange resin changes during the course of the enzyme isolation.

L-*Leucyl-tRNA : Protein Transferase.*[4] The steps involved in the purification are summarized in Table I. All operations are performed at 0–4°. *E. coli* B is grown to mid-exponential phase in a rich broth, washed once with 0.85% NaCl, and frozen. Thawed cells are ground with alumina (2 g/g wet weight) in the presence of 1 mg of pancreatic DNase and the mixture suspended in a solution containing 10 mM Tris–HCl, pH 7.8, 10 mM magnesium acetate, 30 mM KCl, and 5 mM 2-mercaptoethanol (1.5 ml/g cells). After two centrifugations at 30,000 g for 20 min to remove the alumina and unbroken cells, the supernatant fraction is centrifuged at 105,000 g for 90 min. The upper four-fifths of this solution is removed and dialyzed overnight against 20 mM Tris–HCl, pH 7.8, containing 5 mM 2-mercaptoethanol.

The dialyzed supernatant fraction is then diluted 6-fold with 200 mM Tris–HCl, pH 7.8, containing 50 mM 2-mercaptoethanol, and applied to a column (5 × 60 cm) of phosphocellulose (Whatmann P11) equilibrated with the same buffer. The column is washed successively with the starting buffer, starting buffer containing 100 mM KCl, and starting buffer containing 200 mM KCl. The transfer activity is then eluted with a similar buffer containing 500 mM KCl. Fractions containing the enzyme are pooled and concentrated by dialysis against the same Tris-mercaptoethanol buffer saturated with ammonium sulfate. The precipitated material

[4] R. L. Soffer, *J. Biol. Chem.* **248,** 8424 (1973).

TABLE II

PURIFICATION OF L-ARGINYL-tRNA : PROTEIN TRANSFERASE
FROM RABBIT LIVER[a]

Fraction	Vol (ml)	Protein (mg)	Total units	Specific activity
105,000 g supernatant	450	12,000	192	0.016
Ammonium sulfate	150	4,060	171	0.042
pH 5.2 precipitate	100	610	127	0.208
First CM-cellulose eluate	160	100	120	0.200
DEAE-cellulose eluate	72	132	53	0.401
Second CM-cellulose eluate	21	0.175	20	114

[a] R. L. Soffer, *J. Biol. Chem.* **245,** 731 (1970).

in the dialysis sac is then collected by centrifugation and extracted successively with 50% ammonium sulfate, 45% ammonium sulfate, and 40% ammonium sulfate. The enzyme is then removed from the residual pellet by washing with 10% ammonium sulfate in 30 mM Tris–HCl, pH 7.4, with 50 mM 2-mercaptoethanol.

The 40–10% ammonium sulfate fraction is then applied to a column (1.5 × 150 cm) of Sephadex G-100 equilibrated with 20 mM Tris–HCl, pH 7.4, 50 mM 2-mercaptoethanol, and 120 mM ammonium sulfate. The enzyme is eluted with the same buffer. Fractions with the highest activity are pooled, diluted with an equal volume of 20 mM Tris–HCl, pH 7.4, containing 50 mM 2-mercaptoethanol, and applied to a second column (1 × 13 cm) of phosphocellulose equilibrated with 20 mM Tris–HCl, pH 7.4, 50 mM 2-mercaptoethanol, and 60 mM ammonium sulfate. The column is washed with the same buffer and the enzyme eluted with a similar buffer containing 160 mM ammonium sulfate. The enzyme is immediately concentrated by ultrafiltration in a Diaflo apparatus with a UM2 filter and dialyzed overnight against a buffer containing 20 mM Tris–HCl, pH 7.4, 50 mM 2-mercaptoethanol, 120 mM ammonium sulfate, and 50% glycerol. The purified enzyme can be stored at −20° for several months.

L-*Arginyl-tRNA : Protein Transferase.*[5] The steps involved in the purification are summarized in Table II. All operations are performed at 0–4°. Freshly excised rabbit livers are suspended in a solution containing 50 mM Tris–HCl, pH 7.8, 5 mM magnesium acetate, 30 mM KCl, 0.1 mM EDTA, 5 mM 2-mercaptoethanol, and 0.25 M sucrose (2 mg/g wet weight). After homogenization for three 1-min periods in a Waring Blender, the extract is passed through a double layer of cheesecloth, and subjected to two 30-min centrifugations at 20,000 g. The supernatant frac-

[5] R. L. Soffer, *J. Biol. Chem.* **245,** 731 (1970).

tion is then centrifuged for 120 min at 105,000 g. The upper four-fifths of the solution is removed and dialyzed overnight against 20 mM Tris–HCl, pH 7.8, containing 5 mM 2-mercaptoethanol.

A saturated solution of ammonium sulfate in 40 mM Tris–HCl, pH 7.8, with 10 mM 2-mercaptoethanol, is then added to the supernatant fraction to give 35% saturation. After stirring for 10 min, the suspension is centrifuged for 30 min at 20,000 g and the precipitate discarded. An additional amount of the saturated ammonium sulfate solution is then added to the supernatant fraction to give 67% saturation. After centrifugation the precipitate is dissolved in 20 mM Tris–HCl, pH 7.8, containing 5 mM 2-mercaptoethanol and dialyzed overnight against the same buffer.

The dialyzed material is then adjusted to pH 5.2 by the dropwise addition of 1 M acetic acid. After stirring for 15 min, the precipitate is collected by centrifugation at 20,000 g for 30 min and suspended in 20 mM Tris–HCl, pH 9.0, containing 100 mM 2-mercaptoethanol. After homogenization and dialysis for 2 hr against the same buffer, the material is dialyzed overnight against 1 mM potassium citrate, pH 6.5, containing 100 mM 2-mercaptoethanol. The solution is then applied to a column (2 × 26 cm) of carboxymethylcellulose equilibrated with the same buffer. Elution is continued with the same buffer. The protein-containing fractions are combined and applied to a column (3 × 28 cm) of DEAE-cellulose equilibrated with 1 mM potassium citrate, pH 6.5, containing 100 mM 2-mercaptoethanol. After washing the column with this same buffer, the enzyme is eluted with a similar buffer containing 200 mM KCl.

The material from the DEAE-column is then diluted twofold with 1 mM potassium citrate, pH 6.5, 100 mM 2-mercaptoethanol, and applied to a second carboxymethylcellulose column (2 × 26 cm) equilibrated with 1 mM potassium citrate, pH 6.5, 100 mM 2-mercaptoethanol, and 80 mM KCl. After washing with a similar buffer containing 100 mM KCl, the enzyme is eluted with a solution of 1 mM potassium citrate, pH 6.5, 100 mM 2-mercaptoethanol, and 200 mM KCl, Fractions containing transfer activity are pooled and concentrated by dialysis against 20 mM Tris–HCl, pH 9.0, containing 100 mM 2-mercaptoethanol and 30% polyethylene glycol. After a second dialysis against a similar buffer containing 20% glycerol, the enzyme is stored at 0°. The arginine transferase is considerably less stable than the leucine transfer activity, and as much as 50% of the activity is lost in 1 week.

Identification of Substrates

The identification of physiologically significant substrates for the aminoacyl-tRNA : protein transferases poses several problems. Since the residue that is added to a protein in the transfer reactions is an unmodified amino acid, the occurrence of that amino acid (leucine, phenylalanine, methionine, or arginine) at the NH_2-terminus of an isolated protein is not in itself proof that the protein is a substrate. Indeed, it is likely that in most cases, these residues arise as a result of ribosome-dependent protein synthesis and posttranslational cleavage. While methionine is used as the initiating residue in both prokaryotes and eukaryotes, it is commonly removed enzymatically to expose other amino acids. In principle, one could recognize a modified protein containing an added amino acid if one could compare the NH_2-terminal amino acid sequence with the corresponding DNA sequence. However, the sequence data that are available are not yet sufficient to make this feasible. Alternatively, one could identify proteins containing an added amino acid by noting the presence of specific types of heterogeneity in the NH_2-terminal residue (e.g., Leu and Lys, or Arg and Glu). Thus far, this has been demonstrated only in the case of bovine serum albumin.[6] As a result, most studies related to substrate identification have involved either *in vitro* aminoacylation or the characterization of mutants.

In Vitro Aminoacylation. A number of proteins have been identified that can serve as acceptors of ^{14}C-labeled amino acids in one of the transfer reactions. While acceptance *in vitro* does not necessarily imply that the protein is a physiologically significant substrate, it does provide an indication of substrate potential. Two types of *in vitro* aminoacylation experiments can be performed. The first involves addition of a specific purified protein to a reaction mixture containing ^{14}C-labeled aminoacyl-tRNA or a ^{14}C-labeled aminoacyl-tRNA generating system and a partially purified transfer activity. For L-leucyl-tRNA : protein transferase, reactions include in 200 μl: 50 mM Tris–HCl, pH 7.8, 10 mM 2-mercaptoethanol, 150 mM KCl, 10 mM magnesium acetate, 6 mM ATP, 20 μM [14C]phenylalanine (100 μCi/μmol), 10 mg/ml *E. coli* B tRNA, 0.01 units/ml phenylalanyl-tRNA synthetase, 1 unit/ml L-leucyl-tRNA : protein transferase, and up to 3 mg/ml of the protein to be tested.[7] After incubation for 1–4 hr at 37°, reactions are stopped by the addition of 5 mg/ml pancreatic ribonuclease. Incorporation of radioactivity is determined as described above. Under these conditions, acylation is stoichiometric, i.e., proportional to the amount of added acceptor. For characterization of the modified protein, the size of the reaction can be increased 10-fold and the protein recovered from the mixture after aminoacylation. The procedure for testing proteins for acceptance in the arginine transfer reaction is similar, except that the pH of the reaction is 9.0.[8]

[6] M. J. Leibowitz and R. L. Soffer, *J. Biol. Chem.* **246,** 4431 (1971).
[7] M. J. Leibowitz and R. L. Soffer, *J. Biol. Chem.* **246,** 5207 (1971).
[8] R. L. Soffer, *J. Biol. Chem.* **246,** 1602 (1971).

The second type of *in vitro* aminoacylation involves addition of a specific group of proteins or a cellular component to a reaction mixture as the acceptor. After incubation under the reaction conditions described above, the proteins are precipitated with 10% TCA and solubilized in 2% sodium dodecyl sulfate and 5% 2-mercaptoethanol. The proteins can be separated by polyacrylamide gel electrophoresis as described elsewhere in this series (Vol. 22 [39] and Vol. 26 [1]) and the acceptors identified by autoradiography. Using this approach, acceptors of L-leucyl-tRNA : protein transferase have been identified in the soluble fraction and in ribosomes of *E. coli*.[9] Substrates for L-arginyl-tRNA : protein transferase have been found in erythrocyte membranes[10] and mouse chromatin.[11]

Three issues concerning this type of analysis should be mentioned. First, it should be noted that not all proteins with an appropriate acceptor site act as substrates for the transferases *in vitro*. If the NH_2-terminus of the acceptor is buried in the protein's three-dimensional conformation, acceptance may be demonstrable only after denaturation.[12] Second, if an intact cellular component is used as the acceptor, only those proteins with exposed NH_2-termini should be able to act as substrates; the pattern of acceptance therefore may be quite different if the organelle is first disrupted and the proteins extracted. Third, since a protein that is fully modified *in vivo* cannot act as an acceptor *in vitro,* the method is useful for identifying physiologically significant substrates only if there is partial aminoacylation *in vivo* or if the proteins are obtained from cells lacking transfer activity. In the case of *E. coli,* it has been shown that of the soluble and ribosomal proteins labeled *in vitro*, a small fraction is more heavily labeled if they are obtained from a transferase-deficient mutant.[9] This subset of proteins is more likely to represent the true substrates for L-leucyl-tRNA : protein transferases. In the absence of available mutants, an alternative approach is to compare the changing patterns of modification over time. This has been done for arginine acceptance during viral-induced cellular transformation.[13]

Characterization of Transferase-Deficient Mutants. In principle, the observation of specific phenotypic changes in a transferase-deficient strain should provide clues to the identification of proteins that are normally modified *in vivo.* Since one cannot predict in advance which phenotypic traits will be affected, mutants can only be identified by a direct

[9] R. L. Soffer and M. Savage, *Proc. Natl. Acad. Sci. U.S.A.* **71,** 1004 (1974).
[10] H. Kaji, *Biochemistry* **15,** 5121 (1976).
[11] H. Kaji and P. Rao, *FEBS Lett* **66,** 194 (1976).
[12] R. L. Soffer and J. D. Capra, *Nature (London), New Biol.* **233,** 44 (1971).
[13] P. Rao and H. Kaji, *Biochim. Biophys. Acta* **477,** 394 (1977).

assay for the transfer activity. The standard assay methods described above can be used with crude lysates of bacteria or yeasts provided that an antibiotic is included to prevent ribosome-dependent protein synthesis. A transferase-deficient strain of *E. coli* K12 was isolated after mutagenesis with *N*-methyl-*N'*-nitro-*N*-nitrosoguanidine.[9] This mutant has been shown to exhibit a pleiotropic phenotype that includes an abnormal cell morphology, altered specific activities of proline dehydrogenase, tryptophanase, phenylalanyl-tRNA synthetase, and β-galactosidase, and an increased accumulation of enterochelin.[14,15] However, none of the proteins listed has been found to be an acceptor in the leucine transfer reaction; a protein involved in regulating the synthesis of these proteins may therefore be the actual substrate. A mutant of *Saccharomyces cerevisiae* lacking L-arginyl-tRNA : protein transferase has also been isolated after mutagenesis with ethylmethane sulfonate.[16] However, the mutation has not yet been shown to have a clear phenotypic effect. The function of both L-leucyl-tRNA : protein transferase and L-arginyl-tRNA : protein transferase thus still has to be established.

[14] C. E. Deutch, R. C. Scarpulla, E. B. Sonnenblick, and R. L. Soffer, *J. Bacteriol.* **129**, 544 (1976).
[15] L. P. Freedman and C. E. Deutch, *Biochem. Biophys. Res. Commun.* **98**, 693 (1981).
[16] M. Savage, R. L. Soffer, and M. J. Leibowitz, *Curr. Genet.* **7**, 285 (1983).

[19] Identification of N-Terminal Myristyl Blocking Groups in Proteins

By ALASTAIR AITKEN and PHILIP COHEN

A limited number of reports of fatty acids linked covalently to proteins have appeared in the literature. The first was the attachment of fatty acids (mostly palmitate) to the α-amino group of the N-terminal cysteine residue in the murein lipoprotein of the *Escherichia coli* outer membrane.[1] This protein also contains fatty acids in a diglyceride which is linked to the thiol side group of this same N-terminal cysteine residue. The precursor of the diacylglycerol appears to be one of the major phospholipid species[2] and its fatty acid composition reflects that of the phospholipids

[1] K. Hantke and V. Braun, *Eur. J. Biochem.* **34**, 284 (1973).
[2] P. K. Chattopadhyay and H. C. Wu, *Proc. Natl. Acad. Sci. U.S.A.* **74**, 5318 (1977).

from *E. coli*.[1] The composition of the main constituents of the N-terminal amide linked fatty acids was $C_{16:0}$(palmitate), 65%; $C_{16:1}$, 10.9%; $C_{18:1}$, 10.8%; $C_{14:hydroxy}$, 4%, and $C_{14:0}$(myristate), 2.4%.

Recently two proteins have been shown to contain myristic acid covalently attached to the amino-terminus. These are the catalytic subunit of cyclic-AMP-dependent protein kinase (C subunit)[3] and the regulatory subunit of protein phosphatase 2B.[4] The latter is a Ca^{2+}-calmodulin-dependent protein phosphatase that is identical to a protein first isolated from bovine brain, termed calcineurin.[5] The regulatory subunit of protein phosphatase 2B (calcineurin B) binds 4 calcium ions per mole with affinities in the micromolar range[6] and determination of its complete amino acid sequence[6a] has shown extensive homology to calmodulin and other Ca^{2+}-binding proteins such as troponin C.[7] The discovery of myristic acid at the N-terminus of a protein phosphatase and a protein kinase suggests that this unusual blocking group may be involved in the interaction of "converter enzymes" with their protein substrates. Alternatively, the myristyl group could play a role in maintaining the subunit-subunit interactions between the regulatory (B) and catalytic (A) subunits of calcineurin and between the regulatory and catalytic subunits of cyclic-AMP-dependent protein kinase. It is unlikely that the role of myristic acid in either of these proteins is for membrane attachment, in contrast to the fatty acids linked to the *E. coli* murein lipoprotein.[1] The type II isoenzyme of cyclic-AMP-dependent protein kinase is known to bind to the inner surface of the plasma membrance of some cells.[8] However, addition of cyclic-AMP releases the C subunit (with the myristyl blocking group) while the regulatory subunit (containing an *N*-acetyl blocking group) remains membrane bound. It is also possible that the myristyl group is involved in the translocation of the C subunit from the cytoplasm to the nucleus, across the nuclear membrane.[9]

[3] S. A. Carr, K. Biemann, S. Shoji, D. C. Parmelee, and K. Titani, *Proc. Natl. Acad. Sci. U.S.A.* **79**, 6128 (1982).

[4] A. Aitken, P. Cohen, S. Santikarn, D. H. Williams, A. G. Calder, A. Smith, and C. B. Klee, *FEBS Lett.* **150**, 314 (1982).

[5] A. A. Stewart, T. S. Ingebritsen, A. Manalan, C. B. Klee, and P. Cohen, *FEBS Lett.* **137**, 80 (1982).

[6] C. B. Klee, T. H. Crouch, and M. A. Krinks, *Proc. Natl. Acad. Sci. U.S.A.* **76**, 6270 (1979).

[6a] A. Aitken, P. Cohen, and C. B. Klee, *Eur. J. Biochem.*, in press (1984).

[7] W. C. Barker, L. K. Ketcham, and M. O. Dayhoff, eds., "Atlas of Protein Sequence and Structure," Vol. 5, p. 273. Natl. Biomed. Res. Found., Washington, D.C., 1975.

[8] J. D. Corbin, P. H. Sugden, T. M. Lincoln, and S. L. Keely, *J. Biol. Chem.* **252**, 3854 (1977).

[9] H. G. Nimmo and P. Cohen, *Adv. Cyclic Nucleotide Res.* **8**, 145 (1977).

Isolation of Blocked N-Terminal Peptides Containing Myristic Acid

In the study of the N-terminus of calcineurin B, double glass distilled water passed through a Waters C_{18} Sep-pak was used. Acetonitrile (low UV HPLC grade S) and trifluoroacetic acid were from Rathburns, Walkerburn, Peebleshire, Scotland.

The very hydrophobic nature of the blocking groups in this protein was made quickly apparent by the technique of high-performance liquid chromatography (HPLC).[3,4] In these studies, reverse-phase HPLC was employed at ambient temperature, using μBondapak C_{18} columns (30 × 0.39 cm) (Waters) with linear gradients of water/acetonitrile containing 0.1% trifluoroacetic acid. Peptides were detected by absorbance at 210 nm. With a flow rate of 1 ml/min and increase in acetonitrile concentration of 1%/min, N-terminal peptides of calcineurin B, ranging in size from 3 to 21 residues, all eluted as sharp peaks at an apparent concentration of 57% acetonitrile. In the case of the C subunit, the peptides were redissolved in a small volume of 6 *M* guanidine–HCl, and eluted with water/acetonitrile containing 0.1 and 0.08% trifluoroacetic acid, respectively. The flow rate was 2 ml/min with an increase in acetonitrile concentration of about 4%/min. Myristyl peptides were recovered at apparent concentrations of 53% acetonitrile. These concentrations of acetonitrile were much higher than those required to elute much larger peptides that lack the myristyl blocking group.

Identification of Myristyl-Blocking Groups

Mass Spectrometry. Fast atom bombardment (FAB) mass spectrometry has proved the most useful mass spectrometric technique for identification of myristic acid in blocked peptides. With this new technique (reviewed in references 10 and 11) one can in general obtain structural information on peptides up to about 20 residues. Prior chemical derivatization is not necessary. Involatile and thermally labile compounds can be studied by this method and very small amounts of material can be used. Molecular weight determinations may be made with picomolar levels of peptides.[10]

FAB mass spectra of esterified and nonesterified myristyl peptides were obtained on a tripeptide from an *Staphylococcus aureus* proteinase digest and a decapeptide from CNBr cleavage. Peptides were esterified with 15 m*M* methanolic–HCl for 25 hr at ambient temperature.

The mass spectra were recorded on a Kratos MS50 mass spectrometer fitted with a high field magnet. A standard Kratos FAB source was employed to generate a 4 to 6-kV Xenon beam. Samples were dissolved in

[10] D. H. Williams, C. V. Bradley, S. Santikarn, and G. Bojesen, *Biochem. J.* **201,** 105 (1982).

1 μl of a 1 : 1 α-thioglycerol : diglycerol matrix, and the mixture was introduced into the source on a copper probe tip. With this matrix, the sensitivity may be improved by about an order of magnitude as compared to the use of a glycerol matrix, and has proved particularly suitable for nonpolar peptides (Santikarn and Williams, personal communication).

When the number of carboxylic acid groups was determined from the increase in M_r of the esterified peptide, the M_r of the blocking group could be calculated from the known amino acid composition. In the case of calcineurin B and the C subunit this was 211, corresponding to $CH_3(CH_2)_{12}CO-$.

In identifying the myristyl blocking group of the C subunit, Carr et al.[3] also used direct chemical ionisation with ammonia as the reagent gas.

Gas Chromatography. Gas chromatography remains the most widely applicable method for identifying fatty acids despite an increasing number of HPLC methods that have become available in recent years; for example, HPLC of 1-naphthylamine derivatives of fatty acids.[12]

Gas chromatography[4] and gas chromatography coupled to mass spectrometry[3] in fused silica capillary columns were used to confirm the presence of myristic acid. Peptides were hydrolyzed at 110° in sealed glass ampoules for 20[4] or 44 hr[3] in 6 N HCl. The hydrolyzates were extracted 3 times with ether[4] or with chloroform.[3] The ethereal layer was back-extracted with water to remove any HCl taken up by the ether, and the organic layer dried in a stream of nitrogen. The residue was methylated with diazomethane, and in the case of calcineurin B,[4] analyzed by chromatography on a Carbo–Erba 4160 gas chromatograph fitted with an on-column injection system and a 25 m × 0.25 mm glass open tubular column coated with Sil-5 (Chrompak U.K. Ltd).

Discussion

The amino terminal sequences of the 2 proteins known to contain myristyl blocking groups are identical in 3 of the first 4 residues (see the table). This suggests that the acylating enzyme(s) that links myristyl groups to these proteins may recognize the N-terminal sequence Gly-Asn-X-Ala-. The elution of all myristyl peptides at similar concentrations of acetonitrile from reverse phase HPLC columns may facilitate identification of other proteins with this blocking group. The finding that the fatty acid blocking group is almost exclusively myristic acid would indicate

[11] M. Barber, R. S. Bordoli, R. D. Sedgwick, and A. N. Tyler, Nature (London) 293, 270 (1981).

[12] M. Ikede, K. Shimada, and T. Sakaguchi, Chem. Pharm. Bull. 30, 2258 (1982).

AMINO-TERMINAL SEQUENCES OF TWO PROTEINS
KNOWN TO CONTAIN MYRISTYL BLOCKING GROUPS

Protein	Sequence
Calcineurin-B	$CH_3(CH_2)_{12}CO$-Gly-Asn-Glu-Ala-Ser-
C subunit	$CH_3(CH_2)_{12}CO$-Gly-Asn-Ala-Ala-Ala-

that the enzyme(s) acylating these proteins has a high specificity for this fatty acid.

Other proteins known to contain covalently linked fatty acids include viral membrane proteins[13,14] and membrane proteins in fibroblast cell cultures.[15] However, these fatty acids are ester linked, 70–80% being palmitic acid and the rest oleic acid and stearic acid. These fatty acids are readily cleaved from the proteins by incubation at 23° for 1–24 hr with 1.0 M hydroxylamine, or with 0.1 M KOH in 20% methanol. In contrast, amide-linked fatty acids are released by hydrolysis at 85° for 5 hr in 2 M methanolic–HCl.

Sefton *et al.*[16] have recently shown the presence of covalently bound fatty acid in the transforming proteins of Rous sarcoma and Abelson murine leukemia viruses. Both of these are protein kinases with specificity for tyrosine residues.[17,18] In the case of the transforming protein of Rous sarcoma virus, cultures were incubated with [^{3}H]palmitic acid, and the label was amide-linked to an N-terminal fragment (M_r 17,000) known to participate in membrane binding. However, the transforming factor of Harvey sarcoma virus, appeared to contain ester-linked fatty acids. The amount of [^{3}H]palmitate attached to each transforming protein was low, suggesting that the fatty acid incorporated may not be exclusively palmitate. Incubation of cell cultures with [^{3}H]myristic acid may help to identify further proteins with this N-terminal modification.

Note Added in Proof. After the submission of this chapter, a report appeared of a myristylated derivative in the membrane-associated protein P15 from both Rauscher and Moloney murine leukemia viruses.[19] In both proteins the myristylated amino terminal residue is Gly (as in the two

[13] M. F. G. Schmidt, *Virology* **116**, 327 (1982).
[14] U. Klockmann and W. Deppert, *FEBS Lett.* **151**, 257 (1983).
[15] M. J. Schlesinger, A. I. Magee, and M. F. G. Schmidt, *J. Biol. Chem.* **255**, 10021 (1980).
[16] B. M. Sefton, I. S. Trowbridge, and J. A. Cooper, *Cell* **31**, 465 (1982).
[17] M. S. Collet, A. F. Purchio, and R. L. Erikson, *Nature (London)* **285**, 167 (1980).
[18] T. Hunter and B. M. Sefton, *Proc. Natl. Acad. Sci. U.S.A.* **77**, 1311 (1980).
[19] L. E. Henderson, H. C. Krutzsch, and S. Oroszlan, *Proc. Natl. Acad. Sci. U.S.A.* **80**, 339 (1983).

proteins described here), but residues 2–5 (Gln-Thr-Val-Thr) do not match those in the table.

Acknowledgment

The work in the authors' laboratory was supported by a Program Grant from the Medical Research Council, London, U.K.

[20] Detection of an N-Terminal Glucuronamide Linkage in Proteins

By P. E. KOLATTUKUDY

Cutinase, an extracellular enzyme from *Fusarium solani pisi*, catalyzes the hydrolysis of the insoluble biopolyester called cutin.[1,2] This enzyme contains a glucuronic acid molecule attached to the N-terminal amino group of the protein by an amide linkage (Fig. 1).[3–5]

Principle

When D-glucuronic acid is attached via the carboxyl group to the protein, the hemiacetal carbon is available for reduction by NaB^3H_4. The labeled L-gulonic acid thus generated remains attached to the protein because the amide linkage is stable to most conditions used for $NaBH_4$ treatment. On the other hand O-glycosidically attached glucuronic acid, which is also present in cutinase,[5] undergoes β-elimination under basic conditions and the eliminated glucuronic acid is also reduced to gulonic acid by NaB^3H_4. Thus, the O-glycosidically attached glucuronic acid is converted to free labeled gulonic acid whereas the glucuronamide is converted to gulonamide which remains protein bound. Therefore, simple gel filtration methods can be used to remove the free gulonic acid from the macromolecular fraction which would contain the protein-bound gulonic acid. The usual acid hydrolysis of the protein releases the amide-bound gulonic acid which can then be isolated and identified. Alternatively,

[1] P. E. Kolattukudy, *Science* **208**, 990 (1980).
[2] P. E. Kolattukudy, R. E. Purdy, and I. B. Maiti, this series, Vol. 71, Part C, p. 652.
[3] T. S. Lin and P. E. Kolattukudy, *Biochem. Biophys. Res. Commun.* **72**, 243 (1976).
[4] T. S. Lin and P. E. Kolattukudy, *Biochem. Biophys. Res. Commun.* **75**, 87 (1977).
[5] T. S. Lin and P. E. Kolattukudy, *Eur. J. Biochem.* **106**, 341 (1980).

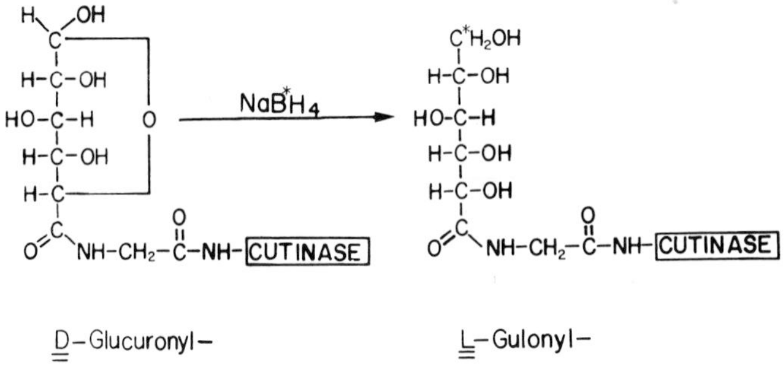

Fig. 1. Reduction of N-terminal D-glucuronic acid to L-gulonic acid by NaB^{3}H$_4$ reduction of cutinase.

cutinase can be treated with NaB^{3}H$_4$ under neutral conditions when only the hemiacetal carbon of the protein-bound uronic acid is reduced and the O-glycosidically attached carbohydrates are not eliminated and therefore are not reduced by NaB^{3}H$_4$. The protein recovered from such a reaction mixture is therefore specifically and exclusively labeled at the amide-linked gulonic acid. Protease can be used to digest such labeled protein and *N*-gulonyl derivative of the N-terminal amino acid residue can be isolated from the digest.

Treatment of Cutinase with NaB^{3}H$_4$

Electrophoretically homogeneous cutinase (10 mg) isolated from the extracellular fluid of cutin-grown *Fusarium solani pisi*[2,6] is treated with 0.2 *M* NaB^{3}H$_4$ in 0.5 ml 0.1 *N* KOH or 0.05 *M* Tris–HCl buffer, pH 7.0, for 216 hr at 4° in the dark. Then 0.2 ml of 0.08 *M* PdCl$_2$ and one drop of *n*-octanol are added and the mixture is kept at room temperature (22°) for 2 hr. The mixture is then acidified with acetic acid and lyophilized. (Since acidification releases ^{3}H$_2$ the process should be done with adequate precautions to avoid release of radioactivity into the laboratory.) Acidification and lyophilization are repeated until essentially all the exchangeable ^{3}H is removed. The residue is dissolved in 1 ml of water and the insoluble material is removed by centrifugation. The solution is subjected to gel filtration on a BioGel P-2 column (1.2 cm diameter, 110 cm long) with water as the solvent. The eluant is monitored for ^{3}H and absorbance at 280 nm (Fig. 2). The protein peak at the void volume contains the protein-bound gulonic acid generated by the reduction of glucuronamide. When neutral conditions are used only the protein is labeled. The fractions

[6] P. E. Kolattukudy, *in* "Lipolytic Enzymes" (B. Borgström and H. Brockman, eds.), p. 471. Elsevier, Amsterdam, 1984.

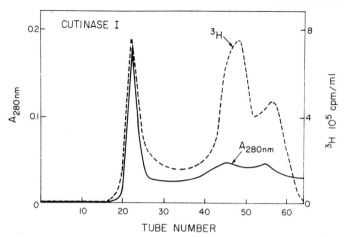

FIG. 2. BioGel P-2 gel filtration of the alkaline NaB³H₄-treated cutinase.

representing the protein peak are pooled and lyophilized. The residue is dissolved in a small volume of water and an aliquot containing adequate amounts of radioactivity is hydrolyzed in 6 N HCl for 24 hr in a sealed tube at 110° as is usually done for the preparation of protein hydrolysates for amino acid analysis. The hydrolyzate is lyophilized, the residue is dissolved in 0.5 ml of water, and relyophilized. After repeating this process twice, the final residue is analyzed using a polystyrene sulfonate column in an amino acid analyzer. The labeled gulonic acid emerges before aspartic acid but after cysteic acid from the analyzer (Fig. 3). Another acidic labeled component emerges just prior to the labeled gulonic acid. This unidentified labeled material is generated when a variety of proteins which do not contain any carbohydrates are subjected to alkaline NaB³H₄ treatment similar to that indicated above.

When the NaB³H₄ treatment is done under basic conditions for the long periods of time required for β-elimination, peptide bond cleavage occurs and a small amount of the amide-bound gulonic acid may be found in molecules which are retarded even in BioGel P-2. The process of β-elimination can be accelerated by doing the NaB³H₄ treatment in a 5 : 4 : 1 mixture of dimethyl sulfoxide, water, and ethanol containing 0.2 M KOH at 40°, and under such conditions β-elimination may be complete within 4 hr.[7] However, in our experience, such conditions appear to promote peptide bond cleavages in cutinase. Therefore, it is advisable to use the

[7] F. Downs, A. Herp, J. Moschera, and W. Pigman, *Biochim. Biophys. Acta* **328**, 182 (1973).

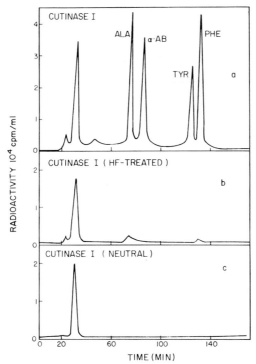

FIG. 3. Analysis of the hydrolyzate of alkaline NaB³H₄-treated cutinase by an amino acid analyzer. The bottom tracing shows the results obtained with cutinase treated with NaB³H₄ under neutral conditions.

low temperature for alkaline NaB³H₄ treatment. However, the sensitivity of peptide bond cleavage might vary with the nature of the protein and therefore a preliminary test with SDS gel electrophoresis of the treated protein is advisable.

When the NaB³H₄ treatment is done under basic conditions, the anhydroamino acid residues generated by β-elimination of the O-glycosidically attached carbohydrates are also generating labeled amino acids. Identification of such labeled amino acid residues showed that cutinase from *F. solani pisi* contains O-glycosidically attached sugars at serine and threonine generating alanine and α-aminobutyric acid by alkaline NaB³H₄ treatment (Fig. 3a). Besides these two residues, which are known to be involved in such linkages in other proteins, O-glycosidic linkages were suggested to be present at β-hydroxyphenylalanine and β-hydroxytyrosine because labeled phenylalanine and tyrosine were also generated by alkaline NaB³H₄ treatment of cutinase (Fig. 3a). These amino acids have not been previously found in any other glycoprotein.[3,5] When cutinase is treated with NaB^{3}H$_4$ under neutral conditions, only the amide linked glucuronic acid is reduced (Fig. 3c). The O-glycosidically linked carbohydrates can be removed by treatment of the protein with anhydrous HF[8] prior to NaB^{3}H$_4$ treatment. Under such conditions also only the protein-bound gulonic acid becomes labeled and therefore amino acid analysis of the hydrolyzate of the labeled protein shows only the radioactive acidic component (Fig. 3b).

Identification of Gulonic Acid

The acidic labeled fraction obtained from the amino acid analyzer is dissolved in 0.01 M formic acid, evaporated to dryness at 60° and the residue is held at this temperature for 3 hr to lactonize gulonic acid. The lactone can be identified by cochromatography of the label with authentic gulonolactone when the material is subjected to (1) thin-layer chromatography on silica gel with methyl ethyl ketone : acetic acid : methanol (3 : 1 : 1) as the solvent, (2) high-performance liquid chromatography with a μBondapak/carbohydrate column (Waters Associates) with 3% H$_2$O in CH$_3$CN as the solvent, and (3) radio gas–liquid chromatography of the trimethylsilyl ether derivative on a 5% OV-1 on 80–100 mesh Gaschrom Q, packed in a stainless-steel column (2.4 m × 6.35 mm outer diameter) held at 240°.[3] A portion of the effluent is passed through a radioactivity monitor.

Isolation of the N-Gulonyl Conjugate of the N-Terminal
  Amino Acid Residue[4]

The NaB^{3}H$_4$ treatment of the protein is done under neutral conditions and the labeled protein is isolated as indicated above. The protein solution, containing about 10 mg protein and adjusted pH to 7.5 with 0.1 N NaOH, is treated with 1.2 mg *Streptomyces griseus* protease at 22° for 24 hr. The digest is passed through a BioGel P-2 column (1.2 × 110 cm) with water as the eluant and aliquots of the fractions are assayed for ^{3}H. If the material eluting at the void volume contains a substantial proportion of the radioactivity, the Pronase treatment should be repeated on the material recovered from the first gel filtration. The labeled material, representing small molecules and containing most of the radioactivity, recovered from the P-2 column is dissolved in 1 ml of 0.1 N HCl and passed through a Dowex 50W-X8 cation exchange column (50–100 mesh, 2.2 × 30 cm) with water as the eluant. The material which is not retained by this

[8] A. J. Mort and T. A. Lamport, *Anal. Biochem.* **82**, 289 (1977).

column contains the bulk of the label contained in the protein. This material, recovered by lyophilization, is dissolved in water, pH is adjusted to 3.5 with HCl and an excess of an etherial solution of diazomethane is added. The aqueous phase containing the methyl ester is lyophilized. The residue is subjected to thin-layer chromatography on silica gel G with n-propanol:water (9:1) as the solvent system. The N-gulonylglycine methyl ester, at an R_f of 0.23, contained the bulk of the label (this material does not react with ninhydrin). This conjugate can be purified by the same thin-layer chromatographic step. Acid hydrolyzate prepared from this material in the usual manner with 6 N HCl (110°, 24 hr) can be analyzed with an amino acid analyzer as indicated above. Cutinase isolated from $F.$ $solani$ $pisi$ gives glycine and labeled gulonate in equimolar amounts (Fig. 4) suggesting that the N-terminal glycine of the enzyme is attached to glucuronic acid via amide linkage.

The labeled N-gulonylglycine obtained from the protease digest of the NaB³H₄-treated cutinase can also be directly compared with synthetic N-gulonylglycine which can be prepared as follows: 3 mmol each of gulono-lactone and glycine methyl ester and 1.7 mmol of anhydrous sodium carbonate are dissolved in 40 ml of absolute methanol and the mixture is heated at 65° under N₂ for 4 hr. After evaporating off the solvent, the residue is dissolved in a minimal volume of water and passed through a Dowex-50W (H⁺) column. From the products eluted with water,

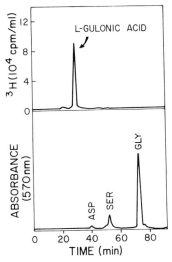

FIG. 4. Analysis of the acid hydrolyzate of the N-gulonylglycine isolated from the protease digest of neutral NaB³H₄-treated cutinase. The analysis was done with an amino acid analyzer.

N-gulonylglycine methyl ester can be purified by thin-layer chromatography using MN-cellulose 300 with ethyl acetate : pyridine : water (1 : 1 : 1) as the developing solvent (R_f, 0.41). The labeled product derived from cutinase can be compared with this synthetic material by[4] (1) thin-layer chromatography with the above solvent systems, (2) radio gas–liquid chromatography of the trimethylsilyl ether derivative as indicated above, and (3) high-performance liquid chromatography using a μBondapak/carbohydrate column as mentioned above.

These methods should be suitable for detecting N-terminal glucuronamide irrespective of the nature of the N-terminal amino acid residue.

Occurrence of N-Glucuronamide in Other Organisms

So far, glucuronamide at the N-terminus was detected in cutinases isolated from *Fusarium solani pisi, Fusarium roseum culmorum, Fusarium roseum sambucinum, Ulocladium consortiale,*[9] and *Colletotrichum gloeosporioides.*[10] It appears that fungal cutinases which contain O-glycosidically linked carbohydrates also contain N-terminal glucuronamide as in the above cases whereas cutinases from *Streptomyces scabies* and *Helminthosporum sativum,* which contain no O-glycosidically attached carbohydrates, do not have glucoronamide at the N-terminus.

Biosynthesis

Cutinase induced by low levels of cutin hydrolysate in glucose-grown cultures also has glucuronamide at the N-terminus.[11] The synthesis of such a structure by *Fusarium solani pisi* grown in defined medium allows convenient biosynthetic studies. The poly(A)$^+$ mRNA isolated from induced cultures can be translated *in vitro* either in the rabbit reticulocyte lysate system or in the wheat germ cell-free system, the latter being more efficient for this fungal mRNA preparation.[12] The IgG fraction isolated from rabbit antiserum prepared against cutinase can be used to isolate the primary translation product of cutinase mRNA. This product is 2100 daltons larger than the mature peptide.[12]

The processing of the precursor to the mature enzyme obviously involves cleavage of peptide bond(s), introduction of the glucuronamide at the N-terminus and glycosylation. Introduction of *O*-glycosyl residue

[9] T. S. Lin and P. E. Kolattukudy, *Physiol. Plant Pathol.* **17,** 1 (1980).
[10] C. L. Soliday and P. E. Kolattukudy, unpublished results (1983).
[11] T. S. Lin and P. E. Kolattukudy, *J. Bacteriol.* **133,** 942 (1978).
[12] W. H. Flurkey and P. E. Kolattukudy, *Arch. Biochem. Biophys.* **212,** 154 (1981).

```
80
MET LYS PHE PHE ALA LEU THR THR LEU LEU ALA ALA THR ALA SER ALA LEU PRO THR SER
ATG AAA TTC TTC GCT CTC ACC ACA CTT CTC GCC GCC ACG GCC TCG GCT CTG CCT ACT TCT


140
ASN PRO ALA GLN GLU LEU GLU ALA ARG GLN LEU GLY ARG THR THR ARG ASP ASP LEU ILE
AAC CCT GCC CAG GAG CTT GAG GCG CGC CAG CTT GGT AGA ACA ACT CGC GAC GAT CTG ATC
                                          ↑
```

FIG. 5. The nucleotide sequence of the cDNA prepared from cutinase mRNA and the derived amino acid sequence. The arrow shows the glycine residue which probably becomes the N-terminal residue of the mature enzyme.

(mannose) is catalyzed by microsomal preparations from *F. solani pisi* induced to synthesize cutinase.[13] However, the cleavage of the peptide bond(s) and introduction of the glucuronamide has not been demonstrated although it has been proposed that glucuronic acid probably forms the amide linkage in a transpeptidization type reaction during the cleavage of the peptide bond.[6]

Structure of the N-Terminal Region Determined by cDNA Cloning and Nucleotide Sequencing

The complete primary structure of cutinase has been determined by nucleotide sequencing of the cloned cDNA[14] and the N-terminal region is shown in Fig. 5. The apparent leader sequence appears to have a composition similar to those of the leader sequences of other extracellular proteins. From the sequence it appears that the first glycine near the N-terminus residue is 31 amino acid residues from the N-terminus of the precursor and therefore it can be tentatively concluded that this glycine residue forms the amide linkage with glucuronic acid and becomes the N-terminal amino acid of mature cutinase.

Acknowledgments

This work was supported in part by Grant PCM-8007908, from the National Science Foundation. I thank Linda Rogers for assistance in preparing this manuscript.

[13] C. L. Soliday and P. E. Kolattukudy, *Arch. Biochem. Biophys.* **197**, 367 (1979).
[14] C. L. Soliday, W. H. Flurkey, T. W. Okita, and P. E. Kolattukudy, *Proc. Natl. Acad. Sci. U.S.A.* (submitted).

[21] Occurrence, Detection, and Biosynthesis of Carboxy-Terminal Amides

By GÜNTHER KREIL

Occurrence

The existence of an amide rather than a free α-carboxyl group at the COOH-terminus of peptides was probably the first posttranslational modification to be discovered. This was independently demonstrated in 1953 by three groups,[1-3] who showed that both oxytocin and vasopressin terminate with glycinamide. The same type of modification has since been found in numerous peptides secreted by tissues of many different organisms. Examples are gastrin, cholecystokinin, substance P, secretin, vasoactive intestinal peptide, thyrotropin-releasing hormone,[3a] corticotropin-releasing factor,[4] etc., from mammalian species, several frog skin peptides like caerulein, bombesin, and its relatives, sauvagine,[5] dermorphin[6] etc. Also, amidated peptides have been found in invertebrates, examples being the bee venom peptides melittin, apamin, MCD peptide, adipokinetic hormone from locusts, red pigment concentrating hormone from prawns, eledoisin from an octopus, cecropins from moths,[7,8] toxin II of scorpion venom,[9] egg laying hormone from *Aplysia*,[10] and conotoxins from the venom of a marine snail.[11] It is currently unknown whether

[1] H. Tuppy and H. Michl, *Monatsh. Chem.* **84**, 1011 (1953).
[2] V. Du Vigneud, C. Ressler, and S. Trippett, *J. Biol. Chem.* **205**, 949 (1953).
[3] R. Acher and J. Chauvet, *Biochim. Biophys. Acta* **12**, 487 (1953).
[3a] Unless indicated otherwise, sequences are taken from M. O. Dayhoff, "Atlas of Protein Sequence and Structure," Vol. 5 (1972) and supplements I(1973), II(1976), and III(1978). National Biomedical Research Foundation, Silver Spring.
[4] J. Spiess, J. Rivier, C. Rivier, and W. Vale, *Proc. Natl. Acad. Sci. U.S.A.* **78**, 6517 (1981).
[5] P. C. Montecucchi and A. Henschen, *Int. J. Pept. Protein Res.* **18**, 113 (1981).
[6] P. C. Montecucchi, R. DeCastiglione, and V. Erspamer, *Int. J. Pept. Protein Res.* **17**, 316 (1981).
[7] D. Hultmark, A. Engström, H. Bennich, R. Kapur, and H. G. Boman, *Eur. J. Biochem.* **127**, 207 (1982).
[8] R. B. Merrifield, L. D. Vizioli, and H. G. Boman, *Biochemistry* **21**, 5020 (1982).
[9] C. Kopeyan, G. Martinez, and H. Rochat, *FEBS Lett.* **89**, 54 (1978).
[10] A. Y. Chiu, M. W. Hunkapiller, E. Heller, D. K. Stuart, L. E. Hood, and F. Strumwasser, *Proc. Natl. Acad. Sci. U.S.A.* **76**, 6656 (1979).
[11] W. R. Gray, A. Luque, B. M. Olivera, J. Barrett, and L. J. Cruz, *J. Biol. Chem.* **256**, 4734 (1981).

METHODS IN ENZYMOLOGY, VOL. 106

unicellular organisms, like bacteria, algae, and yeast as well as higher plants, synthesize amidated peptides.

A comparison of these peptides shows that the nature of the terminal amino acid varies widely so that, with the possible exception of arginine and aspartic acid, all have been found at an amidated COOH end. It is, however, noteworthy that this modification has only been detected in rather small polypeptides, ranging in size from the tripeptide thyrotropin releasing hormone to scorpion toxin II, which contains 64 amino acids.[9] The reason for this size specificity is unclear at present. The biological role of these terminal amide bonds is also a matter of mere speculation. As in the case of modifications at the amino end, a protection against exoproteolytic digestion can be invoked, although, in several instances, it has clearly been demonstrated that the blocked COOH-terminus is essential for biological activity.

Detection

Terminal amides have usually been detected by the change in electrophoretic mobility caused by the modification of the carboxyl end. Intact peptides or one of the proteolytic fragments were found at neutral or weakly acidic pH to have a net charge differing from the one expected from the amino acid composition. Further digestion with proteolytic enzymes of broad specificity or stepwise Edman degradation will then yield the terminal amino acid amide, which can easily be identified by its electrophoretic mobility using standard substances as reference. Total hydrolysis then yields the free amino acid and ammonia in equimolar quantities, which can be quantified using amino acid analyzers.

A more general method to search for amidated peptides has been described by Tatemoto and Mutt.[12] The crude or purified peptide sample is degraded with a protease of broad specificity, like thermolysin, subtilisin, or elastase, and the total digest is subsequently treated with dansyl chloride to yield the dansylated peptides and amino acids. From alkaline solutions, the dansyl derivatives of amidated amino acids and peptides can be preferentially extracted into ethyl acetate. Dansyl derivatives of amino acids can subsequently be identified by two-dimensional chromatography on polyamide sheets.

Using this technique, two new amidated peptides were detected in side fractions obtained during the purification of porcine secretin.[13] These peptides, designated PHI and PYY (P stands for peptide, the other two letters indicate the NH_2- and COOH-terminal amino acid, respectively)

---

[12] K. Tatemoto and V. Mutt, *Proc. Natl. Acad. Sci. U.S.A.* **75**, 4115 (1978).
[13] K. Tatemoto and V. Mutt, *Nature (London)* **285**, 417 (1980).

were subsequently purified and their sequences determined.[13,14] It appears likely that these represent new polypeptide hormones as PHI is similar to glucagon and secretin, while PYY shows homology to pancreatic poly-peptide. More recently, a new amidated neuropeptide was discovered using this approach[15] and was shown to be similar to the two latter pep-tides.

The method does not work for amides of arginine, glutamic, and aspar-tic acid, but these are unusual terminal amino acids. This chemical test for blocked carboxyl ends which in turn serves as a handle for isolating new and potentially interesting peptides of unknown biological action is of obvious interest and should have broad applicability.

Biosynthesis

In view of the diversity of terminal amino acids found in amidated peptides, it is hard to imagine why certain peptides have this modification and others not, even though they may be synthesized in the same cell. Also homologous peptides having either free or blocked COOH-ends have been found. An example are the hormones glucagon and secretin, where the former has a free α-carboxyl group. As peptides are generally derived from larger precursors, it is obvious that the specificity for amida-tion must reside in the structure of these larger forms, most likely in the amino acid(s) adjacent to the terminal residue of the end product. A first clue as to what this "amidation signal" might be came from studies on the biosynthesis of melittin, the main constituent of honeybee venom which terminates with glutaminamide. It could be demonstrated that the precur-sor, prepromelittin, contains an extra glycine adjacent to the glutamine which is amidated in the final venom constituent.[16] A similar situation had earlier been found in corticotropin, which is also the precursor of α-melanotropin. Formation of the latter peptide must involve cleavage and modification of the sequence Val-Gly-Lys-Lys . . . to ultimately yield Val-CONH$_2$.[17] Similar situations, in which a glycine is located next to the terminus of the end product, have since been encountered in the precursors for calcitonin,[18,19] gastrin,[20] caerulein,[21] corticotropin releasing

[14] K. Tatemoto and V. Mutt, *Proc. Natl. Acad. Sci. U.S.A.* **78**, 6603 (1981).

[15] K. Tatemoto, *Proc. Natl. Acad. Sci. U.S.A.* **79**, 5485 (1982).

[16] G. Suchanek and G. Kreil, *Proc. Natl. Acad. Sci. U.S.A.* **74**, 975 (1977).

[17] G. I. Harris and A. B. Lerner, *Nature (London)* **179**, 1346 (1957).

[18] S. G. Amara, D. N. David, M. G. Rosenfeld, B. A. Roos, and R. M. Evans, *Proc. Natl. Acad. Sci. U.S.A.* **77**, 4444 (1980).

[19] J. W. Jacobs, R. H. Goodman, W. W. Chin, P. C. Dee, J. F. Habener, N. H. Bell, and J. T. Potts, Jr., *Science* **213**, 457 (1981).

[20] O. J. Yoo, C. T. Powell, and K. L. Agarwal, *Proc. Natl. Acad. Sci. U.S.A.* **79**, 1049 (1982).

[21] W. Hoffmann, T. C. Bach, H. Seliger, and G. Kreil, *EMBO J.* **2**, 111 (1983).

factor,[22] egg laying hormone,[23] vasopressin,[24] oxytocin,[25] as well as in peptides derived from pro-opiomelanocortin.[26,27]

This would then indicate that a terminal glycine residue exposed during the processing of peptide precursors is the recognition signal for the amidation reaction. Alternatively, it has been proposed that the structure Gly-B-B, where B stands for lysine or arginine, would be the structure recognized by the enzyme catalyzing the formation of terminal amides.[20] This sequence is indeed found in most of the hormone precursors mentioned in the preceding paragraph, with the exception of corticotropin releasing factor,[22] where the glycine is followed by a single lysine residue. In case of melittin, it could, however, be shown that in the nucleotide sequence of the cloned cDNA, a stop codon is present after the codon for the terminal glycine.[28] It is clear that in honeybee venom glands, lysine/ arginine residues play no role in the amidation and this appears also the case for mammalian tissues (see below).

The search for the amidating enzyme has been difficult, since attempts to demonstrate this reaction *in vitro* have been plagued by low yields and poor reproducibility. In principle, two types of reactions could be envisaged: (1) a transamidation, whereby the terminal glycine is replaced by an $-NH_2$ group donated by glutamine or some other amide; and (2) oxidation of the $-NH-CH_2-$ bond of the terminal glycine to yield an unstable $-N=CH-$ double bond. This could be achieved by dehydrogenation or by hydroxylation and subsequent cleavage of water. Hydrolysis of the $-CO-N=CH-COOH$ structure thus formed would then yield the terminal amide $-CONH_2$ and glyoxylic acid.

Evidence presented by Bradbury *et al.*[29] has proved that the second mechanism is the correct one. These authors used a granule fraction prepared from porcine pituitary by differential centrifugation. After solubilization by sonication, the amidating activity could be partly purified by chromatography on Sephadex G-100, where it emerged as a broad peak in

[22] Y. Furutani, Y. Morimoto, S. Shibahara, M. Noda, H. Takahashi, T. Hirose, M. Asai, S. Inayama, H. Hayashida, T. Miyata, and S. Numa, *Nature (London)* **301,** 537 (1983).

[23] R. A. Scheller, J. F. Jackson, L. B. McAllister, B. S. Rothman, E. Mayeri, and R. Axel, *Cell* **32,** 7 (1983).

[24] H. Land, G. Schütz, H. Schmale, and D. Richter, *Nature (London)* **295,** 299 (1982).

[25] H. Land, M. Grez, S. Ruppert, H. Schmale, M. Rehbein, D. Richter, and G. Schütz, *Nature (London)* **302,** 342 (1983).

[26] S. Nakanishi, A. Inoue, K. Kita, M. Nakamura, A. C. Y. Chang, S. N. Cohen, and S. Numa, *Nature (London)* **278,** 423 (1979).

[27] N. G. Seidah, J. Rochemont, J. Hamelin, S. Benjannet, and M. Chrétien, *Biochem. Biophys. Res. Commun.* **102,** 710 (1981).

[28] R. Vlasak, C. Unger-Ullmann, G. Kreil, and A. M. Frischauf, *Eur. J. Biochem.* **135,** 123 (1983).

[29] A. F. Bradbury, M. D. A. Finnie, and D. G. Smyth, *Nature (London)* **298,** 686 (1982).

a position corresponding to a molecular weight of 60,000. Using labeled D-Tyr-Val-Gly as substrate, the slow formation of D-Tyr-valinamide could be observed. The reaction was about 2–3 pmol/hr. Using the same tripeptide containing ^{15}N-labeled glycine, it could furthermore be shown by high-resolution mass spectroscopy that the reaction product retained the ^{15}N isotope. The fact that the nitrogen of the terminal glycine remains in the amidated product proves that the redox reaction described above as alternative 2 is taking place. Without giving any experimental details, Bradbury *et al.* mention that glyoxylic acid has also been identified as the second reaction product. No amidation was observed with D-Tyr-Val-COOH as substrate or with tripeptides of the structure D-Tyr-Val-X, X being alanine, leucine, sarcosine etc. Amidation was also detected with the tripeptide D-Tyr-Phe-Gly, but not with peptides containing lysine, aspartic acid, or D-alanine in position 2. It is clear from these results that a terminal glycine is essential for the amidation, while the penultimate amino acid may have to be one with a neutral side chain.

Similar results have been presented by Husain and Tate[30] with a granule fraction from bovine pituitary prepared by the same procedure, using pGlu-His-Pro-Gly as substrate. This substrate would be the immediate precursor of thyrotropin releasing hormone, pGlu-His-Pro-amide. Formation of the hormone from the tetrapeptide could indeed be observed at a rate of 1–2 nmol/hr/mg protein. This would be considerably higher than the 2–3 pmol/hr observed by Bradbury *et al.*[29] (protein concentration was not given in this paper). The reactions were carried out in an oxygen atmosphere but no mention was made about its influence on the yield.[30]

From these data it is quite clear that peptides containing a terminal glycine can be converted to the amidated peptide whereby the nitrogen of the glycine is retained in the product. The enzyme(s) catalyzing this reaction have so far been poorly characterized. Most important, no mention of any cofactors are made in the two papers dealing with this subject. This is surprising as it would be quite obvious to try the influence of pyridine nucleotides, flavines etc. on the rate of the amidation reaction. The observed reaction rates have been low and in fact some essential cofactor may be limiting. It has been observed in our laboratory[31] that amidation can take place by oxidative cleavage with Cu^{2+} ions and hydrogen peroxide at pH 5.0. This model reaction is, however, not specific for peptides terminating with glycine, yet it may give a clue as to what the mechanism operating *in vivo* might be.

The amidation reaction appears to be quite unstable even in intact

[30] I. Husain and S. S. Tate, *FEBS Lett.* **152**, 277 (1983).
[31] A. Hutticher and G. Kreil, unpublished experiments (1981).

cells. It has been reported[32] that dog pancreatic islets in culture mainly secrete a more acidic form of pancreatic polypeptide, which lacks the terminal amide and instead contains an extension of one or a few amino acids. Honeybee venom glands incubated *in vitro* also make promelittin lacking the terminal amide.[33] All these observations would point toward a complex reaction mechanism.

The tissue distribution of the amidation reaction is currently not known. The activity may well be present only in those cells which secrete amidated peptides. Frog oocytes, which catalyze a variety of posttranslational reactions, apparently cannot synthesize terminal amides. After injection of the mRNA for prepromelittin, promelittin with the extra glycine and a free α-carboxyl group is the stable end product.[34]

It is quite clear that much remains to be done to understand the formation of terminal amides. Apart from the basic aspects, this reaction also has some potential interest for applied research. Synthesis of polypeptides by recombinant DNA techniques in *E. coli* and possibly also in yeast could yield products with a free α-carboxyl group which may be biologically inactive. The subsequent amidation *in vitro* of a product with an extra glycine to obtain the active hormone would clearly be of interest.

[32] T. L. Paquette, R. Gingerich, and D. Scharp, *Biochemistry* **20**, 7403 (1981).
[33] G. Suchanek and G. Kreil, unpublished experiments (1977).
[34] C. D. Lane, J. Champion, L. Haiml, and G. Kreil, *Eur. J. Biochem.* **113**, 273 (1981).

[22] Tyrosine Incorporation in Tubulin

By MARTIN FLAVIN and HIROMU MUROFUSHI

Although probably anticipated by an early report of ribonuclease-resistant incorporation of phenylalanine into protein of a liver extract,[1] tubulin tyrosinolation (the tRNA-independent addition of tyrosine to the C-terminal glutamate of its α chain) was first characterized in Caputto's laboratory, as the fortuitous result of a control experiment carried out while studying developmental fluctuations in tRNAs from brain.[2]

Tubulin-tyrosine ligase can also utilize phenylalanine and Dopa,[3] but no protein substrate other than tubulin has been detected. The reaction

[1] L. Rosen and G. D. Novelli, *Biochim. Biophys. Acta* **145**, 218 (1967).
[2] H. S. Barra, J. A. Rodriguez, C. A. Arce, and R. Caputto, *J. Neurochem.* **20**, 97 (1973).
[3] C. A. Arce, J. A. Rodriguez, H. S. Barra, and R. Caputto, *Eur. J. Biochem.* **59**, 145 (1975).

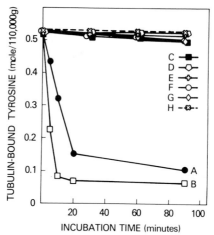

FIG. 1. Requirements for release of tyrosine from tyrosinolated tubulin mediated by tubulin-tyrosine ligase. Incubation mixture (A) contained 0.51 mg/ml of ^{14}C-tyrosinolated tubulin (0.53 mol of C-terminal tyrosine per 110,000 g tubulin); 1 unit/ml of tubulin-tyrosine ligase; 0.1 M K$^+$MES, pH 6.8; 10 μM GTP; 1 mM ADP; 10 mM sodium phosphate; 10 mM MgCl$_2$. The other mixtures were modifications of (A) as follows: (B) 5 units/ml of enzyme; (C) no enzyme; (D) no ADP; (E) no phosphate; (F) no MgCl$_2$; (H) no ADP or phosphate, but containing 1 mM AMP and 10 mM sodium pyrophosphate. All samples were incubated at 37°. Fifty-microliter aliquots were taken at intervals and assayed for protein bound [^{14}C]tyrosine as described under ligase assays.

involves formation of a peptide bond with the α-carboxyl of glutamate at the expense of ATP, is reversible[4] in the presence of ADP + P$_i$ [Eq. (1)] and is formally analogous to the extensively studied reactions of glutamine and glutathione synthesis. Figure 1 shows that tyrosine is not released in the presence of AMP + PP$_i$, and that the reverse reaction can go essentially to completion, presumably with stoichiometric resynthesis of ATP,[5] though this has not been shown.

α-tubulin-Gly-Glu-Glu + L-tyrosine + ATP $\rightleftharpoons$

$$ADP + P_i + \alpha\text{-tubulin-Gly-Glu-Glu-Tyr} \quad (1)$$

---

[4] J. A. Rodriguez, C. A. Arce, H. S. Barra, and R. Caputto, *Biochem. Biophys. Res. Commun.* **54**, 335 (1973).

[5] Assuming the standard free energy change for the reaction shown in Eq. (1) in the forward direction to be in the range of -6000 calories reported for an ATP-dependent peptide synthesis [J. E. Snoke and K. Bloch, *J. Biol. Chem.* **213**, 825 (1955)] calculation shows the reverse reaction could go essentially to completion under the actual conditions of Fig. 1. With the initial concentration of tyrosinolated tubulin 53% of 510 μg/ml or $2.5 \times 10^{-6} M$, $\Delta G = +6000 + RT \ln x^3/(10^{-3} - x)(10^{-2} - x) (2.5 \times 10^{-6} x)$. The value of x at equilibrium, when $\Delta G = 0$, is $2.476 \times 10^{-6} M$, so that the detyrosinolation has gone 99% toward completion.

This type of posttranslational modification is unprecedented. It is rendered even more remarkable by recent sequence analyses of several α chain mRNAs from vertebrate brains, which show them to contain a C-terminal codon for tyrosine.[6,7] Since cytosolic brain α chain has always been isolated as a mixture of tyrosinolated and detyrosinolated species,[8,9] it must be assumed that the initial posttranslational modification is tyrosine removal. An appropriate tubulin detyrosinolating carboxypeptidase (carboxypeptidase-tubulin) has been partially purified from brain.[10,11] This situation might mislead some to believe that the tyrosine present in tubulin is all of genomic origin, and that the machinery for posttranslational tyrosinolation is in some way "artifactual."[12] It is clear, however, from many reports of tyrosine incorporation *in vivo* in various cells and tissues, where protein synthesis has been completely inhibited, that at least a major part of the α chain C-terminal tyrosine has arrived there posttranslationally.[13-16] Further confirmation comes from a report that, by nutritional manipulation, a major part of the C-terminal tyrosine can be replaced by phenylalanine.[17]

Tubulin-Tyrosine Ligase

Assay Method

Principle. The assay is based on the rate at which labeled tyrosine is fixed into a trichloroacetic acid-insoluble form in the presence of purified microtubule protein.[18]

[6] P. Valenzuela, M. Quiroga, J. Zaldivar, M. W. Kirschner, and D. W. Cleveland, *Nature* (*London*) **289**, 650 (1981).
[7] I. R. Lemischka, S. Farmer, V. R. Racaniello, and P. A. Sharp, *J. Mol. Biol.* **151**, 101 (1981).
[8] J. Nath and M. Flavin, *FEBS Lett.* **95**, 335 (1978).
[9] E. Krauhs, M. Little, T. Kempf, R. Hofer-Warbinek, W. Ade, and H. Ponstingl, *Proc. Natl. Acad. Sci. U.S.A.* **78**, 4156 (1981).
[10] M. E. Hallak, J. A. Rodriguez, H. S. Bara, and R. Caputto, *FEBS Lett.* **73**, 147 (1977).
[11] N. Kumar and M. Flavin, *J. Biol. Chem.* **256**, 7678 (1981).
[12] S. N. Timasheff and L. M. Grisham, *Annu. Rev. Biochem.* **49**, 565 (1980).
[13] W. C. Thompson, *FEBS Lett.* **80**, 9 (1977).
[14] J. Nath and M. Flavin, *J. Biol. Chem.* **254**, 11505 (1979).
[15] J. Nath and M. Flavin, *J. Neurochem.* **35**, 693 (1980).
[16] J. Nath and M. Flavin, *J. Cell Biol.* **91**, 232 (1981).
[17] J. A. Rodriguez and G. G. Borisy, *Science* **206**, 463 (1979).
[18] Microtubule protein, the preparation obtained from brain extract by repeated cycles of assembly, is composed of about 80% of tubulin, and 20% of a variety of microtubule-associated proteins.

Stock Solutions

A. Tris–HCl buffer, 0.3 M, pH 7.2, stored at $+4°$
B. KCl, 1.2 M, stored at $+4°$
C. $MgCl_2$, 0.15 M, stored at $+4°$
D. Dithiothreitol (DTT), 0.3 M, stored at $-15°$
E. ATP, 0.05 M, stored at neutral pH at $-15°$
F. L-[^{14}C]Tyrosine, 1.2 mM, 50 μCi/ml, made from stock labeled tyrosine, and 10 mM unlabeled L-tyrosine in 0.1 N HCl. Stored at $-15°$
G. Microtubule protein, stored at about 12 mg/ml at $-70°$, in 100 mM potassium 2-(N-morpholino)ethanesulfonic acid (K$^+$MES), pH 6.8, 1 mM each EGTA and GTP, and 0.5 mM $MgCl_2$. The protein may be prepared from mammalian brain by three cycles of temperature-dependent assembly by the glycerol procedure described elsewhere in this series,[19] or by any of the procedures omitting glycerol which have been described in the literature. It should be stored in small aliquots, and thawed briefly just before use.

Procedure. To a 6 × 50-mm glass tube in an ice bath is added sufficient distilled water to give a final volume of 60 μl, and 5 μl each of (A), (B), and (C), 2 μl of (D), 3 μl of (E), 5 μl of (F), and (the penultimate addition) 15 μl (or sufficient to make 3 mg/ml) of (G). For 1 day's use, all of the stock solutions except (B) and (G) may also be combined. The reaction is started by adding ligase fraction, rapidly mixing with a Vortex-type agitator, and transferring the tubes to a 37° bath. A blank tube without ligase should always be included. Tyrosine fixation is linear for at least 20 min with a ligase concentration of 0.01 to 0.1 unit/ml. For much longer incubations, the tubes may be stoppered with 000 corks. If it is necessary to dilute the ligase, this should be done just before assay in the same mixture used for assay but omitting ATP and tyrosine, and reducing the microtubule protein to 0.2 mg/ml. To stop the reaction 50 μl is transferred (we find a Drummond Wiretrol convenient) to a Whatman 3MM paper disk (2.3 cm diameter) and the disk is dropped into a beaker of cold 10% trichloroacetic acid, about 4 ml per disk. The disks may be washed and dried as described,[20] or more rapidly by 10 min, with rapid rotary agitation at 25°, in each of two changes of 10% trichloroacetic acid; 5 min in the same volume of 5% acid in a bath at about 90°, followed by brief washes in 1 : 1 ethanol–ether, then in ether. The disks are placed in scintillation counter vials with 15 ml of 0.4% (w/v) 2,5-diphenyloxazole in toluene. If

[19] R. E. Williams, Jr. and J. C. Lee, this series, Vol. 85B, p. 364.
[20] R. G. Mans and G. D. Novelli, *Arch. Biochem. Biophys.* **94,** 48 (1961).

[³H]tyrosine is used paper disks should not be used, but the protein should be precipitated and redissolved.

Definition of Unit and Specific Activity. A unit of enzyme is defined as the amount transferring 1 nmol of tyrosine to tubulin in 1 min, under the above conditions. The cpm/nmol are determined by transferring 1 to 3 μl of stock solution (F) to a paper disk, which is air dried without washing, and counted in the same scintillation mixture; free tyrosine does not elute from the disk. Specific activity is defined as units per milligram of protein.

Purification

Preparation of Affinity Columns

SEPHAROSE-SEBACIC ACID HYDRAZIDE ATP. Periodate-oxidized ATP is immobilized to Sepharose gel via a spacer of sebacic acid dihydrazide.[21] Adipic acid dihydrazide should not be used, as it results in a much lower affinity for ligase.

Sebacic acid dihydrazide is prepared from diethyl sebacate and hydrazine hydrate.[21] One hundred milliliters of diethylsebacate, 200 ml of hydrazine hydrate (98%), and 200 ml of ethanol are added to a round-bottom flask with a reflux condenser. Gentle boiling of the solution is carried out for 3 hr with the aid of a mantle heater. The resultant sebacic acid dihydrazide precipitate is collected on a Büchner funnel, washed successively with cold ethanol and ether, and dried in a desiccator.

CNBr-activated Sepharose 4B, 1.2 g, which makes about 4 ml of gel, is washed successively with 100 ml of ice-cold 1 mM HCl and 50 ml of distilled water, immediately followed by the addition of sebacic acid dihydrazide solution (0.4 g dissolved in 8 ml of glacial acetic acid). The gel suspension is gently stirred overnight at room temperature using a small propeller attached to a low-speed motor. The gel is then washed, on a glass filter funnel, successively with 100 ml of 50% (v/v) acetic acid, 200 ml of water, and 100 ml of 1 M sodium acetate (pH 5.0) at room temperature. The washed gel is suspended in 4 ml of the above acetate buffer and chilled on ice. Now, 1.2 ml of 20 mM ATP solution (pH adjusted to 7 with KOH) is mixed with 1.2 ml of 18 mM NaIO$_4$ (freshly dissolved in water in the dark) and the mixture is stirred at 0° for 1 hr in the dark. The resultant oxidized ATP solution is added to the above Sepharose-sebacic acid hydrazide gel and the suspension is gently stirred at 4° for 2.5 hr in the dark. The gel is washed successively with 50 ml each of ice cold 1 M NaCl, 1 mM EDTA, and distilled water, suspended in water, and kept at 4° until used.

[21] M. Wilchek and R. Lamed, this series, Vol. 34, p. 475.

The amount of coupled ATP can be conveniently calculated by the difference between the amount of total ATP added and the unbound ATP which is recovered in the 1 M NaCl wash. Usually 50–65% of the total ATP is coupled. The concentration of the bound ATP in the gel is in the range of 3–4 mM. The ATP-immobilized gel is used for affinity chromatography within a day or two after preparation.

SEPHAROSE-TUBULIN. CNBr-activated Sepharose 4B (1.3 g which makes about 4.5 ml of wet cake) is quickly washed with 100 ml each of ice cold 1 mM HCl, distilled water, and a solution of 50 mM KCl–10 mM K$^+$MES (pH 6.8)–1 mM EGTA–0.5 mM MgCl$_2$, successively. Immediately after the final washing, 30 mg of microtubule protein (about 15 mg/ml in 50 mM KCl–10 mM K$^+$MES (pH 6.8)–1 mM EGTA–0.5 mM MgCl$_2$–0.5 mM GTP) is added to the gel and the gel suspension is transferred into a beaker. A 20 mM GTP solution and a solution of 50 mM KCl–10 mM MES (pH 6.8)–1 mM EGTA–0.5 mM MgCl$_2$ are added to the gel suspension in sufficient amounts to give a final GTP concentration of 1 mM, and a final volume of 9 ml. The gel suspension is gently stirred at 4° overnight. The gel is washed successively with 200 ml of ice cold 0.1 M Tris–HCl (pH 7.0)–0.6 M KCl–1 mM EGTA–0.5 mM MgCl$_2$, and 200 ml of 0.6 M KCl–25 mM K$_+$MES (pH 6.8)–5 mM MgCl$_2$–0.5 mM DTT for 30 min each. Complete washing of unbound proteins is necessary because, if washing is incomplete, the final preparation of tubulin-tyrosine ligase is contaminated with microtubule proteins. The gel is finally washed with 50 ml of ice cold 50 mM KCl–25 mM K$^+$MES (pH 6.8)–5 mM MgCl$_2$–1 mM DTT and transferred to a column for affinity chromatography.

More than 90% of the microtubule protein is routinely coupled, yielding a gel containing 6–7 mg/ml of protein. The microtubule protein that has been used in this preparation was purified with glycerol present during each assembly cycle.[19]

Purification Procedure.[22] All procedures are carried out at 0–4°. The starting material (which can be stored for months at −70°) may be the original extract of brain (porcine, bovine, or ovine have been used). But since microtubule protein is needed, it is convenient to use the first warm supernatant from which microtubule protein has been assembled.[19] In the preparation outlined in Table I, 500 ml of this fraction, from 6 porcine brains, was diluted with 1/2 volume of 25 mM K$^+$MES (pH 6.8)–5 mM MgCl$_2$–0.5 mM DTT (the dilution is needed only when glycerol has been used in the microtubule assembly). Two hundred milliliters of DEAE-cellulose (DE-52) preequilibrated with the above buffer solution is added, and the slurry stirred slowly for 1 hr with a propeller, with care not to

---

[22] H. Murofushi, *J. Biochem. (Tokyo)* **87**, 979 (1980).

TABLE I
PURIFICATION OF TUBULIN-TYROSINE LIGASE

Step	Volume (ml)	Total protein (mg)	Total activity (units)	Specific activity (units/mg)	Yield (%)	Purification (fold)
1. First warm super-natant[a]	500	4400	400	0.091	(100)	(1)
2. DEAE-cellulose	320	840	470	0.55	117	6.1
3. Sepharose-sebacic acid hydrazide-ATP	30	6.5	263	41	66	440
4. Sepharose-tubulin	6.0	0.19	145	740	36	8100

[a] The supernatant obtained after assembling microtubules from an extract of fresh porcine brain.[19]

generate fines. Adding brain extract to a prepoured column would result in very poor flow, due to concentration of protein at the top of the column.

The DEAE-cellulose is collected on a Büchner funnel, suspended in 200 ml of 25 mM K+MES (pH 6.8)–50 mM KCl–5 mM MgCl$_2$–0.5 mM EDTA–0.5 mM DTT and transferred to a column (2.3 cm in diameter). After washing the column with about 70 ml of the above buffer, a linear gradient of KCl (from 0.05 to 0.3 M, total 1000 ml) containing 25 mM K+MES (pH 6.8), 5 mM MgCl$_2$, 0.5 mM EDTA, and 0.5 mM DTT is applied collecting 15 ml fractions. The flow rate is about 70 ml/hr. Tubulin-tyrosine ligase is eluted as a broad peak between 0.1 and 0.2 M KCl (Fig. 2).

Active fractions, indicated by the bracket in Fig. 2, are combined and applied directly to a Sepharose-sebacic acid hydrazide-ATP column (1.2 × 3.5 cm), preequilibrated with 25 mM K+MES (pH 6.8)–5 mM MgCl$_2$–0.5 mM DTT, at a flow rate of 40 ml/hr. With increasing amounts of protein adsorbed to the column, the flow rate may decrease. In this case, the top of the gel may be agitated gently with a glass rod. About 99% of the protein passes through the column. After charging the whole DEAE-cellulose fraction, the column is washed with 100 ml of 0.5 M KCl–25 mM K+MES (pH 6.8)–5 mM MgCl$_2$–0.5 mM DTT to remove materials which are bound nonspecifically. After the A_{280} of the eluate falls below 0.05, the column is washed with 20 ml of 0.1 M KCl–25 mM K+MES (pH 6.8)–5 mM MgCl$_2$–0.5 mM DTT, to lower the ionic strength in the column. Tubulin-tyrosine ligase is eluted with 25 mM ATP–25 mM K+MES (pH 6.8)–0.1 M KCl–25 mM MgCl$_2$–0.5 mM DTT, collecting 10 ml fractions at a flow rate of 15 ml/hr. Active fractions (usually the first to the third) are

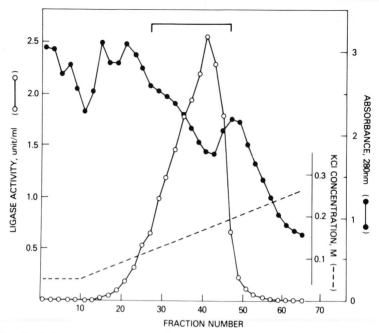

FIG. 2. Elution profile of protein and tubulin-tyrosine ligase activity from DEAE-cellulose.

combined, diluted with an equal volume of 25 mM K$^+$MES (pH 6.8)–5 mM MgCl$_2$–1 mM DTT to reduce the KCl concentration for the next chromatography, and then centrifuged at 15,000 g for 20 min to remove a small amount of precipitate in the enzyme fraction.

The Sepharose–sebacic acid hydrazide–ATP fraction is applied to a Sepharose-tubulin column (1 × 5 cm) preequilibrated with 50 mM KCl–25 mM K$^+$MES (pH 6.8)–5 mM MgCl$_2$–1 mM DTT. The flow rate should be less than 25 ml/hr, otherwise, a part of the ligase activity emerges in the unadsorbed fraction. After the unadsorbed materials are washed out completely with 20 ml of 0.1 M KCl–25 mM K$^+$MES (pH 6.8)–5 mM MgCl$_2$–1 mM DTT, while monitoring the A_{280} of the eluate, the enzyme is eluted with 0.4 M KCl–25 mM K$^+$MES (pH 6.8)–5 mM MgCl$_2$–1 mM DTT, collecting 3 ml fractions at a flow rate of 6 ml/hr. The second fraction of the 0.4 M KCl eluate shows the highest ligase activity and the third fraction the next highest activity. These fractions are divided into tubes and kept at −80°.

Comments on the Purification. Table I illustrates a successful preparation. The 8000-fold purified ligase is homogeneous by SDS–PAGE.[22] It can be stored at $-70°$ with little loss of activity for 2 months. Protein in the final fraction is, however, very dilute, and no method has been devised for concentrating it without loss of activity.

An alternative partial purification has been described, yielding an enzyme fraction with more concentrated protein, and activity completely stable for at least a year at $-70°$.[23] Although the purification was only 250-fold, this preparation is also suitable for the applications described below.

The DE-52 fraction is quite stable at $-70°$, and the pooled fractions from the ATP column may be stored overnight. The purification should be carried out more or less without interruption over 4 to 5 days.

Properties of Tubulin-Tyrosine Ligase

The enzyme is homogeneous by SDS–PAGE (molecular weight 46,000), and the native molecular weight appears to be 37,000 by HPLC gel filtration.[22] In brain extract, however, it exists as a 150,000-dalton stoichiometric complex with tubulin.[24] The pH optima are observed at 8 and 6.5, and 8.5 μM ATP or 30 μM L-tyrosine give half maximal activity.[22] Tubulin dimer appears to be the only substrate, but activity toward small oligomers has not been ruled out. ATP is about 5 times more effective than any other nucleoside triphosphate. Maximal rates are observed with from 3 to 20 mM MgCl$_2$; calcium ions cannot replace, and counteract the activation by, magnesium ions. Potassium ions are a potent activator, with a K_a of 28 mM; NH$_4^+$ is less effective, and Na$^+$ does not activate. At concentrations above 150 mM all 3 become inhibitory. Lithium salts inhibit at all concentrations.[25]

Subcellular localization has not been definitively established.[26] Activity has been detected in every vertebrate tissue examined, but is at least 5 times higher in brain and cultured neuronal cells[14,26] than in other tissues or derived cell cultures. Invertebrates have been little studied; ligase is absent from *Tetrahymena*[27] and present in sea urchin eggs[27] and *Leishmania*.[28]

Analytical Procedures Utilizing Tubulin-Tyrosine Ligase

Tubulin Detyrosinolation. Ligase will catalyze the release of tyrosine from tubulin in the presence of ADP + P$_i$, under the conditions shown in

[23] M. Flavin, T. Kobayashi, and T. M. Martensen, *Methods Cell Biol.* **25B,** 257 (1982).
[24] D. Raybin and M. Flavin, *Biochemistry* **16,** 2189 (1977).
[25] T. M. Martensen and M. Flavin, unpublished results.
[26] D. Raybin and M. Flavin, *J. Cell Biol.* **73,** 492 (1977).
[27] T. Kobayashi and M. Flavin, *Comp. Biochem. Physiol. B* **69B,** 387 (1981).
[28] J. Nath and M. Flavin, *J. Cell Biol.* **91,** 323a (1981).

the legend to Fig. 1. The reaction can be followed by measuring residual fixed labeled tyrosine (Fig. 1) or, more accurately, by precipitating the protein and measuring free tyrosine in the supernatant, as described below in the assay for carboxypeptidase-tubulin. Generally, however, pancreatic carboxypeptidase A (CPA) will be found more convenient for detyrosinolation.

Determination of Proportions of Different Species of Modified Tubulin. Tubulin from every source so far examined has been found to consist, in varying proportions, of 3 species: that which has C-terminal tyrosine, that which does not have but can accept tyrosine, and a third (so far obscure) species that appears not to be a substrate for tubulin-tyrosine ligase *in vitro*.[15,28] The procedure is based on determining the maximum amount of tyrosine which can be enzymatically added to tubulin, before and after incubating the latter with CPA. CPA removes tyrosine without attacking the penultimate glutamate[9] of the α chain. Tubulin suitable for assay may be isolated by ion-exchange columns,[19] by precipitation directly from extracts with vinblastine,[15] or may be microtubule protein purified by assembly cycles[19] but in the latter case may have lost tyrosine during the successive warm incubations.

Aliquots of approximately 50 μg of tubulin in 50 μl of 100 mM K$^+$MES (pH 6.8), 0.5 mM MgCl$_2$, and 1 mM each EGTA and GTP are incubated for 10 min at 37° in duplicate (4 tubes) with and without CPA, which has been diluted as a suspension in water, and is added to give a final concentration of 0.1 μg/ml, or enough to give a maximal effect.[15] The tubes are chilled, and 10 μl of 0.2 M K$^+$ β-phenylpropionate, pH 7.0, is added to those tubes containing CPA, to inhibit further activity. The β-phenylpropionate does not affect ligase activity or change the results when added to tubes without CPA. The following are then added to each tube to give the final concentration indicated (in 100 μl final volume): Tris–HCl, pH 7.2, 25 mM; KCl, 100 mM; MgCl$_2$, 12.5 mM; ATP, 2.5 mM; DTT, 10 mM; [^{14}C]tyrosine (50 μCi/μmol), 0.1 mM. Purified ligase is then added to each pair of duplicate tubes to give final concentrations of 0.5 and 1.0 unit/ml, and the mixtures are incubated for 30 min at 37°. The amount of tyrosine fixed is determined as described above under tubulin-tyrosine ligase assay. If the 2 amounts of ligase give the same amount of tyrosine fixation, this is taken to be the maximum that the tubulin can accept. The incremental fixation elicited by the prior CPA incubation measures the proportion of α chains that were originally tyrosinolated.

There are a number of possible sources of error in the interpretation of this analysis, particularly relating to the possibility that the "nonsubstrate" tubulin might be an analytical artifact. These have been discussed elsewhere,[23] and we believe successfully ruled out.

Preparation of Maximally Tyrosinolated or Detyrosinolated Tubulin.
Microtubule protein purified by 3 assembly cycles is tyrosinolated for 40
min at 37° in a mixture containing 0.9–1.1 unit/ml of purified ligase, microtubule protein at 5–6 mg/ml, and 100 mM K$^+$MES (pH 6.8), 2.5 mM ATP,
0.3 mM GTP, 2 mM DTT, 150 mM KCl, 5 mM MgSO$_4$, and 0.25 mM L-
tyrosine.[29] The reaction is stopped by lowering the temperature to 0°.
After 20 min the mixture is centrifuged for 30 min at +2° at 40,000 g. The
supernatant is equilibrated into 100 mM K$^+$MES (pH 6.8), containing 1
mM EGTA, 0.5 mM MgCl$_2$, and 0.1 mM GTP, by filtration through 10 bed
volumes of preequilibrated Sephadex G-50m. Fractions containing protein[30] are pooled and incubated 30 min at 32° in the presence of 2.5 mM
GTP. Polymerized (fourth cycle) tyrosinolated microtubule protein is collected by centrifugation for 60 min at 30–35° at 100,000 g. The centrifuge
tubes are immersed in liquid nitrogen, and the frozen pellets stored at
−70°.

Detyrosinolation is done by incubating microtubule protein, at 8–10
mg/ml, for 10 min at 37° in 100 mM K$^+$MES (pH 6.8), containing 1 mM
each of EGTA and GTP, 0.5 mM MgCl$_2$, and 0.25 μg/ml of CPA. The
CPA is then inactivated by incubating an additional 10 min at 37° in the
presence of 20 mM DTT, and polymerized detyrosinolated microtubule
protein is isolated as above.

Tubulin is prepared from microtubule protein by phosphocellulose
chromatography.[19] Microtubule protein is applied to a column (1 g moist
cake per 2.5 mg protein; width to height ratio 1 : 6) and eluted at +2° at 10
ml/cm^2 × hr with 100 mM K$^+$MES (pH 6.8), containing 2 mM each DTT
and EGTA, 0.5 mM MgSO$_4$, and 0.1 mM GTP.[29] As each fraction is
collected it is immediately supplemented with MgSO$_4$ and GTP to give
additional final concentrations of 1 and 0.5 mM, respectively. Tubulin is
precipitated from the pooled void volume fractions by adding an equal
volume of saturated ammonium sulfate in 100 mM K$^+$MES (pH 6.8),
containing 2 mM each DTT and EGTA, 0.5 mM MgSO$_4$, and 0.1 mM
GTP. Pellets obtained by centrifugation are dissolved in a small volume
of, and filtered through 10 bed volumes of Sephadex G-50m equilibrated
with, the same buffer but containing 0.5 mM GTP. Fractions containing
tubulin are pooled and stored as small aliquots at −70°.

By the assay described in the preceding section, the percentage of α
chains with C-terminal tyrosine is 45–55 in tyrosinolated, and 0 in detyrosinolated, tubulin.[29]

---

[29] N. Kumar and M. Flavin, *Eur. J. Biochem.* **128,** 215 (1982).
[30] M. Bradford, *Anal. Biochem.* **72,** 248 (1976).

Preparation of Substrate for Carboxypeptidase-Tubulin. Maximally ^{14}C-tyrosinolated microtubule protein is prepared by procedures in the preceding section.[11] Detyrosinolation is followed immediately by tyrosinolation with [^{14}C]tyrosine, 500 μCi/μmol, and a fourth assembly. For most purposes, substrate of lower specific radioactivity is prepared by adding carrier microtubule protein in which a known proportion (usually about 20%) of the α chains have C-terminal tyrosine. Knowing this percentage, one can recalculate the cpm/pmol tyrosine. The diluted microtubule protein is again assembled, and tubulin isolated by phosphocellulose chromatography, as described in the preceding section.

Carboxypeptidase-Tubulin

Assay Method

Principle. The assay is based on the rate of release of trichloroacetic acid soluble radioactivity from ^{14}C-tyrosinolated tubulin. Assembly-incompetent, phosphocellulose-purified tubulin must be used, or conditions employed under which tubulin does not assemble, to obtain consistent results. The assay is not suitable for surveying different tissue extracts, as endoprotease activity would not be excluded, and even in mammalian brain extract the activity is very low.

Procedure. Sufficient 100 mM K$^+$MES (pH 6.8), containing 1 mM each EGTA and GTP and 0.5 mM MgCl$_2$, is added to a 6 $\times$ 50-mm glass tube at 0° to give a final reaction volume of 15 μl. Next phosphocellulose-purified ^{14}C-tyrosinolated tubulin, prepared as in the preceding section, is added to a final concentration of 2 mg/ml (typically 30 μg protein = 45 pmol tyrosine = 5000 cpm). Carboxypeptidase-tubulin (CPT), 10 to 100 units/ml, is added and the reaction started by transferring the tubes to a 37° bath. The reaction is essentially linear with time between 2 and 20 min. The reaction is stopped by successive addition of 600 μl of cold 7% trichloroacetic acid and 40 μl of 1% bovine serum albumin. After vigorous mixing the tubes are centrifuged for 10 min at $+4$° at 2500 rpm in No. 408 adaptors in a Sorvall SS-34 rotor. Five hundred microliters of supernatant is carefully aspirated (we use a small-bore plastic tubing extension on the disposable tip of a Gilson Pipetman pipette) into a vial and counted with 12 ml of Aquasol. Blanks without CPT should be run corresponding to each time interval of incubation.

Definition of Unit and Specific Activity. A unit is defined as the amount of enzyme releasing 1 pmol of tyrosine in 1 min under the above conditions. The cpm/pmol is calculated as described above under "preparation of substrate for CPT." Specific activity is units/mg protein.

Purification Procedure

The starting material may be brain extract or the first warm supernatant after assembling microtubules,[19] but in the latter case glycerol must have been omitted. For the preparation shown in Table II, 925 ml of calf brain first warm supernatant was used.

All procedures are done at 0–4°. Solid ammonium sulfate (259 g) is added, while stirring with a propeller, to 45% of saturation. Stirring is continued for 15 min, followed by centrifugation for 20 min at 27,000 g. The supernatant is brought to 65% saturation (162 g) in the same fashion, and the pellet obtained after centrifugation is dissolved in 1/20 the original volume of 50 mM K$^+$MES (pH 6.7). The solution is dialyzed 17 hr against 2 changes of 3 liters each of the same buffer.

The ammonium sulfate fraction is applied to a 37 ml, 2.2 × 7-cm bed of DEAE-cellulose (DE-52) at a flow rate of 3 ml/min. CPT does not bind to the column. The bed is washed with 20 ml of buffer, and the fractions are combined.

The DE-52 fraction is pumped onto an 18 ml, 1.2 × 14-cm bed of (carboxymethyl)cellulose (CM-52), equilibrated with 50 mM K$^+$MES (pH 6.35), at a flow rate of 1.3 ml/min. The column is sequentially eluted with 33 ml of buffer, 33 ml of buffer + 50 mM KCl, and a linear gradient formed from 82 ml each of 50 mM and 350 mM KCl in buffer, collecting 12 ml fractions.

The 4 active fractions (Fig. 3) are pooled in flexible dialysis sacks, which are placed in the botton of a large beaker containing 250 ml (5 volumes) of 90% saturated ammonium sulfate in 100 mM K$^+$MES (pH

TABLE II
PURIFICATION OF CARBOXYPEPTIDASE-TUBULIN

Step	Volume (ml)	Total protein (mg)	Total activity (units)	Specific activity (units/mg)	Yield (%)	Purification (fold)
1. First warm super-natant[a]	930	7400	~7400	~1[b]	(100)	(1)
2. Ammonium sulfate fraction	80	1700	6200	3.7	84	3.7
3. DEAE-cellulose	75	780	6100	7.8	83	7.8
4. CM-cellulose	50	15	3200	210	43	210
5. Concentrated	0.45	8	1900	230	26	230

[a] The supernatant obtained after assembling microtubules from an extract of fresh calf brain[19]; glycerol must however be omitted from the buffers used.

[b] Not reliably measurable.

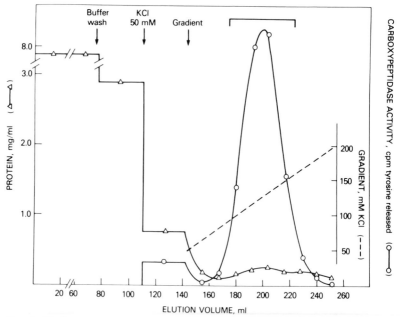

FIG. 3. Elution profile of protein and carboxypeptidase-tubulin activity from (carboxymethyl)cellulose.

6.7) [made by mixing 9 volumes of saturated aqueous ammonium sulfate and 1 of M K$^+$MES (pH 7.0)]. The beaker is gently agitated on a rotary shaker. After 3 hr the volume inside the membrane is reduced to 16 ml, and protein is maximally precipitated. After centrifugation for 15 min at 30,000 g, the pellet is dissolved in 0.3 ml of 50 mM K$^+$MES (pH 6.7), and dialyzed overnight against 2 changes of 200 ml each of the same buffer. After centrifugation to remove a small precipitate, this enzyme is stored at $-70°$ (but appears to be quite stable and may not need this low a temperature). Table II illustrates a typical preparation.

Properties of CPT

The native molecular weight is about 72,000.[31] The strong affinity of CPT for CM-cellulose at low pH (Fig. 3) could be due to affinity for the carboxyl group, but probably also indicates that CPT is a basic protein. If the purified CPT is added in very large amounts to a tubulin solution a precipitate forms, and several small basic proteins have been identified in

[31] T. M. Martensen, Fed. Proc., Fed. Am. Soc. Exp. Biol. 39, 2161 (1980).

the CPT fraction, which aggregate tubulin in much the same fashion that polylysine does.[32,33] The subcellular localization of CPT is not known.

CPT, like CPA, releases only tyrosine from α chain[11] and has no contaminating endoprotease. It has no activity toward aldolase, another protein with C-terminal tyrosine, but is probably not as highly specific as the ligase since denatured tubulin is also slowly detyrosinolated. An interesting property is that it is more active on polymeric and oligomeric forms.[11] That is the reason irregular results may be obtained if assembly-competent tubulin is used for assays; α chain in microtubules is detyrosinolated about 5 times more rapidly than α chain in dimeric tubulin.

Properties of Modified Tubulin

Differences in the relative proportions of tyrosinolated and detyrosinolated tubulin have been reported in different cells and tissues, different subcellular compartments, and in cells in different states, or tissues in different stages of development.[8,14,34–37] However, no real clue to the function of this modification has emerged in the 10 years since its discovery. Marginal but significant effects on *in vitro* polymerization have been reported[29] and a possible role in interaction with membranes has been considered.[8] Some form of coupling between cycles of assembly and tyrosinolation has been suggested by the observation that the enzyme which adds tyrosine acts only on the subunit whereas the enzyme which removes it acts preferentially on the polymer. However, studies of *in vivo* tyrosine turnover in a number of cell types has shown this always to be enhanced when a larger proportion of tubulin is depolymerized.[28]

[32] N. Kumar and M. Flavin, *Biochem. Biophys. Res. Commun.* **106,** 704 (1982).
[33] K. Miyatake and M. Flavin, *J. Cell Biol.* **95,** 341a (1982).
[34] J. Nath, J. Whitlock, and M. Flavin, *J. Cell Biol.* **79,** 294a (1978).
[35] J. A. Rodriguez and G. G. Borisy, *Biochem. Biophys. Res. Commun.* **83,** 579 (1978).
[36] J. A. Rodriguez and G. G. Borisy, *Biochem. Biophys. Res. Commun.* **89,** 893 (1979).
[37] N. Kumar, R. Blumenthal, M. Henkart, J. N. Weinstein, and R. D. Klausner, *J. Biol. Chem.* **257,** 15137 (1982).

[23] Ubiquitination of Proteins

By HARRIS BUSCH

Ubiquitin Structure

The small polypeptide, ubiquitin (Ub), has a molecular weight of 8565 and in its biologically active form contains 76 amino acids.[1,2] It was initially purified by G. Goldstein and his associates in studies on thymus peptides. With a radioimmunoassay it was found that it was widely distributed in plant, animal, yeast, and bacterial cells.[2-4] The sequence of active ubiquitin is

$$
\begin{array}{llll}
1 & 5 & 10 & 15 & 20 \\
\end{array}
$$

Met-Gln-Ile-Phe-Val-Lys-Thr-Leu-Thr-Gly-Lys-The-Ile-Thr-Leu-Glu-Val-Glu-Pro-Ser-

Asp-Thr-Ile-Glu-Asn-Val-Lys-Ala-Lys-Ile-Gln-Asp-Lys-Glu-Gly-Ile-Pro-Pro-Asp-Gln-

Gln-Arg-Leu-Ile-Phe-Ala-Gly-Lys-Gln-Leu-Glu-Asp-Gly-Arg-Thr-Leu-Ser-Asp-Tyr-Asn-

Ile-Gln-Lys-Glu-Ser-Thr-Leu-His-Leu-Val-Leu-Arg-Leu-Arg-Gly-Gly-

Preliminary analysis of the structure of the amino-terminal octapeptide of ubiquitin from celery showed that six of the eight residues were the same as for thymus ubiquitin. The Gly_{75}-Gly_{76} residues are necessary for the biological activity of this structure.[5-9]

The three-dimensional structure of ubiquitin is such that this molecule is extremely resistant to tryptic digestion despite the presence of seven lysine residues and four arginine residues.[2,3,10] It is likely that this structure is tightly coiled in such a way that the basic amino acids are unavailable to the active sites of trypsin, and other proteolytic enzymes. Only after maleylation was the molecule susceptible to the cleavages necessary for structural analysis. Studies with nuclear magnetic resonance show

---

[1] H. Busch and I. L. Goldknopf, *Mol. Cell. Biochem.* **40**, 173 (1981).

[2] G. Goldstein, M. Scheid, U. Hammerling, E. A. Boyse, D. H. Schlesinger, and H. D. Niall, *Proc. Natl. Acad. Sci. U.S.A.* **72**, 11 (1975).

[3] D. H. Schlesinger, G. Goldstein, and H. D. Niall, *Biochemistry* **14**, 2214 (1975).

[4] T. L. K. Low and A. L. Goldstein, *J. Biol. Chem.* **254**, 987 (1979).

[5] I. L. Goldknopf and H. Busch, *Proc. Natl. Acad. Sci. U.S.A.* **74**, 864 (1977).

[6] I. L. Goldknopf and H. Busch, in "The Cell Nucleus" (H. Busch, ed.), Vol. 6, p. 149. Academic Press, New York, 1978.

[7] I. L. Goldknopf and H. Busch, *Biochem. Biophys. Res. Commun.* **96**, 1724 (1980).

[8] M. W. Andersen, I. L. Goldknopf, and H. Busch, *FEBS Lett.* **132**, 210 (1981).

[9] H. D. Wilkinson and T. K. Audhya, *J. Biol. Chem.* **256**, 9235 (1981).

[10] R. E. Lenkinski, D. M. Chen, J. D. Glickson, and G. Goldstein, *Biochim. Biophys. Acta* **494**, 126 (1977).

that ubiquitin has a highly globular, compact, pH- and temperature-resistant conformation; 7 M guanidine hydrochloride is required to denature it. On the basis of the prediction of protein conformation by Chou and Fasman,[11] the 15 NH_2-terminal residues (12%) were suggested to be a pleated sheet. The COOH-terminal (28%) was suggested to be a helical region.[12] Its activity as a hormone is still questionable.[13] The interest in ubiquitin was strongly stimulated by the remarkable finding that it existed as a free polypeptide or in a conjugated form associated with other proteins. When we first proposed that ubiquitin was covalently bound to histone 2A, it was pointed out that there was no previous evidence for proteins linked together by covalent bonds and that this was a most unusual type of linkage. As the evidence for such a structure increased, as indicated previously, it was uncertain whether this was simply a unique event or whether other types of protein–protein conjugations might exist. The further development of information by several related groups on ubiquitination of protein-proteins in red cells prior to proteolysis further extended the evidence for functional roles for these molecules (Table I).

Ubiquitin-Histone 2A (Ub-2A)

Ub-2A was first found in the group of 100 0.4 N H_2SO_4 soluble proteins of nucleoli and nuclei resolved by two-dimensional polyacrylamide gel electrophoresis.[14,15] The presence of ubiquitin in histone 2A was at first unsuspected. It was found after purification of Ub-2A, peptide analysis and amino acid-sequencing procedures.[16–18] Ub-2A had a "Y-shape" with 2 N-terminal and 1 C-terminal amino acids.[16]

The NH_2-terminal sequence[18] of Ub-2A showed it contained the sequence of ubiquitin (Ub).[19] Ubiquitin was covalently linked to the histone

[11] P. Y. Chou and G. D. Fasman, *Biochemistry* **13**, 222 (1974).

[12] P. D. Cary, S. D. King, C. Crane-Robinson, E. M. Bradbury, A. Rabbani, G. H. Goodwin, and E. W. Johns, *Eur. J. Biochem.* **112**, 577 (1980).

[13] T. L. K. Low, G. B. Thurman, M. McAdoo, J. McClure, J. L. Rossio, P. H. Naylor, and A. L. Goldstein, *J. Biol. Chem.* **254**, 981 (1979).

[14] L. R. Orrick, M. O. J. Olson, and H. Busch, *Proc. Natl. Acad. Sci. U.S.A.* **70**, 1316 (1973).

[15] L. C. Yeoman, C. W. Taylor, and H. Busch, *Biochem. Biophys. Res. Commun.* **51**, 956 (1973).

[16] I. L. Goldknopf, C. W. Taylor, R. M. Baum, L. C. Yeoman, M. O. J. Olson, A. W. Prestayko, and H. Busch, *J. Biol. Chem.* **250**, 7182 (1975).

[17] I. L. Goldknopf and H. Busch, *Biochem. Biophys. Res. Commun.* **65**, 951 (1975).

[18] M. O. J. Olson, I. L. Goldknopf, K. A. Guetzow, G. T. James, T. C. Hawkins, C. J. Mays-Rothberg, and H. Busch, *J. Biol. Chem.* **251**, 5901 (1976).

[19] L. T. Hunt and M. O. Dayhoff, *Biochem. Biophys. Res. Commun.* **74**, 650 (1977).

TABLE I

STUDIES IDENTIFYING VARIOUS PROTEINS AS FREE OR CONJUGATED UBIQUITIN

Form and cellular localization	Original nomenclature	Isolation[a]	Structural identification[a]
Cytoplasmic free	UBIP, ubiquitin	Goldstein et al. (1)	Schlesinger et al. (7)[b]
Cytoplasmic free and conjugated	Peptide of bovine Pars Intermedia APFI	Ciechanover et al. (2)	Hershko et al. (8)[c]; Wilkinson et al. (9)[b]
Nuclear free	Ubiquitin	Goldknopf et al. (3)	Goldknopf et al. (3)[b]
	HMG20	Walker et al. (4)	Walker et al. (4)[b];
	Nonhistone protein S		Watson et al. (10)[b]
Nuclear conjugated	A24 = H2A + Ub	Goldknopf et al. (5)	Goldknopf and Busch (11)[c]; Goldknopf and Busch (12)[c]; Olson et al. (13)[b]; Hunt and Dayhoff (14)[d]
	uH2B = H2B = Ub	West and Bonner (6)	West and Bonner (6)[b,c]

[a] Key to references: (1) G. Goldstein, M. Scheid, V. Hammerling, E. A. Boyse, D. H. Schlesinger, and H. D. Niall, *Proc. Natl. Acad. Sci. U.S.A.* **72**, 11 (1975). (2) A. Ciechanover, Y. Hod, and A. Hershko, *Biochem. Biophys. Res. Commun.* **81**, 1100 (1978). (3) I. L. Goldknopf, M. F. French, Y. Daskal, and H. Busch, *Biochem. Biophys. Res. Commun.* **84**, 786 (1978). (4) J. M. Walker, G. H. Goodwin, and E. W. Johns, *FEBS Lett.* **90**, 327 (1978). (5) I. L. Goldknopf, C. W. Taylor, R. M. Baum, L. C. Yeoman, M. O. J. Olson, A. W. Prestayko, and H. Busch, *J. Biol. Chem.* **250**, 7182 (1975). (6) M. H. P. West and W. M. Bonner, *Nucleic Acids Res.* **8**, 4671 (1980). (7) D. H. Schlesinger, G. Goldstein, and H. D. Niall, *Biochemistry* **14**, 2214 (1975). (8) A. Hershko, A. Ciechanover, H. Heller, A. L. Haas, and I. A. Rose, *Proc. Natl. Acad. Sci. U.S.A.* **77**, 1783 (1980). (9) K. D. Wilkinson, M. K. Urban, and A. L. Haas. *J. Biol. Chem.* **255**, 7529 (1980). (10) D. C. Watson, B. Levy-Wilson, and G. H. Dixon, *Nature (London)* **276**, 196 (1978). (11) I. L. Goldknopf and H. Busch, *Biochem. Biophys. Res. Commun.* **65**, 951 (1975). (12) I. L. Goldknopf and H. Busch, *Proc. Natl. Acad. Sci. U.S.A.* **74**, 864 (1977). (13) M. O. J. Olson, I. L. Goldknopf, K. A. Guetzow, G. T. James, T. C. Hawkins, C. J. Mays-Rothberg, and H. Busch, *J. Biol. Chem.* **251**, 5901 (1976). (14) L. T. Hunt and M. O. Dayhoff, *Biochem. Biophys. Res. Commun.* **74**, 650 (1977).

[b] Identified as ubiquitin.

[c] Identified as conjugated ubiquitin.

[d] Identity to ubiquitin determined by amino acid sequence comparison.

2A, specifically on the ε-amine of lysine-119.[5] This remarkable structure led to investigations on similar structures in other histones although functions of such linkages are not yet defined.

Two-dimensional polyacrylamide gel electrophoresis[14] (Fig. 1a) showed that Ub-2A migrated slightly slower than the histones in the urea-acetic acid first dimension and between histone H1 and H2B in the sodium dodecyl sulfate-containing second dimension.[20] Ub-2A was in acid extracts of whole nuclei (Fig. 1b). It was in relatively similar amounts in nuclear, nucleolar, and extranucleolar extracts, which indicated that Ub-2A was present throughout the cell nucleus.[15]

Ub-2A was of interest partly because its nucleolar content markedly decreased in rat liver during thioacetamide administration and in regenerating liver[21–23] (Fig. 1c and d). Since these nucleoli have high levels of rRNA synthesis,[24,25] the decrease in Ub-2A content as well as the presence of free ubiquitin in some active fractions,[26–28] suggested that cleavage of Ub-2A might relate to increased gene activity.

Isolation of Ub-2A

The starting material for purification of Ub-2A[16] was calf thymus chromatin preextracted with 0.35 M NaCl and 0.5 M perchloric acid (Fig. 2a). Ub-2A was in the 0.4 N H_2SO_4 extract along with other histones (Fig. 2a). Ub-2A was separated on Sephadex G-100 (Fig. 2b) and purified on slab gels (Figs. 3 and 4); it migrated in two-dimensional polyacrylamide gel electrophoresis as a single component (Fig. 2c).

The Primary Structure of Ub-2A

The amino acid compositions of Ub-2A and histone 2A (Table II)[17] showed they had similar contents of lysine, histidine, aspartic acid, leucine, and phenylalanine. Ub-2A had a lower molar ratio of arginine and alanine than histone 2A. The content of threonine, glutamic acid, and

[20] I. L. Goldknopf and H. Busch, *Physiol. Chem. Phys.* **5**, 131 (1973).

[21] N. R. Ballal and H. Busch, *Cancer Res.* **33**, 2737 (1973).

[22] N. R. Ballal, I. L. Goldknopf, D. A. Goldberg, and H. Busch, *Life Sci.* **14**, 1835 (1974).

[23] N. R. Ballal, Y.-J. Kang, M. O. J. Olson, and H. Busch, *J. Biol. Chem.* **250**, 5921 (1975).

[24] H. Busch and K. Smetana, "The Nucleolus." Academic Press, New York, 1970.

[25] M. W. Andersen, N. R. Ballal, and H. Busch, *Biochem. Biophys. Res. Commun.* **78**, 129 (1977).

[26] I. L. Goldknopf, M. F. French, Y. Daskal, and H. Busch, *Biochem. Biophys. Res. Commun.* **84**, 786 (1978).

[27] D. C. Watson, B. Levy-Wilson, and G. H. Dixon, *Nature (London)* **276**, 196 (1978).

[28] L. Kuehl, T. Lynes, G. H. Dison, and B. Levy-Wilson, *J. Biol. Chem.* **255**, 1090 (1980).

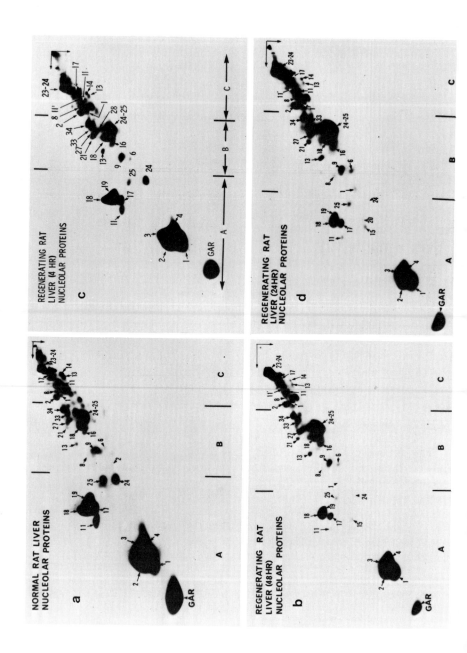

TABLE II

COMPARISON OF Ub-2A AND HISTONE 2A

Residue	Protein A24[a] (mol%)	Histone 2A[b] (mol%)
Trp	0.0	0.0
Lys	11.3	10.9
His	2.4	3.1
Arg	7.4	9.3
Asx	7.3	6.2
Thr	6.5	3.9
Ser	4.5	3.1
Glx	12.3	9.3
Pro	5.6	3.9
Gly	9.2	10.9
Ala	9.6	13.4
Val	4.9	6.2
Met	0.3	0.0
Ileu	5.8	4.7
Leu	10.9	12.4
Try	1.3	2.3
Phe	0.9	0.8
Lys + His + Arg	21.1	23.3
Glx + Asx	19.6	15.5
Glx + Asx/Lys + His + Arg	0.93	0.67
Dansylatable NH₂-terminal	Methionine	Blocked (acetyl-Ser)
Molecular weight	27 000	14 000

[a] Goldknopf et al.[16]

[b] Yeoman et al.[29]

FIG. 1. (a) Two-dimensional polyacrylamide gel electrophoresis of normal rat liver nucleolar acid-soluble proteins. Protein samples (750 μg) were run in the first dimension in gel rods of 10% polyacrylamide/4.5 M urea/0.9 N acetic acid at 120 V for 6 hr. For the second dimension, a 12% polyacrylamide–0.1% sodium dodecyl sulfate slab gel was run for 15 hr at 50 mA per slab. Gels were stained with Coomassie brilliant blue R. (b) Two-dimensional polyacrylamide gel electrophoresis of nucleolar acid-soluble proteins of livers of rats after 48 hr of partial hepatectomy. Conditions were the same as those described in (a). (c) Two-dimensional polyacrylamide gel electrophoresis of nucleolar acid-soluble proteins of livers of rats after 4 hr of partial hepatectomy. Conditions were the same as those described in (a) (d) Two-dimensional polyacrylamide gel electrophoresis of nucleolar acid-soluble proteins of livers of rats after 24 hr of partial hepatectomy. Conditions were the same as those described in (a).

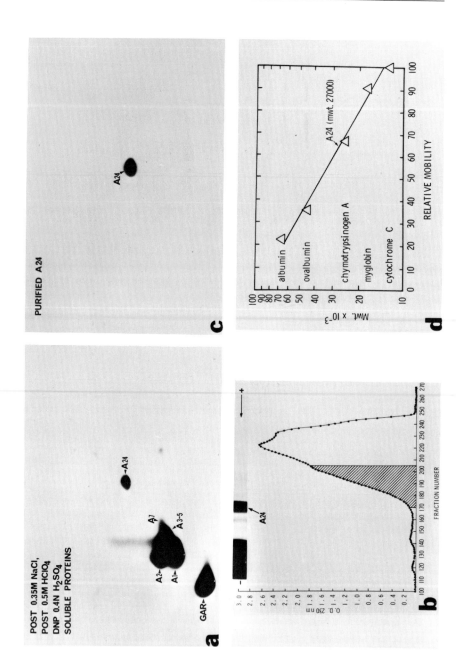

proline was higher in Ub-2A; it also had a higher acidic–basic amino acid molar ratio than histone 2A. Neither protein contained tryptophan while histone 2A contained no methionine. The methionine residue was an amino-terminal amino acid of Ub-2A as shown by dansylation.

Peptide maps of these proteins had striking similarities.[17] The tryptic peptides of histone 2A[29] were present in the corresponding pattern of Ub-2A. Similar results were found with chymotryptic peptides.

Because Ub-2A had a methionine NH_2-terminus and histone 2A contains no methionine and Ub-2A had a molecular weight greater than histone 2A and had additional peptides not in histone 2A upon both tryptic and chymotryptic digestion, it seemed that Ub-2A contained the total amino acid sequence of histone 2A along with an additional structure.[17]

The Terminal Amino Acid Sequences of Ub-2A

Peptide 16 in Ub-2A was not found by the ninhydrin-cadminum or fluorescamine procedures which require a primary amino group (Table III); peptides 16[5] from histone 2A and Ub-2A (Table III) had identical staining characteristics, amino acid composition, and positive tests for the acetyl group. Those results were consistent with the blocked amino terminal tryptic peptide of histone 2A: N-acetyl-Ser-Gly-Arg.

The fact that Ub-2A contained a second NH_2-terminus, a methionine, suggested that it had a nonhistone sequence. This unblocked sequence was amenable to automatic Edman degradation without prior separation of the two polypeptides. The sequence of the first 37 NH_2-terminal residues[18]

5   10   15   20

Met-Gln-Ile-Phe-Val-Lys-Thr-Leu-Thr-Gly-Lys-Thr-Ile-Thr-Leu-Glu-Val-Glu-Pro-Ser-

25   30   35

Asp-Thr-Ile-Glu-Asn-Val-Lys-Ala-Lys-Ile-Gln-Asp-Lys-Glu-Gly-Ile-Pro

was identical to that of ubiquitin.[3,19]

[29] L. C. Yeoman, M. O. J. Olson, N. Sugano, J. J. Jordan, C. W. Taylor, W. C. Starbuck, and H. Busch, *J. Biol. Chem.* **247**, 6018 (1972).

FIG. 2. (a) Two-dimensional gel (see Fig. 1) of rat liver 0.4 *N* H_2SO_4-soluble proteins of the residual DNP after three 0.35 *M* NaCl extractions followed by two 5% perchloric acid extractions. (b) Sephadex G-100 column chromatography of the protein A24-enriched acid-soluble proteins from calf thymus tissues. The fractions containing protein A24 and histones 2B and 3 (indicated by the shaded area of the graph) were pooled and used for preparative electrophoresis as shown in Fig. 3. (c) Two-dimensional gel electrophoresis of purified calf thymus protein A24 by preparative electrophoresis. (d) Estimation of the molecular weight of protein A24. The molecular weight was determined as described using sodium dodecyl sulfate gels and known molecular weight markers.

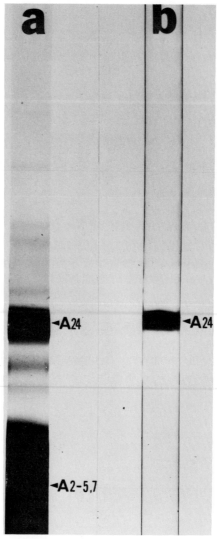

FIG. 3. Purification of protein A24 by preparative electrophoresis. (a) an amido black-stained vertical side strip cut from a preparative 10% polyacrylamide slab gel after electrophoresis of the pooled fractions obtained from column chromatography on Sephadex G-100. The position of protein A24 and histones 2B (A2) and 3 (A3-5,7) are indicated. Such vertical strips from both sides and the center were used as a guide to cut out horizontal sections of the unstained remainder of the slabs which contained the protein A24 band. The protein A24 was then obtained by electrophoresis out of the gel sections into dialysis tubing. (b) Reelectrophoresis of purified protein A24 obtained in (a) on 10% polyacrylamide gels.

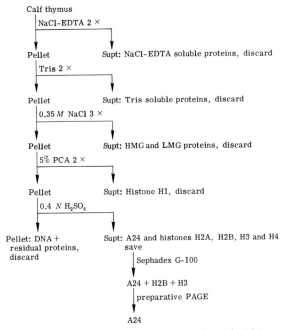

FIG. 4. Flow sheet for preparation of protein A24.

TABLE III
ANALYSIS OF TRYPTIC PEPTIDE 16 OF PROTEIN
A24 AND HISTONE 2A

	Protein A24	Histone 2A
Staining reactions		
Fluorescamine	−	−
Ninhydrin	−	−
Sakaguchi	+	+
Rydon-Smith	+	+
Amino acid ratios		
Ser (= 1.0)	1.00	1.00
Gly	1.09	1.15
Arg	1.12	1.21
Molar yield[a]	0.42	0.46
Acetyl group[b]	+	+

[a] Nanomoles recovered from tryptic peptide
maps/nanomole of protein digested.
[b] Detected as acetyl hydrazide.

TABLE IV
RATIOS OF FREE AMINO ACIDS RELEASED FROM PROTEIN A24 AND
HISTONE 2A DIGESTION WITH CARBOXYPEPTIDASES A AND B

Amino acid	Step 1[a] residues released from		Step 2[b] residues released from	
	Protein A24	Histone 2A	Protein A24	Histone 2A
Lysine	1.0	1.0	3.0	3.0
Histidine	0.1	0.1	0.9	1.0
Alanine	0.1	0.1	0.9	0.7
Glycine	0.1	0.1	0.9	0.6

[a] Carboxypeptidase B: 60-min digestion followed by boiling for 5 min.
[b] Carboxypeptidase A: After step 1, 60-min digestion.

The COOH-terminal amino acid sequence[30] and rate-limiting steps for carboxypeptidase B and A digestions of histones 2A were

-His-His-Lys-Ala-Lys-Gly-Lys-COOH
▲   ▲
A   B

The values in Table IV which equated the nanomoles of lysine to 1.0 after carboxypeptidase B and to 3.0 after carboxypeptidase A digestions were identical for both proteins.[17] Ub-2A had the same COOH-terminal sequence as histone 2A. Quantitative hydrazinolysis indicated that lysine was the sole COOH-terminal amino acid of Ub-2A; its molar yield was identical to that of histone 2A.[5]

The Branched Tryptic Peptide of Ub-2A

The detection of two amino-termini and one carboxyl-terminus[5] indicated that the Ub-2A molecule was branched and that the ubiquitin was linked to histone 2A in a manner which prevented detection of its carboxyl-terminus. Tryptic peptide 17' from Ub-2A had a slightly different electrophoretic mobility from peptide 17 of histone 2A (Fig. 5). The amino acid composition and carboxyl terminus of 17' of Ub-2A (Table V) were the same as that of peptide 17 (residues 119–124) of histone 2A; it had two additional glycine residues.

The first cycle of Edman degradation of the Ub-2A peptide was consistent with a branched structure in which the two glycines were attached to

[30] N. Sugano, M. O. J. Olson, L. C. Yeoman, B. R. Johnson, C. W. Taylor, W. C. Starbuck, and H. Busch, *J. Biol. Chem.* **247**, 3589 (1972).

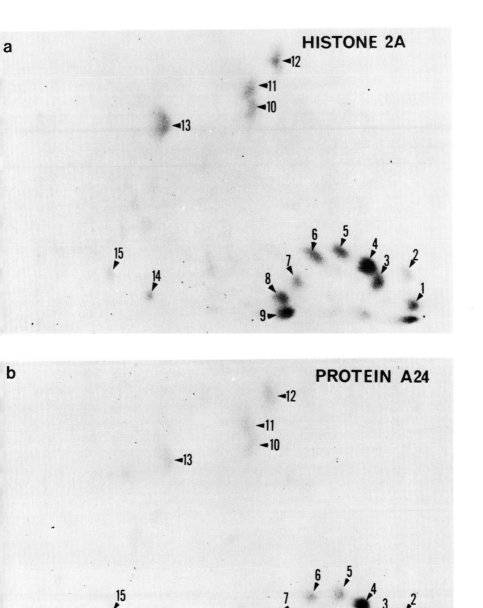

FIG. 5. Two-dimensional tryptic peptide maps of (a) histone 2A and (b) protein A24. Vertical dimension is chromatography and horizontal dimension is electrophoresis. Spots 1–16 are common to both proteins. Note the positions of peptides 16 which were negative to the ninhydrin-cadmium stain used in the maps. Note also the position of peptides 17 of histone 2A and 17′ of protein A24, which differ slightly in electrophoretic mobility.

TABLE V
AMINO ACID RATIOS AND CARBOXYL-TERMINI
OF PEPTIDES 17 OF HISTONE 2A AND 17' OF
PROTEIN A24

	17	17'
Gly		1.64
Lys	1.67	1.68
His	1.71	1.61
Thr	0.91	0.93
Ser (= 1.00)	1.00	1.00
Glu	0.95	0.94
COOH-terminus[a]	Lys (0.57)[b]	Lys (0.64)[b]
Molar yield[c]	0.67	0.74

[a] Data obtained by hydrazinolysis, only lysine was found.
[b] Data in parentheses are molar yield of carboxyl-terminal lysine.
[c] Nanomoles of peptide recovered from tryptic peptide maps/nanomole of protein digested.

the rest of the peptide in an isopeptide linkage to the ε-amino group of lysine (cf. below). Coupling with phenylisothiocyanate occurred at the α-NH$_2$ groups of both lysine and glycine.[5] Cyclization produced two thiazolinones, one containing glycine and one containing lysine with a glycine residue attached to its ε-NH$_2$ group. HI hydrolysis of these thiazolinones released two glycine residues and one lysine residue. Conversion of the thiazolinones to phenylthiohydantoins (PTH) produced sufficient PTH-glycine to account for half of the glycine in the thiazolinones. Gas chromatographic analysis of phenylthiohydantoins showed there was 0.9 nmol of PTH-glycine/nmol of peptide, or approximately one-half of the amount obtained from hydrolysis of thiazolinones. These data and subsequent cycles of Edman degradation,[5] which produced essentially identical results for both peptides (Table VI), supported the branched structure of this peptide and provided the first evidence for the -Gly-Gly isopeptide bond (Fig. 6).

Trypsin treatment of Ub-2A yielded intact ubiquitin[5] as expected from the known resistance of ubiquitin to trypsin.[3] The Arg-Gly-Gly linkage was cleaved under these conditions.[5] Treatment of Ub-2A with the arginine-blocking reagent 1,2-cyclohexanedione,[31] followed by trypsinization,

[31] L. Patthy and E. L. Smith, J. Biol. Chem. 250, 557 (1975).

TABLE VI

SEQUENTIAL EDMAN DEGRADATION OF PEPTIDES
17 AND 17'[a]

Amino acid	Edman cycle			
	1	2	3	4
Peptide 17 of histone 2A				
Glys				
Lys	0.5			
Thr		1.0		
Glu			0.9	
Ser				0.3
Peptide 17' of protein A24				
Gly	1.6	0.5		
Lys	0.4	0.1		
Thr		1.1		
Glu			0.8	
Ser				0.3

[a] Nanomoles of amino acids/nanomoles of peptide
released from thiazolinones after hydrolysis with
H1.

produced[6] a trypsin-resistant polypeptide which included the branched
peptide 17' linked to arginine 74 of the carboxyl-terminus of ubiquitin and
the glycines of peptide 17'. Two-dimensional polyacrylamide gel electrophoretic mobility and amino acid compositions of fragments of Ub-2A

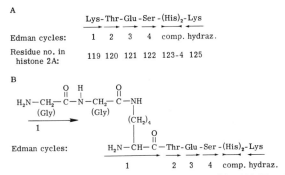

FIG. 6. Structure and outline of proof of structures of peptides (A) 17 and (B) 17'. Note
the identical sequence Lys-Thr-Glu-Ser-His-His-Lys in both peptides and the position number of these amino acid residues in the histone 2A sequence. Comp., amino acid composition; hydraz, hydrazinolysis.

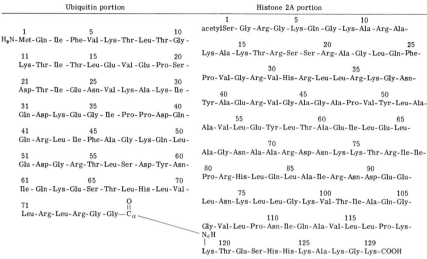

FIG. 7. The complete amino acid sequence of Ub-2A (protein A24) wherein the carboxyl terminal glycine 76 of ubiquitin is attached to the ε-NH$_2$ of lysine 119 of histone 2A by an isopeptide linkage.

from cleavage at tyrosines with N-bromosuccinimide accounted for the whole structure of Ub-2A[7] as shown in Fig. 7.

The novel bifunctional structure of Ub-2A immediately raised questions about other bifunctional proteins. Other proteins contain two functional units; β-lipotropin has both melanocyte-stimulating hormone and β-endorphin,[32–35] but both units are within a single polypeptide chain. The linkage of the histone 2A and ubiquitin by an isopeptide suggested a posttranslational conjugation. Isopeptide linkages have been found in collagen,[36] fibrin,[37,38] peptidoglycans,[39] wheat germ,[40] bovine colostrum,[41]

[32] A. F. Bradbury, D. G. Smyth, and C. R. Snell, *Biochem. Biophys. Res. Commun.* **69**, 950 (1976).

[33] B. M. Cox, A. Goldstein, and C. H. Li, *Proc. Natl. Acad. Sci. U.S.A.* **73**, 1821 (1976).

[34] R. Guillemin, N. Long, and R. Burgus, *C.R. Hebd. Seances Acad. Sci., Ser. D* **282**, 783 (1976).

[35] C. H. Li, D. Chung, and B. A. Doneen, *Biochem. Biophys. Res. Commun.* **72**, 1542 (1976).

[36] G. L. Mechanic and M. Levy, *J. Am. Chem. Soc.* **81**, 1889 (1958).

[37] J. J. Pisano, J. S. Finlayson, and M. Peyton, *Science* **160**, 892 (1968).

[38] S. Matacic and A. G. Loewy, *Biochem. Biophys. Res. Commun.* **30**, 356 (1968).

[39] P. Dezelee and G. D. Shockman, *J. Biol. Chem.* **250**, 6806 (1975).

[40] R. Tkachuk and V. J. Mellish, *Can. J. Biochem.* **55**, 295 (1977).

[41] H. Klostermeyer, K. Rabbel, and E. H. Reimerdes, *Hoppe-Seyler's Z. Physiol. Chem.* **357**, 1197 (1976).

hair medulla protein,[42] chorion of rainbow trout,[43] and *Escherichia coli* ribosomal protein S11.[44] However, histones and nonhistone chromosomal proteins had not been shown to contain such a linkage previously and the ubiquitin–histone 2A bond was the first such conjugate demonstrated.

Approximately 10% of histone 2A is in the form of protein A24,[16,45,46] and this quantitative distribution is the same for all the postsynthetic modification[47,48] and amino acid sequence[45,46] variants of histone 2A.

A ubiquitin adduct of histone 2B, Ub-2B, has been described in which 1–1.5% of histone 2B is conjugated in the carboxyl terminal portion with ubiquitin.[49] This concentration of Ub-2B in the nucleus is approximately one-tenth that of Ub-2A. Both sequence variants of histone 2B are conjugated to the same extent.

Chromosomal Localization—The Presence of Ub-2A in Nucleosomes

Analysis of the distribution of Ub-2A in chromatin[16] showed Ub-2A and the histones were not in the saline-EDTA (0.075 M NaCl/0.025 M EDTA, pH 8.0) (Fig. 8b) or the 0.01 M Tris (Fig. 8c) washes. They were found in the 0.4 N H_2SO_4-soluble proteins of chromatin (Fig. 8d). Ub-2A and the histones were not solubilized when the chromatin was treated with 0.35 M NaCl (Fig. 9a),[16] which extracts the HMG and low-LMG nonhistone chromosomal proteins.[50]

The 0.4 N H_2SO_4-soluble proteins of the chromatin residue contained Ub-2A and the histones (Fig. 9b). After extraction of histone H1 with 0.6 M NaCl or with 0.5 M $HClO_4$, Ub-2A and histones 2A, 2B, 3, and 4 were not extracted (Figs. 2a and 9d, respectively). Ub-2A accounted for 1.9% of the total of histones 1, 2A, 2B, 3, and 4.

When the histones and most nonhistone chromosomal proteins were extracted from chromatin with 3 M NaCl–7 M urea, Ub-2A was also extracted (Fig. 9c). Ub-2A and the histones reassociated with the DNA by one-step or gradient dilution to low ionic strength. Ub-2A had similar binding characteristics to histones 2A, 2B, 3 and 4.

[42] H. W. Harding and G. W. Rogers, *Biochim. Biophys. Acta* **427**, 315 (1976).
[43] H. E. Hagenmaier, I. Schmitz, and J. Fohles, *Hoppe-Seyler's Z. Physiol. Chem.* **357**, 1435 (1976).
[44] R. Chen and U. Chen-Schmeisser, *Proc. Natl. Acad. Sci. U.S.A.* **74**, 4905 (1977).
[45] S. C. Albright, P. P. Nelson, and W. T. Garrard, *J. Biol. Chem.* **254**, 1065 (1979).
[46] M. H. P. West and W. M. Bonner, *Biochemistry* **19**, 3238 (1980).
[47] I. L. Goldknopf, F. Rosenbaum, R. Sterner, G. Vidali, V. Allfrey, and H. Busch, *Biochem. Biophys. Res. Commun.* **90**, 269 (1979).
[48] H. Okayama and O. Hayaishi, *Biochem. Biophys. Res. Commun.* **84**, 755 (1978).
[49] M. H. P. West and W. M. Bonner, *Nucleic Acids Res.* **8**, 4671 (1980).
[50] G. H. Goodwin, C. Sanders, and E. W. Johns, *Eur. J. Biochem.* **38**, 14 (1973).

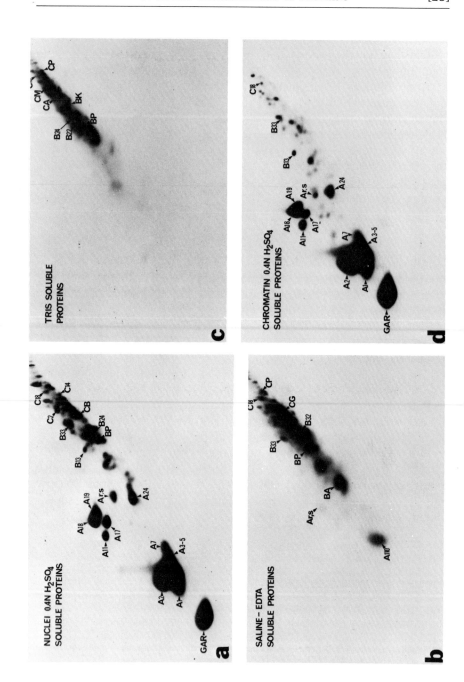

TRIS SOLUBLE
PROTEINS

c

CHROMATIN 0.4N H₂SO₄
SOLUBLE PROTEINS

d

NUCLEI 0.4N H₂SO₄
SOLUBLE PROTEINS

a

SALINE – EDTA
SOLUBLE PROTEINS

b

Accordingly, Ub-2A was solubilized from chromatin with histones 2A, 2B, 3, and 4[16]; two of each of these histones compose the octameric nucleosome core particle.[51] Ub-2A is in a subset of nucleosomes where it substitutes for histone 2A in the core octamers.[52,53] The interactions between histones 2A and other core histones[51,54,55] and histone 1[56–58] are not altered by ubiquitination.[45,53,56] It was speculated that the presence of Ub-2A might affect the flexibility of adjacent nucleosomes[59,60] inhibiting condensation of the chromatin.[59–61] Pro-Ub-2B has also been found in nucleosome cores.[49] Free ubiquitin is not a DNA binding protein[12,62,63]; it has been found among the high mobility group proteins.[64]

Ubiquination and ATP-Dependent Proteolytic Enzymes of Reticulocytes

One of the most studied aspects of ubiquitination of proteins has been the reticulocyte enzyme system which serves to destroy most of the cellular proteins in red cell precursors with the exception of structural ele-

[51] R. D. Kornberg and J. O. Thomas, *Science* **184,** 865 (1974).

[52] I. L. Goldknopf, M. F. French, R. Musso, and H. Busch, *Proc. Natl. Acad. Sci. U.S.A.* **74,** 5492 (1977).

[53] H. G. Martinson, R. True, J. B. E. Burch, and G. Kunkel, *Proc. Natl. Acad. Sci. U.S.A.* **76,** 1030 (1979).

[54] J. A. D'Anna, Jr. and I. Isenberg, *Biochemistry* **13,** 4992 (1974).

[55] H. G. Martinson, R. True, C. K. Lau, and M. Mehrabian, *Biochemistry* **18,** 1075 (1979).

[56] W. M. Bonner and J. D. Stedman, *Proc. Natl. Acad. Sci. U.S.A.* **76,** 2190 (1979).

[57] T. Boulikas, J. M. Wiseman, and W. T. Garrard, *Proc. Natl. Acad. Sci. U.S.A.* **77,** 127 (1980).

[58] J. Allan, P. G. Hartman, C. Crane-Robinson, and F. X. Aviles, *Nature (London)* **288,** 675 (1980).

[59] A. Worcel, *Cold Spring Harbor Symp. Quant. Biol.* **42,** 313 (1977).

[60] I. L. Goldknopf, M. W. Andersen, N. R. Ballal, G. Wilson, and H. Busch, *Eur. J. Cell Biol.* **22,** 95 (1980).

[61] S.-I. Matsui, B. K. Seon, and A. Sandberg, *Proc. Natl. Acad. Sci. U.S.A.* **76,** 6386 (1979).

[62] I. L. Goldknopf, M. F. French, N. R. Ballal, I. Daskal, and H. Busch, *Proc. Am. Assoc. Cancer Res.* **18,** 83 (1977).

[63] J. Jenson, G. Goldstein, and E. Breslow, *Biochim. Biophys. Acta* **624,** 378 (1980).

[64] J. M. Walker, G. H. Goodwin, and E. W. Johns, *FEBS Lett.* **90,** 327 (1978).

FIG. 8. Two-dimensional polyacrylamide gel electrophoretic analyses of the distribution of rat liver nuclear proteins during chromatin preparation. Electrophoresis was from right to left in the first dimension and top to bottom in the second. The gels were stained in Coomassie brilliant blue. The histones are spots GAR (histone 4), A1 (histone 2A), A2 (histone 2B), A3-5,7 (histone 3), and A11, 17-19 (histone 1): (a) 0.4 N H_2SO_4-soluble nuclear proteins; (b) proteins solubilized in two washes of 0.075 M NaCl/0.025 M EDTA, pH 8.0/1 mM PhCH$_2$SO$_2$F; (c) proteins solubilized in two washes of 0.01 M Tris, pH 8.0/1 mM PhCH$_2$SO$_2$F; (d) 0.4 N H_2SO_4 soluble chromatin proteins.

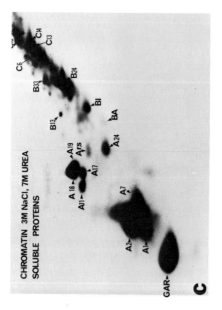

CHROMATIN 3M NaCl, 7M UREA
SOLUBLE PROTEINS

c

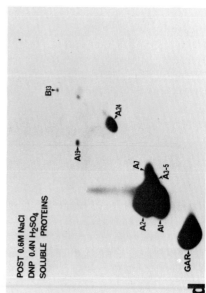

POST 0.6M NaCl
DNP 0.4N H_2SO_4
SOLUBLE PROTEINS

d

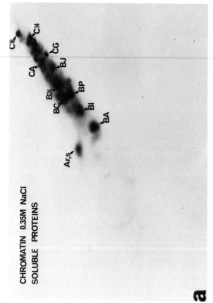

CHROMATIN 0.35M NaCl
SOLUBLE PROTEINS

a

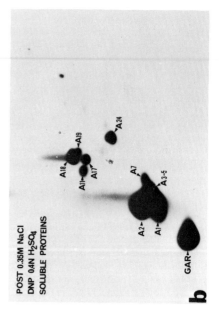

POST 0.35M NaCl
DNP 0.4N H_2SO_4
SOLUBLE PROTEINS

b

ments and hemoglobin.[65] Unlike mature red cells which exhibit little proteolysis, there is marked destruction of many cellular proteins in the
maturing reticulocyte which was shown initially by Schweiger et al.[66] to
be energy dependent. Studies on labeled products had shown that most of
the degraded proteins were synthesized at much earlier stages of differentiation. In addition, inhibitors of ATP formation blocked this proteolysis.
This interesting proteolytic system is obviously supplementary to other
proteolytic systems of cells which include the lysosomal system and other
soluble systems.

Attempts to define the system in biochemical terms[67–74] were undertaken under conditions which solubilized the enzymes; the pH optimum
was mildly alkaline and both Mg^{2+} and ATP were required. Like the intact
reticulocytes, this cell-free system degraded several proteins but not normal hemoglobin; it was inhibited by agents which block protein degradation in the reticulocytes. Resolution of the elements of the system was
undertaken on DEAE cellulose which separated the components into two
fractions, one of which was in the flow-through or void fraction. The other
factor was eluted with high salt concentrations. An interesting feature of

[65] A. Hershko, A. Ciechanover, and I. A. Rose, J. Biol. Chem. 256, 1525 (1981).
[66] H. G. Schweiger, S. Rapoport, and E. Scholzel, Nature (London) 178, 141 (1956).
[67] A. Ciechanover, Y. Hod, and A. Hershko, Biochem. Biophys. Res. Commun. 81, 1100 (1978).
[68] K. D. Wilkinson, M. K. Urban, and A. L. Haas, J. Biol. Chem. 255, 7529 (1980).
[69] A. Ciechanover, S. Elias, H. Heller, S. Ferber, and A. Hershko, J. Biol. Chem. 255, 7525 (1980).
[70] A. Ciechanover, H. Heller, S. Elias, A. L. Haas, and A. Hershko, Proc. Natl. Acad. Sci. U.S.A. 77, 1365 (1980).
[71] A. Hershko, A. Ciechanover, H. Heller, A. L. Haas, and I. A. Rose, Proc. Natl. Acad. Sci. U.S.A. 77, 1783 (1980).
[72] A. Ciechanover, H. Heller, R. Katz-Etzion, and A. Hershko, submitted for publication.
[73] A. Hershko, A. Ciechanover, and I. A. Rose, J. Biol. Chem. 256, 1525 (1981).
[74] A. Hershko, A. Ciechanover, and I. A. Rose, Proc. Natl. Acad. Sci. U.S.A. 76, 3107 (1979).

FIG. 9. Distribution of rat liver chromatin proteins during extraction with various solutions. Electrophoresis conditions were the same as in Fig. 5. (a) proteins solubilized from
chromatin by 3 extractions of 0.35 M NaCl/0.01 M Tris, pH 8.0/1 mM $PhCH_2SO_2F$; (b) 0.4 N
H_2SO_4-soluble proteins of the residual deoxyribonucleoprotein after 0.35 M NaCl extraction. Note the presence of histone 1 (A11, 17-19); (c) proteins dissociated from chromatin in
3 M NaCl/7 M urea/0.05 M sodium acetate, pH 6.0/1 mM $PhCH_2SO_2F$ after the DNA was
pelleted by centrifugation at 214,000 g for 24 hr; (d) 0.4 N H_2SO_4-soluble proteins of the
residual DNP after three (10 ml/g of nuclei) $PhCH_2SO_2F$. Note the markedly reduced
amounts of histone 1 (spots A11,17-19) compared to b.

the system was that neither of the two crude fractions exhibited activity alone but when pooled in the presence of ATP, the activity was restored.

The unbound fraction was ultimately shown to be ubiquitin.[68] It was stable at 90°, appeared to be a polypeptide precipitable with ammonium sulfate, and its activity was lost after incubation with chymotrypsin or pronase. Originally designated APF-1 (ATP dependent proteolysis factor I) it has a molecular weight of 9000 which is similar to that of ubiquitin.

The second DEAE-cellulose fraction contained two components, one a heat-labile high-molecular-weight component stabilized by ATP and the second, another less well defined fraction.[65,67–74] These components were separated by ammonium sulfate fractionation. They and ubiquitin were required by the system. Their binding in the presence of Mg^{2+} and ATP was blocked by N-ethyl maleimide and EDTA. The ubiquitin was covalently bound as indicated by the findings that it was not released with acid, alkali or heating in SDS or 2-mercaptoethanol.

Surprisingly, the ubiquitin was subsequently shown to be covalently bound to many high-molecular-weight proteins by this system. In particular, exogenous substrates such as lysozyme, globin, and lactalbumin were good substrates for the binding of the ubiquitin. Polylysine was also an effective recipient of the ubiquitin moiety. Unlike the case of Ub-2A, several ubiquitin molecules were bound to a single protein.

Analysis of the binding of the ubiquitin to the proteins indicated that an isopeptide linkage formed between the ε-amines of the lysine residues of the proteins and the carboxyls of the glycine termini of the ubiquitin. Chemically, the bonds were stable to 0.1 N NaOH or 1 M hydroxylamine at pH 9.

During the development of these studies, the groups of Hershko, Ciechanover, Rose, Wilkinson, Haas, and others[65,67–74] noted the interesting relationship of these reactions to those for Ub-2A (protein-A24, the ubiquitin–histone 2A complex). The ubiquitin complex to protein 2A was notable because an isopeptide linkage had been proven to exist between two proteins which seemed analogous to the linkage noted between ubiquitin and the protein substrates in the proteolysis reactions. When it was found that the molecular weights and amino acid compositions of the proteolysis factor were similar to ubiquitin, it was a logical conclusion that the factor was indeed the same. The only puzzling point was the presence of six glycine residues in their AFP-1 factor and four in the early reported structure of ubiquitin.

C-Terminal -Gly-Gly in Active Ubiquitin

In our studies (noted above) on the binding of the ubiquitin moiety to histone 2A, we had demonstrated a diglycine "bridge" between the reported structure of ubiquitin and the ε-lysine residue of position 119 of histone 2A. The possibility existed at that time that either the diglycine was added in one or two steps prior to the conjugation with ubiquitin or that the structure proposed for ubiquitin lacked the necessary two germinal glycine residues. Subsequently, Andersen *et al.*[8] found that the "active" conjugating form of ubiquitin contained the two terminal glycine residues and that without these the ubiquitin was inactive or was unconjugated.

Similar experiments were done by Hershko *et al.*[73] who showed that the ubiquitin which was bound to the activating enzyme had a terminal glycine. Wilkinson and Audhya[9] demonstrated that the active form of ubiquitin contains the -Gly-Gly residue and neither the ubiquitin containing one terminal glycine or lacking the terminal glycine was active in the proteolysis system.

In addition, Wilkinson and Audhya[9] found that preparations of ubiquitin which had reduced amounts of -Gly-Gly had reduced activity in the reticulocyte proteolysis system. HPLC analysis coordinated activity and the presence of the -Gly-Gly complex. Accordingly, the enzymes involved must recognize the -Gly-Gly terminal complex. They concluded that the structural requirements of the protein and ubiquitin molecules are identical for formation of protein Ub-2A and for forming the covalent conjugates thought to be intermediates in ATP-dependent protein degradation.

Wilkinson *et al.*[68] confirmed that ubiquitin formed the same conjugates in the reticulocyte system as were noted above. Inasmuch as there was some uncertainty about the precise amino acid of the Arg-Gly-Gly ubiquitin terminus in the formation of the isopeptide linkages, Hershko *et al.*[73] identified the ATP-activated terminal amino acid which they found to be glycine.

Some uncertainties emerged from the sequence studies on calf thymus ubiquitin which had been found by Schlesinger *et al.*[3] and Low and Goldstein[4] to have a C-terminal-Arg-. However, in the studies from our laboratory,[5–7] the linkage of ubiquitin to histone 2A was Arg-Gly-Gly. The question then was whether the Gly-Gly was added later or was an intermediate. In this connection, the studies of Watson *et al.*[27] indicated that in trout nuclear ubiquitin, some termini were Gly, and some Arg- as shown by hydrazinolysis. Hershko *et al.*[73] treated their ubiquitin–conjugate complexes with NaB^3H_4, hydrolyzed the product with acid and released ethanolamine derived from the active acid residue. Thus, they concluded that the carboxyl terminus of the activated ubiquitin was glycine. The active form of ubiquitin in protein conjugation probably has an Arg-Gly-Gly carboxyl-terminus.[5–9] Presumably, the loss of glycines in calf thymus ubiquitin resulted from partial proteolysis by a protease of uncer-

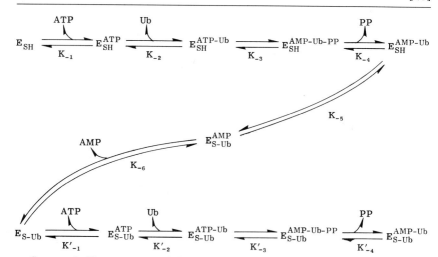

SCHEME 1. The sequence and distribution of enzyme intermediates. From Haas and Rose.[75,76]

tain specificity, although this has not yet been demonstrated. Whether this is a physiological phenomenon or specific processing is uncertain.

Recently, much attention has been focused on the question of the mechanism of formation of the ubiquitin–enzyme complex and the reaction sequence involved. Haas and Rose[75,76] have developed a reaction sequence which is most reasonable for the individual and overall reaction steps involved (Scheme 1):

$$ATP + Ub + E_{SH} \rightleftharpoons E_{SH}^{AMP\text{-}Ub} + PP_i \tag{1}$$
$$E_{SH}^{AMP\text{-}Ub} \rightleftharpoons E_{S\,Ub} + AMP \tag{2}$$
$$E_{S\,Ub} + ATP + Ub \rightleftharpoons E_{S\,Ub}^{AMP\text{-}Ub} + PP_i \tag{3}$$

Deconjugation of Ubiquitin

The lyase (isopeptidase)[77,78] which may be important in the turnover of ubiquitin[79,80] cleaves the Ub-2A to histone 2A and ubiquitin provides a ubiquitin which has two C-terminal glycine residues.[8]

[75] A. L. Haas and I. A. Rose, *J. Biol. Chem.* **257,** 10329 (1982).
[76] A. L. Haas, J. V. B. Warms, A. Hershko, and I. A. Rose, *J. Biol. Chem.* **257,** 2543 (1982).
[77] M. W. Andersen, N. R. Ballal, I. Goldknopf, and H. Busch, *Biochemistry* **20,** 1100 (1981).
[78] S.-I. Matsui, A. A. Sandberg, S. Negoro, B. K. Seon, and G. Goldstein, *Proc. Natl. Acad. Sci. U.S.A.* **79,** 1535 (1982).
[79] I. L. Goldknopf, S. Sudhakar, F. Rosenbaum, and H. Busch, *Biochem. Biophys. Res. Commun.* **95,** 1253 (1980).
[80] R. S. Wu, D. Nishika, P. Patarzis, M. H. P. West, and W. M. Bonner, *J. Cell Biol.* **87,** 46A (1980).

Functions of Ubiquitin Conjugates of Protein

In the proteolysis system, the ubiquitin conjugate serves as a recognition site for the protease.[81] Ubiquitin, which is remarkably insensitive to proteases,[3] is either an allosteric binding site or a portion of the protein configuration necessary for the reaction site on the enzyme surface. The protein A24 (Ub-2A) was decreased in nucleoli of cells stimulated for cell division (regenerating liver) or for hypertrophy of the nucleoli (thioacetamide treatment).[21,22]

The finding that Ub-2A was decreased on active ribosomal genes has been confirmed recently by Levinger et al.[82] who showed that they are deficient in Ub-2A. The decrease of the ubiquitin-conjugates makes sense in relation to the displacement of nucleosomes from the rDNA.

One in four of all nucleosomes are ubiquinated in nonsynchronized cultured Drosophila cells.[83] In the satellite DNA nucleosomes which are not transcribed, fewer than one in 25 nucleosomes contained Ub-2A. In the transcribed copia and heat shock genes, the concentration of Ub-2A was very high, i.e., one of the two nucleosomes contained the Ub-2A complex. Accordingly, in this system the Ub-2A was associated with active genes.[83]

Matsui et al.[61] reported that metaphase chromosomes, which are essentially collapsed chromatin, contained no Ub-2A conjugates and hence, ubiquitin residues were lost. In addition, the relatively inactive mature chicken erythrocytes did not contain significant amounts of Ub-2A but in the related transcriptionally active "reticulocytes" the Ub-2A adduct was present in higher concentration.[84] Accordingly, a dichotomy exists in that in the transcriptionally active chromatin for RNA polymerase II Ub-2A was present. For RNA polymerase I readouts the conjugate is decreased.

Recent studies by Seale[85] indicated that the half-life of the ubiquitin moiety was only 90 min in HeLa cells. Ub-2A and free ubiquitin are lost from nuclei during chicken erythropoiesis[84] when transcription stops.[86]

The cleavage of Ub-2A itself is tied to cell activity. In the chicken reticulocyte, the endogenous lyase (isopeptidase) activity is brisk, but in the mature, inactive erythrocyte, the lyase activity ceased and little Ub-2A was detected.[84]

In a mouse "ts" mutant defective in H1 histone phosphorylation, under nonpermissive conditions (39°) Ub-2A disappeared from the chro-

[81] U. Scheer, Cell 13, 535 (1978).
[82] L. Levinger, J. Barsoum, and A. Varshavsky, J. Mol. Biol. 146, 287 (1981).
[83] L. Levinger and A. Varshavsky, Cell 28, 375 (1982).
[84] I. L. Goldknopf, G. Wilson, N. R. Ballal, and H. Busch, J. Biol. Chem. 225, 10555 (1980).
[85] R. L. Seale, Nucleic Acids Res. 9, 3151 (1981).
[86] A. F. Williams, J. Cell Sci. 10, 27 (1972).

matin.[87–90] Ub-2A reappeared in the shiftdown to the permissive temperature. Clearly a sorting out of the various signals ubiquitination provides will be an area of exciting research from the points of view of both proteolysis and gene function.

Acknowledgments

The original studies from our laboratories were supported by grants from U.S. Public Health Services, Cancer Research Grants, The Bristol-Myers Fund, The Michael E. DeBakey Medical Foundation, The Pauline Serne Wolff Memorial Foundation, The Finger Fund, the Davidson Fund, and The William Farrish Fund.

[87] I. L. Goldknopf, S. Cheng, M. W. Andersen, and H. Busch, *Biochem. Biophys. Res. Commun.* **100,** 1464 (1981).

[88] T. Marunouchi, H. Yasuda, Y. Matusmoto, and M. Yamada, *Biochem. Biophys. Res. Commun.* **95,** 126 (1980).

[89] H. Yasuda, Y.-I. Matsumoto, T. Marunouchi, S. Mita, and M.-A. Yamada, submitted for publication.

[90] T. Marunouchi, S. Mita, Y. Matsumoto, and H. Yasuda, submitted for publication.

Section V

Protein Alkylations/Dealkylations (Arylations)

[24] A Simple Analysis of Methylated Proteins

By Woon Ki Paik

Protein methylation is a posttranslational modification reaction which occurs in many proteins. This reaction involves side chains of various amino acids, such as lysine, arginine, histidine, and glutamine for N-methylation and glutamic and aspartic acid for O-methylesterification (see the table).

One of the most useful approaches in investigating the biological significance of protein methylation is to isolate a protein whose amino acid side chain(s) is substituted with a methyl group, and to compare its characteristics with the unmethylated form of the protein. However, the methyl substitution for the hydrogen atom of the amino or guanidino group of a protein molecule results in only a slight change in steric configuration or charge. In addition, most of the *in vivo* or *in vitro* enzymatically methylated proteins contain less than an equimolar amount of methylated amino acid residue per mole of the protein, making the physicochemical difference extremely subtle. Thus separation of the methylated protein from its unmethylated counterpart species is not easily accomplished. However, application of modern purification techniques such as ion-exchange chromatography, high-performance liquid chromatography, isoelectrofocusing, and two-dimensional chromatography should be able to resolve these protein species from each other. Presently, only a single example of successful resolution of the methylated protein from its counterpart has been reported; separation of methylated and unmethylated cytochrome *c* by ion-exchange column chromatography[1,2] or isoelectrofocusing.[3]

As described above, it is an extremely useful approach to prepare the methylated protein in large quantity and to compare it with the unmethylated counterpart with respect to physicochemical as well as biological function of the protein. It is equally important to manipulate and analyze methylated protein species among the multitude of unmethylated proteins without resorting to an extensive purification. A simple double-radiolabeling method to investigate methylated proteins is described below.

[1] M. Foucher, J. Verdiere, F. Lederer, and P. Slonimski, *Eur. J. Biochem.* **31,** 139 (1972).
[2] W. A. Scott and H. K. Mitchell, *Biochemistry* **8,** 4282 (1969).
[3] C.-S. Kim, F. Kueppers, P. DiMaria, J. Farooqui, S. Kim, and W. K. Paik, *Biochim. Biophys. Acta* **622,** 144 (1980).

METHODS IN ENZYMOLOGY, VOL. 106

NATURAL OCCURRENCE OF VARIOUS METHYLATED AMINO ACIDS

Methylated amino acid	Proteins
ε-N-Monomethyllysine	Flagella protein, histones, myosin, actin, ribosomal proteins, opsin, tooth matrix protein, EF 1α (fungus *Mucor*)
ε-N-Dimethyllysine	Flagella protein, histones, myosin, actin, opsin, ribosomal proteins, EF 1α (fungus *Mucor*)
ε-N-Trimethyllysine	Histones, cytochrome c (*Ascomycetes*), myosin, actin, ribosomal proteins, calmodulin, EF-Tu, citrate synthase (porcine heart), heat-shock protein, EF 1α (fungus *Mucor*)
N^G-Monomethylarginine[a]	Histones, acidic nuclear protein, AI basic protein, heat-shock protein
N^G,N^G-Dimethylarginine	Histone, AI basic protein, myosin, HnRNP protein, ribosomal proteins, tooth matrix protein, HMG-1 and HMG-2 protein, nucleolar protein C23
N^G,N'^G-Dimethylarginine	Histone, AI basic protein, tooth matrix protein, myosin, actin, opsin
3-N-Methylhistidine	Myosin, actin, histone (avian), opsin
δ-N-Methylglutamine	Ribosomal protein
Glutamyl or aspartyl methyl ester	Membrane proteins of *Escherichia coli*, *Salmonella typhimurium* and erythrocyte

[a] N^G refers to the nitrogen of the guanidino group of arginine. N^G,N^G-Dimethylarginine is dimethyl substituted on the same nitrogen in the guanidine group. Thus, this compound is also called asymmetric dimethylarginine. In the same terminology, N^G,N'^G-dimethylarginine is dimethyl substituted at two different nitrogen atoms and, therefore, called symmetric dimethylarginine.

Simple Identification Method for Methylated Proteins

In order to study the methylation pattern, the rate of turnover, or demethylation of previously labeled protein, double-labeling of synchronized HeLa S-3 cells with both L-[*methyl*-³H]methionine and L-[1-¹⁴C]methionine or L-[U-¹⁴C]lysine was carried out.[4] The rationale behind this approach is that there is an equal probability that both isotopes will be incorporated into the backbone of newly synthesized proteins and, in addition, the [³H]methyl-labeled group derived from L-[*methyl*-³H]methionine by way of *S*-adenosyl-L-[*methyl*-³H]methionine will appear in the methylated amino acid residues. Thus, the ratio of ³H to ¹⁴C will indicate the extent of protein methylation.

This strategy was successfully applied in the identification of methylated ribosomal proteins in *Escherichia coli* and *Saccharomyces cerevi-*

[4] T. W. Borun, D. Pearson, and W. K. Paik, *J. Biol. Chem.* **247,** 4288 (1972).

siae.[5,6] Chang *et al.*[5] grew *E. coli* in a medium containing both L-[*methyl-*
^{3}H]methionine and L-[1-^{14}C]methionine. After isolation of 70 S ribosomes,
the individual ribosomal proteins were resolved by two-dimensional
polyacrylamide gel electrophoresis. After staining the gels, the regions of
the individual proteins were cut from the gels and the radioactivity was
measured. Among more than 55 ribosomal proteins analyzed, only six
proteins were methylated as indicated by the ratio of ^{3}H to ^{14}C.

Analytical Procedure

Escherichia coli cells were grown in 10 ml of a Tris-buffered medium
[0.1 M tris(hydroxymethyl)aminomethane(Tris) hydrochloride, 1 mM
$MgCl_2$, 0.01 mM $FeCl_2$, 0.1 mM $CaCl_2$, 1 mM KH_2PO_4, and 0.32 mM
Na_2SO_4, adjusted to a final pH of 7.2] supplemented with 0.3% glucose, 2
μg of thiamin/ml, and 20 μg/ml of each of 20 amino acids except
methionine. A 15 μCi amount of L-[1-^{14}C]methionine (specific activity, 54
mCi/mmol) and 150 μCi of L-[methyl-^{3}H]methionine (specific activity, 2.5
Ci/mmol) were then added. The final concentration of methionine was 6
μg/ml. After the cells were harvested at late log phase, 250 mg of carrier
Q13 cells (purchased from General Biochemical Inc., Chagrin Falls, Ohio)
was added.

The cells were disrupted with a French press in a buffer containing
0.01 M Tris–HCl (pH 7.8), 10 mM $MgCl_2$, and 0.05 M KCl. By a modification of the Nierenberg and Matthaei procedure[7] 70 S ribosomes were
prepared; the final ribosomal pellet was resuspended in the above buffer.
Sixty absorbancy units (260 nm) of the 70 S ribosomes in 0.5 ml of buffer
were then adjusted to 0.1 M $MgCl_2$, and the ribosomal RNA was precipitated by the rapid addition of two volumes of glacial acetic acid according
to the method of Hardy *et al.*[8] The supernatant solution was lyophilized to
dryness, and the proteins were redissolved in 0.15 ml of the sample gel
solution used for the first-dimensional run of the two-dimensional polyacrylamide gel electrophoresis (procedure of Kaltschmidt and Wittmann[9]). After electrophoresis and staining of the gels, regions of the
individual proteins were cut from the gels and analyzed for radioactivity
by the procedure of Nashimoto *et al.*[10]

[5] F. N. Chang, C. N. Chang, and W. K. Paik, *J. Bacteriol.* **120,** 651 (1974).
[6] M. Cannon, D. Schindler, and J. Davies, *FEBS LETT.* **75,** 187 (1977).
[7] M. W. Nierenberg and J. H. Matthaei, *Proc. Natl. Acad. Sci. U.S.A.* **47,** 1588 (1961).
[8] S. J. S. Hardy, C. G. Kurland, P. Voynow, and G. Mora, *Biochemistry* **8,** 2897 (1969).
[9] E. Kaltschmidt and H. G. Wittmann, *Anal. Biochem.* **36,** 401 (1970).
[10] H. Nashimoto, W. Held, E. Kaltschmidt, and M. Nomura, *J. Mol. Biol.* **62,** 121 (1971).

Comments

The above double-labeling method is highly useful for preliminary identification of methylated proteins among the group of unmethylated species, studies on the rate of methyl turnover, and in the investigation on the extent of protein methylation. It should also be pointed out that L-[^{35}S]methionine could replace ^{14}C-labeled methionine or lysine, and that ^{14}C-labeled arginine or histidine could also be employed in combination with L-[*methyl*-^{3}H]methionine in investigation of the metabolism of methylated arginine or histidine derivatives, analogous to lysine methylation.

The above described method necessitates simultaneous presence of two radioisotopic compounds. However, these sets of two radioisotopic compounds could be employed independently to quantitate their incorporation into proteins. For example, while investigating the methylation of heat shock proteins, chicken fibroblast cells were grown in presence of [^{35}S]methionine and the proteins were resolved by two-dimensional gel electrophoresis.[11] Radioactivity incorporation patterns in various proteins were visualized by radiofluorography. Independently, another set of chicken fibroblast cells was labeled with L-[*methyl*-^{3}H]methionine, and the extent of radioactivity incorporation into various proteins was compared. This method provides a simplified means to identify methylated proteins compared with the double-labeling experiments described earlier.

Finally, the following points should be taken into consideration when evaluating the results of the double-labeling experiment. With high methionine content in protein, a low level of methylation (as indicated by the ratio of ^{3}H to ^{14}C) will be obscured. On the other hand, with a low methionine content in the protein, the high ratio of ^{3}H to ^{14}C does not necessarily indicate the high degree of methylation.

[11] C. Wang, R. H. Gomer, and E. Lazarides, *Proc. Natl. Acad. Sci. U.S.A.* **78**, 3531 (1981).

[25] S-Adenosylmethionine : Protein (Arginine) N-Methyltransferase (Protein Methylase I) (Wheat Germ)

By Martin Tuck and Woon Ki Paik

S-Adenosylmethionine : protein (arginine) N-methyltransferase or protein methylase I [EC 2.1.1.23] was first isolated from calf thymus by Paik and Kim.[1] The enzyme is mainly found in the cytosol; however,

[1] W. K. Paik and S. Kim, *J. Biol. Chem.* **243**, 2108 (1968).

METHODS IN ENZYMOLOGY, VOL. 106

trace amounts have also been shown in the nucleus. Purified protein methylase I forms *in vitro* three products which have been identified as N^G-mono-, N^G,N^G-di- (asymmetric), and $N^G,N^{'G}$-di- (symmetric) methylarginine.[2] Many proteins are known to contain methylated arginines. Included in this class are histones, myelin basic protein (AI basic protein), HMG-1 and HMG-2 protein, HnRNP protein, ribosomal proteins, and heat shock proteins. The enzyme is found in many vertebrate tissues, although it is most abundant in brain,[3] thymus, testis, and spleen.[4] A form of protein methylase I has also just recently been isolated from wheat germ,[5] which is the first report of the enzyme being present in the plant kingdom.

Published reports have shown that protein methylase I has been successfully purified from Krebs II ascites cells,[6] rat and calf thymus,[1] rat and calf brain,[7,8] and chicken embryo.[9] In the following section the purification and properties of wheat germ protein methylase I are discussed. Wheat germ protein methylase I has several distinctive properties from animal protein methylase I's. These differences include optimum pH, relative ratio of *in vitro* reaction products, protein substrate specificity, inhibition by adenosine, and the requirement of a peptide cofactor for full enzyme activity.

Assay Method

Principle. The formation of methylated arginines is followed by the incorporation of [^{14}C]- or [^{3}H]methyl groups into the substrate proteins from S-adenosyl-L-[*methyl*-^{14}C or -^{3}H]methionine.[10] Acid hydrolysis of

---

[2] N^G- refers to the nitrogen of the guanidino group of arginine. N^G,N^G-Dimethylarginine is dimethyl substituted on the same nitrogen atom in the guanidino group. Thus, this compound is also called asymmetric dimethylarginine. In the same terminology, $N^G,N^{'G}$-dimethylarginine is dimethyl substituted at two different nitrogen atoms and, therefore, called symmetric dimethylarginine.

[3] W. K. Paik and S. Kim, *J. Neurochem.* **16**, 1257 (1969).

[4] W. K. Paik and S. Kim, *Adv. Enzymol.* **42**, 227 (1975).

[5] A. Gupta, D. Jensen, S. Kim, and W. K. Paik, *J. Biol. Chem.* **257**, 9677 (1982).

[6] P. Casellas and P. Jeanteur, *Biochim. Biophys. Acta* **519**, 243 (1978).

[7] G. M. Jones and P. R. Carnegie, *J. Neurochem.* **23**, 1231 (1974).

[8] H. W. Lee, S. Kim, and W. K. Paik, *Biochemistry* **16**, 78 (1977).

[9] J. Enouf, F. Lawrence, C. Tempete, M. Robert-Gero, and E. Lederer, *Cancer Res.* **39**, 4497 (1979).

[10] S-Adenosyl-L-[*methyl*-^{14}C]methionine is the preferable isotope if products are to be analyzed after 6 N HCl hydrolysis, even though the specific radioactivity is low. When S-adenosyl-L-[*methyl*-^{3}H]methionine is used and hydrolysis products are analyzed, many products other than methylated arginines appear with radioactivity. The ^{3}H isotope is good for routine assay purposes since the specific radioactivity of S-adenosyl-L-[*methyl*-^{3}H]methionine is high and it is less expensive.

the methylated proteins shows that only methylated arginine residues are formed. Protein methyl esters produced by contaminating S-adenosylmethionine : protein-carboxyl O-methyltransferase [protein methylase II; EC 2.1.1.24] are removed by alkaline hydrolysis at elevated temperatures.

Reagents

1 *M* glycine–NaOH buffer, pH 9.0
0.1 *M* dithiothreitol (DTT) in 30% glycerol, 0.1 *M* glycine–NaOH buffer, pH 9.0
S-Adenosyl-L-[*methyl*-^{14}C]methionine (25 nmol/0.1 ml; 90–100 cpm/pmol)
Calf thymus histone type II-AS (3 mg/0.1 ml; obtained from Sigma Chemical Co.)
15% trichloroacetic acid (TCA)
0.5 *M* phosphate buffer, pH 8.0
95% ethanol

Procedure. Protein methylase I activity is assayed as described[8] with minor modifications. Assay mixture contain 0.1 ml of 1 *M* glycine–NaOH buffer, pH 9.0, 0.1 ml of 0.1 *M* DTT, 0.1 ml of S-adenosyl-L-[*methyl*-^{14}C] methionine, 0.1 ml of protein substrate (usually calf thymus histone type II-AS), and 0.1 ml of enzyme preparation in a final volume of 0.7 ml. The incubation mixture without S-adenosyl-L-[*methyl*-^{14}C]methionine is first preincubated at 37° for 3 min, and the reaction is started by the addition of the radioactive methyl donor compound. After 10 min of incubation, the reaction is terminated by the addition of 3 ml of 15% TCA. The acid-soluble fraction, nucleic acids and phospholipids are removed by heating the assay tubes at 90° for 15 min, followed successively by two washes with 15% TCA and one treatment with ethanol at 70° for 10 min.[11] The protein-methyl esters produced by the action of contaminating protein methylase II are removed by alkaline hydrolysis; 1.0 ml of 0.5 *M* phosphate buffer, pH 8.0, is added to the above treated samples and the mixture is heated to 80° for 10 min. Approximately 3 ml of 15% TCA is added and the precipitates are collected by centrifugation. The precipitates are treated once with ethanol and transferred quantitatively into scintillation vials containing 5 ml of scintillation solution Formula 963 (New England Nuclear Corp.). The enzyme activity is expressed as picomoles of *methyl*-^{14}C group incorporated per minute per milligram of enzyme protein.

---

[11] E. Durban, H. W. Lee, S. Kim, and W. K. Paik, *Methods Cell Biol.* **19**, 59 (1978).

Purification Procedures

Step 1. Homogenization. Sixty grams of wheat germ (obtained from General Mills, Vallejo, CA) is suspended in 180 ml of cold deionized water and stirred for 30 min using a magnetic stirrer at 4°. All of the following procedures are carried out at 0–4°. The homogenate is then centrifuged at 39,000 g for 30 min and the supernatant is obtained by passage through glass wool to remove the lipid layer.

Step 2. First Ammonium Sulfate Precipitation. To the above supernatant (136 ml) 54 ml of saturated ammonium sulfate solution (saturated at room temperature) is added slowly while stirring. The suspension is allowed to stand for 10 minutes and then it is centrifuged at 39,000 g for 15 min. The supernatant is collected while the precipitate is discarded. To the supernatant another 54 ml of saturated ammonium sulfate is added. The precipitate is again collected by centrifugation at 39,000 g for 15 min.

Step 3. Calcium Phosphate Gel Treatment. The precipitate in Step 2 is dissolved in 68 ml of water, and 204 ml of calcium phosphate gel (suspension of 17 mg solid/ml) is added to the enzyme suspension under gentle stirring. The suspension is then allowed to stand for 20 min. After centrifugation at low speed (about 3000 g), the sedimented gel is washed twice with a large volume of water and then resuspended in 30 ml of 0.3 M phosphate buffer, pH 7.2, prepared in buffer A (buffer A; 20 ml of 0.5 M phosphate buffer, pH 7.2, containing 10% glycerol and 0.7 ml of 2-mercaptoethanol diluted to a total volume of 1000 ml). The gel suspension is centrifuged at 39,000 g for 10 min and 40 ml of eluate is obtained.

Step 4. Second Ammonium Sulfate Precipitation. The eluate in Step 3 is treated with 40 ml of saturated ammonium sulfate solution and the precipitate is collected at 39,000 g for 10 min. The collected precipitate is dissolved in 50 ml of buffer A. To this preparation 12 ml of 1.5% protamine sulfate is added and the preparation is centrifuged after standing for 15 min.

Step 5. DEAE-Sephadex Chromatography. The supernatant from Step 4 is loaded onto a DEAE-Sephadex column (4.0 × 8.0 cm), which has been equilibrated with buffer A. The column is first washed with 300 ml of buffer A, and then eluted batchwise with 150 ml of 1.2 M KCl in buffer A. The eluted enzyme is precipitated by adding an equal amount of saturated ammonium sulfate solution.

Step 6. CM [(Carboxymethyl)cellulose]-Sephadex Chromatography. The enzyme preparation is dissolved in 25 ml of buffer A and then applied onto a CM-Sephadex column (2.5 × 8.0 cm), which has previously been equilibrated with buffer A. The column is washed continuously with buffer A until a total volume of 100 ml run-off eluate is collected. The

SUMMARY OF PURIFICATION PROCEDURE

Purification steps	Total protein (mg)	Total enzyme activity (pmol/min)	Specific activity (pmol/min/mg enzyme protein)	Yield (%)	Fold purification
Whole homogenate	8520	1320	0.155	—	—
Supernatant	7527	2343	0.311	177.5	2.1
Ammonium sulfate precipitate	2550	1650	0.640	125.0	4.1
Calcium phosphate gel	853	3145	3.69	236.7	23.8
DEAE-Sephadex	312	2725	8.73	206.0	56.3
CM-Sephadex	150	2101	14.0	159.0	90.3

enzyme is precipitated again with an equal volume of saturated ammonium sulfate solution. Finally, the precipitated enzyme is dissolved in 15 ml of buffer A and stored frozen. The concentration of protein is estimated by the method of Lowry et al.,[12] using bovine serum albumin as a standard. The overall purification is 90-fold with a yield of 160%. Data pertinent to each purification step are presented in the table.

Properties

The various properties of the wheat germ protein methylase I are widely different from the calf brain enzyme.

pH Optimum. The pH optimum for the enzyme is about 9.0, which is significantly higher than the 7.2 observed for the calf brain enzyme.

Methylated Arginines Produced. Product analysis reveals another major difference between the wheat germ and calf brain protein methylase I. The wheat germ enzyme forms only two methylated products compared to three for the calf brain enzyme: The two products formed are N^G-monomethylarginine and N^G,N'^G-dimethylarginine at the ratio of $75:25$, whereas the ratio of N^G-mono-, N^G,N^G-di, and N^G,N'^G-dimethylarginine produced by calf brain enzyme is $50:40:5$.[8]

K_m Value. The K_m values of the wheat germ and calf brain enzymes for S-adenosyl-L-methionine are 5.7×10^{-6} and 2.1×10^{-6} M, respectively. The K_m values for histone with the wheat germ and calf brain enzymes are 2.5×10^{-5} and 5.5×10^{-4} M, respectively.

[12] O. H. Lowry, N. J. Rosebrough, A. L. Farr, and R. J. Randall, *J. Biol. Chem.* **193**, 265 (1951).

Peptide Cofactor. Another major difference between wheat germ and calf brain protein methylase I is that the former requires a low-molecular-weight cofactor for its enzyme activity. The cofactor is dialyzable and heat labile, and the stimulatory activity is destroyed by trypsin treatment, indicating that the cofactor is peptide in nature. The activity of the co-factor is observed by applying the enzyme preparation to a Sephadex G-100 column after CM-Sephadex chromatography. At this step the majority of the proteins are eluted in the void volume but no enzyme activity is detected in any of the fractions. When the pooled protein peak is mixed with various other pooled fractions from the column, the enzyme activity is restored (Fig. 1).

Inhibitors. Another unusual property of wheat germ protein methylase I is that the enzyme has a natural inhibitor. This is observed during the purification of the enzyme when an unusually high yield of activity is obtained (see the table). The wheat germ enzyme appears to be much more sensitive to the inhibitor when compared with that of calf brain.

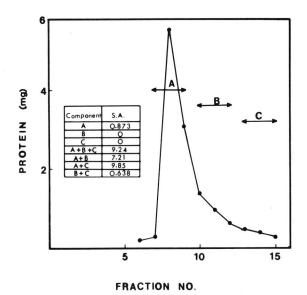

FRACTION NO.

Fig. 1. Presence of a cofactor for wheat germ protein methylase I. The enzyme preparation after CM-Sephadex chromatography is charged on a Sephadex G-100 column. Eluent fractions are divided into three portions, A containing fractions 7–9, B containing 10–12, and C containing 13–15. For reconstitution of the enzyme, equal amounts of each portions are mixed and precipitated with saturated ammonium sulfate. The precipitates are then resuspended in the original volume of buffer A for assay. ●——●, protein concentration. S.A., specific activity.

The purified natural inhibitor appears to be identical to adenosine, since the two compounds have identical UV absorption spectra and retention time on a high-performance liquid chromatography column.[5] Both the natural inhibitor and adenosine also demonstrate identical pI values (6.54) when determined by isoelectrofocusing.[5]

Wheat germ protein methylase I is also inhibited by many S-adenosyl-L-homocysteine analogs.[5] S-Adenosyl-L-homocysteine, A9145C, sinefungin, and S-inosyl-L-(2-hydroxy-4-methylthio)butyrate are the most powerful inhibitors of the enzyme demonstrating K_i values of 10^{-6} to 10^{-8} M. ATP, ADP, and AMP are inactive as inhibitors, as well as other nucleosides and bases.

Protein Substrate Specificity. In contrast to the calf brain protein methylase I, the wheat germ enzyme is highly specific toward substrate protein.[5] Histones appear to be the only substrate for the wheat germ enzyme *in vivo* as well as *in vitro*. On the other hand, various proteins such as histones, myelin basic protein, ribonuclease, polyarginine, and fibrinogen serve *in vitro* as substrates for the calf brain protein methylase I.

[26] Enzymatic Methylation and Demethylation of Protein-Bound Lysine Residues

By WOON KI PAIK and PETER DiMARIA

S-Adenosylmethionine : Protein (Lysine) N-Methyltransferase (Protein Methylase III) (*Neurospora crassa*)

Introduction

Our consideration of S-adenosylmethionine : protein (lysine) N-methyltransferase (protein methylase III; EC 2.1.1.43; the enzyme which catalyzes the transfer of the methyl group from S-adenosyl-L-methionine to the ε-amino groups of lysyl residues in protein) is complicated by the fact that a whole class of enzymes of this description exists. Their most remarkable characteristics are undoubtedly a high specificity both toward a particular protein as a substrate and toward particular lysyl residues within that protein (see Table I). The enzymes of capable of methylating histone or cytochrome c have been most intensively studied, and purified

METHODS IN ENZYMOLOGY, VOL. 106

TABLE I
CURRENTLY KNOWN PROTEIN METHYLASES III

Source	Substrate specificity	Residue modified	Subcellular location
Calf thymus[a,b]	Histone H4	Res-20	Nuclei
Calf brain[c]	Histone	n.d.[l]	Chromatin
Chicken embryo[d]	Histone	n.d.	Nuclei
N. crassa[e]	Cytochrome c (vertebrate)	Res-72	Cytosol
S. cerevisiae[f]	Cytochrome c (vertebrate)	Res-72	Cytosol
Wheat germ[g]	Cytochrome c (vertebrate)	Res-72	n.d.
Crithidia oncopelti[h]	Cytochrome c-557	Res-3	
E. coli[i]	L11 ribosomal protein	n.d.	Ribosome-bound
Physarum polycephalum[j]	Histone	n.d.	Cytosol
Rat brain[k]	Calmodulin	n.d.	Cytosol

[a] P. Sarnow, I. Rasched, and R. Knippers, *Biochim. Biophys. Acta* **655,** 349 (1981).
[b] W. K. Paik and S. Kim, *J. Biol. Chem.* **245,** 6010 (1970).
[c] J. C. Wallwork, D. P. Quick, and J. A. Duerre, *J. Biol. Chem.* **252,** 5977 (1977).
[d] P. J. Greenaway and D. Levine, *Biochim. Biophys. Acta* **350,** 374 (1974).
[e] E. Durban, S. Nochumson, S. Kim, W. K. Paik, and S.-K. Chan, *J. Biol. Chem.* **253,** (1978).
[f] P. DiMaria, E. Polastro, R. J. DeLange, S. Kim, and W. K. Paik, *J. Biol. Chem.* **254,** 4645 (1979).
[g] P. DiMaria, S. Kim, and W. K. Paik, *Biochemistry* **21,** 1036 (1982).
[h] J. Valentine and G. W. Pettigrew, *Biochem. J.* **201,** 329 (1982).
[i] F. N. Chang, L. B. Cohen, I.J. Navickas, and C. N. Chang, *Biochemistry* **14,** 4994 (1975).
[j] M. Venkatesan and I. R. McManus, *Biochemistry* **18,** 5365 (1979).
[k] A. Sitaramayya, L. S. Wright, and F. L. Siegel, *J. Biol. Chem.* **255,** 8894 (1980).
[l] Not determined.

to varying extents from a variety of sources in each case. To date, the cytochrome *c* methylases have provided the clearest model of protein-lysyl methylation enzymology and will be considered in this chapter with particular emphasis on the enzyme characterized from *Neurospora crassa*.

Assay Method

Principle. The enzymatic transfer of radioactive methyl groups from *S*-adenosyl-L-[*methyl*-^{14}C]methionine into trichloroacetic acid (TCA)-precipitable material (including cytochrome *c*) is determined under conditions which have been generally found to be favorable for the methylation of ε-amino groups of lysyl residues (pH 9.0).

Reagents

0.5 M Tris–HCl buffer, pH 9.0

Absolute ethanol

Horse heart cytochrome c (Sigma type II); 30 mg/ml in H_2O

0.2 mM S-adenosyl-L-[*methyl*-^{14}C]methionine (specific activity, 55 mCi/mmol in 0.01 M HCl)

7.5% trichloroacetic acid (TCA)

0.2 M NaOH

1 M HCl

30% H_2O_2

Procedure. The reaction is carried out in a 12 ml conical centrifuge tube in a total volume of 0.25 ml. The incubation mixture consists of 50 μl of 0.5 M Tris–HCl buffer, 50 μl of the horse heart cytochrome c solution, 50 μl of S-adenosyl-L-[*methyl*-^{14}C]methionine, and the remaining 100 μl consisting of the added enzyme solution and/or H_2O. First, all components except the enzyme and S-adenosyl-L-[*methyl*-^{14}C]methionine are added to the tube. The enzyme is then added and the tube is placed in a 37° water bath. After a preincubation period of 5 min, the S-adenosyl-L-[*methyl*-^{14}C]methionine is added to initiate the methylation reaction and the incubation is allowed to proceed for 10 min. The reaction is stopped by the addition of 3 ml of 7.5% TCA and the tubes are cooled on ice for 15 min. The tubes are then heated at 90° in a water bath for 15 min. This treatment serves to solubilize endogenous nucleic acids which may contain transferred methyl groups. The tubes are cooled on ice for 5 min and then centrifuged in a clinical table-top centrifuge at top speed for 10 min. The supernatant is discarded and the pellet is washed twice with 10 ml of 7.5% TCA followed by centrifugation. These washes serve to remove the unreacted S-adenosyl-L-[*methyl*-^{14}C]methionine and other small molecules which may have been methylated. In these washes, the pellet is usually dispersed easily by vortexing, but if tightly packed, it may be necessary to use a tapered glass rod. The pellet is then resuspended in 10 ml of absolute ethanol and heated at 70° for 10 min. This treatment will solubilize lipids including phospholipids which may contain radioactive methyl groups. The tubes are cooled on ice and then centrifuged. After decantation of the supernatant, 0.5 ml of 0.2 M NaOH is added to the tube and the tube is placed in a boiling water bath. The heating is continued for 1 hr whereby via evaporation the volume is considerably reduced. This alkaline treatment will labilize protein-carboxyl methyl esters (product of protein methylase II) and the methyl groups on the guanidino moiety of arginyl residues (product of protein methylase I) forming methanol and methylamine, respectively. These components are volatile and will be driven off by evaporation.[1] After alkali treatment, 0.1 ml of 1 M HCl is added to the tubes to neutralize the NaOH. Since the cytochrome c produces a greenish color which could cause quenching during scintillation counting, 0.1 ml of 30% H_2O_2 is added and the tube is heated for 2 to 5 min at 80–90° to decolorize the mixture. Ten milliliters of scintillation cocktail (Formula 963, New England Nuclear Corp.) is then added to the tube and following vortexing the clear solution is transferred to a vial for scintillation counting.

In the routine use of this assay, several controls should be run. To control for possible endogenous activities whose reaction products are not removed by the protocol described above, control tubes lacking the added cytochrome c substrate should be run. Also, to control for nonenzymatic methylation and nonspecific trapping of radioactivity, control tubes containing heat-inactivated enzyme (5 min at 100°) should also be run. Finally, when highly purified enzyme is assayed, the alkaline hydrolysis can be eliminated and the pellet following the ethanol treatment can be dissolved and decolorized directly by the addition of 0.1 ml of 30% H_2O_2.

Enzyme activity is expressed as picomoles of [^{14}C]methyl incorporated per minute per milligram of enzyme protein (specific activity).

Growth of Neurospora crassa

The general method of culture of *N. crassa* which we have used has been described in an earlier volume of *Methods in Enzymology* and will not be detailed here.[2] The cultures are grown in 20-liter carbuoys containing 10 liters of media with cotton plugs at the mouth. Aeration is accomplished by forced air filtered through cotton which is allowed to bubble from a glass tube extending into the culture vessel. To inoculate the culture, 3 ml of sterile water is added to an agar slant containing a *N. crassa* culture and is vortexed. About 1 ml of the resulting spore- and mycelium-laden suspension is aseptically pipetted into the carbuoy. The cultures are grown at room temperature. After about 5 days, the *N. crassa* are harvested on a Buchner funnel with suction. While on the funnel, the *N. crassa* are washed with deionized water. The thick pad of

---

[1] Although in our work with various *N. crassa* enzyme preparations, we have observed only ε-N-methylated lysines to remain after the hot alkali treatment, one should be aware of the fact that other methylations could survive this treatment. These are methylations occurring at the following: the δ-nitrogen of arginine, the 1- or 3-nitrogen of histidine, the nitrogen of proline, and the α-amino group of the N-terminal amino acid residues.

[2] R. H. Davis and F. J. DeSerres, this series, Vol. 17, Part A, p. 79.

mycelial material is then cut into small chunks (about 1 cm in diameter) and then lyophilized. The freeze-dried material is stored at $-20°$.

Purification Procedures

Step 1. Preparation of the Crude Extract and Initial Purification. Throughout the course of the purification, all procedures are carried out at 0–4°. The lyophilized *N. crassa* (100 g) is suspended in 2 liters of deionized water and then homogenized in a conical ground glass homogenizer at 1200 rpm until no mycelial fragments are apparent. The thick homogenate is centrifuged at 39,000 *g* for 30 min to obtained a clear orange-colored supernatant. To this supernatant, saturated ammonium sulfate solution (saturated at room temperature) is added dropwise with stirring until the final saturation is 45% (add 45 ml of the ammonium sulfate to 55 ml of the supernatant). The suspension is allowed to stand for 30 min to ensure precipitation and then is centrifuged at 39,000 *g* for 10 min. The supernatant is discarded and the pellet is resuspended in a total volume of 500 ml of 10 mM sodium phosphate buffer, pH 7.2.

Step 2. Adsorption onto Calcium Phosphate Gel and Batch Elution. To the above suspension 600 ml of calcium phosphate gel suspension (17 mg solid/ml suspension) is added with stirring. After allowing to adsorb to the gel for 5 min with occasional stirring, the gel is collected by centrifugation at 3000 *g* for 5 min. The gel is washed once with 10 mM sodium phosphate buffer, pH 7.2 and then the enzyme is eluted from the gel by suspending it in a total volume of 100 ml of 30 mM sodium phosphate buffer, pH 7.2 (calcium phosphate gel is found to vary from lot to lot in terms of molarity of phosphate buffer necessary to elute the enzyme; from 30 to 60 mM; therefore, before using the new batch of calcium phosphate gel, a test experiment should be carried out to determine this parameter). After about 10 min the gel suspension is centrifuged at 39,000 *g* for 30 min. The supernatant containing the eluted enzyme is then dialyzed against 10 mM Tris–HCl (pH 7.0) containing 10% glycerol and 1 mM 2-mercaptoethanol (TGM buffer) for 2 hr in preparation for the subsequent step.

Step 3. DEAE-Cellulose Chromatography. The dialyzed calcium phosphate gel eluate is loaded onto a DEAE-cellulose column (1 × 20 cm) which has previously been equilibrated with TGM buffer. After charging the sample, the column is subsequently eluted with TGM buffer containing 0.1 M KCl. A large peak of A_{280}-absorbing material is eluted. Following the elution of this peak, the column is eluted with a 400 ml linear KCl gradient from 0.1 M to 0.7 M in TGM buffer. The protein methylase III is eluted at a position centering at 0.225 M KCl. This activity peak (fractions 96–105 with 2 ml in each fraction) is pooled and is brought to 60% saturation with ammonium sulfate (60 ml of saturated ammonium sulfate per 40 ml of pooled enzyme). In this form, the enzyme is stable for at least 1 week at 4°.

Step 4. Preparative Gel Electrophoresis. Further purification is accomplished by discontinuous preparative gel electrophoresis utilizing a Prep-Disc apparatus (Canalco Co.) essentially as described by Shuster.[3] The three reagents (1) Tris buffer (24 ml 1 M HCl, 18.2 gm Tris, 0.23 ml TEMED (N,N,N',N'-tetramethylethylenediamine)/100 ml, pH 8.9), (2) 30% acrylamide plus 0.8% bisacrylamide, (3) freshly prepared 0.14% ammonium persulfate, are respectively mixed in a 1:1:2 ratio in order to cast a 7.5% polyacrylamide gel of 4 cm in height and 3.2 cm^2 in cross-sectional area.

The enzyme preparation obtained in Step 3 is centrifuged at 39,000 g for 10 min and then is dissolved in 1 ml of the pH 8.5 buffer (30 ml 1 M HCl, 22.7 g Tris, 100 ml glycerol, 0.1 ml 2-mercaptoethanol in a final volume of 1,000 ml). The sample is dialyzed for 2 hr against this buffer and then after the addition of several drops of glycerol is carefully applied to the top of the gel. The sample is carefully overlaid with pH 8.3 electrode buffer 3.0 g Tris and 14.4 g glycine/liter). This buffer is also used for the lower electrode as well. The electrophoresis is carried out at 15 mA and the pH 8.5 buffer described above is used for elution during the run. The flow rate of the elution buffer is set at 1 ml/min and 3 ml fractions are collected. The enzyme is eluted after 2 hr. The fractions containing the enzyme activity peak (42–44) are pooled and brought to 50% ammonium sulfate by the addition of an equal volume of saturates ammonium sulfate solution. The enzyme suspension is then stored at −20° until use. The entire purification is summarized in Table II. The overall purification of the enzyme is 3500-fold with a yield of 4.1%. In the routine use of the enzyme, an aliquot of the ammonium sulfate suspension is dialyzed 12 hr against TGM buffer.

Properties

Purity. When the 3500-fold purified enzyme is run on sodium dodecyl sulfate–polyacrylamide gels which have been subsequently stained with Coomassie blue and scanned at 560 nm, only two bands are apparent on the scanning profile with one comprising four times the staining intensity of the other. Assuming the major band to be the enzyme, we judge the preparation to be 80% homogenous.

Stability. The enzyme is stable for several months when stored in the presence of 50% saturated ammonum sulfate at 0–4°.

[3] L. Shuster, this series, Vol. 22, p. 412.

TABLE II
SUMMARY OF PURIFICATION STEPS

Purification step	Total protein (mg)	Specific activity (pmol/min/mg)	Total activity (pmol/min)	Yield (%)	Purification (fold)
Homogenate	20,680	8.5	176,400	100	1.0
Supernatant	9,420	10.8	101,940	57.8	1.3
Ammonium sulfate (45%)	3,350	24	80,400	45.6	2.8
Calcium phosphate gel	109	331	36,080	20.4	39
DEAE-cellulose	6.1	2,380	14,500	8.2	280
Preparative gel electrophoresis	0.246	29,500	7,250	4.1	3,470

Molecular Properties. The molecular weight of the enzyme as determined by gel filtration on a Sephadex G-200 column is 120,000. The isoelectric point (p*I*) is 4.8.

Catalytic Properties. The K_m value for the methyl donor, S-adenosyl-L-methionine, is 19 μM and for native horse heart cytochrome c, the protein substrate routinely used, the K_m is 1.7 mM. S-Adenosyl-L-homocysteine is a strong product inhibitor of the enzymatic reaction with a K_i of 2 μM. The mode of this compound's inhibition is determined to be competitive with respect to S-adenosyl-L-methionine and uncompetitive with respect to cytochrome c.

Enzymatic Products. In the transfer of methyl groups to cytochrome c, the enzyme produces a mixture of ε-N-mono-, di-, and trimethylated lysyl residues in the protein as shown by amino acid analysis of the hydrolyzed methylated cytochrome c. The ratio of mono-, di-, and trimethylated lysine produced after a 30 min incubation is determined using the enzyme at each of the various purification stages. The ratio varies little with respect to the state of purification and for all stages averages $11.5 \pm 2.8 : 35 \pm 6.8 : 53 \pm 9.4$ for mono : di : tri. The lack of ratio variation during the purification indicates that a single enzyme or enzyme complex is responsible for the stepwise methylation of the lysyl residue.

Protein–Substrate Specificity. Among a variety of proteins tested, only cytochromes c of various species serve as substrates. Because histone is among the nonsubstrates, the enzyme can be differentiated from the methyltransferases that have been described which recognize histones or histone subfractions as a substrate (see Table I). The variety of cytochromes c recognized as substrates are all unmethylated *in vivo* and vary only moderately in substrate capability (40 to 166% relative to horse heart cytochrome c). Included in this class are cytochromes c from mammals, birds, fishes, amphibians, and insects. An exception is the cytochrome c from *Candida krusei* which is methylated *in vivo* but also serves as a fair substrate for the enzyme. Yeast cytochrome c, which is *in vivo* trimethylated at residue 72 is a nonsubstrate.

Residue Site Specificity. The purified enzyme is shown to methylate native horse heart cytochrome c exclusively at residue 72 which resides in the

$$\underset{70\quad71\quad72\quad73\quad74}{\text{Asn-Pro-Lys-Lys-Tyr-}}$$

sequence. As mentioned below, the residues 1–65 from CNBr cleavage of horse heart cytochrome c also serves as a substrate. The site of methylation in this polypeptide has been localized to a Lys-Lys sequence at residues 7–8. Because both of these sites are Lys-Lys sequence, it is suggested by us that this sequence is a necessary although not a sufficient factor in site recognition. (Other Lys-Lys sequences exist in the molecule which are not *in vitro* methylated.)

Sequence and Structural Determinants for Methylation. In lieu of the recognition of residue 72 in native horse heart cytochrome c as the methylation site, the hemoprotein is treated in various ways and is also cleaved into fragments by CNBr. The K_m and V_{max} values for each of the modified cytochromes c and fragments are determined and are shown in Table III. Ethanol-denatured cytochrome c (K_m, 0.20 mM) is only a slightly better

TABLE III
KINETIC PARAMETERS OF VARIOUSLY TREATED CYTOCHROMES
c AND CNBr-CLEAVAGE FRAGMENTS

Substrate	K_m (mM)	V_{max} (nmol methyl group/min/mg enzyme protein)
Cytochrome c (horse heart)		
Native	0.32	24
Ethanol-denatured	0.20	29
Heme-free	0.03	7
CNBr-peptide		
Peptide I (residues 1–80)	0.007	16
Peptide II (residues 1–65)	0.04	27
Peptide III (residues 66–104)	0.04	2.4
Peptide IV (residues 81–104)	N.A.[a]	N.A.
Peptide V (residues 66–80)	N.A.	N.A.

[a] Not active.

substrate than the native hemoprotein (K_m, 0.32 mM). Apocytochrome c is a much better substrate than either of these (K_m, 0.03 mM) and residues 1–80 generated by CNBr treatment is the best substrate overall (K_m, 7 μM). These data have led us to propose that *in vivo* methylation occurs when the protein is nascent or before heme attachment which occurs in the intramembrane space of the mitochondria. The residues 66–104 peptide is a good substrate (K_m, 0.04 mM) while the residues 66–80 peptide is a nonsubstrate even though it contains the residue 72 methylation site. This result alone with the mentioned substrate capability of residues 1–80 suggests that sequence on either side of the 66–80 region is necessary for substrate recognition. It is not known whether this requirement reflects a minimal peptide length or the presence of specific recognition sequences. Although a nonsubstrate, the residues 66–80 peptide is a strong inhibitor of the methylation [at 5 μM, it inhibits the methylation of native cytochrome c (100 μM) by 50%]. The residues 1–65 peptide is a fairly good substrate despite lacking the residue 72 site. As mentioned earlier, the site recognized in this region is lysine 7 and/or 8. This site, however, is not methylated when native, ethanol-denatured, or apocytochrome c is used as substrate.

Enzymatic Demethylation of Protein-Bound Methyllysine (ε-Alkyllysinase)

Introduction

The *in vivo* turnover of ε-N-methyl groups in protein-bound methyllysine residues has been described in several systems. Methyl groups have been observed to turn over in histone in HeLa S-3 cells and perfused cat kidney.[4,5] Similarly, the phenomenon was observed in the iso-1 cytochrome c of yeast.[6]

Only one demethylating enzyme which is capable of removing protein-bound lysyl methyl groups has been described.[7] The enzyme is present in mammalian tissues, and is especially abundant in the kidney. This enzyme has been demonstrated utilizing chemically lysyl-methylated histone. With a high degree of certainty, this enzyme has been shown to be identical to a free ε-N-methyllysine demethylase which was originally

[4] T. W. Borun, D. Pearson, and W. K. Paik, *J. Biol. Chem.* **247**, 4288 (1972).
[5] K. Hempel, G. Thomas, G. Roos, W. Stoecker, and H.-W. Lange, *Hoppe-Seyler's Z. Physiol. Chem.* **360**, 869 (1979).
[6] J. Farooqui, S. Kim, and W. K. Paik, *J. Biol. Chem.* **255**, 4468 (1980).
[7] W. K. Paik and S. Kim, *Arch. Biochem. Biophys.* **165**, 369 (1974).

described by us (ε-alkyllysinase or ε-alkyl-L-lysine : oxygen oxidoreductase; EC 1.5.3.4).[8] The enzyme utilizes molecular O_2 and the cofactor FAD to oxidatively demethylate the methyllysines to form formaldehyde.

Assay Method

Principle. When the free ε-N-methyllysine demethylase was originally described, it was assayed by manometrically determining lysine formed with lysine decarboxylase.[8] A more facile radiometric assay was then developed in which radioactive formaldehyde released from ε-N-L-[*methyl*-14C]lysine (either free or protein-bound in histone) is trapped as a formaldemethone derivative and counted for radioactivity.[7] This method is detailed below.

Preparation of Substrate. One gram of histone (Sigma Chemical Co., type II-A) is dissolved in 13 ml of water, and 6 ml of 0.2 M borate buffer (pH 9.0) and 10 mg of NaBH$_4$ are added. Radioactive [14C]formaldehyde in 0.01-ml aliquots is added at 5 min intervals at 0° until a total of 0.05 ml is added. The formaldehyde used is prepared by diluting [14C]formaldehyde (specific activity, 59 Ci/mol) with an equal volume of nonradioactive 37% formaldehyde. Subsequent to the last addition, the reaction is allowed to proceed for another 5 min, at which time the reaction mixture is dialyzed against 6 liters of water for 24 hr with 3 changes during this time. The dialyzed sample is then lyophilized. At this point, the [*methyl*-14C]histones can be used directly as a substrate for the enzyme. It is pertinent to mention that the above chemical methylation is specific towards the ε-amino group of lysyl residues.[9,10] If it is desired that free ε-N-[*methyl*-14C]-L-lysine be used as a substrate, the histone can be hydrolyzed in 6 M HCl *in vacuo* at 110°. The resulting hydrolysate is evaporated under reduced pressure to remove HCl, and the radioactive material is added to a 0.05 M solution of nonradioactive ε-N-monomethyl-L-lysine to give a specific activity of 1800 cpm/μmol of ε-N-monomethyl-L-lysine.

Reagents

Phosphate buffer, 0.5 M (pH 7.2)
Semicarbazide, 0.1 M (pH around 7)
Flavin adenine dinucleotide (FAD), 1 mM
Phenazine methosulfate, 1%
ε-N-L-[*methyl*-14C]Lysine, 0.05 M (approximately 1800 cpm/μmol)
Ethanol

[8] S. Kim, L. Benoiton, and W. K. Paik, *J. Biol. Chem.* **239**, 3790 (1964).
[9] G. E. Means and R. E. Feeney, *Biochemistry* **7**, 2192 (1968).
[10] J. LaBadie, W. A. Dunn, and N. Aronson, Jr., *Biochem. J.* **160**, 85 (1976).

Formaldehyde, 0.74% (commercially available solution is diluted 50 times)
5,5-Dimethyl-1,3-cyclohexanedione (dimedon), 0.4%
NaOH, 2 M
Acetic acid, 2 M
Trichloroacetic acid (TCA), 50%
NH_4OH-NH_4Cl buffer, 1.0 M (pH 9.7)
[methyl-^{14}C]Histone (0.9 mg/0.15 ml)

Procedure

Free ε-N-[methyl-^{14}C]Lysine Demethylation. The reaction mixture contains 0.2 ml of 0.05 M ε-N-[methyl-^{14}C]lysine, 0.1 ml of phosphate buffer, 0.1 ml of semicarbazide, 0.1 ml of FAD, 0.1 ml of phenazine methosulfate, and an appropriate amount of enzyme (4–8 mg of protein) in a total volume of 1.5 ml. The blank contains everything as in the reaction mixture, except that the enzyme is inactivated by boiling for 5 min. The mixture is incubated for 2 hr at 37°, at which time 1.5 ml of a 50% solution of TCA is added to stop the reaction. The mixture is then centrifuged in a clinical centrifuge (table top) for 10 min and the supernatant is added to 60 ml of dimedon in 250-ml flask. One milliliter of 0.74% formaldehyde is added as a carrier, giving 78 mg of formaldemethone as the theoretical yield. About 3.5 ml of 2 M NaOH and 1.7 ml of 2 M acetic acid are then added to form a salt. The precipitate is then collected in a Büchner funnel (i.d. 3 cm) after standing 30 min at room temperature, washed several times with water, and then dried under an infrared lamp. Drying is completed within 30 min with the lamp 25 cm above the sample. A portion of the sample is weighed and transferred to a scintillation vial containing 10 ml of scintillation solution (Formula 963, New England Nuclear Corp.). From the radioactivity obtained, the total radioactivity based on 78 mg of formaldemethone is calculated. The enzyme activity is expressed as micromoles of formaldehyde formed/hour/milligram of enzyme protein.

[methyl-^{14}C]Histone Demethylation. The incubation mixture for the demethylation activity of [methyl-^{14}C]histone is similar to that above except for the following modifications: 0.15 ml 0f 1.0 M NH_4OH-NH_4Cl buffer is substituted for phosphate buffer and 0.15 ml of a solution containing 0.9 mg of [methyl-^{14}C]histone (8770 cpm) is substituted for the ε-N-[methyl-^{14}C]lysine. The procedure up to the crystallization and drying of the [^{14}C]formaldemethone is as described above. However, the following additional steps are carried out. The dried material is washed from the

Büchner funnel into a large (1.5 × 15 cm) test tube with 10 ml of cold ethanol. The material is put into solution with heating and then is filtered into a smaller test tube containing several boiling chips. The tube is boiled at 90–95° until the volume is reduced to 1.5–2.0 ml. After cooling, the formed crystals are collected and counted for radioactivity as described above. The enzyme activity is expressed as picomoles of formaldehyde formed/hour/milligram of enzyme protein.

Purification Procedures

Step 1. Triton X-100 Extraction. Throughout the course of the purification all procedures are carried out at 0–4°. Five grams of rat kidney from adult animals is homogenized in 9 volumes of 0.25 M sucrose–6 mM CaCl$_2$ using an electrically driven Teflon-glass homogenizer. The homogenate is filtered through a double layer of cheesecloth and centrifuged at 39,000 g for 20 min. The precipitate is extracted with 20 ml of 0.5% Triton X-100 for 10 min in a ice-cold Teflon-glass homogenizer. The suspension is centrifuged at 39,000 g for 30 min and the supernatant is saved.

Step 2. Calcium Phosphate Gel Treatment. Forty milliliters of calcium phosphate gel (17 mg of solid/ml, product of Sigma Chemical Co.) is added to the Triton X-100 extract from Step 1, and the mixture is centrifuged at 700 g for 5 min. The precipitate is washed twice with a large volume of cold deionized water and the adsorbed enzyme removed by the addition of 10 ml of 0.3 M phosphate buffer (pH 7.2). The enzyme is found in the supernatant after removing the calcium phosphate gel by centrifugation at 39,000 g for 10 min.

Step 3. Ammonium Sulfate Precipitation. Saturated ammonium sulfate (saturated at room temperature) is added to the gel eluate from Step 2 in a ratio of 6 : 10 [(NH$_4$)$_2$SO$_4$: eluate]. After about 10 min, the precipitate is collected by centrifugation at 39,000 g for 10 min. If frozen at this stage the enzyme activity remains unchanged for at least 2 weeks.

Step 4. Heat Treatment. The precipitate is dissolved in 1.8 ml of 5 mM ε-N-monomethyl-L-lysine, and the suspension is heated at 55° for 6 min in a small test tube. After cooling in ice, the sample is centrifuged at 105,000 g for 20 min and the supernatant saved.

Step 5. Sephadex G-200 Chromatography. The supernatant from Step 4 is applied to a Sephadex G-200 column (1 × 93 cm) equilibrated with 5 mM phosphate buffer (pH 7.2) containing 10% glycerol. The activity peak (fractions 29–32 with 1 ml each fractions) is pooled. At this stage, the enzyme has been purified 15-fold with a 5% yield. The entire purification is summarized in Table IV.

TABLE IV
PURIFICATION SCHEDULE OF ε-ALKYLLYSINASE

Purification step	Volume (ml)	Protein (mg/ml)	Specific activity (μmol/hr/mg protein)	Total activity (μmol/hr)	Yield (%)	Purification (fold)
Whole homogenate	40.0	16.9	0.162	109.3	100.0	1.00
Triton X-100 extract	20.0	7.14	0.303	43.3	39.6	1.86
Ca$_3$(PO$_4$)$_2$ gel eluate	12.8	5.13	0.400	26.3	24.1	2.47
(NH$_4$)$_2$SO$_4$ precipitate	1.80	16.23	0.711	20.8	19.0	4.39
Heated at 55° for 6 min	1.80	3.36	1.43	8.65	7.91	8.82
Sephadex G-200	5.00	0.376	2.60	4.89	4.47	16.0

Properties

Identity of ε-N-Methyllysine and Histone Demethylating Enzymes. At each stage of purification the ε-N-L-[*methyl*-^{14}C]lysine and [*methyl*-^{14}C]histone demethylating activities are monitored. As shown in Table V, the ratio of these two activities remains relatively constant during the course of the purification. This indicates that the same enzyme is responsible for both activities. In addition, the two activities decrease at identical rates upon heating of the partially purified enzyme preparation at 55°, also indicating a common enzyme.

pH Optimum. The enzyme shows optimal activity for demethylating ε-N-[*methyl*-^{14}C]lysine at pH 7.2 using phosphate buffer.

TABLE V
DEMETHYLATION ACTIVITY DURING ENZYME PURIFICATION, USING [*methyl*-^{14}]HISTONE
AND ε-N-L-[*methyl*-^{14}C]LYSINE AS SUBSTRATE[7]

Purification step	With [*methyl*-^{14}C]histone (enzyme activity, pmol/hr/mg protein)	With ε-N-L-[*methyl*-^{14}C]lysine (enzyme activity, μmol/hr/mg protein)	Ratio of column 2 to column 3
Whole homogenate	6.5	0.155	42
Triton X-100 extract	16.3	0.356	46
(NH$_4$)$_2$SO$_4$ precipitate	24.7	0.711	35
Sephadex G-200 eluate	63.7	1.70	38

Kinetic Parameters. The enzyme exhibits classical Michaelis–Menten kinetics and has a K_m of 1.05 mM for ε-N-monomethyl-L-lysine.

Substrate Specificity. The enzyme will also demethylate ε-N-di-methyl-L-lysine and α-keto-ε-methylaminocaproic acid. The ornithine analog, δ-N-methyl-L-ornithine is not demethylated.

The Effect of Various Compounds on Enzyme Activity. The divalent cations Ni^{2+}, Zn^{2+}, and Co^{2+} are found to be inhibitory at 1.2 mM concentration. At 1.3 mM, KCl reduces the enzyme activity to 40% and 2,6-dichlorophenol is also found to be inhibitory toward the partially purified enzyme.

[27] Methylated Lysines and 3-Methylhistidine in Myosin: Tissue and Developmental Differences

By GABOR HUSZAR

Methylated Amino Acids in Myosin, Tissue-Specific Occurrence

Four different methylated amino acids have been identified so far in myosins: 3-methylhistidine (3-MeHis), ε-N-monomethyllysine (MeLys), ε-N-dimethyllysine (Me$_2$Lys), and ε-N-trimethyllysine (Me$_3$Lys).[1-3] All these unusual residues are in the heavy chain of subfragment 1, the globular head of myosin that carries the sites for ATP hydrolysis and actin combination.[4,5] While 2 mol of Me$_3$Lys per myosin heavy chain is found in all the mammalian cardiac and skeletal myosin investigated, the occurrence and amounts of the other methylated amino acids show a high degree of tissue, species, and developmental variation. MeLys and 3-MeHis are constituents of myosin of fast skeletal muscles of adult animals, but they are not present in myosin of slow and cardiac muscles, and in fast muscles of unborn or newborn animals. During early development, e.g., in the first 28 days of life in rabbits, the contents of 3-MeHis and MeLys increase to that of the adult level. There are trace amounts of Me$_2$Lys in myosins of several species, but significant amounts of this

---

[1] G. Huszar and M. Elzinga, *Nature (London)* **223**, 834 (1969).

[2] W. M. Kuehl and R. S. Adelstein, *Biochem. Biophys. Res. Commun.* **37**, 59 (1969).

[3] I. P. Trayer, C. I. Harris, and S. V. Perry, *Nature (London)* **217**, 452 (1968).

[4] G. Huszar and M. Elzinga, *Biochemistry* **10**, 229 (1971).

[5] M. F. Hardy and S. V. Perry, *Nature (London)* **223**, 300 (1969).

METHODS IN ENZYMOLOGY, VOL. 106

residue have been found only in cat soleus and lobster myosins.[6] Actin, the other major muscle protein, has one fully methylated 3-MeHis in all muscles and species investigated. The actin in *Acanthamoeba castellanii* also contains MeLys and Me$_2$Lys.[7]

The distribution of methylated amino acids was investigated by isolating and sequencing the peptides which contain the 3-MeHis and the two Me$_3$Lys in rabbit fast skeletal muscle myosin. Both 3-MeHis and Me$_3$Lys were present as fully methylated and trimethylated amino acids in the myosin heavy chain, and there were no partially methylated Lys and His residues. The 3-MeHis containing tryptic peptide contained 13 amino acid residues, including the methylated histidine.[4]

The question arises why myosins in cardiac and fetal muscles lack 3-MeHis. To distinguish between the various possibilities (e.g., the absence of histidine in this particular position of the myosin sequence, major structural differences in the various myosins, or the lack of histidine methylating enzyme in the heart and developing muscle tissues), the peptides homologous to the 3-MeHis peptide were isolated and sequenced in myosins of rabbit and bovine heart and of fast muscle of 21-day-old developing rabbits (50% histidine methylation[3]). In comparing the 13-member peptides in skeletal and cardiac myosins of the rabbit (Fig. 1), 4 amino acid replacements were found, and there was an unmethylated histidine residue in cardiac myosin in the position corresponding to the 3-MeHis. In the peptide of bovine cardiac myosin there was also the unmethylated histidine, but there was only one amino acid difference between the peptides of rabbit and bovine hearts.[8] These differences in the structure of skeletal and cardiac muscle myosins were the first demonstration of tissue and species differences in myosin isozymes. The peptide of the baby rabbits contained 50% 3-MeHis and 50% histidine, and there was another position with a 50% Ala/Gly heterogeneity. Thus, another isozymic change from fetal-type to adult-type myosin was also demonstrated.[9]

The two Me$_3$Lys were also fully methylated residues in fast skeletal muscle myosin, but in nonidentical peptides.[10] One of them had 10 residues in a sequence -Ala-Thr-Asp-Thr-Ser-Asn-Phe-Me$_3$Lys-Lys-Lys-; the other Me$_3$Lys residue has been isolated by tryptic and thermolytic digestion in the dipeptide -Tyr-Me$_3$Lys-. Methylated lysines are constituents of several proteins (cytochromes, histones, calmodulin, ribosomal and flagellar proteins), but 3-MeHis has so far only been demonstrated in

[6] W. M. Kuehl and R. S. Adelstein, *Biochem. Biophys. Res. Commun.* **39**, 956 (1970).
[7] R. E. Weihing and E. D. Korn, *Nature (London)* **227**, 1263 (1970).
[8] G. Huszar and M. Elzinga, *J. Biol. Chem.* **247**, 745 (1972).
[9] G. Huszar, *Nature (London), New Biol.* **240**, 260 (1972).
[10] G. Huszar, *J. Biol. Chem.* **247**, 4057 (1972).

Rabbit skeletal: Leu-Leu-Gly-Ser-Ile-Asp-Val-Asp- 3-methyl histidine -Gln-Thr-Tyr-Lys

Rabbit foetal: Leu-Leu-Ala-Ser-Ile-Asp-Ile-Asp- histidine -Gln-Thr-Tyr-Lys

Rabbit cardiac: Leu-Leu-Ser-Ser-Leu-Asp-Ile-Asp- histidine -Gln-Asn-Tyr-Lys

FIG. 1. The sequence of the myosin peptides with 3-MeHis and His. The homologous peptide in bovine cardiac myosin showed a Ser → Gly replacement in the third position as compared to the sequence of rabbit cardiac myosin. Comparison of the three sequences: ■, different in one peptide; ▨, different in all three peptides (from Huszar[9]).

the contractile proteins, actin and myosin. The amino acid sequence around the 3-MeHis in rabbit skeletal myosin has been confirmed subsequently as a part of a 93-member CNBr peptide.[11]

Determination of Methylated Basic Amino Acids

The methylated amino acids, like other amino acids, are analyzed in the HCl hydrolyzates of proteins. Hydrolysis in 6 N HCl for 24 or 48 hr does not cause a loss of methyl groups. There are two technical problems associated with the determination of methylated amino acids in myosin. One is the relative amount of these residues. In the myosin heavy chain there are about 1900 amino acids and about 160 lysines and 30 histidines for each 3-MeHis residue. A program with special resolution power is necessary to quantify methylated lysine or histidine next to 100-fold higher lysine or ammonia peaks. The other technical difficulty is the resolution of various methylated derivatives of lysine from each other.

Determination of methylated lysines in myosin, c cytochrome, calmodulin, etc., is still carried out by conventional methods using amino acid analyzers.[12,13] In analysis of 3-MeHis there are other alternatives which include gas–liquid chromatography,[14] HPLC,[15,16] and a recently developed photometric method.[17]

Amino acid analysis programs designed to determine methylated basic amino acids are based on the use of the high-resolution amino acid analyzer resins (e.g., Beckman AA-15, Bio-Rad-A-5, Beckman W-3H). Usu-

[11] M. Elzinga and J. H. Collins, *Proc. Natl. Acad. Sci. U.S.A.* **74,** 4283 (1977).
[12] G. Zarkadas, *Can. J. Biochem.* **53,** 96 (1975).
[13] P. DiMaria, E. Polastro, R. J. DeLange, S. Kim, and W. K. Paik, *J. Biol. Chem.* **254,** 4645 (1979).
[14] T. W. Larsen and R. F. Thornton, *Anal. Biochem.* **109,** 137 (1980).
[15] S. J. Wassner, J. L. Schlitzer, and J. B. Li, *Anal. Biochem.* **104,** 284 (1980).
[16] Z. Friedman, H. W. Smith, and W. S. Hancock, *J. Chromatogr.* **182,** 414 (1980).
[17] E. Radha and S. P. Bessman, *Anal. Biochem.* **121,** 170 (1982).

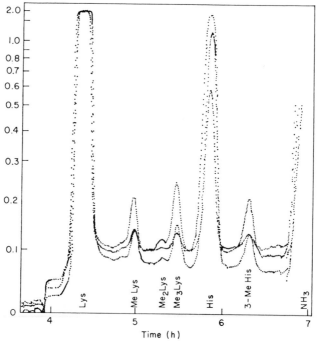

FIG. 2. Amino acid analysis of the basic amino acids in a peptide mixture of chicken pectoralis muscle myosin. Elution conditions are described below.

ally, a 30- to 40-cm-long column is used. It is eluted at 26–30° with a buffer of 0.3–0.4 M Na citrate with pH between 5.0 and 6.0 and flow rates of 25–35 ml. The conditions used in our program with a Beckman model 119 automated amino acid analyzer (Fig. 2) were as follows: a column of 38 × 0.9 cm filled with AA-15 resin was developed with 0.35 M sodium citrate, pH 5.85, at 28°, with a flow rate of 30 ml/hr. The operating pressure was about 200 lb/in.[2] Elution order and times of the amino acids were as follows: tyrosine, 111 min; phenylalanine, 120 min; lysine, 258 min; MeLys, 291 min; Me$_2$Lys, 310 min; Me$_3$Lys, 320 min; histidine, 344 min; 3-MeHis, 367 min; and NH$_3$, 412 min. The use of methylated lysine standards is recommended, as the methyl groups influence the color yields.

Biosynthesis of Methylated Amino Acids

Similarly to other posttranslational amino acid side-chain modifications, lysines and histidines are enzymatically methylated after they are incorporated into the myosin peptide chain. The methyl donor is *S*-adenosyl-L-methionine (SAM). Methylation of myosin with SAM donors has been carried out in various experimental designs: in muscle homogenates and cultures,[5,18-21a] in perfused hearts,[22] and by *in vitro* methylation of nascent myosin chains still attached to polyribosomes.[23]

Experiments were directed to the mechanisms underlying the tissue specific myosin methylation pattern: 3-MeHis and MeLys are absent in cardiac myosin, while the presence of 2 Me₃Lys residues are common in both skeletal and cardiac myosins. The sequencing data demonstrated (Fig. 1) that cardiac myosin contains the unmethylated histidine, and that the peptides with histidine or 3-MeHis are homologous. We have made the hypothesis that the tissue-specific occurrence of methylated amino acids is due to variations in the presence of methylating enzymes, rather than to the structural differences in the myosins. This hypothesis was tested by *in vitro* methylation of nascent myosin chains.[23]

Skeletal muscle myosin was synthesized with skeletal myosin polyribosomes and with the S-150 supernatant enzymes from heart in the presence of amino acid mixtures and [³H]SAM. Subsequently, cardiac myosin was synthesized on cardiac polysomes using skeletal muscle enzymes. Two alternative explanations were tested. If the tissue specificity of methylation is based on the presence of methylating enzymes, cardiac or skeletal myosin should be methylated to the same extent with skeletal supernatant. If, however, the methylases are constituents of both skeletal and cardiac muscle tissues, and some structural features of cardiac myosin prevent the biosynthesis of MeLys, then skeletal muscle myosin would contain the monomethylated residues after incubation with either supernatant.

The S-150 supernatant enzymes from skeletal muscles catalyzed MeLys biosynthesis in both skeletal and cardiac myosin, while the cardiac supernatant generated only Me₃Lys residues in both myosins. The myosin peptides with methylated lysines were identified by ion-exchange chromatography of the tryptic digest of the nascent myosin. The peptides with [³H]methyl-labeled groups were HCl hydrolyzed, and the methylated lysines were identified by amino acid analyses. Methylation of denatured cardiac myosin with the skeletal S-150 enzymes was unsuccessful.

[18] B. Krzysik, J. P. Vergnes, and I. R. McManus, *Arch. Biochem. Biophys.* **146,** 34 (1971).
[19] K. Laki and E. Wilson, *Physiol. Chem. Phys.* **1,** 536 (1969).
[20] M. Reporter, *Arch. Biochem. Biophys.* **158,** 577 (1973).
[21] J. Wikman-Coffelt and D. T. Mason, *Res. Commun. Chem. Pathol. Pharmacol.* **27,** 319 (1980).
[21a] W. K. Paik and S. Kim, *Adv. Enzymol.* **42,** 227 (1975).
[22] C. A. Watkins and H. E. Morgan, *J. Biol. Chem.* **254,** 693 (1979).
[23] G. Huszar, *J. Mol. Biol.* **94,** 311 (1975).

The formation of MeLys in cardiac myosin incubated with skeletal supernatants indicates that both skeletal and cardiac muscle myosins contain a site for monomethylation. Only nascent myosin chains of <50,000 daltons were methylated, which indicates that there are steric considerations in the accessibility of the lysines to be methylated. These data and the homologous sequences suggest that the tissue-specific occurrence of methylated amino acids is due to the presence of the methylating enzymes in the cell rather than to differences in myosin structure.

Mechanism of Lysine Methylation

At the present time, the nature of the recognition site for lysine methylation is uncertain. The site is most likely a group of amino acids around the lysines to be methylated, or amino acids also may contribute that are close the the lysine in the tertiary structure, but far apart in the unfolded myosin peptide chain. This question was examined in further experiments in which myosin biosynthesis and lysine methylation were carried out using polyribosomes from skeletal muscle and S-150 enzymes from wheat germ. The rationale of these experiments was based on the homology in amino acid sequences around Me$_3$Lys in myosin (-Asn-Phe-Lys-Me$_3$Lys-Lys-Lys-) and in c cytochromes of various plants, including the wheat germ (-Leu-Asn-Pro-Me$_3$Lys-Lys-Tyr-).[24] Based on the methylation of myosin by the wheat germ supernatant enzymes, we suggested that the short common sequence (-Asn-X-Lys-Lys-) is likely to be a part of the recognition site for lysine methylation.[25]

Functional and Research Role of Methylated Amino Acids

Apart from c cytochromes and the few above mentioned proteins, only myosin and amoeba actin[7] were reported to contain methylated lysines while 3-MeHis is present only in actin and myosin. This unusual distribution suggests that they may have a special role in the contractile process. Other ideas with respect to function have emphasized the possible role of the very hydrophobic Me$_3$Lys as a signal site or recognition point. Muscle turnover studies measuring myosin and actin biosynthesis and methylation have demonstrated no rapid changes in the extent of histidine and lysine methylation; thus, it makes unlikely a regulatory role in muscle function similar to phosphorylation and dephosphorylation.

In the past decade, several metabolic and protein turnover studies

[24] A. N. Glazer, R. J. De Lange, and R. S. Martinez, *Biochim. Biophys. Acta* **188**, 164 (1969).
[25] G. Huszar, *Fed. Proc., Fed. Am. Soc. Exp. Biol.* **33**, Abstr. 2008 (1974).

took advantage of the fact that 3-MeHis is only present in actin and myosin. As 3-MeHis does not get demethylated, 3-MeHis excretion in the urine follows muscle protein breakdown. With each mole of actin and myosin catalyzed, 1 mol of 3-MeHis appears in the urine. Urinary creatinine excretion is proportional to the total muscle mass; thus, the ratio of 3-MeHis and creatinine in the urine reflects the rate of protein degradation over the total muscle mass. This ratio has been very useful in studying protein metabolism and nutrition. Several publications have appeared employing this approach including various metabolic studies,[26] diabetes,[27,28] muscular dystrophy,[29] and metabolism of premature infants.[30]

Techniques Used in the Study of Structure and *in Vitro* Methylation of Myosin

Myosin Polysome Preparation

Leg muscles from 14-day-old chick embryos are freed from skin and bones and homogenized. Hearts of chick embryos of 14 to 20 days old are used for preparation of heart polysomes. The muscle of hearts are cut into small pieces with sharp scissors. The pieces are washed free of blood with 10 mM Tris–HCl, pH 7.6. Homogenization is carried out in 2 volumes of buffer (0.25 M KCl, 10 mM magnesium acetate, 10 mM Tris–HCl, pH 7.5) in a Dounce glass homogenizer with a loose pestle (average, 7 strokes). The homogenate is centrifuged at 10,000 g for 10 min, and the supernatant is layered on 42% sucrose (containing KCl, magnesium acetate, and Tris buffer in concentrations equal to that of the homogenizing buffer) in a Spinco 42 rotor tube. The tubes are then spun at 35,000 rpm (96,000 g) for 180 min. The pelleted material is the myosin polysomes, which usually have an $A_{260/280}$ ratio of about 1.7.[23]

Preparation of Supernatant Enzymes

Homologous Muscle System. Twenty-one-day old chick embryos are used. The leg muscles or hearts are cleaned and cut into small pieces which are extensively homogenized with a tight Dounce glass homogenizer in 3 volumes of buffer (100 mM KCl, 2 mM CaCl$_2$, 3 mM magnesium acetate, and 10 mM Tris–HCl pH 7.6). The homogenate is clarified by

[26] V. R. Young and H. N. Munro, *Fed. Proc., Fed. Am. Soc. Exp. Biol.* **37,** 2291 (1978).
[27] A. F. Nakhooda, C. Wei, and E. B. Marliss, *Metab., Clin. Exp.* **29,** 1272 (1980).
[28] G. Huszar, V. Koivisto, E. Davis, and P. Felig, *Metab. Clin. Exp.* **31,** 188 (1982).
[29] R. C. Griggs, R. T. Moxley, and G. B. Forbes, *Neurology* **30,** 1262 (1980).
[30] J. H. Seashore, G. Huszar, and E. Davis, *Metab., Clin. Exp.* **30,** 959 (1981).

centrifugation at 35,000 g for 30 min, and the middle two-thirds of the supernatant is passed through a 2.5 × 28-cm Sephadex G-25 column. The early fractions with the most tubidity (ribosomes) are pooled and centrifuged at 175,000 g for 60 min (Spinco 65 Ti rotor at 55,000 rpm). The resulting clear postribosomal supernatant (about 8 to 10 mg/ml protein) is used subsequently in the cell-free system. The supernatant is stored in liquid nitrogen.

Heterologous Wheat Germ System. Dry wheat embryos are ground thoroughly with a small amount of sand in a precooled mortar in 10 volumes buffer (100 mM KCl, 2 mM CaCl$_2$, 1 mM magnesium aetate, 10 mM Tris-maleate) with the pH rigorously maintained between 6.4 and 6.8. The slurry is then centrifuged for 10 min at 25,000 g, and the supernatant under the upper lipid layer is removed with a Pasteur pipette. Prior to use, the supernatant is dialyzed against 100 mM KCl, 3 mM magnesium acetate, 3 mM 2-mercaptoethanol, and 10 mM Tris–HCl, pH 7.6. The resulting wheat germ cell-free system may be stored in liquid nitrogen.[31]

In Vitro Methylation of Myosin

The polysome pellet is resuspended in a buffer of 0.25 M KCl, 5 mM dithiothreitol, 50 mM Tris–HCl, pH 7.5. S-Adenosyl-L-[*methyl-*^{3}H] methionine and amino acid mixture (cold or radioactive, depending on the purpose of the experiment) is combined with the gently resuspended polysomes (approx. 1 μCi/14 A_{260} units of polysomes). Addition of supernatant enzymes (about 100 μl/14 A_{260} units of polysomes) completes the assay. Protein incorporation or the methylation is monitored by precipitation of small aliquots with trichloroacetic acid. The aliquots are filtered ice cold after incubation at 80°, dried on Millipore filters, and counted in toluene-based scintillation solution. To purify the nascent myosin chains methylation is stopped by cooling the assay to 0°; the reaction mixture (about 2 to 3 ml) is layered on 42% sucrose–salt solution and spun in the 65 Ti rotor for 60 min at 150,000 g. The pellet contains the methylated myosin chains that are associated with the polysomes. The myosin is then subjected to various analytical procedures, e.g., determination and characterization of the incorporated radioactivity, SDS–gel electrophoresis, tryptic digestion, and ion-exchange chromatography.

An important modification has been introduced by running the myosin synthesis and methylation assay at 27°. Under these conditions, the completed chains are not released from the polysomes; thus the purification of the nascent myosin chains becomes possible by sucrose centrifugation.[23]

[31] A. Marcus, D. Efron, and D. W. Weeks, this series, Vol. 30, p. 749.

Strategy for Isolation of Peptides with Methylated Amino Acids in Myosin

These methods have been primarily developed for the isolation of 3-MeHis peptides. Myosin is prepared[4] and the purity is checked with SDS electrophoresis. The myosin chains are subjected to S-alkylation by NEM in the presence of 5 M guanidine hydrochloride, 100 mM EDTA, 0.2% 1-mercaptoethanol, and 100 mM Tris at pH 8. The myosin is then extensively dialyzed, dissolved in formic acid, and cleaved with CNBr.[4] At this point, the CNBr peptides may be isolated or the CNBr–peptide mixture can be further cleaved by trypsin after blockage of the lysine residues with citraconic anhydride.[32] The resulting Arg peptides (the Arg 3-MeHis peptide in myosin is about 45 residues long) are fractionated by ion-exchange chromatography on SP-Sephadex or on Dowex-50. The peptide mixture enriched in methylated amino acids is further chromatographed by Sephadex, BioGel or by HPLC.[33]

[32] H. B. F. Dixon and R. N. Perham, *Biochem. J.* **109**, 312 (1968).
[33] S. W. Tong and M. Elzinga, *J. Biol. Chem.* **24**, 13100 (1983).

[28] S-Adenosylmethionine : Protein-carboxyl O-Methyltransferase (Protein Methylase II)

By SANGDUK KIM

S-Adenosylmethionine : protein-carboxyl O-methyltransferase (protein methylase II) transfers the methyl group of S-adenosyl-L-methionine (SAM) to free carboxyl groups of protein or polypeptide substrates, yielding protein-carboxylmethyl esters according to the following scheme:

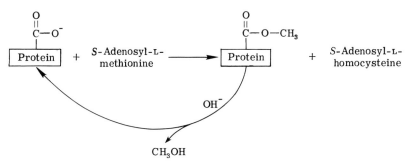

The protein-methyl ester is unstable in aqueous alkaline solution, and yields methanol and the original protein upon hydrolysis. The chemical stability of the ester varies greatly depending on the substrate modified. Esters in soluble proteins are generally more labile than those in membrane-associated proteins.

The enzyme is ubiquitous in nature and has been purified from various mammalian organs,[1-6] wheat germ,[7] and bacteria.[8,9] The substrate specificity and molecular properties of the enzyme vary greatly depending on the origin (Table I); however, enzymes from mammalian organs appear to have similar properties.

According to the systematic nomenclature suggested by the Enzyme Commission, the enzyme is S-adenosylmethionine : protein O-methyltransferase (EC 2.1.1.24). However, other names have been used to describe the enzyme, causing some confusion. Initially, the term "methanol-forming enzyme"[10] was used in the literature until the enzymatic reaction mechanism was better understood; i.e., elucidation of the immediate enzymatic product as an alkali-labile protein-carboxylmethyl ester and not methanol. In our laboratory, protein methylase II is used to distinguish it from protein methylases I and III. However, protein carboxymethylase, protein-carboxylmethyltransferase, CheR methyltransferase (*Salmonella typhimurium* and *Escherichia coli*), and methyl-accepting chemotaxis protein methyltransferases I and II (*Bacillus subtilis*) have been also used to describe the enzyme.

Assay Method

Principle. Protein methylase II activity can be determined by two different approaches utilizing radiomethyl-labeled SAM. The first approach is to directly determine the amount of radioactivity transferred from [*methyl*-^{14}C]SAM into trichloroacetic acid (TCA)-precipitable substrate protein.[11] The second approach takes advantage of the alkali lability of the enzymatically formed ester bond; the reaction product is hydro-

[1] S. Kim, *Arch. Biochem. Biophys.* **157,** 476 (1973).
[2] S. Kim, *Arch. Biochem. Biophys.* **161,** 652 (1974).
[3] E. J. Diliberto, Jr. and J. Axelrod, *Proc. Natl. Acad. Sci. U.S.A.* **71,** 1701 (1974).
[4] M. Iqbal and T. Steenson, *J. Neurochem.* **27,** 605 (1974).
[5] E. T. Polastro, M. M. Deconinck, M. R. Devoel, E. L. Mailier, Y. B. Looza, A. G. Schneck, and J. Leonis, *Biochem. Biophys. Res. Commun.* **81,** 920 (1978).
[6] S. Kim, S. Nochumson, W. Chin, and W. K. Paik, *Anal. Biochem.* **84,** 415 (1978).
[7] L. Trivedi, A. Gupta, W. K. Paik, and S. Kim, *Eur. J. Biochem.* **128,** 349 (1982).
[8] W. R. Springer and D. E. Koshland, Jr., *Proc. Natl. Acad. Sci. U.S.A.* **74,** 533 (1977).
[9] A. Burgess-Cassler, A. H. J. Ullah, and G. W. Ordal, *J. Biol. Chem.* **257,** 8412 (1982).
[10] J. Axelrod and J. Daly, *Science* **150,** 892 (1965).
[11] S. Kim, L. Wasserman, B. Lew, and W. K. Paik, *J. Neurochem.* **24,** 625 (1975).

TABLE I

PROPERTIES OF PROTEIN METHYLASE II ISOLATED FROM VARIOUS SOURCES

Sources of enzyme	Substrate protein specificity	K_m for SAM ($M \times 10^{-6}$)	K_i for SAH ($M \times 10^{-6}$)	Molecular weight	Comments
Mammalian[a] Calf brain Bovine pituitary gland Erythrocytes of human, rat	ACTH, calmodulin, γ-globulin, histone, ovalbumin, ribonuclease, erythrocyte membrane protein[b,c,d]	0.87	0.9	25,000	Broad protein substrate specificity
Plant Wheat germ[e]	ACTH, histone,[d,e] ribonuclease	5.0	1.5	41,000	Various histones are the major endogenous substrates
Bacteria Salmonella typhimurium[f] Escherichia coli	Membrane proteins of S. typhimurium and E. coli	10		38,000	Membrane proteins from CheR⁻ cell are poor substrates. Sensitive to salts[g]
Bacillus subtilis	Membrane proteins from B. subtilis	2	0.2	44,000	MCP methyltransferase I. Ca^{2+} is competitive inhibitor
subtilis[h,i]	Membrane proteins from B. subtilis. Absent in chemotaxis mutant	5	0.2	30,000	MCP methyltransferase II. Ca^{2+} and Mg^{2+} are activators

[a] W. K. Paik and S. Kim, in "Protein Methylation" (A. Meister, ed.), p. 202. Wiley, New York, 1980.
[b] P. Galletti, W. K. Paik, and S. Kim, Eur. J. Biochem. **97**, 221 (1979).
[c] C. A. Janson and S. Clarke, J. Biol. Chem. **255**, 11640 (1980).
[d] Endogenous methyl-acceptor proteins.
[e] L. Trivedi, A. Gupta, W. K. Paik, and S. Kim, Eur. J. Biochem. **128**, 349 (1982).
[f] W. R. Springer and D. E. Koshland, Jr., Proc. Natl. Acad. Sci. U.S.A. **74**, 533 (1977).
[g] S. Clarke, K. Sparrow, S. Panasenko, and D. E. Koshland, Jr., J. Supramol. Struct. **13**, 315 (1980).
[h] A. H. J. Ullah and G. W. Ordal, Biochem. J. **199**, 795 (1981).
[i] A. Burgess-Cassler, A. H. J. Ullah, and G. W. Ordal, J. Biol. Chem. **257**, 8412 (1982).

lyzed in an alkaline solution and the liberated radioactive methanol is extracted using isoamyl alcohol[3,12] or collected by distillation under reduced pressure and moderate temperature.[13]

The direct method is preferable to assess the activity in crude tissue extract, because relatively high blank values (0-time activity or heated enzyme) are often observed when assayed by indirect method. The indirect method is best suited for assaying the enzyme activity when using small peptide or protein that are not precipitated by TCA. The methanol extraction method is simple and rapid, and therefore is an ideal method for the assay of large numbers of samples.

Reagents

  Citrate-phosphate buffer cocktail, pH 6.0; prepared by mixing 6 ml of 0.75 *M* citric acid, 10 ml of 1.5 *M* Na_2HPO_4, 8 ml of 0.06 *M* EDTA, and 0.08 ml of 2-mercaptoethanol. The pH of the mixture is readjusted to 6.0 with citric acid

  S-Adenosyl-L-[*methyl*-^{14}C]methionine; specific activity, 60 mCi/mmol; 0.2 m*M* in 0.01 *M* HCl. Stored at −20°

  Substrate protein[14]; γ-globulin (100 mg/ml), dissolve crystals slowly in 0.01 *M* HCl and adjust pH to 6.0 with 1 *M* HCl

  Procedure. In a 12-ml thick-wall Pyrex conical centrifuge tube immersed in an ice bucket, 0.025 ml of the citrate-phosphate buffer cocktail, 0.05 ml of substrate solution, and an appropriate amount of the enzyme preparation are mixed in a final volume of 0.20 ml. After incubating the tubes at 37° for 3 min, the reaction is initiated by adding 0.05 ml of [*methyl*-^{14}C]SAM. The reaction is then allowed to proceed for 10 min. Two sets of duplicate tubes are prepared, one of which serves as a control. After the incubation, one of the following methods can be used to quantitate the amount of methyl label incorporated into the substrate protein.

Method I. TCA Precipitation Method (Direct Assay): Reagents

  TCA, 15%[15]

  Chloroform : ether : ethanol mixture (1 : 2 : 2 by v/v)

[12] M. Jamaluddin, S. Kim, and W. K. Paik, *Biochemistry* **15**, 3077 (1976).
[13] S. Kim, and W. K. Paik, *Anal. Biochem.* **42**, 255 (1971).
[14] γ-Globulin is used as a standard methyl acceptor, because the protein is readily available at a moderate cost and also the TAC precipitate is easy to handle. However, many other proteins (ovalbumin, histones etc.) can equally be used as the methyl acceptor substrate for the enzyme assay.
[15] The use of 15% TCA ensures complete precipitation of the substrate protein after the reaction. However, depending on the protein a lower percentage of TCA solution can be used.

Ethanol, 98%

Sodium phosphate buffer, 0.2 M (pH 7.4)

Procedure. The reaction is terminated by addition of 5 ml TCA. The mixture is carefully overlayered with ethanol (about 0.4 ml) and centrifuged for 15 min in a table-top clinical centrifuge. The supernatant is decanted, and the precipitate is successively washed with 5 ml of TCA, once with chloroform : ether : ethanol mixture, and once with ethanol. The precipitates in one set of tubes are transferred quantitatively into scintillation vials containing 5 ml of scintillation solution and counted in a scintillation counter. One ml of 0.2 M phosphate, pH 7.4, is added to the precipitates in the second set of tubes (serving as controls), which are then placed in a boiling water bath for 5 min.[16] This treatment decomposes the protein methyl ester. The proteins are precipitated by the addition of TCA, then washed once with TCA and once with ethanol. Finally, the radioactivity in the precipitate is counted. The difference in the radioactivity between the heated and unheated tubes is taken as the radioactivity due to the protein-methyl esters formed by protein methylase II.

Method II. Methanol Extraction Extraction Method (Indirect Assay): Reagents

Sodium borate, 0.25 M, pH 11; heating to 60° may be necessary to completely dissolve the salt

Isoamyl alcohol (3-methyl-1-butanol)

Procedure. The enzyme reaction is stopped by the addition of an equal volume (0.25 ml) of sodium borate.[17] Incubation is continued for another 2 min at 37° and then cooled in an ice bucket. Twenty volumes of the original incubation mixture of isoamyl alcohol (5 ml) is added and the mixture vortexed vigorously for 15 sec. The layers are separated by centrifugation for 2 min using a clinical centrifuge. Two-milliliter aliquots of the organic layer are transferred to two scintillation vials. Scintillator is added to one, the detected counts representing the total extracted radioactivity. The solution in the second vial is evaporated to dryness on a steam bath for 30 min. Scintillator is then added and the sample counted. The difference in radioactivity in these two vials is a measure of the protein-methyl ester formed in the reaction mixture. When using a purified enzyme preparation, the control tube can be prepared with heated enzyme (100° for 15 min), Thus obviating the steam bath evaporation step.

---

[16] The hydrolysis of the soluble protein-carboxylmethyl ester can also be carried out at pH 8, 37° for 60 min.

[17] If the volume of reaction mixture is reduced, all the reagents used in subsequent methanol extraction method should be reduced proportionally.

Quantitative Analysis of [^{14}C]Methanol Derived from Protein-Carboxylmethyl Ester by Derivatization as 3,5-Dinitrobenzoate

Since the enzymatic product of protein methylase II, the protein-carboxylmethyl ester, is unstable in alkaline solution yielding methanol, this lability can be exploited for a quantitative identification of the product. The radioactive methanol released from the modified protein under mild alkaline conditions can be converted into methyl ester of 3,5-dinitrobenzoate according to the following reaction:

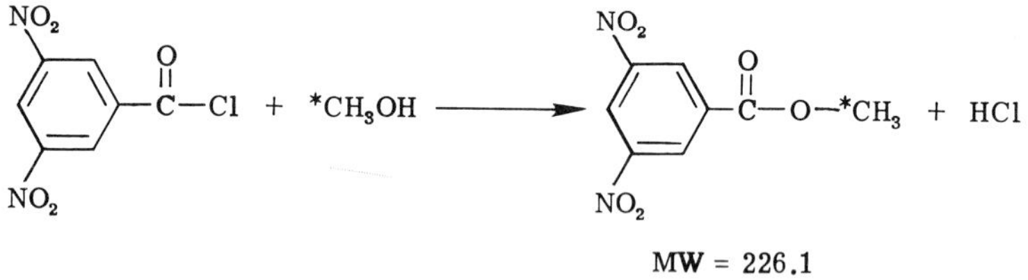

MW = 226.1

By repeated crystallization of this derivative to a constant radiospecific activity and sharp melting point it is possible to determine quantitatively the amount of methyl groups derived from the methylated protein.

Preparation of Sample. Carboxyl[*methyl*-^{14}C]methylated protein must be free of unreacted radioactive SAM before subjecting it to alkaline hydrolysis. This can be achieved by repeated washing and precipitation of the methylated protein from the assay mixture with TCA as described under Method I. Traces of ethanol in the protein precipitates introduced by the final washing step are removed by suspending the precipitate in anhydrous ether followed by air drying. Alternatively, gel filtration of the methylated protein mixture on Sephadex G-25 (or G-10) can be used to remove unreacted SAM, or dialysis of the sample against 0.1 M acetic acid, if it is applicable.

The washed radioactive protein is then hydrolyzed at pH 7.4 (0.01 M sodium phosphate or another buffer of low ionic strength) for 5 min at 100°, or alternatively, at pH 8.0 for 60 min at 37°. The mixture is cooled and adjusted to 5% TCA (higher concentration interferes with derivatization). The resulting supernatant[18] containing the radioactive methanol is used for derivatization.

Alternatively, distillation of the total incubation mixture (after adjusting the pH to 7.4) under reduced pressure and temperature can be used to

[18] If the isoelectric point of substrate protein is known, radioactive methanol released after the hydrolysis from the methylated protein can be separated by isoelectric precipitation of the protein; for example, precipitation of histones at pH 10.

obtain [*methyl*-^{14}C]methanol from protein-carboxylmethyl ester. Details of the distillation method have been reported elsewhere.[13]

Reagents

Sample of radioactive methanol obtained as above, 1–2 ml containing 10,000 cpm or more
3,5-Dinitrobenzoyl chloride, mp 66–68°
10 M NaOH
Pyridine
Methanol, 100%, as carrier

Procedure.[19] In a 30 ml beaker, 1–2 ml of sample, 0.05 ml of carrier methanol (1.25 nmol) and 0.5–1.0 ml of 10 M NaOH are well mixed and cooled in an ice bath. To the mixture 0.5 g of 3,5-dinitrobenzoyl chloride (finely powdered) is added and triturated vigorously with a glass stirring rod for about 1 min and then occasionally for the next 10 min. The mixture is then diluted with an equal volume of cold water. The solid is collected on a filter using suction, washed thoroughly with cold water, and air dried. The methyl ester of 3,5-dinitrobenzoate thus formed is recrystallized by dissolving the solid in a minimum amount of hot pyridine (about 1 ml) followed by the addition of 20 ml of cold water. The yield after the first recrystallization is about 15%. Recrystallization is repeated until constant radiospecific activity is obtained. Usually the first and the second recrystallizations gave the same specific activity with a sharp melting point of 106–107°. Total radioactive methanol formed is calculated based on the theoretical yield of methyl 3,5-dinitrobenzoate (282.5 mg) from 1.25 nmol of carrier methanol.

Definition of Unit and Specific Activity

An enzyme unit is defined as the amount of enzyme that catalyzes the transfer of one picomole of [^{14}C]methyl to the methyl acceptor substrate per minute at pH 6.0 and 37°. Specific activity is expressed as units of enzyme per milligram of protein.

Purification of Protein Methylase II from Mammalian Tissues

Several methods have been used to purify protein methylase II from mammalian tissues, utilizing ion-exchange or molecular sieve chromatography in combination with affinity chromatography. The affinity chromatographic method utilizing *S*-adenosyl-L-homocysteine (SAH), a reaction product inhibitor, covalently linked to Sepharose beads is a very

---

[19] A modified procedure based on A. C. Neish, this series, Vol. 3, p. 255.

effective means to obtain highly purified enzyme. A few representative purification procedures mainly developed in our laboratory will be described below.

Preparation of SAH-Sepharose 4B[6]: Reagents

Sepharose 4B-1,6-diaminohexane, Sigma Chemical Co.

S-Adenosyl-L-homocysteine, 100 mg/20 ml of 0.1 M sodium carbonate, pH 9. The compound is easily dissolved in warm temperature (60°)

Bromoacetic acid

N-Hydroxysuccinimide

Dicyclohexylcarbodiimide

Dioxane

2-Mercaptoethanol, 0.2 M

Sodium phosphate buffer, 0.1 M, pH 7.5

Sodium carbonate buffer, 0.1 M; 0.01 M, pH 9.0

NaCl, 0.1 M; 0.2 M

Procedure. In a 250 ml beaker, 5 g of Sepharose 4B-1,6-diaminohexane is suspended in 0.1 M sodium phosphate buffer, pH 7.5, and washed with the buffer a few times to remove contaminating lactate. Meanwhile, O-bromoacetyl-N-hydroxysuccinimide is prepared as follows. In a 25 ml flask, 8 ml of dioxane, 1.0 mmol (139 mg) of bromoacetic acid and 1.2 mmol (138 mg) of N-hydroxysuccinimide are mixed. To this solution, 1.1 mmol (267 mg) of dicyclohexylcarbodiimide is added and dissolved. After allowing the reaction to proceed for 70 min at room temperature the dicyclohexyl-urea precipitate is removed by filtration. The filtrate containing O-bromoacetyl-N-hydroxysuccinimide is then added to the above washed 1,6-diaminohexane-Sepharose suspension at 4°. After standing for 30 min, the Sepharose is washed with 2 liters of cold 0.1 M NaCl and then equilibrated in 0.1 M sodium carbonate buffer, pH 9.0. The moist packed gel (bromoacetyl-1,6-diaminohexane-Sepharose) is mixed with 20 ml of SAH solution and the mixture is incubated for 3 days at room temperature with occasional stirring. The gel is then washed with about 1 liter of 0.2 M NaCl and resuspended in 20 ml of 0.01 M sodium carbonate buffer. The residual unreacted bromoacetyl groups are then blocked by reacting the gel with an equal volume of 0.2 M 2-mercaptoethanol in 0.01 M sodium carbonate buffer for 2 hr at room temperature. The gel is finally washed with approximately 2 liters of 0.2 M NaCl.

To quantitate the amount of SAH covalently attached to the gel, the total A_{260} of the SAH solution before and after the Sepharose reaction is calculated. Routinely about 75 μmol of SAH is found to be covalently linked per 10 ml bed volume of the gel.

Affinity Purification (from Calf Brain)[6]

All procedures are performed at 4°.

Step 1. Preparation of Supernatant. Three hundred grams of frozen calf brain is homogenized in 4 volumes of cold 0.3 M sucrose, and cytosol is prepared by centrifuging the homogenate at 105,000 g for 60 min.

Step 2. pH 5.1 Treatment. The supernatant obtained from Step 1 (1350 ml) is adjusted to pH 5.1 with dropwise addition of cold 0.5 M acetic acid. Care should be taken to avoid overacidification since the enzyme is easily inactivated at a lower pH. The precipitate is dissolved after centrifugation of the mixture at 39,000 g for 30 min.

Step 3. Ammonium Sulfate Precipitation. The pH 5.1 supernatant is brought to 70% $(NH_4)_2SO_4$ saturation at 4° by the addition of solid $(NH_4)_2SO_4$. The precipitate is then recovered by centrifugation at 39,000 g for 30 min, dissolved in 65 ml of buffer B (5 mM sodium borate–5 mM EDTA–2.4 mM 2-mercaptoethanol, pH 9.3), and dialyzed overnight against the same buffer. Use of this buffer B prevents inactivation of the enzyme during the storage and dialysis.

Step 4. Affinity Chromatography. A column of SAH-Sepharose-4B (1.9 × 19 cm) is equilibrated with buffer A (5 mM sodium phosphate–5 mM EDTA–2.4 mM 2-mercaptoethanol, pH 6.2), having a flow rate of 30 ml/hr. The dialyzed enzyme preparation obtained in Step 3 is carefully adjusted to pH 6.2 by dropwise addition of cold 0.5 M acetic acid immediately before applying it to the affinity column. The column is then washed with buffer A until no more protein is eluted as detected by A_{280} (about 20 times the column volume). The affinity bound methylase is then eluted with buffer B containing 0.02 mM SAM, pH 6.2. Fractions (4.5 ml) are collected into test tubes containing 0.5 ml of 50 mM sodium borate–50 mM EDTA, pH 9.3. The enzyme peak is located by assaying 0.02 ml of each fraction (Fig. 1). The fractions containing the enzyme activity are pooled and concentrated under mild vacuum in long dialyzing tubing (1 cm in flat width) or alternatively by immersible membranes (Immercible CX; Millipore, Cat. No. PTGC 11K 25) to about 10 ml. This enzyme solution is then brought to 20% glycerol and further concentrated to 1.0 ml in a dialyzing tubing which is surrounded by dried Sephadex G-200. These methods of concentrating the highly diluted enzyme solution result in little inactivation of the enzyme.

Step 5. Sephadex G-100 Chromatography. The sample obtained from Step 4 is applied to a column of Sephadex G-100 (1.0 × 160 cm) previously equilibrated with buffer B. Elution is carried out with the same buffer at a flow rate of 30 ml/hr. One-milliliter fractions are collected, and 0.01-ml aliquots are assayed for the enzyme activity. The active fractions (frac-

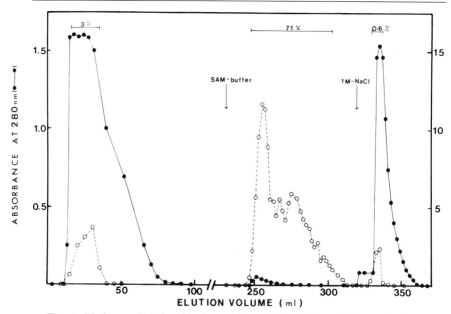

Fig. 1. Elution profile of calf brain protein methylase II from S-adenosyl-L-homocysteine-Sepharose 4B column (6). SAM buffer contains 0.02 mM S-adenosyl-L-methionine in 5 mM sodium phosphate–5 mM EDTA–2.4 mM 2-mercaptoethanol, pH 6.2. CPM represents enzyme activity.

tions 65–86) are pooled, concentrated as described above, and stored in 20% glycerol at −20°. Table II provides a summary of all the purification steps.

Affinity Purification (from Human Erythrocytes)

Although the general procedure for purification of the enzyme from erythrocytes utilizing SAH-Sepharose is similar to that described in the previous section, the bulk of the hemoglobin must be removed from the erythrocyte lysate in order for the affinity gel to provide successful purification. In this section, a simple but effective procedure for removing hemoglobin is described.

Procedure.[20] All procedures are performed at 4°. Freshly drawn heparinized human blood (about 120 ml) is centrifuged at 2000 g for 10 min to remove the plasma and then washed three times with 5 volumes of cold 0.15 M NaCl. The washed leukocyte-free erythrocytes (50 ml) are lysed in

[20] S. Kim, J. Choi, and G.-J. Jun, *J. Biochem. Biophys. Methods* **8**, 9 (1983).

TABLE II
PURIFICATION OF PROTEIN METHYLASE II FROM CALF BRAIN

			Protein		Enzyme activity[a]		
Step	Fraction	Total volume (ml)	mg/ml	Total (mg)	Units/ mg	Total units	Yield (%)
	Whole homogenate	1,500	14.8	22,200	10	222,000	100
1	Supernatant at 105,000 g	1,350	3.15	4,250	31	133,200	60
2	pH 5.1 treatment	1,320	1.90	2,500	40	99,900	45
3	$(NH_4)_2SO_4$, 70%	65	22.5	1,460	48	70,100	31.6
4	SAH-Sepharose 4B chromatography	1.6	1.75	2.8	17,200	48,160	21.7
5	Sephadex G-100 chromatography	22	0.065	1.4	30,000	42,000	18.9

[a] γ-Globulin was used as the methyl acceptor protein.

10 volumes of 5 mM sodium phosphate buffer, pH 8.0. After centrifugation at 27,000 g for 30 min, the resulting supernatant is brought to 60% $(NH_4)_2SO_4$ saturation by the addition of the solid over a period of 30 min. The addition of the crystals can be rapid in the beginning, but must be slowed near the end, otherwise a large amount of hemoglobin will precipitate. The solution is then stirred for another 30 min. The precipitate was recovered by centrifugation at 27,000 g for 30 min, resuspended in 500 ml of a 55% $(NH_4)_2SO_4$ saturated solution in buffer B, and stirred for 30 min. The precipitate recovered as before is resuspended in a minimum volume of buffer B and dialyzed overnight against the same buffer. The dialyzed material is then applied on a column of Sephadex G-75 (4 × 130 cm) preequilibrated in buffer B. Seven milliliter fractions are collected and 0.05-ml aliquots from each fraction are assayed for protein methylase II activity. The active fractions (fractions 65–96) are pooled, adjusted carefully to pH 6.2 with cold acetic acid, and applied to a SAH-Sepharose 4B column (2 × 5 cm). The subsequent procedures are the same as Steps 4 and 5 of the previous section, except that a Sephadex G-75 column (1.0 × 120 cm) is used (see Table III).

Alternative Purification Procedure (from Calf Thymus)[1]

All the procedures are carried out at 4°.

Step 1. Preparation of Supernatant. Five hundred grams of freshly frozen, defatted calf thymus is thawed and homogenized in 5 volumes of

TABLE III
PURIFICATION OF PROTEIN METHYLASE II FROM HUMAN ERYTHROCYTES

Purification step	Volume (ml)	Protein		Enzyme activity[a]		Y
		mg/ml	Total mg	Units/mg	Total units	
Whole lysate (50 ml erythrocyte)	550	22.0	12,000	2.00	24,200	1
Supernatant at 27,000 g	480	19.0	9,120	1.85	16,900	
60% $(NH_4)_2SO_4$ precipitates	560	4.8	2,690	4.25	11,400	
55% $(NH_4)_2SO_4$ wash	4.5	40.4	182	60.0	10,900	
Sephadex G-75 chromatography (4.0 × 130 cm)	220	0.095	21	390	8,200	
SAH-Sepharose 4B (2.0 × 5.0 cm)	0.6	0.410	0.246	17,000	4,180	
Sephadex G-75 (1.0 × 120 cm)	0.8	0.138	0.110	27,000	3,000	

[a] γ-Globulin was used as the methyl acceptor protein.

sucrose solution containing 0.25 M sucrose–5 mM Tris(hydroxymethyl)aminoethane–3 mM $CaCl_2$–1 mM EDTA–2.4 mM 2-mercaptoethanol, pH 7.0. The homogenate is centrifuged at 39,000 g for 60 min, and the supernatant is passed through a funnel plugged with glass wool.

Step 2. pH 5.1 Treatment. Approximately 2000 ml of the supernatant obtained in Step 1 is adjusted to pH 5.1 by dropwise addition of 0.2 M acetic acid. The suspension is stirred for 10 min and then centrifuged at 39,000 g for 30 min.

Step 3. Ammonium Sulfate Fractionation. The pH 5.1 supernatant is brought to 45% saturation of $(NH_4)_2SO_4$ by the addition of saturated solution over a period of 30 min [$(NH_4)_2SO_4$ is saturated at 20°]. After stirring for another 15 min, the precipitate is recovered by centrifugation at 39,000 g for 30 min. The precipitate is suspended in about 100 ml of the buffer B (5 mM sodium borate–5 mM EDTA–2.4 mM 2-mercaptoethanol, pH 9.3) and the pH of the suspension is immediately adjusted to 9.3. The solution is then dialyzed overnight against the same buffer.

Step 4. DEAE-Sephadex Batch Concentration. DEAE-Sephadex A-50 is swollen and is washed several times with buffer B. To the enzyme preparation obtained in Step 3, an equal volume of the washed gel is added and the mixture is stirred slowly for 30 min (about 15 mg protein/ml preswollen gel). Unadsorbed protein is collected either by centrifugation or by using a sintered glass funnel, and saved. The gel is washed four times with 50-ml aliquots of the buffer B containing 0.02 M NaCl. All the washings and the unadsorbed protein are combined, lyophilized, and dialyzed against the buffer B.

Step 5. QAE-Sephadex Chromatography. A column of QAE-Sephadex A-50 (2.5 × 20 cm) is equilibrated with the buffer C (10 mM monoethanolamine–2.5 mM EDTA–2.4 mM 2-mercaptoethanol, pH 9.7). The dialyzed sample obtained in Step 4 is adjusted to pH 9.7 and applied to the column. The column is washed with 50 ml of buffer C before the elution of the enzyme is carried out with a NaCl gradient using 5 chambers of a Buchler Varigrad Instrument; the first and second chamber contain 100 ml each of buffer C, the third chamber 100 ml of 0.1 M NaCl in the buffer, the fourth 100 ml of the buffer, and the fifth 100 ml of 0.25 M NaCl in the buffer, pH 9.7. Four milliliters each fractions are collected at a flow rate of 20 ml/hr. If the flow rate of the column slows down, a peristaltic pump can be placed between the gradient and the column. Two-hundredth milliliter of each fraction is assayed for protein methylase II activity. The active fractions are pooled and concentrated.

Step 6. Chromatography on Sephadex G-75. The sample obtained in Step 5 is dissolved in 1 ml of buffer B and applied to a column of Sephadex G-75 (1.0 × 160 cm) previously equilibrated with the buffer B. Elution is carried out with the same buffer at a flow rate of 20 ml/hr. One milliliter fractions are collected, and 0.01-ml aliquots are assayed for enzyme activity. The protein peak at this stage of purification does not coincide with the enzyme activity peak. Repetition of Steps 5 and 6 will considerably improve the purity of the enzyme. The summary of purification scheme is listed in Table IV.

Properties

Stability. The enzyme in the crude form may be stored at −20° for a few years without appreciable loss of the activity. The activity of purified enzyme is very susceptible toward acidic pH,[1] air oxidation, high temperature, and low protein concentration. However, stored at −20° at a concentration of 0.5–1.0 mg/ml in 20–50% glycerol containing 5 mM sodium borate–5 mM EDTA–2.4 mM 2-mercaptoethanol, pH 9.3, the enzyme retained full activity for over a year.

Molecular Properties. The highly purified enzyme is about 95% pure as determined using SDS–polyacrylamide gel electrophoresis. The major species of the enzyme from various mammalian tissues is estimated to be 25,000 daltons[2,5] by gel filtration, although aggregates of various size classes have been noted. The highly purified enzyme preparations show

TABLE IV
PURIFICATION OF PROTEIN METHYLASE II FROM CALF THYMUS

Step	Fraction	Volume (ml)	Protein		Enzyme activity[a]		Yield (%)
			mg/ml	Total mg	Units/mg	Total units	
	Whole homogenate	2,500	16.8	42,000	2.0	84,000	100
1	Supernatant at 39,000 g	2,400	5.0	12,000	7.3	87,600	104
2	pH 5.1 treatment	2,350	2.3	5,400	12.3	66,400	79
3	(NH₄)₂SO₄, 45%	70	18.5	1,300	38.2	49,700	59
4	DEAE-Sephadex A-50	34	6.1	207	211	43,700	52
5	QAE-Sephadex A-50	40	0.83	33	682	22,500	26
6	Sephadex G-75	5	0.75	3.75	2,020	7,560	9
	Rechromatography by Sephadex G-75	6	0.13	0.78	4,300	3,350	4

[a] Histone was used as the methyl acceptor protein.

several peaks on isoelectrofocusing; pI values of 4.9, 5.5, and 6.0 from rat erythrocytes,[2] 5.5, 6.0, and 6.2 from calf brain, and 4.85 from calf thymus.[1]

Catalytic Properties. The optimum pH of the reaction varies depending on the methyl acceptor substrate; γ-globulin,[2] histone,[21] and ovalbumin exhibit at around pH 6, whole serum[2] at around 7 and gelatin[22] at pH 7.4. The kinetic mechanism of the calf thymus enzyme follows a sequential random Bi Bi mechanism.[23] The K_m value of the enzyme for SAM is 0.87 μM. The K_m values for methyl acceptor substrates depend greatly on the nature of the polypeptides: 0.1 mM for histone, 0.34 mM for pancreatic ribonuclease,[23] 46 μM for γ-globulin, and 7.7 μM for follicle-stimulating hormone.[12]

Specificity. The enzyme is highly specific for SAM as the methyl donor. *S*-Adenosyl-L-ethionine shows only a 2.4% donor efficiency when compared to the methionine derivative.[24] Deaminated SAM analog, *S*-inosyl-L-(2-hydroxy-4-methylthio)butyrate and *S*-inosyl-L-methionine, and the decarboxylated derivative, *S*-adenosyl-(5′)3-methylthiopropy-

[21] S. Kim and W. K. Paik, *J. Biol. Chem.* **245,** 1806 (1970).
[22] S. Kim and W. K. Paik, *Biochim. Biophys. Acta* **252,** 526 (1971).
[23] M. Jamaluddin, S. Kim, and W. K. Paik, *Biochemistry* **14,** 694 (1975).
[24] S. Kim and W. K. Paik, unpublished result.

lamine, do not all act as methyl donors.[25] The demethylated SAM, SAH, is a competitive inhibitor with a K_i value of 0.9 μM. The D isomer of SAH, however, is inactive as an inhibitor.[25] SAH analogs and antifungal antibiotics, sinefungin and A1945C, are potent competitive inhibitors[7,26] with K_i values of 2.6 and 0.4 μM, respectively.

The specificity for polypeptide substrates (methyl acceptor) is rather broad. The best substrate so far reported in the literature is ACTH[3,27] followed by calmodulin,[28] acetylcholine receptor,[29] β-subunit of lutropin (or follicle-stimulating hormone), and β-lipotropin. γ-Globulin, ovalbumin, histones, and native bovine pancreatic ribonuclease are moderately active as substrate and are often used as a model methyl acceptor for the enzyme assay. As a rule, denatured proteins are far better methyl acceptors; for example, gelatin,[13] F-P-100,[21] and oxidized pancreatic ribonuclease.[30] The copolymer of L-glutamic acid and L-tyrosine, although not a methyl acceptor, exhibits a noncompetitive inhibitory activity (with corticotropin as a substrate) with K_i value of 0.11 μM.[25] Polyglutamic and polyaspartic acids are weak inhibitors; 50% inhibition at the concentration of 0.166 and 0.22 mM, respectively.[25]

Activators and Inhibitors. The enzyme does not require any cofactor. Various metals such as Ca^{2+}, Mn^{2+}, Cu^{2+}, Zn^{2+}, Co^{2+}, Fe^{2+}, Fe^{3+}, and Mg^{2+} at 2 mM have no effect on enzyme activity.[21] p-Chloromercuribenzoate at 0.4 mM inhibits the enzyme activity by 40%; however, this inhibition can be reversed by 12 mM 2-mercaptoethanol. The enzyme activity is completely lost in 2 M urea and 0.5 M guanidinium chloride; however, the activity can be restored upon dialysis of the denaturants.[1] The enzyme activity is not inhibited with 40% Triton X-100.

[25] A. Oliva, P. Galletti, V. Zappia, W. K. Paik, and S. Kim, *Eur. J. Biochem.* **104,** 595 (1980).

[26] R. T. Borchardt, E. E. Lee, B. Wu, and C. O. Rutledge, *Biochem. Biophys. Res. Commun.* **89,** 919 (1979).

[27] S. Kim and C. H. Li, *Int. J. Pept. Protein Res.* **13,** 282 (1979).

[28] C. Gagnon, S. Kelly, V. Manganiello, M. Vaughan, C. Odya, W. Strittmatter, A. Hoffman, and F. Hirata, *Nature (London)* **291,** 515 (1981).

[29] Y. Kloog, D. Flynn, A. R. Hoffman, and J. Axelrod, *Biochem. Biophys. Res. Commun.* **97,** 1474 (1980).

[30] S. Kim and W. K. Paik, *Biochemistry* **10,** 3141 (1971).

[29] The Protein Carboxylmethyltransferase Involved in *Escherichia coli* and *Salmonella typhimurium* Chemotaxis

By JEFFREY B. STOCK, STEVEN CLARKE, and
DANIEL E. KOSHLAND, JR.

General Properties

In *Escherichia coli* and *Salmonella typhimurium* a set of membrane proteins is reversibly methylated during chemotaxis.[1-3] The enzyme responsible for this modification catalyzes the transfer of methyl groups from *S*-adenosylmethionine (SAM) to specific glutamyl residues in the methylated 60,000-dalton membrane proteins.[4,5] The reaction is catalyzed by a single transferase which is encoded by the *cheR* gene.[6,7] Mutants in this gene are defective in transferase activity, and show severely impaired sensory motor coordination.[6,8] The protein appears to have a molecular weight of 31,000 from electrophoresis on polyacrylamide gels containing sodium dodecyl sulfate.[9-11] Under native conditions a molecular weight close to this value has been obtained.[7,11]

The apparent K_m of the transferase for *S*-adenosylmethionine is approximately 12 μM. The transferase shows a broad pH optimum between 6.5 and 8.5 (Fig. 1) and its activity depends on the ionic composition of the assay medium (Fig. 2). The enzyme appears to be completely specific for the methyl-accepting chemotaxis proteins (MCPs). No other proteins, polypeptides or small molecules have been found to act as substrates.[6,7] It acts only on glutamic acid residues.[12,13]

[1] M. S. Springer, M. F. Goy, and J. Adler, *Nature* (*London*) **280**, 279 (1979).
[2] D. E. Koshland, Jr., *Annu. Rev. Biochem.* **50**, 765 (1981).
[3] A. Boyd and M. Simon, *Annu. Rev. Physiol.* **44**, 501 (1982).
[4] A. Boyd, M. Kendall, and M. Simon, *Nature* (*London*) **301**, 623 (1983).
[5] A. F. Russo and D. E. Koshland, Jr., *Science* **220**, 1016 (1983).
[6] M. R. Springer and D. E. Koshland, Jr., *Proc. Natl. Acad. Sci. U.S.A.* **74**, 533 (1977).
[7] S. Clarke, K. Sparrow, S. Panasenko, and D. E. Koshland, Jr., *J. Supramol. Struct.* **13**, 315 (1980).
[8] J. B. Stock, A. M. Mareris, and D. E. Koshland, Jr., *Cell* **27**, 37 (1981).
[9] M. Silverman and M. Simon, *J. Bacteriol.* **130**, 1317 (1977).
[10] A. L. DeFranco and D. E. Koshland, Jr., *J. Bacteriol.* **147**, 390 (1981).
[11] A. Russo, C. Bibus, and D. E. Koshland, Jr., unpublished results.
[12] P. Van der Werf and D. E. Koshland, Jr., *J. Biol. Chem.* **252**, 2793 (1977).
[13] M. L. Kleene, M. L. Toews, and J. Adler, *J. Biol. Chem.* **252**, 3214 (1977).

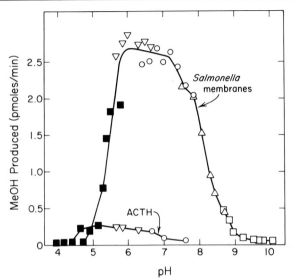

Fig. 1. Dependence of carboxylmethyltransferase activity on pH.[7] Equal parts of a mixture of a ST1038 membrane fraction and an ST1 soluble fraction in 10 mM NaPO$_4$, 0.2 M NaCl, pH 7.0, were diluted into 0.1 M buffers at the indicated pH containing S-adenosyl[*methyl*-^{3}H]methionine. The final concentration of soluble protein was 2.2 mg/ml and the incubation was carried out for 30 min at 37°. Methyltransferase activity was determined by measuring the amount of methanol extracted into the organic solvent after base hydrolysis as described in the text. Buffers included the sodium salt of citric acid (■), maleic acid (▽), phosphoric acid (○), glycylglycine (△), and glycine (□). The small activity shown with ACTH as a substrate is not due to the *cheR* methyltransferase because extracts of *cheR*⁻ strains catalyzed this reaction at the same rate.

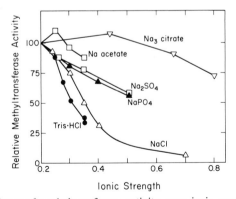

Fig. 2. Dependence of methyltransferase activity upon ionic conditions in the assay medium.[7] Crude soluble extracts of *Salmonella* were assayed in 0.1 M NaPO$_4$, 1 mM EDTA, as described in the text, in the presence of additional salts at pH 7.0. Methyltransferase activity relative to that measured in control experiments, ionic strength, 210 mM, is shown as a function of the total ionic strength.

Methyltransferase Assay

Assays of the enzyme can be performed by a variety of methods *in vivo, in situ,* or *in vitro.* Since different techniques have proven useful on different occasions, the components will be discussed individually and then evaluated at the end.

In Vivo Assay. Transferase activity may be assessed *in vivo* by incubating cells in the presence of chloramphenicol and [*methyl-*^{3}H]methionine which is rapidly transported into cells and converted into *S*-adenosyl[^{3}H-*methyl*]methionine. Labeling in chloramphenicol has the advantage that relatively low concentrations and high specific activities of [^{3}H]methionine may be used. At cell densities of approximately 5×10^7 cells/ml, 10 μM [^{3}H]methionine is not depleted for up to 1 hr at 30°. There are two drawbacks of the chloramphenicol procedure: (1) glutamyl methyl esters which do not turn over are never labeled,[14] and (2) at cell densities greater than 10^8 cells/ml overall levels of methylation seem to decline with time of incubation, and the rate and extent of methyl ester turnover increases.[15] The latter effects may be explained if cells in chloramphenicol excrete a repellent. These problems are obviated if cells are simply grown in [^{3}H]methionine in the absence of chloramphenicol.[14] Using this procedure the microdistillation or vapor phase equilibrium quantitation methods (see below) must be used to distinguish glutamyl [γ-^{3}H]methyl esters from [^{3}H]methionine incorporated directly into protein. Approximately 10-fold higher concentrations of [^{3}H]methionine must also be supplied or the cells will exhaust the extracellular [^{3}H]methionine. Endogenous methionine synthesis seems to be effectively turned off by exogenous methionine.[16] Thus in cells grown in [^{3}H]methionine the specific activity of [^{3}H]methyl groups corresponds to the specific activity of exogenous label. For more rigorous measures of methyl incorporation, *metF* mutants may be used.[14]

Three methods have been used to stop the *in vivo* methylation/demethylation reaction at the end of an incubation interval: (1) The cell suspension is made 10% in trichloroacetic acid (TCA) by addition of an appropriate volume of 100% TCA.[14] The cells are pelleted by centrifugation at 12,000 *g* for 10 min and residual TCA is removed by washing the pellet with cold acetone. (2) The suspension is diluted 1 : 10 into ice cold acetone.[17] The cells are pelleted by centrifugation at 12,000 *g* for 10 min, and washed with 1 volume of ice cold 50% acetone : water to remove

[14] J. B. Stock and D. E. Koshland, Jr., *J. Biol. Chem.* **256,** 10826 (1981).
[15] J. B. Stock and D. E. Koshland, Jr., unpublished observations.
[16] D. Smith, *Adv. Genet.* **16,** 141 (1971).
[17] N. F. Paoni and D. E. Koshland, Jr., *Proc. Natl. Acad. Sci. U.S.A.* **76,** 3693 (1979).

salts. (3) The suspension is made 10% in formalin, and the cells are pelleted by centrifugation at 12,000 g for 10 min at 4°.[18] All these methods give essentially identical results.

In Situ Assay. Toluene treatment allows free access of S-adenosylmethionine to the cell interior so that transferase may be assayed *in situ.*[17] The extent of methyl group incorporation depends on the exact conditions of toluene treatment. The major variables seem to be the degree of cell disruption during the permeabilization process and the extent of protein demethylation prior to addition of S-adenosyl[*methyl*-³H]methionine. In wild type *S. typhimurium* LT2, maximal rates are obtained under the following conditions[17]: cells are grown to mid-log phase (approximately 5 × 10⁸ cells/ml) in minimal medium (Vogel–Bonner citrate[19] containing 1% glycerol), harvested by centrifugation at 12,000 g for 10 min at 4°, and washed twice by resuspension in minimal media followed by centrifugation. The cell pellets are resuspended in minimal media containing 9.5 mM EDTA, pH 7.0 to a density of 1.6 × 10⁹ cells/ml. Toluene is added to a concentration of 0.14% (v/v) and the suspension is vortexed 2 min. The cells are incubated in a shaking water bath at 30° for 10 min, washed once with minimal medium containing 9.5 mM EDTA, and finally resuspended in fresh medium. S-Adenosyl[³H-*methyl*] methionine, 500 mCi/mmol, is added to these cells (final concentration, 20 μM) and the suspension is incubated at 30°. Aliquots of 0.2 ml are removed at various intervals, and the reaction is quenched by addition to 10 ml of 10% TCA. The [³H]methyl-labeled toluene-treated cells in 10% TCA can be assayed by the butanol extraction assay or by simply filtering on Whatman GF/F glass fiber filters, washing with 0.1 M NaCl, and using the diffusion or microdistillation assays described below.

In Vitro Assay. Transferase activity may be assayed *in vitro* by combining a cytoplasmic preparation containing the enzyme with membranes containing the methyl accepting proteins, and S-adenosyl[*methyl*-³H]methionine. There are 3 parts to this procedure: preparation of the membranes, incubation with cytoplasm, and determination of radioactive methyl groups incorporated into membrane proteins.

Membranes are generally prepared from a strain which is deficient in transferase activity such as *S. typhimurium* ST1038 *cheR* (ATCC29596).[7] These cells are grown to early stationary phase in nutrient broth, harvested by centrifugation at 12,000 g for 20 min, washed by resuspension in 0.1 M sodium phosphate, 1.0 mM EDTA, pH 7.0, centrifuged again, and finally resuspended in 3 ml of sodium phosphate–EDTA buffer/g, wet

[18] M. F. Goy, M. S. Springer, and J. Adler, *Proc. Natl. Acad. Sci. U.S.A.* **74,** 4964 (1977).
[19] H. Vogel and D. Bonner, *J. Biol. Chem.* **218,** 97 (1956).

weight, cells. The cells are disrupted by sonication, whole cells and large debris are removed by centrifugation at 12,000 g for 15 min, and the supernatant is centrifuged at 100,000 g for 90 min. The pellet is resuspended in sodium phosphate–EDTA buffer, centrifuged again at 100,000 g for 90 min, and the washed membranes are finally resuspended in a minimal volume of sodium phosphate–EDTA. Cytoplasmic extracts are generally prepared by sonication in 0.1 M NaPO$_4$, 1.0 mM EDTA, pH 7.0, as described above. Whole cell debris and membrane fractions are removed by centrifugation at 236,000 g for 30 min at 4°. This soluble extract, at a protein concentration of 15–20 mg/ml, is stored at −20° in small aliquots. Under these conditions, methyltransferase activity is stable for months.

For the accurate determination of methyltransferase activity it is essential that S-adenosylhomocysteine (SAH) hydrolase be present. SAH, one product of the methyltransferase reaction, is a potent competitive inhibitor of the $E.$ $coli$ and $Salmonella$ methyltransferases.[17] Cell extracts prepared from enteric bacteria contain a highly active SAH hydrolase which, under the conditions of the in $vitro$ transferase assay, completely degrades micromolar concentrations of SAH within a few seconds.[20] If transferase is purified away from this degradative enzyme, SAH is not degraded and transferase activity may be significantly inhibited. To preclude this possibility an aliquot of crude extract from a transferase-deficient strain is routinely added to the in $vitro$ incubation mixtures if it is not already present.

The incubations of membranes with varying amounts of transferase are carried out at 30° in a total volume of 0.1 ml containing 100 μM S-adenosyl[^{3}H-$methyl$]methionine (200 cpm/pmol), ST1038 membranes prepared as described above (1.4 mg protein), and 0.1 M sodium phosphate, 1 mM EDTA, buffer at pH 7.0. The reaction is terminated at 5 to 30-min time points by removal of 20-μl aliquots onto 1-cm squares cut from Whatman 3MM filter paper.[14] These are immediately immersed into a large volume of 10% TCA. After 10 min the TCA is decanted and the filters are washed twice with fresh 10% TCA, then twice in methanol. The filters are then analyzed by the microdistillation or vapor phase equilibrium methods (below). Alternatively, the reaction is terminated by the addition of 2 ml of 10% TCA, proteins are precipitated by centrifugation for 10 min at 4000 g at 4°, the supernatant is removed by aspiration, the pellet is resuspended once more in 2 ml 10% TCA, centrifuged, and it is assayed by procedures described below.

[20] K. Walsh and D. E. Koshland, Jr., unpublished observations.

Quantitation of Methyl Ester Groups

Several methods are available to assay methyl ester groups. One is an organic *extraction procedure*.[7] The trichloroacetic acid pellet containing [3H]-methylated proteins is dissolved in 0.5 ml of 0.2 M NaOH, 0.4% sodium dodecyl sulfate, and 1% methanol. Samples are incubated 30 min at 30° during which time the protein-bound esters are hydrolyzed to form [3H]methanol. This product is extracted with 6.0 ml of a 3 : 2 (v/v) mixture of toluene/3-methyl-1-butanol. Samples are vortexed in 15 ml Corex centrifuge tubes for 15 sec and centrifuged 5 min at 3000 g. Control experiments performed with [14C]methanol indicate that 65% of the total methanol partitions into the organic phase under these conditions. A portion of the organic phase (2.0 ml) is counted directly in 10 ml of a water-miscible scintillation cocktail. Another 2.0-ml aliquot is evaporated to dryness in a 20-ml plastic scintillation vial at 50°, then counted after the addition of 2 ml of the toluene/3-methyl-1-butanol mixture and 10 ml of scintillation fluor. The radioactivity of these two samples is corrected for the volumes counted, the efficiency of extraction, and a small amount (4%) of quenching contributed by the organic phase. The difference when related to the original specific activity of the S-adenosyl[3H-*methyl*]methionine, is taken as the number of methyl groups transferred to the protein.

A second assay method is the *microdistillation procedure*.[14] A sample containing [3H]-methylated membrane proteins is placed into a 13 × 100-mm test tube containing 0.1–0.2 ml of 1 M Na$_2$CO$_3$. The test tube is closed with a rubber stopper through which have been inserted two 1-ml syringes each containing 1 ml of methanol. A third insert is attached to a tube which leads to a 5-ml syringe barrel immersed in a liquid scintillation vial containing 10 ml of a water-miscible liquid scintillation fluid. The test tube with its various attachments is placed into an 80° water bath. After a 30 min incubation (sufficient to completely hydrolyze the carboxylmethyl esters) the 1 ml of methanol from one syringe is injected into the tube. After 15 min the methanol from the second tube is dispensed, and 15 min later (after all the cold and [3H]methanol has distilled into the scintillation fluid) the scintillation vials are removed, capped, and radioactivity is assayed in a liquid scintillation spectrometer.

A third method is the *vapor-phase equilibrium procedure*.[21] A sample containing [3H]methyl-labeled MCPs is placed into a 1.5-ml Eppendorf microfuge tube containing 0.1–0.2 ml of 1 M Na$_2$CO$_3$ or NaOH. This tube is placed onto a 7-ml liquid scintillation vial containing 2.4 ml of scintilla-

[21] J. E. Campillo and S. J. H. Ashcroft, *FEBS Lett.* **138**, 71 (1982).

tion fluid. The vial is closed and incubated overnight at 37°. The vials are then placed in the scintillation counter without shaking (to avoid introducing scintillation fluid in the microfuge tube). Samples on filter paper or in solution can also be assayed as long as the pH is greater than 10. The average recovery of methanol can be as high as 90%. The exact yields, however, depend on the precise conditions of assay and should be calibrated with internal standards.

Comments on Analytical Procedures. The microdistillation procedure is occasionally useful for rapid screening. The vapor-phase equilibration method is less laborious and is the method of choice in most situations.

In some cases it is desirable to assay proteins after separation by gel electrophoresis on polyacrylamide gels.[22] To do this the regions of interest are identified (in this case in the 60,000-dalton region). Gels are generally stained with Coomassie, destained overnight in 10% acetic acid, and dried. The 60,000-dalton region is then excised and analyzed for carboxylmethyl groups by the microdistillation or vapor phase equilibration assays described above. Under labeling conditions in which it has previously been determined that only carboxylmethyl groups are labeled (e.g., in chloramphenicol), the total radioactivity in electrophoretogram segments may be determined directly. The carboxylmethyl-containing regions of the dried gel are cut into 1–6 mm slices, placed in scintillation vials, and hydrated with 20–50 μl of H_2O. After 5 min, 10 ml of 3% Protosol/0.4% Omnifluor (New England Nuclear) in toluene is added and the slices are incubated overnight at 37° before analyzing total radioactivity in a liquid scintillation spectrophotometer. Methods of gel solubilization which involve incubations of small volumes for several hours at elevated temperatures have given irreproducible results because of loss of [³H]methanol.

Identification of Glutamic Acid γ-Methyl Ester Residues as a Product of the *cheR* Methyltransferase

Membranes containing radiolabeled methyl esters (prepared as described above) are enzymatically digested by either the protocol of Van der Werf and Koshland[12] or Kleene *et al.*[13] Glutamic acid γ-methyl esters can be identified in the digestion mixtures by either of the methods described in those papers, or by the specific type of protocol described for aspartic acid β-methyl ester in this volume.[23] The latter procedure involves an initial ion-exchange chromatography step on amino acid ana-

[22] V. K. Laemmli, *Nature (London)* **227**, 680.
[23] S. Clarke, P. N. McFadden, C. M. O'Connor, and L. L. Lou, this volume [31].

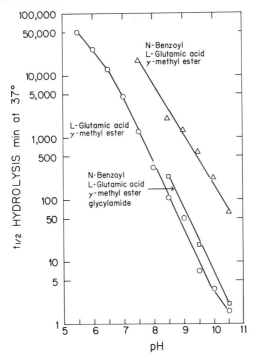

FIG. 3. Rate profile of the hydrolysis of glutamic acid γ-methyl ester and derivatives with blocked α-amino and α-carboxyl groups as a function of pH at 37°.[25]

lyzer resin, followed by desalting on a Sephadex G-15 column. The identity of the radioactive methyl ester is then confirmed by thin-layer chromatography[24] and by the comparison of the pH hydrolysis rate profile of the isotopically labeled ester with that of standard glutamic acid γ-methyl ester. Free glutamic acid γ-methyl ester hydrolyzes much more rapidly than derivatives with a blocked amino group, presumably as a result of an intramolecular reaction resulting in the formation of the cyclic pyroglutamic acid derivative. This property is a useful feature for confirming the assignment of glutamic acid γ-methyl ester (Fig. 3).[25] The stereochemistry of the isolated glutamic acid γ-methyl ester has not been established, but it is presumed to be in the L-configuration based on the relative ease of proteolytic digestion and the stoichiometric levels of this modified residue in chemotaxis proteins. This situation is different from

[24] C. A. Janson and S. Clarke, *J. Biol. Chem.* **255,** 11640 (1980).
[25] T. C. Terwilliger and S. Clarke, *J. Biol. Chem.* **256,** 3067 (1981).

that observed with human erythrocyte methylated proteins, where the isolated aspartic acid β-methyl ester has the D-configuration.[23]

Effects of Attractants

The primary methylated chemotaxis proteins in *E. coli* and *Salmonella* function as chemoreceptors for serine and aspartate.[26,27] Saturating concentrations of these attractants cause roughly twofold increases in levels of receptor methylation *in vivo*.[14,18] These changes are accompanied by large transient increases in transferase activity. The stimulation elicited by attractants may be seen in toluene-treated cells.[17] *In vitro,* however, the effects of attractants are more difficult to detect.[7,27,28] The table summarizes levels of methylation in various chemotaxis mutants and the effect of attractants on these levels.

Differences between *E. coli* and *S. typhimurium* Methyl Transferases

There is a general genetic homology between the chemotaxis systems of *E. coli* and *Salmonella*.[10,29] *E. coli cheR* mutants may be complemented by a wild-type *cheR* gene from *S. typhimurium* and vice versa[29]; and *in vitro* assays indicate that *E. coli* and *Salmonella* transferase work equally with analogous receptors from either species. Dramatic differences are observed, however, in the relative activities of the two enzymes. The *Salmonella* transferase is roughly 10 times more active than the *E. coli* enzyme when the activities are compared in toluene-treated cells or in cell-free extracts. The cloned *Salmonella cheR* gene expressed in an *E. coli* strain exhibits the enhanced activity characteristic of the *Salmonella* transferase. The difference therefore appears to reflect a characteristic difference between the transferase enzymes rather than a secondary influence of some other cellular constituent.

Partial Purification of the Methyl Transferase Enzymes

The methyltransferase from *S. typhimurium* has been partially purified and characterized. The purification involves three steps and yields a preparation which is approximately 70% pure.[11] The first step takes advantage of the strong interaction ($K_D \sim 1 \mu M$) between the enzyme and its substrates, the methyl-accepting chemotaxis receptors in the inner mem-

[26] S. Clarke and D. E. Koshland, Jr., *J. Biol. Chem.* **254,** 9695 (1979).
[27] E. A. Wang and D. E. Koshland, Jr., *Proc. Natl. Acad. Sci. U.S.A.* **77,** 7157 (1980).
[28] S. J. Kleene, A. C. Hobson, and J. Adler, *Proc. Natl. Acad. Sci. U.S.A.* **76,** 6309 (1979).
[29] A. DeFranco, J. S. Parkinson, and D. E. Koshland, Jr., *J. Bacteriol.* **139,** 107 (1979).

RECEPTOR CARBOXYLMETHYLATION IN VARIOUS MUTANTS

Relevant genotype	Strain	S. typhimurium[a] Methyl-MCP[b] (pmol/mg protein) −Attractants	+Attractants	Increase (%)	Strain	E. coli[a] Methyl-MCP[b] (pmol/mg protein) −Attractants	+Attractants	Increase (%)
Wild type	ST1	114 ± 14	218 ± 42	91	RP437	62 ± 12	130 ± 42	110
cheA	ST1008	219 ± 30	405 ± 70	85	RP4303	42 ± 10	133 ± 21	217
cheW	ST1024	104 ± 30	179 ± 43	72	RP4305	46 ± 9	89 ± 17	93
cheR	ST1038	1 ± 5	10 ± 15	—	RP4306	24 ± 19	33 ± 14	—
	ST36	0 ± 2	ND[c]	—	RP4080	18 ± 13	ND	—
cheB	ST410	211 ± 34	217 ± 52	3	RP4310	77 ± 11	110 ± 14	43
					RP4479	243 ± 7	337 ± 45	39
cheY	ST112	136 ± 34	239 ± 21	76	RP4315	60 ± 13	216 ± 35	260
	ST1001	138 ± 4	222 ± 25	61				
cheZ	ST171	280 ± 19	555 ± 11	98	RP4318	81 ± 5	162 ± 21	100
					RP4742	152 ± 19	305 ± 31	101
fla	ST426	1 ± 2	2 ± 4	—	MS6104	0 ± 5	6 ± 12	—

[a] The mutants examined are closely related to their corresponding wild-type parents, but they are not entirely isogenic outside the che region. Except in cheR and fla mutants, different absolute levels of receptor methylation are probably more reflective of varying levels of receptor production than differences in the activity of the methylation enzymes. S. typhimurium strains are from the collection of D. E. Koshland, Jr.; E. coli strains were obtained from J. S. Parkinson, University of Utah, except MS6104 which was obtained from M. Simon, University of California, San Diego.

[b] Cells were grown at 30° in Vogel-Bonner Medium containing 1% glycerol plus 0.1 mM [^{3}H-$methyl$]methionine (specific activity, 100–150 cpm/pmol) plus, for S. typhimurium strains, 50 μg/ml of histidine and thymine; and, for E. coli strains, 50 μg/ml histidine, threonine, and leucine, and 1 μg/ml thiamine. At culture densities of approximately 5×10^8 cells/ml, the attractants serine and aspartate (pH 7.0) were added to half of each culture (final concentrations, 10 mM). After 1 hr continued growth at 30°, three aliquots were removed from each half culture, added to ice-cold acetone, washed with 50% acetone : water, solubilized in 2% sodium dodecyl sulfate, and subjected to slab gel electrophoresis according to the method of Laemmli.[22] The 60,000 molecular weight region of each gel lane was excised and analyzed for [^{3}H]carboxylmethyl groups by the microdistillation procedure. Results are given ± the standard deviation of the mean of the results from the triplicate samples.

[c] ND, Not determined.

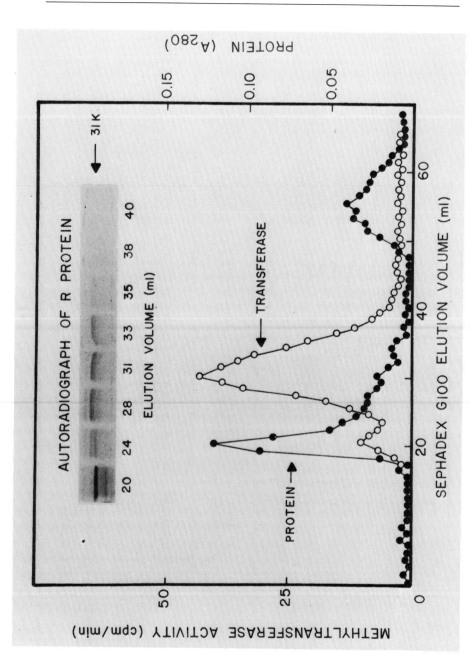

Fig. 4. Coelution of methyltransferase and *cheR* protein during Sephadex G-100 column chromatography. ^{35}S-Labeled *Salmonella cheR* protein (expressed by the cloned *cheR* gene as described previously[10]) was mixed with 0.5 ml of the 100,000 *g* supernatant fraction from an *S. typhimurium* ST 1 cell extract. This material was then subjected to Sephadex G-100 column chromatography (1.2 × 50 cm column; elution buffer, 0.1 *M* sodium phosphate, 1 m*M* EDTA, pH 7.3). Fractions (1 ml) were collected and assayed for A_{280} (●) and transferase activity (○). In addition, the indicated fractions were analyzed for the *cheR* protein by dodecyl sulfate–polyacrylamide gel electrophoresis and fluorography. It has been shown that the *cheR* product is the labeled protein which migrates at 31,000 molecular weight.[10]

brane.[20] The activity can thus be precipitated in the membrane fraction and then eluted by 2 *M* KCl into a soluble fraction. The eluted material is then bound to a hexyl agarose column and eluted by decreased ionic strength. The last step exploits the extremely basic nature of the enzyme, as it is bound to a carboxymethyl cellulose column at pH 7.00 and eluted with 150 m*M* KCl. The enzyme appears to have a p*I* of greater than 9 under native and denaturing isoelectric focusing conditions. The purified enzyme has the same apparent size, 31,000 daltons, as the cloned *cheR* gene product, and the cheR protein copurifies with methyltransferase activity (Fig. 4).

[30] Carboxylmethyl Esterase of Bacterial Chemotaxis

By Mark A. Snyder, Jeffrey B. Stock, and
Daniel E. Koshland, Jr.

Methylation of a set of transmembrane proteins, originally termed methyl-accepting chemotaxis proteins (MCPs)[1] and now known to be receptors,[2] play a central role in bacterial chemotaxis.[3–5] The enzyme responsible for the demethylation of these proteins has been shown to be the product of the *cheB* gene in *Escherichia coli* and *Salmonella typhimurium*.[6,7] Mutants in this gene are severely deficient in chemotactic ability.

[1] E. N. Kort, M. F. Goy, S. H. Larsen, and J. Adler, *Proc. Natl. Acad. Sci. U.S.A.* **72,** 3939 (1975).
[2] E. Wang and D. E. Koshland, Jr., *Proc. Natl. Acad. Sci. U.S.A.* **77,** 7157 (1980).
[3] M. S. Springer, M. F. Goy, and J. Adler, *Nature (London)* **280,** 279 (1979).
[4] D. E. Koshland, Jr., *Annu. Rev. Biochem.* **50,** 765 (1981).
[5] A. Boyd, G. Mandel, and M. I. Simon, *Annu. Rev. Physiol.* **44,** 501 (1982).
[6] J. B. Stock and D. E. Koshland, Jr., *Proc. Natl. Acad. Sci. U.S.A.* **75,** 3659 (1978).
[7] M. A. Snyder and D. E. Koshland, Jr., *Biochimie* **63,** 113 (1981).

The residues which are methylated and demethylated have been shown to be glutamyl residues.[8,9] The purification, assay, and properties of the enzyme are described below.

Assay Method

Principle. Bacterial membranes are prepared with protein methyl groups ^{3}H-labeled either *in vitro* or *in vivo*. Esterase activity results in the production of [^{3}H]methanol, which is distilled and counted by liquid scintillation spectrometry.

Reagents

Buffers used were buffer I, 0.1 M potassium phosphate, pH 7.0 containing 1 mM EDTA; buffer II, 20 mM potassium phosphate, pH 7.0, containing 2.0 M KCl and 1 mM EDTA; buffer III, 50 mM MES, pH 6.5, containing 50 mM NaCl and 1 mM EDTA.

Strains useful in the assay procedures are *S. typhimurium* ST1038 (ATCC29596), a strain deficient in methyltransferase and whose MCPs are thus unmethylated *in vivo*. This strain is used as the source for esterase substrate. *S. typhimurium* ST89pGK2, a plasmid-bearing strain which overproduces esterase and transferase is used as esterase source. *S. typhimurium* ST89pGK3, a plasmid-containing strain which overproduces methyltransferase, is the source for this enzyme.

All cells are grown in L broth at 30°, collected when the cell density reaches an A_{650} of 0.7 to 0.9 (0.5–1.0 × 10^9 cells/ml), and frozen until use. Additionally, plasmid-bearing strains were grown in the presence of 25 μg/ml ampicillin.

The amount of tritiated methanol produced is determined by a variety of methods described previously[6,7,10] and in detail in the description of assays for the protein methyltransferase of chemotaxis.[11]

Procedure. In a typical procedure for the preparation of esterase substrate, 2 to 3 g of ST1038 cells is suspended with 10 ml of buffer I and sonically disrupted until >95% of the cells are broken. The sonicate is centrifuged at 10,000 g to remove whole cells, and then at 100,000 g for 60 min to collect membranes. The membrane pellet is resuspended with a glass homogenizer in 1 ml of buffer I to a final protein concentration of 20–40 mg/ml. In a separate tube, 200 μCi of S-adenosyl-L-[^{3}H]methionine and 400 nmol of unlabeled material (in 0.5 N HCl) are taken to dryness by

[8] S. J. Kleene, M. L. Toews, and J. Adler, *J. Biol. Chem.* **252**, 3214 (1977).
[9] P. Van der Werf and D. E. Koshland, Jr., *J. Biol. Chem.* **252**, 2793 (1977).
[10] J. B. Stock and D. E. Koshland, Jr., *J. Biol. Chem.* **256**, 10826 (1981).
[11] J. B. Stock, S. Clarke, and D. E. Koshland, Jr., this volume [29].

lyophilization. To this tube is then added 0.8 ml ST1038 membranes, 2.1 ml of ST89pGK3 cytosol (20 mg/ml protein; prepared by sonically disrupting ST89pGK2 in twice the cell volume of buffer I and collecting the 100,000 g supernatant), L-aspartate and L-serine to 1 mM, and enough buffer I to bring the final volume to 4.0 ml. The mixture is incubated at 30° for 30 min with occasional shaking. Following the labeling period, the mixture is diluted to 95 ml with buffer II and centrifuged at 100,000 g for 60 min. The resulting membrane pellet is resuspended in 95 ml of buffer I, centrifuged as above, and the final membrane pellet resuspended with 1–4 ml buffer I (9–35 mg/ml protein). The membranes were either used immediately or made 10% in glycerol, divided into small aliquots, and frozen. Aliquots, when thawed, were used within 4 hr and any remaining membranes discarded.

For the assay of esterase esterase and [³H]ST1038 membranes (final concentration 0.1 mg/ml protein) are diluted into 1 ml buffer III and incubated at 30°. At various times, 100 μl is withdrawn into a microdistillation tube (see above) containing 1 ml methanol. The methanol is then distilled directly into 10 ml scintillation fluid. The [³H]methanol so collected is then counted and the amount of glutamyl ester hydrolyzed is calculated based upon the known specific activity of the S-adenosyl-L-[³H]methionine used in the substrate preparation.

Unit Definition. One unit of activity hydrolyzes 1 fmol of methyl ester/min. While the concentration of substrate in the above assay condition is well below the K_m (see below), we have observed that, if hydrolysis is limited to a maximum rate of 20% available methyl ester/10 min, the initial rate of hydrolysis changes proportionately with enzyme concentration. Specific activity is expressed as the number of enzyme units per mg protein.

Purification Procedure

Unless otherwise indicated, all procedures are carried out at 4°.

Step 1. Extraction. Sixty grams of ST89pGK2 is resuspended in 120 ml of buffer I. Cells are sonically disrupted for 20–30 min, at which time greater than 95% of the cells are broken as determined visually. The sonicate is centrifuged at 10,000 g for 10 min to remove whole cells, and the resulting supernatant centrifuged at 100,000 g for 60 min to remove membranes.

Step 2. Removal of DNA. A 10% solution of Polymin P (pH 7.0) is added to the high-speed supernatant to a final concentration of 0.5%. The mixture is stirred for 5 min, then centrifuged at 30,000 g for 15 min, and the pellet discarded.

Step 3. Ammonium Sulfate Fractionation. A saturated ammonium sulfate solution (4°) is added to the Polymin P supernatant to a final concentration of 35% saturation. The solution is slowly stirred for 1 hr and then centrifuged as above. The supernatant is brought to 50% ammonium sulfate and slowly stirred for 6 hr. Following centrifugation as above, the supernatant is discarded and the pellet is resuspended with 40 ml buffer I and dialyzed overnight against 2 liters of buffer I, with two changes.

Step 4. Mercurial Agarose Chromatography. The dialysate is applied to a 40 ml (1.5 × 23 cm) mercurial agarose column at a rate of 15 ml/hr. The column is then washed with 10 mM sodium phosphate, pH 7.0 containing 90 mM NaCl and 0.1 mM EDTA (buffer IV) until the absorbance of the effluent at 280 nm is less than 0.08. Elution is then carried out in two steps. First, the column is developed with 80 ml buffer IV containing 5 mM dithiothreitol at a rate of 3.3 ml/hr. The eluate, which contains the bulk of the eluted protein and a small amount of esterase activity, is discarded. The column is then unpacked and repoured into a 2.8 × 6.5 cm column. Then, once per day, for 5–9 days, 40 ml of elution buffer is passed through the column at a rate of 15 ml/hr and 10 ml fractions are collected. Fractions containing activity (usually 5–7 days worth) are pooled and, if necessary, concentrated to 200 ml volume using an immersible ultrafiltration unit.

Step 5. Ocytl Sepharose Chromatography. The pooled mercurial agarose eluate is applied to a 40 ml (2.8 × 6.5 cm) octyl-Sepharose column equilibrated with buffer IV, at a flow rate of 60 ml/hr. The column is then washed with 60 ml of 5 mM sodium phosphate, pH 7.0, containing 45 mM NaCl, 0.1 mM EDTA and 20% ethylene glycol. Esterase is then eluted with this buffer by increasing the ethylene glycol concentration to 55%. Elution buffer was applied at a rate of 10 ml/hr. Four milliliter fractions are collected and those containing activity are pooled.

Step 6. CM-Cellulose [(carboxymethyl)cellulose] Chromatography. The octyl-Sepharose pool is diluted with distilled water to an ionic strength of 32. mmho, adjusted to pH 6.5 with 0.1 N HCl, and applied to a 7 ml (1.0 × 10 cm) CM-cellulose column equilibrated in 4.5 mM sodium phosphate, pH 6.5 containing 45 mM NaCl and 0.05 mM EDTA (buffer V), at a flow rate of 5 ml/hr. The column is washed with 15 ml of buffer V and developed with 10 mM sodium phosphate, pH 7.0 plus 300 mM NaCl plus 0.01 mM EDTA. Fractions of 0.5 ml are collected, and those containing activity are pooled; the pool volume is approximately 3 ml.

Step 7. Gel Filtration. The CM-cellulose pool is applied to a 100 ml (1.0 × 120 cm) BioGel P-60 column equilibrated with 10 mM sodium phosphate, pH 7.0 containing 200 mM NaCl and 0.1 mM EDTA and the column developed with this buffer at a flow rate of 4 ml/hr. One milliliter fractions are collected and activity is measured across the elution volume. Active fractions are then pooled, made 10% in glycerol and 0.04% in sodium azide, and stored at 4°. The enzyme so prepared is stable for at least several months.

The purification procedure yields are summarized in Table I.

Purity and pI. The enzyme isolated from the above protocol is quite pure, as judged by SDS–polyacrylamide gel staining. Figure 1 shows the pattern obtained from 10 μg of purified protein. One band is visible, with a molecular weight of 37,000, in agreement with previous results. Two-dimensional gels reveal the pI of this protein to be 9.3.

Effect of pH and Ionic Conditions in Activity. Activity is maximum at pH 5.5, being some threefold higher than at pH 7.0. Optimum activity was achieved using MES [2-(N-morpholino)ethanesulfonic acid] as buffering agent; substitution of phosphate or Tris for MES decreased activity by 50 and 25%, respectively. Activity observed with varying buffer salts and pH is shown in Fig. 2A. Activity is also sensitive to ionic strength, as shown in Fig. 2B. In 50 mM MES, maximum activity was observed with no added salt; addition of NaCl to 100 mM decreased activity by 50%. Whether these factors act directly on the esterase or its membrane-bound substrate or both is unknown.

In general, stability of the esterase is inversely related to its activity. Maximum stability and low activity is observed in phosphate buffer at pH 7 with 50–100 mM added NaCl, while MES and low salt afford minimum protection against inactivation but high enzyme activity. For this reason it is often best to assay esterase at less than optimal activity conditions, as the enzyme is stable for longer periods.

TABLE I
PURIFICATION OF CHEMOTACTIC METHYLESTERASE[a]

Step	Total protein (mg)	Total units	Specific activity units/mg	Yield (%)
Extraction	17,100	146,000	8.6	100
Ammonium sulfate fractionation	9,170	110,000	12	75
Mercurial agarose pool	1,110	93,400	84	64
Octyl-Sepharose pool	269	67,200	249	46
CM-cellulose pool	10.0	51,100	5070	35
Gel filtration pool	5.2	36.500	7020	25

[a] One unit hydrolyzes 1 fmol glutamyl methyl ester per min under the assay conditions described in the text.

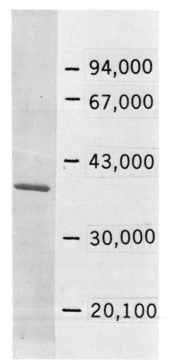

FIG. 1. Purity and molecular weight of the *cheB* esterase. Ten micrograms of purified esterase were electrophoresed on 10% polyacrylamide-SDS gels, and stained as described previously.[11]

Substrate Specificity. The esterase appears to be specific for the glutamyl esters of the MCPs. A variety of peptidyl and naphthyl esters are not hydrolyzed, as evidenced by their failure to competitively inhibit hydrolysis of the natural substrate. Similar specificity has been observed for the methyltransferase, the enzyme which transfers the methyl group to the glutamyl residues from S-adenosylmethionine.[12]

Amino Acid Composition. The amino acid composition of the esterase is given in Table II. A comparison of these values to the overall amino acid composition of bacterial proteins reveals that only histidine is significantly different (fivefold higher) than the average. With a p*I* of 9.3, it is estimated that some 47% of the aspartyl and glutamyl residues are in their amide form.

[12] S. Clarke, K. T. Sparrow, S. Panasenko, and D. E. Koshland, Jr., *J. Supramol. Struct.* **13,** 315 (1980).

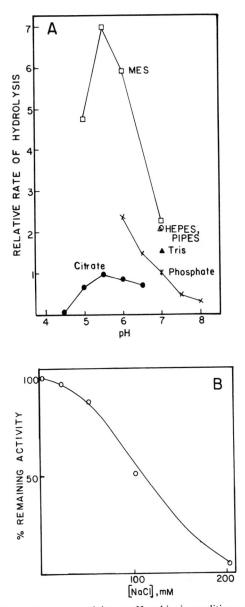

FIG. 2. Dependence of esterase activity on pH and ionic conditions. (A) Assay mixtures contain 50 mM buffer + 1 mM EDTA at the indicated pH, esterase (4.5 μg/ml) and [³H]ST1038 membranes (0.27 mg/ml). At various times, aliquots were withdrawn and assayed for distillable counts. Points shown are initial rates during the first 8 min of hydrolysis. Rates shown are relative, with phosphate at pH 7.0 = 1.0. (B) Hydrolysis of [³H]ST1038 membranes (0.27 mg/ml) by esterase (4.5 μg/ml) in 50 mM MES, pH 6.5 + 1 mM EDTA + indicated concentrations of NaCl. Activity was calculated from initial rates over the first 8 min of hydrolysis.

TABLE II
AMINO ACID COMPOSITION OF THE ESTERASE

Amino acid	Residues per molecule[a]	Rel frequency (Ala = 100)	E. coli relative freq. (Ala = 100)
Asx	26.12	69	76
Cys	4.09	11	14
Glx	32.28	85	83
Pro	20.17	53	35
Gly	28.77	76	60
Ala	37.85	100	100
Val	23.60	62	46
Leu	30.40	80	60
Tyr	3.96	10	17
Phe	6.94	18	25
His	8.97	24	5
Lys	15.07	40	54
Arg	19.97	53	41
Thr	17.01	45	35
Ser	22.12	58	46
Met	19.60	52	29
Ile	19.72	52	34
Trp[b]	4.5	12	8

[a] Calculated on the basis of a molecular weight of 37,000. Values are not corrected for loss during HCl hydrolysis.

[b] Value for tryptophan based on absorbance of esterase (in 10 mM sodium phosphate, pH 7.0 + 300 mM NaCl + 1 mM EDTA) at 290 nm, where the contribution of tyrosine to absorbance is comparatively negligible.

Titration with the sulfhydryl reagent, 5,5'-dithiobis(2-nitrobenzoate) revealed that, of the four cysteines present in the molecule, two are present as free sulfhydryls and two are in a disulfide bridge. The same values were obtained whether the titration was done in phosphate buffer at pH 7 or in buffer plus 8 M urea.

Sensitivity of Activity to Inhibition. Activity is inhibited by a variety of residue-specific reagents. The results of some of these assays are shown in Table III.

The results with sulfhydryl-specific reagents are less clear, however. Methyl methanothiosulfonate completely eliminates activity. Addition of 2-mercaptoethanol completely reverses this inhibition, suggesting that a sulfhydryl residue is involved in some stage of catalysis. Similarly, 5,5'-dithiobis(2-nitrobenzoate) (DTNB) completely and reversibly eliminates activity. However, N-ethylmaleimide, iodoacetate, and iodoacetamide

TABLE III
SENSITIVITY OF THE ESTERASE TO MODIFICATION REAGENTS

Reagent	Conc. of reagent (mM)	Activity remaining[a] (%)
None		100
Methyl methanethiosulfonate	0.08	5
Methyl methanethiosulfonate/2-mercaptoethanol[b]	0.08/100	100
2-Mercaptoethanol	100	100
5,5'-Dithiobis(2-nitrobenzoate)	0.3	5
2-Hydroxy-5-nitrobenzyl bromide	0.5	5
N-Acetylimidazole	1.5	11
Diethyl pyrocarbonate	0.15	20
Diethyl pyrocarbonate	5	5
2,4,6-Trinitrobenzene sulfonate	1	5
Phenylglyoxal	10	5

[a] Mixtures containing 0.1 M sodium phosphate, pH 7.0, purified esterase (0.1 mg/ml), and the indicated reagent were incubated for 2 hr at 4°. Excess reagent was then removed by dialysis [methyl methanethiosulfonate and 5,5'-dithiobis(2-nitrobenzoate)], by addition of 100 mM 2-mercaptoethanol at 4° 2 hr prior to assay (N-acetylimidazole), or by addition of 20 mM glycine (2,4,6-trinitrobenzene sulfonate and phenyl glyoxal). [³H]ST1038 *cheR* membranes (0.1 mg/ml) were then added and initial rates of methyl ester hydrolysis were determined.

[b] Esterase was incubated with 0.08 mM methyl methanethiosulfonate for 2 hr, 100 mM 2-mercaptoethanol was added, and the enzyme assayed 1 hr later with no intervening manipulations.

produce only slight inhibition under a variety of conditions. Serine esterase inhibitors such as PMSF (phenylmethylsulfonyl fluoride) have no effect. Previous treatment of the enzyme with iodoacetamide does not block sensitivity to DTNB. This suggests that the free cysteines on the esterase are differentially reactive to sulfhydryl reagents. The basis for this is not understood; however, such selectivity has been observed and commented on before.[13–17]

The small molecules associated with the methylation reaction have essentially no effect on the hydrolysis reaction *in vitro* (methionine at 1 mM, S-adenosylmethionine at 1 mM, and S-adenosylhomocysteine at 1 mM).

[13] J. A. Thoma, D. E. Koshland, Jr., R. Shinke, and J. Ruscica, *Biochemistry* **4,** 714 (1965).
[14] G. Runca, C. Bauer, and C. A. Rossi, *Eur. J. Biochem.* **1,** 434 (1967).
[15] M. W. Bitensky, K. L. Yielding, and G. L. Tomkins, *J. Biol. Chem.* **240,** 663 (1965).
[16] M. W. Bitensky, K. L. Yielding, and G. L. Tomkins, *J. Biol. Chem.* **240,** 1037 (1965).
[17] S. G. Withers, D. S. C. Yang, N. B. Madson, and R. J. Fletterick, *Biochem. Biophys. Res. Commun.* **86,** 1044 (1979).

Kinetic Parameters. The K_m of the esterase for its protein substrates prepared as described above is approximately 15 μM; this value is not far from the *in vivo* concentration of approximately 0.1 μM. Saturation kinetic studies revealed that the turnover number for this enzyme is 0.24 mol methanol/mol esterase/min, assuming all of the esterase present in the assays is active. While the rate of catalysis is extremely low, it must be remembered that the MCP substrates are not present in more than a 10-fold stoichiometric excess over the esterase *in vivo,* and changes in methylation levels normally involve only a fraction of the receptors. Under wild-type conditions saturation of the enzyme *in vitro* is not observed at membrane concentrations as high as 20 mg membrane protein/ml.

[31] Isolation of D-Aspartic Acid β-Methyl Ester from Erythrocyte Carboxyl Methylated Proteins

By STEVEN CLARKE, PHILIP N. MCFADDEN, CLARE M. O'CONNOR, and LILLIAN L. LOU

Enzymes which catalyze the reversible carboxyl methylation of cellular proteins have been found in every eukaryotic tissue examined.[1] A major class of such protein carboxyl methyltransferases is characterized by a rather nonspecific requirement for a methyl-accepting substrate and by a marked substoichiometric methylation of the protein substrate.[2,3] The isolation of methanol as a hydrolysis product indicates that a protein methyl ester is formed in the reaction, and the hydrolysis rates of the esters are consistent with aspartyl, glutamyl, or α-carboxyl linkages.[4] This eukaryotic enzyme can be distinguished from the bacterial *cheR* protein carboxyl methyltransferase which is specific for its chemoreceptor protein substrate and which catalyzes a stoichiometric reaction.[5–7]

[1] W. K. Paik and S. Kim, "Protein Methylation," p. 202. Wiley, New York, 1980.
[2] Y. Kloog, D. Flynn, A. R. Hoffman, and J. Axelrod, *Biochem. Biophys. Res. Commun.* **97,** 1474 (1980).
[3] D. W. Aswad and E. A. Deight, *J. Neurochem.* **40,** 1718 (1983).
[4] T. C. Terwilliger and S. Clarke, *J. Biol. Chem.* **256,** 3067 (1981).
[5] S. Clarke, K. Sparrow, S. Panasenko, and D. E. Koshland, Jr., *J. Supramol. Struct.* **13,** 315 (1980).
[6] J. B. Stock, S. Clarke, and D. E. Koshland, Jr., this volume [29].
[7] J. B. Stock and D. E. Koshland, Jr., *J. Biol. Chem.* **256,** 10826 (1981).

As a result of the lability of protein methyl esters in strong acids and bases, it is not possible to isolate free amino acid esters by chemical hydrolysis. However, enzymatic digestion has been successfully used to isolate glutamic acid γ-methyl ester in good yield from bacterial membranes which had been methylated *in vitro* and *in vivo* by the chemotaxis-specific bacterial enzyme.[6,8,9] It has been more difficult to isolate and identify the residue or residues which are methylated by the eukaryotic nonspecific protein carboxyl methyltransferases. We have been able, however, to isolate aspartic acid β-methyl ester from proteolytic digests of methylated human erythrocyte proteins.[10] This modified amino acid is unusual in that it has the D stereoconfiguration.[11] Although the yield of ester isolated is low, it is likely that this residue represents a major product of the erythrocyte enzyme(s). Because these enzyme(s) share most of the properties of the other eukaryotic protein carboxyl methyltransferases characterized so far, we would predict that many of these other nonspecific enzymes are also D-aspartyl protein carboxyl methyltransferases. The physiological role of these enzymes is not established but it appears that they function in the metabolism of age-racemized proteins.[11-13]

The problem of isolating eukaryotic amino acid methyl esters in reasonable yield is probably related to two factors. In the first place, a large fraction of the methyl esters of these proteins are very susceptible to hydrolysis even in the neutral pH range. This lability is probably a consequence of the intramolecular attack of the amide nitrogen of the carboxyl peptide bond on the β-methyl ester of modified aspartyl residues. This reaction results in the formation of a 5-membered imide ring that is subsequently rapidly hydrolyzed.[4,14] Second, the D configuration of the aspartyl residue may limit its release from the polypeptide by proteases since these enzymes are generally specific for hydrolyzing peptide bonds adjacent to L-residues.

We describe here one successful approach for isolating and identifying D-aspartic acid β-methyl ester from human erythrocytes and discuss possible artifacts that can arise during such preparations from erythrocytes and other eukaryotic tissues.

[8] P. Van Der Werf and D. E. Koshland, Jr., *J. Biol. Chem.* **252**, 2793 (1977).
[9] S. J. Kleene, M. L. Toews, and J. Adler, *J. Biol. Chem.* **252**, 3214 (1977).
[10] C. A. Janson and S. Clarke, *J. Biol. Chem.* **255**, 11640 (1980).
[11] P. N. McFadden and S. Clarke, *Proc. Natl. Acad. Sci. U.S.A.* **79**, 2460 (1982).
[12] J. R. Barber and S. Clarke, *J. Biol. Chem.* **258**, 1189 (1983).
[13] S. Clarke and C. M. O'Connor, *Trends Biochem. Sci.* **8**, 391 (1983).
[14] S. A. Bernhard, A. Berger, J. H. Carter, E. Katchalski, M. Sela, and Y. Shalitin, *J. Am. Chem. Soc.* **84**, 2421 (1962).

Preparation of ^{3}H-Methylated Proteins

Because the extent of eukaryotic protein methylation is on the order of 1–100 pmol/mg protein, direct nonisotopic methods for analyzing methylated residues have not been successful. Therefore, in our studies we have utilized either L-[*methyl*-^{3}H]methionine or S-adenosyl-L-[*methyl*-^{3}H] methionine to radiolabel methyl groups on erythrocyte proteins. The former isotope has been used to label methyl esters in intact cells,[12,15] since it is taken up and converted into the active methyl group donor, S-adenosyl[*methyl*-^{3}H]methionine, by a cellular synthetase activity.[16] Since mature erythrocytes do not synthesize protein, these results are not complicated by incorporation of the labeled methionine into the protein backbone. Alternatively, S-adenosyl[*methyl*-^{3}H]methionine can be used to label erythrocyte membrane proteins in cell lysates[4] or in a system utilizing a purified methyltransferase.[17] We have found that the use of isotopes containing up to 3 tritium atoms per methyl group (60–90 Ci/nmol) gives the highest yield of aspartic acid β-methyl ester in subsequent digestions, although the reasons for this apparent isotope effect are unclear. ^{3}H-Methylated proteins can be readily separated from unreacted isotope by centrifugation in the case of membrane species or by gel filtration in the case of soluble proteins.

Similar procedures can be used to label protein methyl esters in other types of cells and tissues. To minimize the hydrolysis of these labile esters, it is important to maintain mild acid conditions (pH 2–6) during the isolation of these proteins.

Isolation of Aspartic Acid β-Methyl Ester from Erythrocyte Membranes

Protein Digestion. The enzymatic digestion procedure described here was optimized for ^{3}H-methylated membranes prepared from intact erythrocytes, but it has also been useful for other erythrocyte membrane and soluble protein preparations. A typical proteolytic digestion is performed in a 400 μl plastic microfuge tube containing 20 μl of membranes (5 mg protein/ml, 160 pmol of [^{3}H]methyl groups/ml) and 80 μl of a Sigma or Worthington preparation of bakers' yeast carboxypeptidase Y (2 mg protein/ml, 100 units/mg protein, in 8 mg/ml pH 5 citrate buffer; stable at $-20°$).[11] The final pH of the digestion mixture is adjusted to pH 5–6 to minimize ester hydrolysis and the mixture is then incubated for 16–25 hr at 37°. Analysis of the digestion mixture by dodecyl sulfate gel electropho-

---

[15] C. Freitag and S. Clarke, *J. Biol. Chem.* **256,** 6102 (1981).
[16] K. Oden and S. Clarke, *Biochemistry* **22,** 2978 (1983).
[17] C. M. O'Connor and S. Clarke, *J. Biol. Chem.* **258,** 8485 (1983).

resis indicates that all membrane polypeptides are digested to small fragments under these conditions. The reaction is terminated by the addition
of 20 μl of 1 M HCl and 250 μl of pH 2.20 sodium citrate sample dilution
buffer (0.2 M in Na$^+$) containing 2% thiodiglycol and 0.1% phenol
(Pierce). Two micromoles each of standard amino acid esters are added to
the mixture from 0.2 M stock solutions and the pH is adjusted to approximately 2 if necessary with HCl as measured with indicator paper. The
stock solutions of L-aspartic acid β-methyl ester · HCl (Vega Biochemi-
cals), D-aspartic acid β-methyl ester · HCl,[18] and L-glutamic acid γ-methyl
ester (Sigma) are prepared in water and stored at $-20°$.

Ion-Exchange Chromatography. Insoluble materials are removed
from the proteolytic digest by centrifugation in a Beckman Microfuge B
for 8 min at 9600 g. An aliquot of the digest supernatant (generally 350 μl)
is applied directly to a 0.9 × 30-cm column of sulfonated divinyl benzene
cross-linked polystyrene amino acid analyzer resin (Durrum DC-6A,
Dionex Chemical Corp.) equilibrated in pH 3.25 citrate buffer (0.2 M in
Na$^+$) at 50°.[10] Fractions are eluted from the column in the same buffer at a
flow rate of 66–70 ml/hr using a pump operating at 200–400 psi. After 70
min, the elution is continued with 0.2 M NaOH for an additional 20 min.
Fractions of 2 min are collected and analyzed for radioactivity from
methyl groups by counting a portion in 7 to 10 volumes of ACS II counting
fluid (Amersham) and for the standard amino acid esters by a manual
ninhydrin procedure modified from that of Moore.[19] Here, 0.7 ml of each
fraction is mixed with 0.3 ml of ninhydrin reagent made from 0.40 g
ninhydrin (Pierce) and 0.06 g hydrindantin dihydrate (Pierce) dissolved in
15 ml of dimethyl sulfoxide (Mallinckrodt Analytical Reagent) and then
mixed with 5 ml of pH 5.20 acetate buffer (4.0 M in Li$^+$, Pierce). The
mixture of sample and reagent is heated in a boiling water bath for 30 min,
cooled on ice, and the absorbance of the samples at 570 nm read versus a
water blank. We have found that the color yield of aspartic acid β-methyl
ester is dependent upon the time of heating and is less than the corresponding molar color yield of aspartic acid (Fig. 1) or glutamic acid γ-
methyl ester (data not shown).

Under these conditions, aspartic acid β-methyl ester elutes from the
column at about 32–36 min and glutamic acid γ-methyl ester elutes at 56–
60 min. An additional ninhydrin peak is generally seen at the elution
position (20–24 min) of aspartic acid, which contaminates the aspartic
acid β-methyl ester standards. When digests of large amounts of protein

[18] D-Aspartic acid β-methyl ester is not commercially available at this time. It is synthesized
according to the Protocol in Ref. 11.
[19] S. Moore, *J. Biol. Chem.* **243,** 6281 (1968).

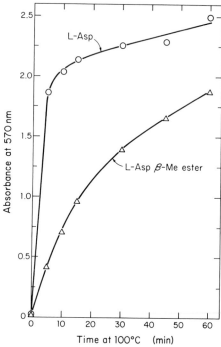

Fig. 1. Ninhydrin color yield of aspartic acid and aspartic acid β-methyl ester as a function of heating time. Samples of each amino acid (100 nmol) were assayed by the manual ninhydrin assay described in the text except that the time of the 100° heating step was varied. Comparison of the spectra of the reaction mixtures of aspartic acid samples with those of aspartic acid β-methyl ester samples revealed that both contained major absorption peaks at 405 and 570 nm while the ester spectrum displayed an additional absorbing species at 460 nm. This suggests that the ester is not only slow to form the major 570 nm chromatophore but is also capable of forming a ninhydrin side product with a maximal absorance of about 460 nm.

are applied to the column, ninhydrin peaks can also be detected at the characteristic elution positions of other free amino acids. Figure 2 is a representative chromatograph obtained by this procedure. In this case, 4.0% of the initial radioactivity was recovered as aspartic acid β-methyl ester. Up to 15% of the total has been recovered in other preparations. The radioactive peak at 12 min elutes at the position of methanol and consists largely of [3H]methanol from esters that are hydrolyzed during the digestion. The peak of radioactivity in the aspartic acid β-methyl ester fraction elutes about 1.5 min before the corresponding ninhydrin peak. This results from an isotope effect due to the presence of up to three tritium rather than hydrogen atoms in the methyl group; its magnitude is

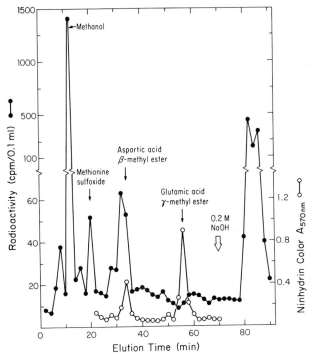

Fig. 2. Separation of ³H-labeled amino acid esters on amino acid analyzer resin. A proteolytic digest of ³H-labeled methylated erythrocyte membrane proteins was prepared and chromatographed on a 0.9 × 30 cm column of Durrum DC-6A resin as described in the text. The elution positions of standard compounds are indicated with arrows. In this particular experiment, the digestion mixture also included 2.45 mg/ml papain and 280 mM 2-mercaptoethanol and was carried out at room temperature for 16.5 hr. Internal standards included 1 μmol of L-aspartic acid β-methyl ester and 1 μmol of glutamic acid γ-methyl ester and the elution positions of these standards are indicated with arrows. The expected elution positions of methanol and methionine sulfoxide are also indicated. The ninhydrin reactions were carried out for 20 min with 0.2 ml of each fraction.

similar to those described previously for amino acids on high-resolution ion-exchange columns.[9,20]

For larger digestion samples, we have also performed this ion-exchange chromatography on the standard "long" column of a Beckman Model 120C amino acid analyzer. Using a 0.9 × 53 cm column of AA-15 resin equilibrated and eluted at 70 ml/hr at 56° in pH 3.25 citrate buffer (0.2 M Na$^+$), aspartic acid β-methyl ester is eluted at 53 min and glutamic acid γ-methyl ester is eluted at about 83 min.

[20] P. D. Klein and P. A. Szczepanik, *Anal. Chem.* **39,** 1276 (1967).

Characterization of Fractionated Aspartic Acid β-[³H]Methyl Ester by Gel Filtration, Thin-Layer Chromatography, and pH Hydrolysis Rate Profiles

Because it is possible that radioactive materials other than aspartic acid β-methyl ester, including undigested peptides, may comigrate with the non-isotopically labeled standard ester (see below), it is necessary to further characterize the radioactive product isolated from the ion-exchange chromatography step. Our standard procedure is to perform a gel filtration step which functions not only to desalt the sample but also to separate the aspartic acid β-methyl ester from other radioactive components. The pooled fraction containing aspartic acid β-methyl ester from the ion exchange column (up to 7 ml) are applied to a 1.5 × 85 cm column of Sephadex G-15 resin, equilibrated and eluted with 0.1 M acetic acid at room temperature. Fractions of approximately 3 ml are collected and aliquots assayed for radioactivity and ninhydrin-reactive material as described above. Aspartic acid β-methyl ester elutes at approximately 78 ml in this system and is well separated from the citrate salts.[10] For the erythrocyte membrane samples, essentially all of the radioactivity coincides with the ninhydrin peak of standard aspartic acid β-methyl ester (cf. Fig. 3); for other ³H-methylated proteins we have been able to separate con-

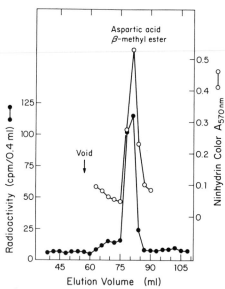

FIG. 3. Purification and desalting of aspartic acid β-[³H]methyl ester by Sephadex G-15 gel filtration. The pooled fractions (32–34 min) containing aspartic acid β-methyl ester from ion-exchange chromatography (Fig. 2) were chromatographed as described in the text.

taminating radioactive peaks from the ester standard. The pooled fractions are lyophilized overnight. Confirmation of the radioactive product
as aspartic acid β-[^{3}H]methyl ester is established by thin-layer chromatography as described, or by measuring the chemical rate of hydrolysis.[10] As
shown in Fig. 4, the pH rate profile of aspartyl ester hydrolysis is very
dependent upon substituents at either the α-carboxyl or the α-amino
group. The hydrolysis rate for aspartic acid β-methyl ester is relatively
constant in the pH range of 8.5 to 10.5. This characteristic feature is due
to the compensating effects of the increasing OH$^-$ concentration and the
formation of the less reactive unprotonated α-amino species as the pH is
increased in this range.[4] Such a compensation is not seen when the α-
amino group is involved in an amide linkage (for example in N-benzoyl-L-
aspartic acid β-methyl ester in Fig. 4). Significantly, when the carboxyl
group is involved in a peptide linkage (for example in N-benzoyl-L-aspar-
tyl β-methyl ester glycylamide), the rates of hydrolysis are approximately
500-fold faster. The latter effect is presumably due to the intramolecular

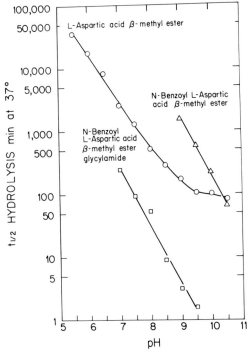

FIG. 4. Chemical hydrolysis rates of free aspartic acid β-methyl ester and of model
peptides containing aspartyl β-methyl ester residues. Data are taken from Terwilliger and
Clarke.[4]

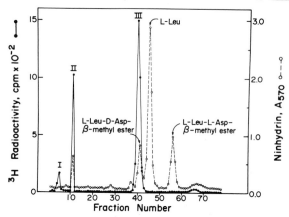

FIG. 5. Demonstration that aspartic acid β-[^{3}H]methyl ester from erythrocyte membranes is in the D stereoconfiguration. L-Leucyl dipeptides were synthesized and separated on an amino acid analysis column as described in the text. The volume of each five minute fraction used for the radioactivity determination was 1.0 ml; the volume taken for the ninhydrin assay was 0.7 ml.

imide formation which increases the rate of hydrolysis as discussed above.[4,14]

Determination of the Configuration of Isolated Aspartic Acid β-[^{3}H]Methyl Ester

Preparation of L-*Leucyl Dipeptide Diastereomers.* The most convenient procedure in use in our laboratory to determine the optical configuration of isolated aspartic acid β-[^{3}H]methyl ester involves the synthesis of dipeptide leucyl diastereomers which can be separated on amino acid analysis resin after the procedure of Manning and Moore.[21,22] The sample of radioactive ester, usually in lyophilized form after the Sephadex G-15 step, is dissolved in 1.2 ml of ice-cold 0.45 M sodium borate, pH 10.2. Additional nonradioactive standards of D- and L-aspartic acid β-methyl esters are added, if necessary, to bring the total amount to 2 μmol of each. This solution is immediately transferred to a tube containing a 2-fold molar excess of solid L-leucine N-carboxyanhydride [4-(2-methylpropyl)-2,5-oxazolidinedione].[23] The tube contents are vigorously mixed on a vor-

---

[21] J. M. Manning and S. Moore, *J. Biol. Chem.* **243**, 5591 (1968).

[22] J. M. Manning, this series, Vol. 25, p. 9.

[23] This reagent is presently unavailable commercially. It can be conveniently synthesized by the procedure of D. Konopinska and I. Z. Siemion [*Angew. Chem., Int. Ed. Engl.* **6**, 248 (1967)]. We have stored this material desiccated at −70°C for several months with no detectable loss of reactivity.

texer at 4° for 2 min, and the coupling reaction is then terminated by the addition of 0.48 ml of 1 *M* HCl. The quenched reaction mixture is mixed with an equal volume of pH 2.2 citrate sample buffer (see above) and is applied to the 0.9 × 53 cm column of the Beckman Model 120C amino acid analyzer as described above. The column is equilibrated and eluted at 70 ml/hr at 56° with pH 3.25 citrate buffer (0.2 *M* in Na$^+$). Fractions (5 min) are collected and aliquots are analyzed for radioactivity and ninhydrin color as above. L-Leucyl-D-aspartyl β-methyl ester elutes at about 190 min, while L-leucyl-L-aspartyl β-methyl ester elutes at about 270 min (Fig. 5). Once again, a small isotope effect is seen where the tritiated dipeptide elutes about 3 min before the non-isotopically labeled standard. It is possible to confirm that the radioactivity eluting at this position is in fact the dipeptide by subsequent gel filtration and thin-layer chromatography as shown in Fig. 6.

Use of D- *and* L-*Specific Amino Acid Oxidases.* We have also been successful in using stereospecific amino acid oxidases to demonstrate that isolated aspartic acid β-[³H]methyl ester has the D configuration.[11] For L-amino acid oxidase treatment, the radioactive methyl ester is mixed with 4 μmol of either L- or D-aspartic acid β-methyl ester standard and 200 μg of enzyme from *Crotalus adamanteus* venom (Sigma type IV, 8.1 units/

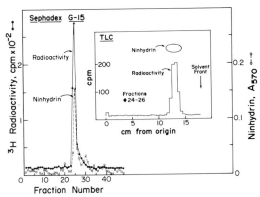

FIG. 6. Cochromatography of L-leucyl-D-aspartic acid β-[³H]methyl ester derived from red blood cell ³H-methylated membranes with synthetic L-leucyl-D-aspartic acid β-methyl ester. Gel filtration was performed as described in the text; the radioactive peak (Fractions 24–26) was lyophilized and applied to a cellulose thin-layer plate and chromatographed with 4/1/1/ (v/v/v) 1-butanol/acetic acid/H₂O. The position of the standard was found by spraying with a solution containing 1% (w/v) ninhydrin and 0.1% cadmium acetate in 4/1 (v/v) acetone/acetic acid. Radioactivity was determined from 0.5 cm sections scraped into scintillation vials containing 100 μl of H₂O and 1.0 ml of NCS tissue solubilizer (Amersham). After incubation for 4 hr at room temperature, 4.0 ml of OCS counting fluid (Amersham) was added and the vials counted.

mg protein) in a total volume of 0.17 ml containing a final concentration of 0.2 M Tris · HCl buffer at pH 7.5. This mixture is incubated at 37° for 3.5 hr. The reaction is quenched with 10 μl of 1 M HCl and 220 μl of pH 2.2 sodium citrate sample buffer and chromatography is performed on the 30 cm amino acid analysis column as described above. No detectable oxidation of the radioactive species is seen by the L-specific oxidase.[11]

For the D-amino acid oxidase treatment, radioactive samples mixed with standards (40 μl) are mixed with 200 μl of enzyme from porcine kidney (Sigma, 5 mg/ml, 17 units/mg protein, dialyzed overnight at 4° against 500 volumes of 100 mM sodium pyrophosphate, 3 mM EDTA, 25 mM NaCl at pH 8.3) for 5 hr at room temperature. The reaction is quenched with 150 μl of 8% 5-sulfosalicyclic acid (Pierce) and the products are analyzed as described above for the L-amino acid oxidase sample. In this case, the D-specific oxidase completely degrades the radioactive ester derived from erythrocyte proteins.[11] Control experiments demonstrated that there is no oxidation or hydrolysis of L-aspartic acid β-methyl ester by the D-specific oxidase, and no breakdown of D-aspartic acid β-methyl ester by the L-specific oxidase.

Alternative Proteolytic Digestion Procedures

For methylated proteins from human erythrocytes, digestion with carboxypeptidase Y alone (which may be contaminated with small amounts of endopeptidase activities) gave the highest yields of aspartic acid β-[^{3}H]methyl ester. The reasons for this are not entirely clear, but several factors appear to be involved. In the first place, the enzyme is active at acidic pH values (pH 5 was used here); this limits the chemical hydrolysis of methyl esters during the digestion. Second, this enzyme does not appear to have any substantial esterase activity toward either free aspartic acid β-methyl ester or aspartyl esters in peptides. Finally, the enzyme is able to cleave peptide bonds on both sides of D-amino acid residues, although at a very slow rate (cf. Hayashi et al.[24]). The rate of cleavage of the peptide bonds immediately adjacent to the D-aspartic acid β-methyl ester residue is likely to be the rate limiting step in the digestion outlined here; this may explain the requirements that we observe for high protease concentration and long digestion times. The results shown in the table indicate that the addition of a variety of proteolytic enzymes does not result in any marked increase in the yield of aspartic acid β-methyl ester from red cell membrane digests.

[24] R. Hayashi, Y. Bai, and T. Hata, *J. Biochem. (Tokyo)* **77**, 69 (1975).

EFFECT OF ADDITIONAL PROTEASES ON THE RELEASE OF ASPARTIC ACID β-METHYL
ESTER FROM ERYTHROCYTE MEMBRANES[a]

Carboxypeptidase Y (mg/ml)	Additional enzymes[b]	Percentage of initial radioactivity recovered as	
		Aspartate β-Me ester	Methanol
0	Control zero-time incubation	0	7.1
0	None	0	14.0
1.4	None	3.2	17.3
1.0[c]	None	2.2	10.9
1.0[c]	Trypsin	3.8	12.5
1.0[c]	Thermolysin	3.7	11.0
1.0[c]	Protease V8 from *Staphylococcus aureus*	5.7	11.9
1.0[c]	Carboxypeptidase A	2.7	12.4
0.8	Pepsin	3.3	15.1
0.8	Chymotrypsin	3.6	17.1
1.2	Pronase	3.1	33.2
1.0[c]	Proteinase K	4.3	33.7
0.8	Subtilisin	2.1	36.1

[a] [3H]-Methylated membranes from intact erythrocytes labeled with [*methyl-*
^{3}H]methionine were digested for a total of 1 hr at 37° in 0.1 *M* sodium acetate pH 5.5
with the indicated addition of proteolytic enzymes. Calcium chloride was included at 20
m*M* when required. The reaction products were analyzed by ion-exchange chromatography as shown in Fig. 2.
[b] All were present at 1.0 mg/ml with the exception of protease V8 which was present at
0.2 mg/ml.
[c] After 30 min incubation, sodium dodecyl sulfate was added to a final concentration of
2.0 mg/ml and the incubation continued another 30 min.

Isolation of Aspartic Acid β-Methyl Ester from Various Erythrocyte Fractions and Other Eukaryotic Cells

We have isolated D-aspartic acid β-[^{3}H]methyl ester from membrane[11]
and cytosolic[25] proteins which had been labeled in intact erythrocytes by
incubation with [*methyl-*^{3}H]methionine. The ester can also be successfully isolated from band 3-enriched membrane preparations from similarly
labeled cells.[26] Using S-adenosyl[*methyl-*^{3}H]methionine as the methyl do-

[25] C. M. O'Connor and S. Clarke, *J. Biol. Chem.* **259,** 2570 (1984).
[26] L. Lou and S. Clarke, unpublished.

nor, we have also labeled membrane proteins in both a lysed cell system[4] and in a defined system containing a purified preparation of erythrocyte methyltransferase.[17] After such labeling, it has been possible to isolate D-aspartic acid β-[³H]methyl ester from whole membranes[10,17] and a band 3-enriched membrane fraction[26] by the digestion procedure described here. At this point it is unclear to what extent this specific digestion protocol will be effective for other eukaryotic methylated polypeptides. For these proteins other combinations of proteolytic enzymes may be useful. In any case, once the methylated amino acid is released from the peptide chain, the analytical procedures described above can be used to determine the specific methyl-accepting residue (aspartate or glutamate) and its stereoconfiguration.

Possible Sources of Artifacts in the Isolation of Radiolabeled Methyl Esters from Proteolytic Digests

Peptide Esters. When proteolytic enzymes are used to digest ³H-methylated proteins, incomplete digestion products can accumulate, possibly because the D-aspartyl β-methyl ester residue prevents or slows the cleavage of adjacent peptide bonds. Many of these peptides elute on the 30 cm ion-exchange column described here in the flow-through fraction (6–8 min) and in the NaOH fraction. In a number of instances, however, radioactive products of proteolytic digests of ³H-methylated proteins have been found to elute near or with the aspartic acid β-methyl ester or glutamic acid γ-methyl ester standards. Nevertheless, the subsequent Sephadex G-15 gel chromatography step effectively separated the radioactive products from the standards and we have concluded that the radioactivity was present on small, incompletely digested peptides. Even if a methylated peptide were to coelute with the amino acid ester standard on gel filtration chromatography, such contamination could be readily detected by measuring the rate of ester hydrolysis for the radioactive fractionated material. As shown in Fig. 4, the hydrolysis rates of free and peptidyl aspartyl esters are very characteristic and can be distinguished from each other. Similar results have been shown for free and peptidyl glutamyl esters.[4]

Contaminants Derived from Isotopic Precursors. The incoporation of methyl groups into eukaryotic proteins is markedly substoichiometric and thus large amounts of high specific activity radioisotopes are required in order to obtain sufficient radioactivity in methylated proteins. Therefore minor contaminants in the isotope, or metabolic byproducts of the isotopes, can cause substantial problems if they elute coincidentally with amino acid esters on ion exchange columns. In some cases, it is possible to separate these radioactive products from the amino acid esters by gel filtration and thin layer chromatography as described above. Generally, however, we have found that a parallel digest of control radioactive samples in which inhibitors of the protein carboxyl methyltransferase were present during the labeling is most useful for verifying the presence of contaminants. For *in vitro* preparations, this inhibition is generally accomplished by adding S-adenosyl-L-homocysteine (K_i is about 1 μM^1) to the incubation mixture of enzyme, substrate protein, and S-adenosyl[*methyl*-^{3}H]methionine. Methylation in intact cells ([*methyl*-^{3}H]methionine labeled) can be inhibited by the addition of adenosine and homocysteine thiolactone to the incubation medium, which results in the accumulation of intracellular S-adenosylhomocysteine.[12]

A sometimes troublesome contaminant is found when cells are labeled with [*methyl*-^{3}H]methionine, and this is especially marked in cases where nonerythroid cells are labeled in the absence of protein synthesis inhibitors. In the latter case, L-[*methyl*-^{3}H]methionine incorporated into proteins is released upon proteolytic digestion and a significant portion of this isotope can then be oxidized to L-[*methyl*-^{3}H]methionine sulfoxide(s). This material normally elutes well before aspartic acid β-methyl ester (20 versus 34 min on the 30 cm preparative analyzer column). However, we have found that a considerable amount of the sulfoxide trails into the aspartic acid β-methyl ester peak. One source of this material may be from the slow oxidation of methionine on the chromatography column itself. Since methionine normally migrated much more slowly than aspartic acid β-methyl ester on the analyzer column (eluting at about 105 min), the net effect of the oxidation is a steady release of the more rapidly migrating sulfoxide which then contaminates all of the intermediate fractions from 20 to 105 min. Methionine sulfoxides can be partially resolved from aspartic acid β-methyl ester on Sephadex G-15 chromatography (it elutes about 6 ml earlier on the column described here) and its presence can be confirmed by the detection of its product after reaction with L-leucyl-N-carboxyanhydride. The L-leucyl-L-methionine sulfoxide(s) elutes at about 160 min on the 53 cm amino acid analysis column and is well separated from L-leucyl-D-aspartic acid β-methyl ester which elutes at about 190 min.

Do L-Glutamyl γ-Methyl Esters and L-Aspartyl β-Methyl Esters Exist in Eukaryotic Tissues?

Our studies in human erythrocytes do not show the presence of either of these components in proteolytic digests. We do not think that this absence represents inefficiency of the protease digestion protocols because it has not been difficult to isolate L-glutamyl γ-methyl ester from bacterial membrane proteins,[6,8,9] or L-aspartic acid β-methyl ester from synthetic poly-L-aspartic acid β-methyl ester.[11] Nevertheless, because we have not been able to isolate D-aspartic acid β-methyl ester in 100% yield, we cannot rule out the possibility that the L-esters exist in erythrocytes or other eukaryotic cells.

Acknowledgments

This work was supported by grants from the National Institutes of Health and the American Heart Association. C. M. O'Connor was supported by an Advanced Research Fellowship from the American Heart Association, and P. N. McFadden and L. L. Lou were supported by USPHS training grants.

[32] Hypusine Formation: A Unique Posttranslational Modification in Translation Initiation Factor eIF-4D

By HERBERT L. COOPER, MYUNG HEE PARK, and J. E. FOLK

Hypusine [N^ε-(4-amino-2-hydroxybutyl)lysine][1] is an unusual amino acid formed *in vivo* by a chemical modification of the side chain of a lysine residue.[2,3] The modifying butylamine group is derived from the polyamine, spermidine, and is subsequently hydroxylated.[2,3] Hypusine formation has been detected, in intact animal cells, in only a single protein,[2–4] which has been identified as a translation initiation factor (eIF-4D).[5]

This factor has been purified from rabbit reticulocytes by several groups.[6–8] In cell-free systems, eIF-4D has been shown to promote translation of polyuridylic acid to form polyphenylalanine, and to enhance formation of methionylpuromycin.[7–9] However, to date, the factor has not

[1] J. Shiba, H. Mizote, T. Kaneko, T. Nakajima, Y. Kakimoto, and I. Sano, *Biochim. Biophys. Acta* **244**, 523 (1971).

[2] M. H. Park, H. L. Cooper, and J. E. Folk, *Proc. Natl. Acad. Sci. U.S.A.* **78**, 2869 (1981).

[3] M. H. Park, H. L. Cooper, and J. E. Folk, *J. Biol. Chem.* **257**, 7217 (1982).

[4] H. L. Cooper, M. H. Park, and J. E. Folk, *Cell* **29**, 791 (1982).

[5] H. L. Cooper, M. H. Park, J. E. Folk, B. Safer, and R. Braverman, *Proc. Natl. Acad. Sci. U.S.A.* **80**, 1854 (1983).

[6] W. C. Merrick, this series, Vol. 60, p. 101.

[7] T. Staehlin, B. Erni, and M. Schreier, this series, Vol. 60, p. 136.

[8] R. Benne, M. Brown-Leudi, and J. Hershey, this series, Vol. 60, p. 15.

[9] W. M. Kemper, K. W. Berry, and W. C. Merrick, *J. Biol. Chem.* **251**, 5551 (1976).

$$
\begin{array}{ccc}
 & & NH_2 \\
 & & | \\
 & & (CH_2)_2 \\
 & NH_2 & | \\
 & | & CHOH \\
 & (CH_2)_4 & | \\
NH_2 & | & CH_2 \\
| & NH & | \\
(CH_2)_4 & | & NH \\
| & (CH_2)_4 & | \\
---NHCH & | & (CH_2)_4 \\
| & ---NHCH & | \\
O{=}C--- & | & ---NHCH \\
 & O{=}C--- & | \\
 & & O{=}C---
\end{array}
$$

Peptide-bonded   Peptide-bonded   Peptide-bonded
lysine   deoxyhypusine   hypusine

been demonstrated to influence translation of native globin mRNA under cell-free conditions.[7] Therefore, the physiological function of eIF-4D remains uncertain, although it is possible that it acts in the formation of the first peptide bond.[8,10]

The hypusine-containing protein is continuously formed in all growing animal cells, of every tissue type and species examined,[4] and is also observed in phorbol ester-treated HL-60 promyelocytic leukemia cells which are growth arrested but maintain high rates of protein synthesis.[11] However, hypusine formation is virtually undetectable in quiescent (nongrowing) peripheral blood lymphocytes, which exhibit low basal rates of protein synthesis. The substrate protein, however, is continuously synthesized and turned over in these cells.[4] Following growth stimulation of lymphocytes hypusine formation is stimulated within the first few hours and rises in parallel to the rate of protein synthesis.[4] This relationship between protein synthetic rate and degree of hypusine formation raises the possibility that this posttranslational modification modulates the activity of eIF-4D as an initiation factor, thereby contributing to the regulation of translation under physiological conditions.

Properties of Hypusine

The following characteristics of hypusine form the basis of methods for its detection and identification: (1) The amino acid is highly basic and elutes from ion-exchange resins under conditions used for study of polyamines and their derivatives. (2) Hypusine is stable both to proteolysis and to acid hydrolysis. Therefore hypusine can be recovered from protein digests produced by either of these methods. This characteristic distinguishes hypusine from other basic materials, such as γ-glutamylspermi-

[10] A. Thomas, R. Benne, and H. O. Voorma, *FEBS Lett.* **128,** 175 (1982).
[11] N. Feuerstein and H. L. Cooper, unpublished results.

dines, which exhibit similar chromatographic behavior but are cleaved by acid hydrolysis. (3) Hypusine may be oxidatively cleaved according to the following reaction: radioactivity derived from polyamines is recovered in the β-alanine fragment, that from lysine in the lysine fragment.

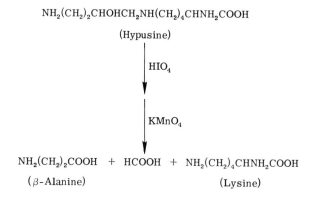

$$NH_2(CH_2)_2CHOHCH_2NH(CH_2)_4CHNH_2COOH$$

(Hypusine)

HIO_4

$KMnO_4$

$$NH_2(CH_2)_2COOH \quad + \quad HCOOH \quad + \quad NH_2(CH_2)_4CHNH_2COOH$$

(β-Alanine)   (Lysine)

Production of Radiolabeled Hypusine in Growing Human Peripheral Lymphocytes

Hypusine can be radiolabeled through biological synthesis in growing cells, or in cells manifesting high rates of protein synthesis, by addition of radioactive spermidine, or its intracellular precursor, putrescine, to cell culture fluids. Care must be exercised in adding polyamines to cell cultures, since amine oxidases present in serum, particularly in fetal bovine serum, may degrade the added polyamines.[12] This is particularly troublesome with spermidine. Our studies with human lymphocytes are generally conducted using human serum or plasma, which do not contain high levels of amine oxidases. Since endogenous polyamine pools of intact cells are large and turn over slowly, extended period of labeling—up to 48 hr—may be required. Hypusine may also be labeled with radioactive lysine. However, polyamine labeling is specific for hypusine and for eIF-4D, while lysine will label most proteins.

Human lymphocytes are prepared and placed in short-term culture with a growth stimulating mitogen, as described.[13] Radiolabeled polyamine (2–5 μCi/ml, [2,3-^{3}H]putrescine · 2HCl, or [*terminal methylenes*-^{3}H]spermidine · 3HCl, specific activity usually >20 Ci/mmol) is added and the cells are incubated for 4–48 hr. Cells are harvested by centrifugation

---

[12] J. C. Allen, C. J. Smith, J. I. Hussain, J. M. Thomas, and J. M. Gaugas, *Eur. J. Biochem.* **102**, 153 (1979).

[13] H. L. Cooper, this series, Vol. 32, p. 633.

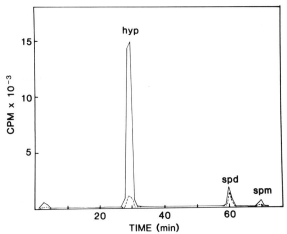

Fig. 1. Chromatographic separation of hypusine from lymphocyte protein digests. Lymphocyte cultures containing 2.5×10^6 cells were incubated for 48 hr with 2.5 μCi/ml [³H]spermidine and with or without mitogen (PHA) (5 μg/ml). Acid hydrolyzates were prepared and analyzed by ion-exchange chromatography, as described. Hyp, Position of hypusine marker. Spd, Spm, Spermidine and spermine, present as contaminants despite extensive washing of TCA precipitate. (——), Radioactivity from growing lymphocytes. (---), Radioactivity from quiescent lymphocytes. From Cooper et al.[4] by permission.

and washed with a large volume of 0.15 M NaCl/0.01 M Na-phosphate, pH 7.4. After resuspension and washing in cold 5% trichloroacetic acid (TCA), the acid-insoluble residue is washed with ether and subjected either to exhaustive proteolytic digestion[14] or to hydrolysis in 6 N HCl at 106° for 18 hr, *in vacuo*. The digestion or hydrolysis products are analyzed by ion-exchange chromatography, as described.[14,15] The elution pattern of radioactivity from such an analysis is shown in Fig. 1.

We have identified radioactive material recovered from the chromatographic procedure as hypusine by the following criteria:

1. Reference to a standard preparation of purified hypusine.[1,2] Since this material is not generally available, identification in most cases will rely on the following criteria. A secondary standard may be derived from hydrolysis products of electrophoretically purified, radiolabeled eIF-4D, as described in the following section.

[14] J. E. Folk, M. H. Park, S. I. Chung, J. Schrode, E. P. Lester, and H. L. Cooper, *J. Biol. Chem.* **255,** 3695 (1980).
[15] M. H. Park, H. L. Cooper, and J. E. Folk, this series, Vol. 94, p. 458.

2. Resistance to acid hydrolysis, i.e., a component which shows the same behavior on ion-exchange chromatography whether derived from proteolytic digestion or from acid hydrolysis.

3. Susceptibility to oxidative cleavage with production of β-alanine and lysine. Details of oxidative cleavage and thin-layer chromatography of products have been described.[15]

Identification of eIF-4D as the Unique Hypusine-Containing Protein

When growing lymphocytes are radiolabeled with polyamines, as in the preceding section, analysis of total cell proteins by standard methods of two-dimensional polyacrylamide gel electrophoresis and fluorography[4,16] reveals only a single radioactive peptide, with M_r ~17,000 and pI ~5.1[2,4] (Fig. 2). Identification of hypusine as the radioactive component of this protein is accomplished as follows.

Using the radiofluorograph as a guide, the labeled protein is located on the dried gel and carefully excised. The gel spot is placed in 1 ml of 6 N HCl and subjected to acid hydrolysis, as in the preceding section. After removal of debris by centrifugation, the sample is analyzed by ion-exchange chromatography, as in the preceding section. Virtually all of the radioactivity in this preparation elutes at the position of hypusine.

Identification of the uniquely radiolabeled peptide as eIF-4D is performed by coelectrophoresis in 2 dimensions with eIF-4D purified from rabbit reticulocytes.[6] Comigration of the single labeled protein, identified by fluorography, and the purified initiation factor, identified by Coomassie blue staining, is strong evidence of identity. The presence of hypusine in purified eIF-4D has been documented by ion-exchange chromatography of acid hydrolysates of the protein subjected to fluorometric analysis.[5] Such analysis shows that eIF-4D contains 1 mol of hypusine per mol of protein. Approximately 0.5 nmol of pure protein is required for this analysis (Fig. 3).

Identification of eIF-4D among total cell proteins which are visualized on electrophoretic gels, either by protein stains or by labeling with radioactive amino acids, is accomplished by reference to eIF-4D labeled with polyamines. Cells are labeled by exposure to radioactive amino acids (e.g., [^{3}H]leucine, 150 μCi/ml for 2–6 hr). Proteins are analyzed by two-dimensional electrophoresis in parallel with a sample of polyamine-labeled material. Provided the electrophoretic technique produces reproducible patterns, fluorographs of the two gels may be superimposed and eIF-4D identified among leucine-labeled proteins. Greater certainty may be obtained by a double-label technique in which material from cells

[16] P. H. O'Farrell, *J. Biol. Chem.* **250**, 4007 (1975).

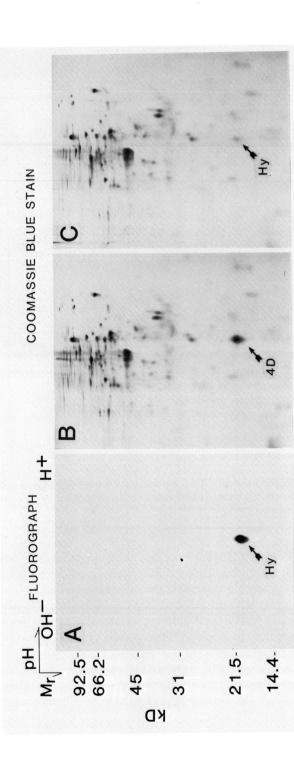

FIG. 2. Identity of two-dimensional electrophoretic mobility of eIF-4D and the single hypusine-containing lymphocyte protein. Growing lympho-cytes (5×10^7 cells) were incubated with [³H]spermidine ($5 \mu Ci/ml$, 31 Ci/mmol) for 40 hr. C shows two-dimensional electrophoresis of whole-cell proteins, Coomassie blue stain. B shows Coomassie blue stain of a mixture of the same spermidine-labeled lymphocyte proteins as in C, plus 2 μg of purified rabbit reticulocyte eIF-4D, indicated by the intensely staining spot labeled 4D. A shows the radiofluorograph of the gel seen in B, revealing the single radiolabeled lymphocyte protein (Hy). The radioactivity in this protein was shown to be in hypusine by ion-exchange chromatography. Spots Hy and 4D corresponded exactly when the fluorograph and the gel were superimposed. From Cooper et al.[5] by permission.

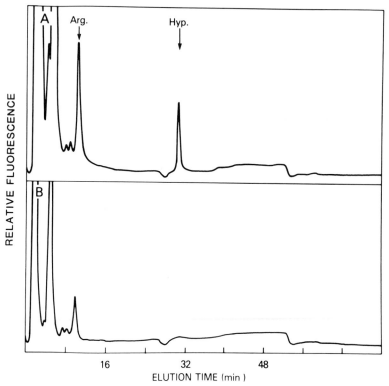

Fig. 3. Fluorometric analysis of acid hydrolysis products of purified rabbit reticulocyte eIF-4D separated by ion-exchange chromatography. Approximately 10 μg of purified eIF-4D were hydrolyzed in acid and analyzed. Elution positions of hypusine and arginine are indicated. (A) eIF-4D. (B) eIF-4C, another purified initiation factor analyzed as a control. From Cooper *et al.*[5] by permission.

labeled with [3H]-labeled polyamines is coelectrophoresed with [35S]methionine-labeled proteins. Direct fluorography provides an image of all labeled proteins. Radioautography of the same gel with interposition of a carbon paper layer between gel and film[17] reveals only [35S]Met-labeled material. By comparison and superimposition of the two films an identification of the amino acid-labeled protein corresponding to eIF-4D is made.

By these methods, the biosynthesis of an eIF-4D-like protein in non-growing lymphocytes, where hypusine formation is virtually undetectable, has been observed.[4] This protein is presumed to be a hypusine-free

[17] K. H. Choo, R. G. Cotton, and D. M. Danks, *Anal. Biochem.* **103**, 33 (1980).

precursor to eIF-4D. Surprisingly, this protein exhibits the same two-dimensional electrophoretic mobility as eIF-4D despite the fact that conversion of lysine to hypusine causes a change in net charge which should be detectable in the isoelectric focusing dimension.

The characteristic electrophoretic mobility of eIF-4D, its two-dimensional position relative to other proteins, and its abundance, make eIF-4D fairly simple to identify and study in such preparations. To date, these characteristics of eIF-4D and the uniqueness of the posttranslational hypusine modification, have been consistent in all animal systems studied.[4] The ability to identify either polyamine- or amino acid-labeled eIF-4D permits detailed study of the intracellular metabolism of this protein, its subcellular distribution and associations, and the role of posttranslational modification by hypusine formation in modulating its biological activity.

[33] Histidinoalanine: A Naturally Occurring Cross-Linking Amino Acid

By RONALD L. SASS and MARY E. MARSH

Occurrence

Histidinoalanines, 3-(τ-histidino)alanine and 3-(π-histidino)alanine, are members of the class of cross-linking compounds postulated to form *in vitro* when proteins are treated with strong alkali or high temperature.[1] Under these conditions dehydroalanine is formed from serine, phosphoserine, or cysteine residues through β-elimination. Dehydroalanine reacts with an available nucleophilic residue to complete the cross-link. Of the possible cross-linking compounds formed in this manner, the one most often observed is lysinoalanine, which occurs in cooked or processed food protein. The formation of histidinoalanine under these conditions has been postulated but not demonstrated.

Histidinoalanine has been identified as a naturally occurring cross-link in bone and dentin collagen and in the calcium-binding phosphoprotein particles derived from the extrapallial fluid of certain bivalve molluscs. Of the two possible isomers, N^τ- and N^π-histidinoalanine, only the former has been observed in collagen preparations, where it represents less than

---

[1] J. W. Finley and M. Friedman, *in* "Protein Crosslinking" (M. Friedman, ed.), p. 123. Plenum, New York, 1977.

METHODS IN ENZYMOLOGY, VOL. 106

0.04% of the amino acid residues.[2] Formation of the cross-link in collagen is apparently a slow, nonspecific and nonenzymatic process. It is observed principally in collagen derived from older individuals of long-lived species.

Both N^τ- and N^π-histidinoalanine are components of the phosphoprotein particles,[3] where they occur at a relative abundance of 3 : 1. In particles isolated from *Rangia cuneata,* phosphoserine, histidine, and histidinoalanine represent 29, 35, and 6% of the amino acid residues, respectively. Thus, the cross-link is probably derived from phosphoserine and histidine residues. Histidinoalanine cross-links monomeric phosphoprotein subunits into discrete particles approximately 40 nm in diameter with a molecular weight in excess of 30 million. *In situ* the particles sequester a protected pool of calcium and phosphate ions,[4] so the cross-links are functionally significant. The degree of cross-linking is species specific but independent of the animal's age. In particles derived from *Mercenaria mercenaria* only 1% of the residues (based on leucine equivalents) are histidinoalanine. The cross-linking reaction is apparently quite specific, but whether or not it is enzymatically mediated at any level is unknown.

Isolation Procedure

Extrapallial fluid is withdrawn from *Rangia cuneata* through a syringe needle inserted between the mantle and shell at the pallial attachment. About 30 ml of fluid can be obtained from a 250 g specimen. Alternately, a mixture of extrapallial fluid and hemolymph can be obtained by collecting all of the fluids which drain from the animal when the adductor muscles are cut and the clam removed from its shell. With this procedure approximately 40 ml of fluid is obtained, about 75% of which is extrapallial fluid. Generally, 600 ml of fluid is collected to obtain about 1 g of phosphoprotein particles. First the fluid is centrifuged 10 min at 600 g to remove the cells. The supernatant fluid is passed successively through 5.0- and 0.45-μm filters to remove particulate matter. The particles are sedimented from the opalescent filtrate by centrifugation at 100,000 g for 30 min, resuspended in water, and sedimented again under the same conditions.

One gram of the isolated particles is hydrolyzed in 25 ml of 6 N HCl in sealed tubes for 24 hr at 105°. HCl is removed under reduced pressure at room temperature. The residue is suspended in 10 ml of water and filtered

[2] D. Fujimoto, M. Hirama, and T. Iwashita, *Biochem. Biophys. Res. Commun.* **104,** 1102 (1982).
[3] R. L. Sass and M. E. Marsh, *Biochem. Biophys. Res. Commun.* **114,** 304 (1983).
[4] M. E. Marsh and R. L. Sass, *Biochemistry,* in press (1984).

through a 0.45-μm filter. The filtrate is applied to a 0.8 × 25 cm column of Dowex 50W-X8 (200–400 mesh; H$^+$ form). The column is washed with 50 ml 3 N HCl to elute the monofunctional amino acids. Then the two histidinoalanine isomers are eluted with 30 ml of 4 N HCl. HCl is removed as before, then the isomers are separated from each other by chromatography on the same resin previously equilibrated with 0.2 N sodium citrate, pH 3.25. The 0.8 × 25 cm column is eluted with 0.38 N sodium citrate, pH 5.35, at 50° and 1.0 ml fractions are collected. N^τ-Histidinoalanine elutes in fractions 21 through 30, and N^π-histidinoalanine elutes in fractions 32 through 48. The appropriate fractions are pooled and desalted on 0.8 × 25 cm columns of Dowex 50W-X8 (H$^+$ form). Citrate ions are eluted with 25 ml water, and sodium ions with 35 ml 2 N HCl. The histidinoalanines are eluted with 4 N HCl, evaporated to dryness and taken up in 0.5 ml water. Contaminants derived from ion-exchange resins are removed by gel filtration chromatography on Sephadex G-10 columns (1.6 × 100 cm) equilibrated and eluted with water.

Using a Glenco amino acid analyzer equipped with a 0.32 × 30 cm column of DC-4A resin and Pico-Buffer System II (Pierce Chemical Co., Rockford, IL), N^τ-histidinoalanine elutes after 71 min and N^π-histidinoalanine elutes after 72 min compared with lysine which elutes after 75 min.

Identification

N^τ- and N^π-Histidinoalanine (Fig. 1) can be identified by their nuclear magnetic resonance spectra. The ^{13}C proton decoupled NMR chemical shifts and proton coupled multiplicities are given in Table I. The spectra were recorded in D$_2$O at pD 7. Chemical shifts are in parts per million downfield from the internal standard sodium 3-trimethylsilylpropionate (TSP). The chemical shifts are very sensitive to pD and care must be taken using comparison spectra recorded at different values of pD. The proton-coupled multiplicity of the C-2 resonance is obscured because of

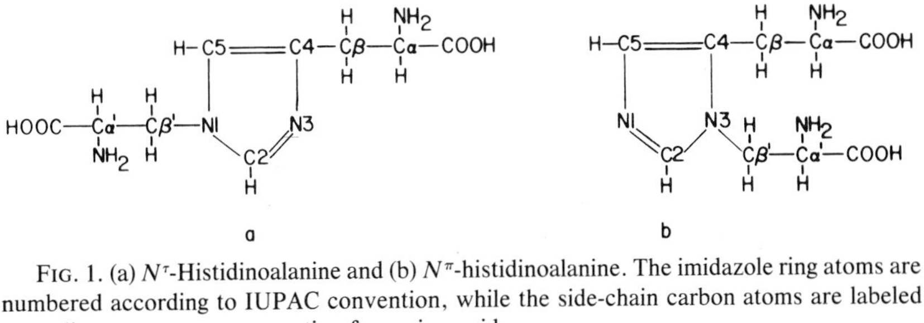

FIG. 1. (a) N^τ-Histidinoalanine and (b) N^π-histidinoalanine. The imidazole ring atoms are numbered according to IUPAC convention, while the side-chain carbon atoms are labeled according to common convention for amino acids.

TABLE I
^{13}C NMR SPECTRA OF N^τ- AND N^π-HISTIDINOALANINE

Atomic assignment	N^τ-Histidinoalanine δppm	N^π-Histidinoalanine δppm	H-coupled multiplicity
C-β	32.39	29.80	3
C-β'	50.78	51.46	3
C-α, C-α'	58.44, 58.73	57.46, 58.68	2, 2
C-5	122.28	124.87	2
C-4	139.45	134.98	1
C-2	142.44	142.28	a
COOH	175.41	174.92	1
COOH	177.65	176.48	1

^{a} H-coupled multiplicity is obscured by H–D exchange during data recording.

H–D exchange during the several hours required to take the spectra. The ^{1}H NMR spectra are presented in Table II using the same experimental conditions. The chemical shifts are given in δppm of the multiplet centroid followed by the integrated number of hydrogen atoms, the mutiplet structure, and the J value when appropriate. The spectra of the hydrogen atoms associated with C-α and C-α' form a broad multiplet with peaks between the values given. Both isomers have similar spectra. The most distinctive differences are in the chemical shifts of C-β and C-4 in the ^{13}C spectra and of the C-β and C-2 hydrogen atoms in the ^{1}H spectra.

N^τ- and N^π-Histidinoalanine have similar mass spectra as might be expected. The desorption chemical ionization positive ion spectra of both show a molecular ion at m/e 243 corresponding to (M + H) where M is the molecular weight consistent with the elemental composition $C_9H_{14}N_4O_4$.

TABLE II
^{1}H NMR SPECTRA OF N^τ- AND N^π-HISTIDINOALANINE

Atomic assignment	N^τ-Histidinoalanine δppm	N^π-Histidinoalanine δppm
(C-β)H$_2$	3.16 (1H, d, J = 9 Hz) 3.18 (1H, d, J = 5 Hz)	3.32(2H, d, J = 6 Hz)
(C-α')H, (C-α)H	3.96–4.19 (2H, m)	3.96–4.16 (2H, m)
(C-β')H$_2$	4.55 (2H, d, J = 5 Hz)	4.55 (2H, d, J = 7 Hz)
(C-5)H	7.09 (1H, s)	7.09 (1H, s)
(C-2)H	7.71 (1H, s)	7.87 (1H, s)

The molecular fragment corresponding to the loss of a side chain at m/e 156 is the only other spectral line typical of amino acids. These compounds are very polar and are difficult to desorb from the probe before thermal decomposition. Other observed spectra lines at m/e 199, 175, and 95 can be accounted for by thermally generated fragments. The methyl ester trifluoroacetyl derivatives of both isomers give both positive and negative ion desorption chemical ionization mass spectra. The positive ion spectra exhibit a parent peak at m/e 463 (M + H). Other mass peaks at m/e 293, 294, 280, and 266 are all consistent with the loss of side chain neutral fragments. The negative ion spectra exhibit major peaks at m/e 461 corresponding to the parent ion (M − H) and m/e 575 which is due to the adduct ion (M + CF$_3$COO). Other mass peaks are observed at m/e 556 and 442 corresponding to the loss of a fluorine atom from the adduct ion and the parent ion, respectively.

[34] S^β-(2-Histidyl)cysteine: Properties, Assay, and Occurrence

By Konrad Lerch

S^β-(2-Histidyl)cysteine has been isolated during amino acid sequence analysis of tyrosinase (EC 1.14.18.1) from *Neurospora crassa*.[1-3] The hitherto unknown structure is thought to arise from a posttranslational modification of the tripeptide Cys-Thr-His (amino acid residues 94–96 of *Neurospora* tyrosinase).[3] This chapter discusses the physicochemical properties of S^β-(2-histidyl)cysteine and describes an analytical procedure for its detection in peptides and proteins.

Properties

The chemical structure of the tripeptide Cys-Thr-His [residues 94–96 of *Neurospora* tyrosinase[3] containing S^β-(2-histidyl)cysteine] is shown in Fig. 1. It is characterized by an intramolecular thioether linking the S^β-cysteine sulfur to the C-2 imidazole carbon of the histidine moiety.[3] Acid hydrolysis (6 N HCl) of the tripeptide yields Thr and a ninhydrin-positive

[1] K. Lerch, *Proc. Natl. Acad. Sci. U.S.A.* **75**, 3635 (1978).
[2] K. Lerch, C. Longoni, and E. Jordi, *J. Biol. Chem.* **257**, 6408 (1982).
[3] K. Lerch, *J. Biol. Chem.* **257**, 6414 (1982).

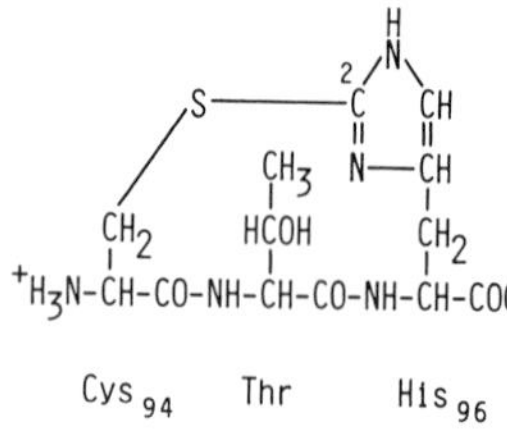

FIG. 1. Chemical structure of the tripeptide Cys-Thr-His (amino acid residues 94–96 of *Neurospora* tyrosinase) containing S^β-(2-histidyl)cysteine. From Lerch.[3]

species identified as 2-thiolhistidine (Fig. 2). The presence of a thioether in this peptide is also substantiated by a positive chloroplatinic acid test and by the cleavage with Ag_2SO_4 resulting in cysteic acid after performic acid oxidation.[3–5]

As shown in Fig. 3, the covalent addition of a sulfur to the C-2 position of the imidazole leads to a drastic change in the absorption spectrum of the histidine moiety. With an extinction coefficient ε^{255} of 1.44×10^4 at pH 2.0, the absorption properties of S^β-(2-histidyl)cysteine are strikingly similar to those reported for 2-thiolhistidine.[3,6] The characteristic absorption features of S^β-(2-histidyl)cysteine are also reflected in the absorption ratio $A_{280/260}$ of native *Neurospora* tyrosinase.[3] Despite the high content of aromatic amino acids in this protein,[1] the calculated $A_{280/260}$ ratio is distinctly higher than the one measured in the presence of 6 M guanidine hydrochloride.

Assay

Principle. The determination of S^β-(2-histidyl)cysteine is based on the diagonal procedure of Brown and Hartley[7] using ion-exchange chromatography. The first step involves cleavage of the thioether by acid hydrolysis and subsequent isolation of 2-thiolhistidine. In a second step, 2-thiolhistidine is converted to histidine with hydrogen peroxide. Histidine is then quantitated by amino acid analysis.

The procedure outlined below has been developed for the determination of S^β-(2-histidyl)cysteine in *Neurospora* tyrosinase or peptides derived from the protein.[3] However, it is basically applicable to any protein or peptide supposedly containing this thioether structure.

[4] G. Toennies and J. J. Kolb, *Anal. Chem.* **23**, 823 (1951).
[5] S. Sano and K. Tanaka, *J. Biol. Chem.* **239**, 3109 (1964).
[6] H. Heath, A. Lawson, and C. Rimington, *Nature (London)* **166**, 106 (1960).
[7] Ch. Rüegg, D. Ammer, and K. Lerch, *J. Biol. Chem.* **257**, 6420 (1982).

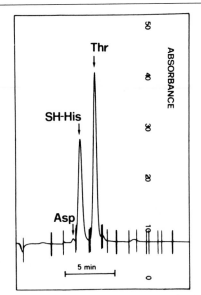

FIG. 2. Amino acid elution profile of the acid-hydrolyzed tripeptide (Fig. 1) obtained on a Durrum D-500 amino acid analyzer. SH-His denotes 2-thiolhistidine. From Lerch.[3]

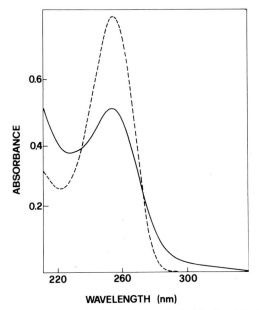

FIG. 3. UV absorption spectra of 2-thiolhistidine (dashed line) and the tripeptide (Fig. 1) (solid line). 2-Thiolhistidine (50 μM) and the tripeptide (34.5 μM) were dissolved in 10 mM HCl. From Lerch.[3]

Procedure

A sample of 20–50 mmol protein or peptide is hydrolyzed in an evacuated, sealed ampoule containing 1 ml of 6 *N* HCl + 0.1% SnCl$_2$ at 110°. The hydrolyzate is evaporated to dryness and then subjected to amino acid analysis (Durrum D-500 amino acid analyzer) by disconnecting the separation column from the reaction coil. Fractions of 50 μl eluting between the positions of aspartic acid and threonine (the elution position of 2-thiolhistidine; see text and Fig. 2) are collected and pooled (400–500 μl). The pooled sample is treated with 5 μl H$_2$O$_2$ (30% w/v) for 10 min at 25° to allow conversion of 2-thiolhistidine to histidine. The amount of histidine is determined by amino acid analysis in the normal running mode. The overall yield of histidine from S^β-(2-histidyl)cysteine ranges between 40 and 50%.

Occurrence

S^β-(2-Histidyl)cysteine has been detected in three different allelic forms of *Neurospora* tyrosinase.[8] The structure, however, was found to be absent in tyrosinase from the prokaryote *Streptomyces glaucescens*[3,9] as well as in *Neurospora* laccase and *Cancer pagures* hemocyanin, two copper proteins containing a similar binuclear copper active site to *Neurospora* tyrosinase.[3,10]

Very little is known about the mechanism of formation of S^β-(2-histidyl)cysteine in *Neurospora* tyrosinase. It has been proposed that the posttranslational generation of the thioether structure might be related to the activation of tyrosinase (conversion of protyrosinase to tyrosinase).[1,11] In this context it is of interest to note that *Neurospora crassa* is capable of synthesizing ergothioneine,[12] an N-trimethylated derivative of 2-thiolhistidine. From biosynthetic studies it has been shown that ergothioneine is derived from histidine and that cysteine is the principal source of the sulfur in ergothioneine.[13] Hence, possibly related enzymatic processes might be responsible both for parts of the synthesis of ergothio-

[8] J. R. Brown and B. S. Hartley, *Biochem. J.* **101**, 214 (1966).
[9] K. Lerch and L. Ettlinger, *Eur. J. Biochem.* **31**, 427 (1972).
[10] K. Lerch, *in* "Metal Ions in Biology" (H. Sigel, ed.), Vol. 13, p. 143. Dekker, New York, 1981.
[11] A. S. Fox and J. Burnett, *in* "Pigment Cell Biology" (M. Gordon, ed.), p. 249. Academic Press, New York, 1959.
[12] A. Meister, "Biochemistry of the Amino Acids," p. 838. Academic Press, New York, 1965.
[13] D. B. Melville, S. Eich, and N. L. Ludwig, *J. Biol. Chem.* **224**, 871 (1957).

neine and for the generation of S^β-(2-histidyl)cysteine in *Neurospora* tyrosinase.

Acknowledgments

This work was supported by Swiss National Science Foundation Grants 3.420-0.78 and 3.709-1.80.

[35] S^β-(Bilin)cysteine Derivatives: Structures, Spectroscopic Properties, and Quantitation[1]

By ALEXANDER N. GLAZER

Open-chain tetrapyrroles, also called bile pigments or bilins, occur as covalently bound prosthetic groups in several important types of proteins. The most abundant of such proteins are the diverse phycobiliproteins which serve as light-harvesting photosynthetic accessory proteins in cyanobacteria (blue-green algae), and in three groups of eukaryotic algae, the red algae, the cryptomonads, and certain dinoflagellates.[2] Phytochrome, the most important photoreceptor for photomorphogenesis in higher plants, carries a bilin prosthetic group.[3] Finally, bilin-carrying proteins function as coloring pigments in the skin of certain sea fishes.[4] In the phycobiliproteins and phytochrome, the bilins are linked to the polypeptide chain through a thioether bond to cysteinyl residues.

Structure

Several different bilins function as covalently attached prosthetic groups in biliproteins.[2,5] Various methods have been employed to break the bilin–peptide linkage(s). These include treatment with refluxing meth-

---

[1] Preparation of this contribution was supported in part by Grant PCM82-08158 from the National Science Foundation and Grant GM28994 from the National Institutes of Health, U.S. Public Health Service.
[2] A. N. Glazer, *in* "The Biochemistry of Plants" (M. D. Hatch and N. K. Boardman, eds.), Vol. 8, p. 51. Academic Press, New York, 1981.
[3] W. Rüdiger, *Struct. Bonding (Berlin)* **40**, 101 (1980).
[4] L. Abolins and W. Rüdiger, *Experientia 22*, 298 (1966).
[5] W. Rüdiger, *Ber. Dtsch. Bot. Ges.* **92**, 413 (1979).

anol,[6] 12 N HCl,[7] hydrogen bromide in trifluoroacetic acid,[8,9] or mercuric or silver chloride at pH 4.0.[10] The yields of free tetrapyrrole obtained by these procedures are in general not quantitative and vary from one biliprotein to another. Information on the structure of the protein-bound bilins was also obtained from studies of the products of chromic acid degradation of biliproteins.[3] For three of the bilins, phycocyanobilin, phycoerythrobilin, and phytochromobilin, the structures have been determined for both the free and the polypeptide-bound pigment (Fig. 1). Degradation and synthetic studies leading to the determination of the structures of the free bilins have been described in several recent reviews.[3,5,11] Various lines of evidence have established for certain bilin peptides that these tetrapyrroles are attached to cysteinyl residues through the 3-ethyl substituent at C-3', as shown in Fig. 1, structures **IV–VI**.

The thioether linkage between the bilins and the polypeptide is stable under the conditions used for automated sequential Edman degradation of proteins. In a number of studies, it has been shown that at the step of the degradation that removes the bilin-bearing residue the color extracts into the organic solvent wash (e.g., butyl chloride). In fact, such an observation has been used in some sequence determinations on biliproteins and bilipeptides as presumptive evidence of the location of bilin-linked cysteinyl residues with no further proof. Procedures have been reported for the isolation of bilipeptides after cleavage of biliproteins either with CNBr or proteolytic enzymes. Bilipeptides can be purified to homogeneity by high-performance liquid chromatography with no alteration in the spectroscopic properties of the bilins.[12–14] Amino acid sequence determination and ^{1}H NMR analysis of such peptides provide the most direct data on the structure and mode of linkage of these prosthetic groups. Structures of bilipeptides determined in this manner are shown in Fig. 1 (**IV–VI**).[12–14] Whereas a cysteine residue appears indispensable for bilin attachment, no general obvious patterns of residues in the regions of sequence flanking the attachment site have yet been observed.

[6] H. W. Siegelman, B. C. Turner, and S. B. Hendricks, *Plant Physiol.* **41**, 1289 (1966).
[7] C. O'hEocha, *Arch. Biochem. Biophys.* **73**, 207 (1958).
[8] H. H. Kroes, *Meded. Landbouwhogesch. Wageningen* **70-18**, 1 (1970).
[9] B. L. Schram and H. H. Kroes, *Eur. J. Biochem.* **19**, 581 (1971).
[10] V. P. Williams and A. N. Glazer, *J. Biol. Chem.* **253**, 202 (1978).
[11] A. Gossauer and H. Plieninger, *in* "The Porphyrins" (D. Dolphin, ed.), Vol. 6A, p. 585. Academic Press, New York, 1979.
[12] J. C. Lagarias, A. N. Glazer, and H. Rapoport, *J. Am. Chem. Soc.* **101**, 5030 (1979).
[13] J. C. Lagarias and H. Rapoport, *J. Am. Chem. Soc.* **102**, 4821 (1980).
[14] R. W. Schoenleber, S.-L. Leung, D. J. Lundell, A. N. Glazer, and H. Rapoport, *J. Am. Chem. Soc.* **105**, 4072 (1983).

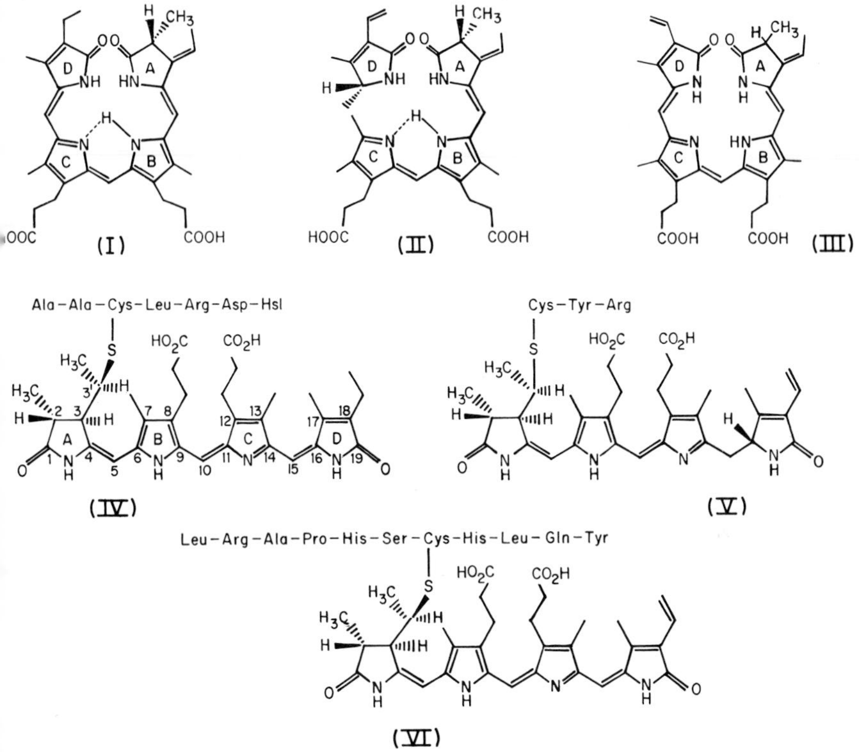

FIG. 1. (**I**) Phycocyanobilin; (**II**) phycoerythrobilin; and (**III**) phytochromobilin. Structures of the bilipeptides **IV–VI** were established in each instance by amino acid sequence determination and ^{1}H NMR analysis of the intact chromopeptide. (**IV**) Phycocyanobilin-peptide resulting from CNBr cleavage of the β subunit of *Synechococcus* 6301 C-phycocyanin.[10,12] Hsl is homoserine lactone. (**V**) Tryptic phycoerythrobilin-peptide from the α subunit of *Porphyridium cruentum* B-phycoerythrin.[14] (**VI**) Peptic phytochromobilin-peptide from oat phytochrome.[13]

Spectroscopic Properties

Polypeptide-linked bilins have distinctive visible absorption spectra, dependent on the chemical nature of the bilin, which can be exploited to determine the number and type of bilins present. Since the absorption spectra of these tetrapyrroles are strongly conformation dependent, complete denaturation of the proteins is a prerequisite to spectroscopic determination of bilin composition. Polypeptide-bound bilins are most stable at pH 3, or below. Absorption spectra of biliproteins, measured immediately after exposure of the protein to either 8 M urea or 6 M guanidine HCl at pH 1.9–3.0, have provided reliable data on the bilin content of numerous

TABLE I
LONG-WAVELENGTH ABSORPTION MAXIMA OF BILIPROTEIN CHROMOPHORES IN
BILIPROTEINS DENATURED AT ACID pH

Bilin	Occurrence	λ_{max} (nm)	ε_{max} $(cm^{-1} M^{-1})$	References[a]
Phycocyanobilin	Cyanobacterial, red algal, and cryptomonad biliproteins	662	35,400[b]	1
Phycoerythrobilin	Cyanobacterial, red algal, and cryptomonad biliproteins	555	43,300[b] 45,500[c]	2 3
Phycourobilin	Cyanobacterial and red algal biliproteins	495	104,000[b]	4
Bilin-600	Cyanobacterial phycoerythro-cyanin; β subunit of cryptomonad phycocyanin	590	43,000[b]	5, 6
Bilin-694	α Subunit of cryptomonad phycocyanin from *Hemiselmis virescens* Millport 64	694	—	6
Phytochromobilin	Phytochrome P_R	657–680	36,400[d]	7
	Phytochrome P_{FR}	610–615	—	8

[a] Key to references: (1) A. N. Glazer and S. Fang, *J. Biol. Chem.* **248**, 659 (1973). (2) A. N. Glazer and C. S. Hixson, *J. Biol. Chem.* **250**, 5487 (1975). (3) G. Muckle and W. Rüdiger, *Z. Naturforsch. C: Biosci.* **32C**, 957 (1977). (4) A. N. Glazer and C. S. Hixson, *J. Biol. Chem.* **252**, 32 (1977). (5) D. A. Bryant, A. N. Glazer, and F. A. Eiserling, *Arch. Microbiol.* **110**, 61 (1976). (6) A. N. Glazer and G. Cohen-Bazire, *Arch. Microbiol.* **104**, 29 (1975). (7) T. Brandlmeier, H. Scheer, and W. Rüdiger, *Z. Naturforsch. C: Biosci.* **36C**, 431 (1981). (8) S. Grombein, W. Rüdiger, and H. Zimmerman, *Hoppe-Seyler's Z. Physiol. Chem.* **356**, 1709 (1975).
[b] Determined in 8 M urea adjusted to pH 3.0 with HCl.
[c] A value of 47,100 $cm^{-1} M^{-1}$ at 550 nm per bilin was determined for a tryptic digest of a C-phycoerythrin (3), and a value of 45,500 $cm^{-1} M^{-1}$ calculated from model compound studies (3).
[d] Determined in 1% (v/v) methanolic H_2SO_4.

biliproteins.[2,15] The absorption maxima and molar extinction coefficients of polypeptide-bound bilins are presented in Table I.

Table II lists the prosthetic group composition of the major biliproteins which serve as photosynthetic accessory proteins in cyanobacteria and red algae. Where available, compositions determined by sequence analysis agree closely with those obtained by spectroscopy. References to both sources of data are included in Table II. Phytochrome carries one phytochromobilin per polypeptide chain.[13,16,17]

[15] A. N. Glazer, *Annu. Rev. Microbiol.* **36**, 173 (1982).
[16] K. T. Fry and F. E. Mumford, *Biochem. Biophys. Res. Commun.* **45**, 1466 (1971).
[17] T. Brandlmeier, H. Scheer, and W. Rüdiger, *Z. Naturforsch. C: Biosci.* **36C**, 431 (1981).

TABLE II

DISTRIBUTION OF BILIN PROSTHETIC GROUPS AMONG THE SUBUNITS OF VARIOUS CYANOBACTERIAL AND RED ALGAL PHYCOBILIPROTEINS

Protein subunit		Number and type of bilins[a]	References[b]	Protein subunit		Number and type of bilins	References
Allophycocyanin B	α	1 PCB	1, 2	C-Phycoerythrin	α	2 PEB	7, 11, 12
Allophycocyanin	α	1 PCB	3, 4		β	3–4 PEB	7, 11, 12
	β	1 PCB	3, 4	B-Phycoerythrin	α	2 PEB	13
C-Phycocyanin	α	1 PCB	3, 5, 6		β	3 PEB[c]	13
	β	2 PCB	3, 5, 6		γ[d]	2 PEB, 2 PUB	13
R-Phycocyanin	α	1 PCB	7, 8	R-Phycoerythrin[e]	α	2 PEB	13–15
	β	1 PCB, 1 PEB	7, 8		β[d]	2 PEB, 1 PUB	13–15
Phycoerythrocyanin	α	1 PXB	8–10		γ[d]	4 (PEB + PUB)	13–15
	β	2 PCB	8–10				

[a] PCB, Phycocyanobilin; PEB, phycoerythrobilin; PUB (phycourobilin) and PXB are bilins of undetermined structure.

[b] Key to references: (1) A. N. Glazer and D. A. Bryant, *Arch. Microbiol.* **104,** 15 (1975). (2) D. J. Lundell and A. N. Glazer, *J. Biol. Chem.* **256,** 12600 (1981). (3) A. N. Glazer and S. Fang, *J. Biol. Chem.* **248,** 659 (1973). (4) W. Sidler, J. Gysi, E. Isker, and H. Zuber, *Hoppe-Seyler's Z. Physiol. Chem.* **362,** 611 (1981). (5) V. P. Williams and A. N. Glazer, *J. Biol. Chem.* **253,** 202 (1978). (6) G. Frank, W. Sidler, H. Widmer, and H. Zuber, *Hoppe-Seyler's Z. Physiol. Chem.* **359,** 1491 (1978). (7) A. N. Glazer and C. S. Hixson, *J. Biol. Chem.* **250,** 5487 (1975). (8) D. A. Bryant, C. S. Hixson, and A. N. Glazer, *J. Biol. Chem.* **253,** 220 (1978). (9) D. A. Bryant, A. N. Glazer, and F. A. Eiserling, *Arch. Microbiol.* **110,** 61 (1976). (10) H. Zuber, *in* "Photosynthetic Prokaryotes. Cell Differentiation and Function" (G. C. Papageorgiou and L. Packer, eds.), p. 23. Elsevier/North-Holland Biomedical Press, Amsterdam, 1983. (11) G. Muckle and W. Rüdiger, *Z. Naturforsch. C: Biosci.* **32C,** 957 (1977). (12) G. Muckle, J. Otto, and W. Rüdiger, *Hoppe-Seyler's Z. Physiol. Chem.* **359,** 345 (1978). (13) A. N. Glazer and C. S. Hixson, *J. Biol. Chem.* **252,** 32 (1977). (14) M.-H. Yu, A. N. Glazer, K. G. Spencer, and J. A. West, *Plant Physiol.* **68,** 482 (1981). (15) M.-H. Yu, Ph.D. Thesis, University of California, Berkeley (1981).

[c] The β subunit of *Porphyridium cruentum* B-phycoerythrin has been shown by sequence studies to carry three phycoerythrobilin prosthetic groups. One of these is linked to the polypeptide chain through two thioether linkages to cysteinyl residues (D. J. Lundell and A. N. Glazer, unpublished observations).

[d] The bilin content of these subunits has not been rigorously established (see ref. 13).

[e] The relative proportion of PEB and PUB in R-phycoerythrins from different organisms is variable. See ref. 14, and also A. N. Glazer, J. A. West, and C. Chan, *Biochem. Syst. Ecol.* **10,** 203 (1982); D. A. Bryant, G. Cohen-Bazire, and A. N. Glazer, *Arch. Microbiol.* **129,** 190 (1981).

Determination of the Number of Thioether-Linked Cysteine Residues in Biliproteins

The following methodology has been successfully applied to the determination of thioether-linked cysteine residues in biliproteins and in cytochromes c.[18] Hydrolysis of proteins in 6 N HCl containing 0.2 M dimethyl sulfoxide at 110° for 24 hr results in a quantitative conversion of cysteinyl and cystinyl residues to cysteic acid.[19] Cysteine residues linked through thioether bonds to bilins or to heme, as in cytochromes c, are converted to cysteic acid in > 94% yield under these hydrolysis conditions. A number of amino acid residues are destroyed, but aspartic, glutamic, alanine, valine, isoleucine, and leucine residues are stable. Methionyl residues and methionine itself are partially converted to homocysteic acid in ~10% yield.[18] For amino acid analyses performed under conditions where homocysteic acid elutes together with cysteic acid, a correction for the homocysteic acid contribution is required. Calculation of the correction factor naturally requires independent determination of the methionine content of the protein in a hydrolyzate prepared in 6 N HCl. The recovery of cysteic plus homocysteic acid in the HCl–dimethyl sulfoxide hydrolyzate is calculated on the basis of an average one-residue figure determined from the recoveries of aspartic acid, glutamic acid, alanine, valine, isoleucine, and leucine, and the known amino acid composition of the protein.

The number of non-thioether-linked half-cystine residues is determined after conversion to the S-carboxymethyl derivative by reduction and carboxymethylation by conventional procedures,[20] and the total half-cystine plus thioether-linked cysteine residues by hydrolysis in HCl–dimethyl sulfoxide as described above. The difference between these values yields the number of thioether-linked cysteine residues.

Hydrolysis in the Presence of Dimethyl Sulfoxide[18]

Protein sample (~0.05 μmol) is lyophilized in a hydrolysis tube of ~4 ml capacity. It is then dissolved in 2.0 ml 6 N HCl and 30 μl of reagent grade dimethyl sulfoxide added. (Amounts and volumes may be scaled down proportionally as desired.) The sample is frozen, evacuated briefly, and sealed. Hydrolysis is performed for 22 hr at 110°. The hydrolyzates are then dried in a heated vacuum dessicator at 50° over NaOH pellets. Amino acid analysis is performed by conventional procedures.

[18] A. N. Glazer, C. S. Hixson, and R. J. DeLange, *Anal. Biochem.* **92,** 489 (1979).
[19] R. L. Spencer and F. Wold, *Anal. Biochem.* **32,** 185 (1969).
[20] A. M. Crestfield, S. Moore, and W. H. Stein, *J. Biol. Chem.* **238,** 622 (1963).

[36] N-Terminal Glyceride-Cysteine Modification of Membrane Penicillinases in Gram-Positive Bacteria

By J. OLIVER LAMPEN *and* JENNIFER B. K. NIELSEN

Several penicillinases (β-lactamase; EC 3.5.2.6) secreted by gram-positive bacteria are also present in the cells as amphipathic membrane proteins that require detergent for solubilization.[1,2] The most thoroughly studied example is the membrane penicillinase of *Bacillus licheniformis* strain 749.[3,4] The secreted enzymes are hydrophilic globular proteins, relatively resistant to protease action and readily renatureable. The amphipathic character of the membrane forms derives from a lipophilic N-terminal segment not present in the exoenzyme. These forms can be processed to yield the exoenzyme but, at least for the 749 penicillinase, the membrane enzyme is not an obligate precursor.[5]

Nature of Lipophilic Modification

Initial purifications yielded preparations containing tightly complexed phospholipids[5a] and this led to the erroneous suggestion that the membrane form was a phospholipoprotein. With improved procedures for purification and SDS–polyacrylamide gel electrophoresis, the membrane enzyme was shown to contain a hydrophilic N-terminal extension of 16 residues. The terminal cysteine carries a diglyceride in thioether linkage and a long-chain fatty acid in amide linkage.[3,4] This structure is identical to the glyceride-cysteine modification on the major outer membrane lipoprotein of *Escherichia coli*.[6] The *E. coli* lipoprotein, a small amphipathic molecule, is a structural component of the membrane, but in the gram-positive bacteria the lipoprotein appears to be part of a membrane protein: secreted protein pair. The lipophilic modification serves as a membrane

[1] J. B. K. Nielsen and J. O. Lampen, *J. Biol. Chem.* **257**, 4490 (1982).
[2] J. B. K. Nielsen and J. O. Lampen, *J. Bacteriol.* **152**, 315 (1982).
[3] J. B. K. Nielsen, M. P. Caulfield, and J. O. Lampen, *Proc. Natl. Acad. Sci. U.S.A.* **78**, 3511 (1981).
[4] J.-S. Lai, M. Sarvas, W. J. Brammar, K. Neugebauer, and H. C. Wu, *Proc. Natl. Acad. Sci. U.S.A.* **78**, 3506 (1981).
[5] L. J. Crane, G. E. Bettinger, and J. O. Lampen, *Biochem. Biophys. Res. Commun.* **50**, 220 (1973).
[5a] S. Yamamoto and J. O. Lampen, *J. Biol. Chem.* **251**, 4102 (1976).
[6] K. Hantke and V. Braun, *Eur. J. Biochem.* **34**, 284 (1973).

METHODS IN ENZYMOLOGY, VOL. 106

anchor while the hydrophilic extension and the major globular portion appear to be free in the aqueous environment of the periplasm.

The structure of the N-terminal segment of the *E. coli* lipoprotein and the stages in its synthesis will not be discussed here. When the gene (*penP*) for *B. licheniformis* 749 penicillinase is cloned in *E. coli*, only the membrane lipoprotein is produced[3,4] (this is often the most convenient source of large amounts of the lipoprotein). Lipophilic modification precedes removal of the signal sequence by a lipoprotein-specific signal peptidase[7,8] and in the presence of the antibiotic globomycin, which inhibits the lipoprotein-specific enzyme, the modified full-length translation product accumulates.[8a] In *B. licheniformis* the surface proteases can cleave the peptide chain beyond the modified Cys residue. As a result, in the presence of globomycin the exocellular forms accumulate[3] instead of the modified translation product. This shift of products in *Bacillus* cultures treated with globomycin is seen with all of the gram-positive glyceride-cysteine penicillinases examined.

The identified lipoproteins from gram-positive bacteria are the membrane penicillinases of *B. licheniformis* strains 749 and 6346,[3,4] the γ-penicillinase of *B. cereus* 569/H (now termed β-lactamase III),[9,10] and the type A and C β-lactamases of *Staphylococcus aureus*.[1] *B. licheniformis* 749 produces substantial quantities of a 63-kilodalton membrane protein which on brief treatment with trypsin yields a hydrophilic 60-kilodalton product that is present as well in the culture fluid.[2] Its production is increased in high phosphate medium,[2] but it has not been identified further. *B. licheniformis, B. subtilis, B. cereus*, and *S. aureus* produce sets of membrane lipoproteins ranging in size between 25 and 65 kilodaltons.[2,11] *B. cereus* is particularly rich in lipoproteins.

Criteria for Identifying Glyceride-Cysteine Lipoproteins

Incorporation of [9,10-³H]Palmitic Acid, L-[³⁵S]Cysteine, and [2-³H] Glycerol into the Membrane Form. Procedures for growing the cultures and carrying out the labeling, preparation of protoplasts, extraction of membranes, and isolation of the products by immunoprecipitation and

[7] M. Hussain, S. Ichihara, and S. Mizushima, *J. Biol. Chem.* **257**, 5177 (1982).
[8] M. Tokunaga, J. M. Loranger, P. B. Wolfe, and H. C. Wu, *J. Biol. Chem.* **257**, 9922 (1982).
[8a] M. Hussain and J. O. Lampen, *FEBS Lett.* **157**, 31 (1983).
[9] A. K. Connolly and S. G. Waley, *Biochemistry* **22**, 4647 (1983).
[10] J. B. K. Nielsen and J. O. Lampen, *Biochemistry* **22**, 4652 (1983).
[11] J. B. K. Nielsen, unpublished data (1982).

electrophoresis on SDS–polyacrylamide gels are presented by Nielsen.[12] The lipoprotein modification includes two ester-linked (O-acyl) and one amide-linked (N-acyl) palmitate residues. The ester bond is cleaved in 0.1 N NaOH at 37° while the amide bond is stable.

Cultures should be labeled with palmitate in a rich medium to minimize randomization of the label by conversion to acetate units and to ensure efficient incorporation into both the ester-linked and the amide-linked residues. In a lean medium incorporation into the amide-linked residue is low. To determine if both types of residues are present, it is convenient to elute a palmitate-labeled band from an SDS gel and expose it to 0.1 N NaOH at 37°. Samples are removed at intervals, neutralized, and rerun on an SDS gel. The percentage of label migrating with the 32-kilodalton band is calculated from a scan of an autoradiograph of the enhanced dried gel. With a typical lipoprotein the two-thirds of the palmitate in ester linkage is removed; the amide-linked material remains.

Formation of Glyceryl Cysteine Sulfone by Oxidation of Gel-Purified [35S]Cysteine-Labeled Protein. Formation of this oxidation product directly demonstrates the presence of a thio ether linkage. The procedure has been previously described.[3] For any palmitate-labeled protein, one can distinguish in this manner the diglyceride thioether type from the O-acyl type reported in eukaryotic cells by Schlesinger.[13]

Expression in E. coli as a Membrane-Bound Lipophilic Form. If the gene for the putative thioether lipoprotein is cloned in *E. coli,* it should be expressed as a membrane-bound lipophilic form. Furthermore, exposure of the cells to globomycin will cause the accumulation of a precursor, usually the full length translation product, carrying the diglyceride modification.[8a] The only known action of globomycin is to inhibit the lipoprotein-specific signal peptidase. Nevertheless, the concurrent derangement of the synthesis of other membrane lipoproteins can produce activation of proteases and the resultant degradation of the nascent precursor. Pretreatment of the cells with a high level of globomycin for a short time appears to provide the most suitable conditions for demonstration of the lipid-modified precursor.

Accessibility to Trypsin on the Surface of Protoplasts. The characterized gram-positive lipoproteins are located on the outer side of the plasma membrane and are readily released from protoplasts by trypsin as soluble proteins. The N-terminal lipid-containing peptide is cleaved off and presumably remains attached to the membrane. Other unidentified lipopro-

[12] J. B. K. Nielsen, this series, Vol 97, p. 153.
[13] M. J. Schlesinger, *Annu. Rev. Biochem.* **50,** 193 (1981).

teins behaved similarly.[2] In the usual procedure, palmitate-labeled protoplasts, prepared as described[3] are suspended in osmotically supported
medium and incubated with and without 40 μg of trypsin/ml for 30 min at
37°. Trypsin action is terminated by the addition of diisopropyl fluoro-
phosphate (1 mM). Membrane extracts are prepared, electrophoresed,
and fluorographed as described.[2] Typical lipoproteins will be absent from
SDS electropherograms of the membranes of the trypsin-treated protoplasts because the globular protein segment is released and the short
chain containing the palmitate label migrates off the gel.

The released proteins are relatively resistant to further proteolysis and
usually can readily be detected. Adjustment of the level of trypsin and the
incubation time may be required in some instances. If desired, protoplasts
labeled with [^{35}S]methionine can be treated in parallel with the palmitate-
labeled sample to simplify detection of the released proteins.

Sequence Required for Lipophilic Modification

The amino acid residues preceding and following the cysteine residue
that undergoes modification are conserved to a considerable extent. The
sequence can be given as

$$\downarrow$$
Lys-Ala-Gly-Cys-a-Ser-Asn

where **a** represents a neutral or nonpolar residue. The three residues
underlined appear to be those most conserved. Cleavage by the lipoprotein-specific peptidase occurs just before the modified Cys ($\downarrow$). If the Cys
is replaced by a Ser residue[14] or if Ala-Lys-Ala-Gly-Cys is removed,[15]
modification does not occur to a significant extent. In the signal sequences
that do undergo lipophilic modification, the Cys is located near the junc-
tion between the hydrophobic stretch of amino acids and the hydrophilic
stretch that precedes the mature enzyme. These segments may orient the
cysteine residue at the surface of the bilayer where it is accessible to the
enzyme transferring the glyceride moiety.[1]

[14] H. C. Wu, M. Tokunaga, S. Hayashi, and C.-Z. Giam, *J. Cell. Biochem.* (in press).
[15] P. S. F. Meźes, W. Wang, Y. Q. Yang, and E. C. Yeh, *J. Biol. Chem.* **258,** 11211 (1983).

[37] Covalent Attachment of Flavin to Flavoproteins: Occurrence, Assay, and Synthesis[1]

By THOMAS P. SINGER and WILLIAM S. MCINTIRE

Occurrence

The first covalently bound flavin to be discovered was the flavin component of mammalian succinate dehydrogenase.[2,3] Proof that its structure is 8α-[N(3)-histidyl]-FAD came 15 years later[3,4] (Fig. 1). At this writing five different structures have been identified in this group of enzymes and chemically synthesized; four are adducts formed with the 8α-methyl group of the flavin ring system, and the fifth is a substituent at C(6) of the aromatic ring of the flavin. Some 25 different enzymes, whose sources range from bacteria to the highest vertebrates, have been shown to contain covalently linked flavin (Table I).[5-40]

[1] The unpublished studies summarized here were supported by the National Institutes of Health (HL-16251), the National Science Foundation (PCM 81-19609), and the Veterans Administration.

[2] E. B. Kearney and T. P. Singer. *Arch. Biochem. Biophys.* **60**, 225 (1955).

[3] T. P. Singer, E. B. Kearney, and V. Massey, *in* "Enzymes: Units of Biological Structure and Function" (O. H. Gaebler, ed.), p. 417. Academic Press, New York, 1956.

[4] W. H. Walker and T. P. Singer, *J. Biol. Chem.* **245**, 4224 (1970).

[5] W. H. Walker, T. P. Singer, S. Ghisla, and P. Hemmerich, *Eur. J. Biochem.* **26**, 279 (1972).

[6] W. C. Kenney, D. E. Edmondson, and R. L. Seng, *J. Biol. Chem.* **251**, 5386 (1976).

[7] D. E. Edmondson, W. C. Kenney, and T. P. Singer, *Biochemistry* **15**, 2937 (1976).

[8] W. C. Kenney and T. P. Singer, *J. Biol. Chem.* **252**, 4767 (1977).

[9] W. C. Kenney, D. E. Edmondson, T. P. Singer, H. Nakagawa, A. Asano, and R. Sato, *Biochem. Biophys. Res. Commun.* **71**, 1194 (1976).

[10] K. Kiuchi, M. Nishikimi, and K. Yagi, *Biochemistry* **21**, 5076 (1982).

[11] W. C. Kenney, D. E. Edmondson, T. P. Singer, M. Nishikimi, E. Nuguchi, and K. Yagi, *FEBS Lett.* **97**, 40 (1979).

[12] W. C. Kenney, T. P. Singer, M. Fukyama, and Y. Miyake, *J. Biol. Chem.* **254**, 4689 (1979).

[13] W. C. Kenney and A. Kröger, *FEBS Lett.* **73**, 239 (1977).

[14] G. Oestreicher, S. Grossman, J. Goldenberg, E. B. Kearney, D. E. Edmondson, T. P. Singer, and J. P. Lambooy, *Comp. Biochem. Physiol.* **67B**, 395 (1980).

[15] H. Möhler, M. Brühmüller, and K. Decker, *Eur. J. Biochem.* **29**, 152 (1972).

[16] D. J. Steenkamp, W. C. Kenney, and T. P. Singer, *J. Biol. Chem.* **253**, 2812 (1978).

[17] J. T. Pinto and W. R. Frisell, *Arch. Biochem. Biophys.* **169**, 483 (1975).

[18] M. Sato, N. Ohishi, and K. Yagi, *Biochem. Int.* **3**, 89 (1981).

[19] A. J. Wittner and C. Wagner, *Proc. Natl. Acad. Sci. U.S.A.* **77**, 4484 (1980).

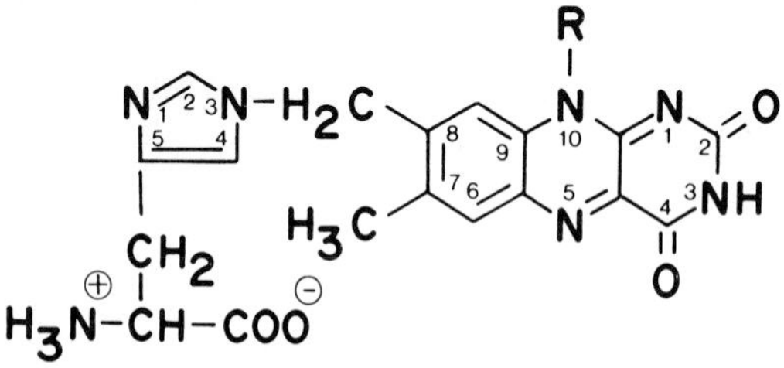

FIG. 1. Structure of 8α-[N(3)-histidyl]-FAD. R denotes the rest of the FAD molecule.

Recognition that an enzyme contains covalently bound flavin has become a relatively simple task. The conventional means is to precipitate and wash the protein with trichloroacetic acid, digest it for a few hours with proteolytic enzymes (usually trypsin and chymotrypsin), and look

[20] A. J. Wittner and C. Wagner, *J. Biol. Chem.* **256,** 4109 (1981).

[21] E. Shinagawa, K. Matsushita, O. Adachi, and M. Ameyama, *Agric. Biol. Chem.* **45,** 1079 (1981).

[22] Unpublished studies from this laboratory.

[23] M. Ohta-Fukuyama, Y. Miyake, S. Emi, and T. Yamano, *J. Biochem. (Tokyo)* **88,** 197 (1980).

[24] N. Onishi and K. Yagi, *Biochem. Biophys. Res. Commun.* **86,** 1084 (1979).

[25] N. Mori, Y. Tahi, H. Yamada, and R. Hayashi, *Agric. Biol. Chem.* **45,** 539 (1981).

[26] J. H. Weiner and P. Dickie, *J. Biol. Chem.* **254,** 8950 (1979).

[27] S. Hayashi, J. Nakamura, and M. Suzuki, *Biochem. Biophys. Res. Commun.* **96,** 924 (1980).

[28] M. Suzuki, *J. Biochem. (Tokyo)* **89,** 599 (1981).

[29] K. Matsushita, E. Shinagawa, O. Adachi, and M. Ameyama, *J. Biochem. (Tokyo)* **85,** 1173 (1981).

[30] J. I. Salach and K. Detmer, *in* "Monoamine, Oxidase: Structure, Function and Altered Function" (T. P. Singer, R. W. von Korff, and D. L. Murphy, eds.), p. 121. Academic Press, New York, 1979.

[31] E. B. Kearney, J. Salach, W. H. Walker, R. L. Seng, W. C. Kenney, E. Zeszotek, and T. P. Singer, *Eur. J. Biochem.* **24,** 321 (1971).

[32] W. H. Walker, E. B. Kearney, R. L. Seng, and T. P. Singer, *Eur. J. Biochem.* **24,** 328 (1971).

[33] J. I. Salach, M. Minamiura, K. T. Yasunobu, and M. B. H. Youdim, *in* "Flavins and Flavoproteins" (T. P. Singer, ed.), p. 605. Elsevier, Amsterdam, 1976.

[34] W. C. Kenney and T. P. Singer, *J. Biol. Chem.* **252,** 4767 (1977).

[35] W. C. Kenney, W. McIntire, and T. Yamanaka, *Biochim. Biophys. Acta* **483,** 467 (1977).

[36] D. J. Steenkamp, *Biochem. Biophys. Res. Commun.* **88,** 244 (1979).

[37] D. J. Steenkamp, W. McIntire, and W. C. Kenney, *J. Biol. Chem.* **253,** 2818 (1978).

[38] S. Ghisla, W. C. Kenney, W. R. Knappe, W. McIntire, and T. P. Singer, *Biochemistry* **19,** 2537 (1980).

[39] K. Matsushita, E. Shinagawa, and M. Ameyama, this series, Vol. 89, p. 187.

[40] M. Ameyama and E. Shinagawa, this series, Vol. 89, p. 141.

TABLE I
ENZYMES CONTAINING COVALENTLY BOUND FLAVIN

Enzyme	Source	Reference
8α-N(1)-Histidyl-FAD		
Thiamin dehydrogenase	Unidentified soil bacterium ATCC 25589	6
β-Cyclopiazonate oxidocyclase	Penicillium cyclopium	7, 8
L-Gulono-γ-lactone oxidase	Mammalian liver	9
	Chicken liver	10
L-Galactonolactone oxidase	Saccharomyces cerevisiae	11
Cholesterol oxidase	Schizophyllum commune	12
8α-N(3)-Histidyl-FAD		
Succinate dehydrogenase	Mammalian heart mitochondria	4, 5
	Vibrio succinogenes	13
	Saccharomyces cerevisiae mitochondria	14
6-Hydroxy-D-nicotine oxidase	Arthrobacter oxidans	15
D-Gluconolactone dehydrogenase	Penicillium cyaneofulvum	16
Sarcosine dehydrogenase	Pseudomonas sp. WRF	17
	Rat liver mitochondria	18–20
Dimethylglycine dehydrogenase	Rat liver mitochondria	18–20
2-Keto-D-gluconate dehydrogenase	Gluconobacter melanogenus	21, 22
Choline oxidase	Alcaligenes sp.	23
	Arthrobacter globiformis	24
	Cylindrocarpon didymum M-1	25
Fumarate reductase	E. coli	26
Sarcosine oxidase[a]	Corynebacterium	27, 28
D-Gluconate dehydrogenase	Pseudomonas aeuginosa	22, 29
8α-S-Cysteinyl-FAD		
Monoamine oxidase A	Human placenta	30
Monoamine oxidase B	Bovine liver mitochondria	31, 32
	Porcine brain mitochondria	33

(continued)

TABLE I (continued)

Enzyme	Source	Reference
Cytochrome c-552	Chromatium strain D	34
Cytochrome c-553	Chlorobium thiosulfatophilum	35
6-S-Cysteinyl-FMN		
Trimethylamine dehydrogenase	Hyphomicrobium X	36
	bacterium W3A1	37, 38
Dimethylamine dehydrogenase	Hyphomicrobium X	36
Proteins with covalently bound flavin of unknown linkage		
D-Gluconate dehydrogenase	Pseudomonas fluorescens	22, 39
	Klebsiella pneumoniae	
	Serratia marcescens	
D-Sorbitol dehydrogenase	Gluconobacter melanogenus	22, 40

a Reported to contain 1 mol each of covalent and noncovalent flavin per mol of enzyme.

for flavin-type fluorescence at acid pH (~pH 3) before or after oxidation with performic acid. A simpler method is to subject the protein to acrylamide gel electrophoresis, which separates the non-covalently bound flavins and look for a yellowish protein band or, preferably, for flavin type fluorescence at acid pH, before and after spraying the gels with performic acid. The latter treatment is necessary to overcome the very intense internal quenching in thio ether-linked flavins and in tyrosyl flavins.

Comparison of the amino acid sequences around the flavin reveals relatively little homology among enzymes containing the same type of flavin structure (Table II,[41–44] underlined sequences), except between enzymes of closely related function (mammalian succinate dehydrogenase and bacterial fumarate reductase). An exception to this is the presence of

[41] W. C. Kenney, D. E. Edmondson, and T. P. Singer, Eur. J. Biochem. 48, 449 (1974).
[42] W. C. Kenney, W. H. Walker, and T. P. Singer, J. Biol. Chem. 247, 4510 (1972).
[43] S. T. Cole, Eur. J. Biochem. 122, 479 (1982).
[43a] M. Brühmüller and K. Decker, Eur. J. Biochem. 37, 256 (1973).
[44] W. C. Kenney, W. McIntire, D. J. Steenkamp, and W. F. Benisek, FEBS Lett. 85, 137 (1978).

a short homologous sequence in bacterial choline oxidase and 6-hydroxy-D-nicotine oxidase.

Qualitative Identification and Assay

Because of the many different types of covalently bound flavin present in biological material, their overlapping properties, and the modification of their chemical properties (e.g., pK values, fluorescence) by interaction with other amino acids in flavin peptides, determination of the type of flavin present is not always a simple task. Figure 2 is intended to present several schemes applicable to all known flavins of this type. One commonly starts out with digestion of the denatured and well-washed protein, which should contain no non-covalently bound flavin. It is essential to purify the flavin peptide first in order to remove materials which interfere in subsequent tests, such as hemes originating from cytochromes. We usually adsorb the flavin on a Florisil column, wash it extensively, and then elute it with 5% (v/v) pyridine. In order to eliminate losses by having the flavin present in several phosphorylated forms, as a result of breakdown during manipulation, it is desirable to convert the crude peptide to the dephosphorylated form at this point, by sequential treatment with nucleotide pyrophosphatase and alkaline phosphatase. The peptide is then purified on thin-layer cellulose plates, using either n-butanol–acetic acid–H_2O (12 : 3 : 5, v/v) or pyridine–n-butanol–acetic acid–H_2O (15 : 10 : 3 : 12, v/v). Advantage may be taken of the different R_f values of flavin peptides in these systems for the monophosphorylated and diphosphorylated forms, so that the procedure may be carried out both before and after digestion with the phosphatases.

While the flavin peptide is seldom pure at this point it may be sufficiently clean for unambiguous identification. For this purpose it is digested with aminopeptidase M to liberate all but the flavin-linked amino acid (which resists digestion by peptidases). The aminoacyl flavin is then further purified by ion-exchange chromatography, HPLC, thin-layer chromatography (TLC) under the conditions mentioned, or high-voltage electrophoresis. These procedure may also identify the type of flavin by comparison with synthetic standards.

Preliminary identification of the flavin is commonly found on the difference in the quantum yield of fluorescence between pH < 3.5 and neutral pH and in the enhancement of fluorescence on oxidation with cold performic acid (Fig. 2). 6-S-Cysteinylflavin is distinguished from other covalently bound flavins by its abnormal absorption spectrum, while 8α-O-tyrosylflavin becomes fluorescent only after reductive cleavage with dithionite and reoxidation by air. The tests outlined in the boxes marked

TABLE II

AMINO ACID SEQUENCES OF FLAVIN PEPTIDES FROM VARIOUS ENZYMES

Flavin	Source	Sequence	Reference
8α-O-Tyrosyl	p-Cresol methylhydroxylase	FAD | Tyr-Asn-Trp-Arg-Gly-Gly-Gly- (Gly,Ser,Met)	22
8α-S-Cysteinyl	Monoamine oxidase	FAD | Ser-Gly-Gly-Cys-Tyr	31
	Cytochrome c-552	FAD | Tyr-Thr-Cys-Tyr	41
	Cytochrome c-553	FAD | Val-Thr-Cys-Pro-Phe-Ser-Asn	35

8α-N(3)-Histidyl	Succinate dehydrogenase	FAD | Ser-His-Thr-Val-Ala-Ala-Glx-Gly-Gly Ile-Asx-Leu-Ala-Ala-Gly-<u>Asx</u>	42
	Fumarate reductase (bacterial)	FAD | Pro-Met-Arg-Ser-His-Thr-Val-Ala-Ala-Glu-Gly-Gly- Ser-Ala-Ala-Val-Ala-Glu-<u>Asp</u>	43
	Choline oxidase	FAD | Asp-Asn-Pro-Asn-(His,Ser,Arg)	23
	6-Hydroxy-D-nicotine oxidase	FAD | Ser-Gly-Gly-Gly- Asn-<u>Asn</u>-Pro-Asp-His-Tyr-Gln-Pro-Ala	43a
6-S-Cysteinyl	Trimethylamine dehydrogenase	FMN | Cys-Ile-Gly-Ala-Gly-Ser-Asp-Lys- Pro-Gly-Phe-Gln	44

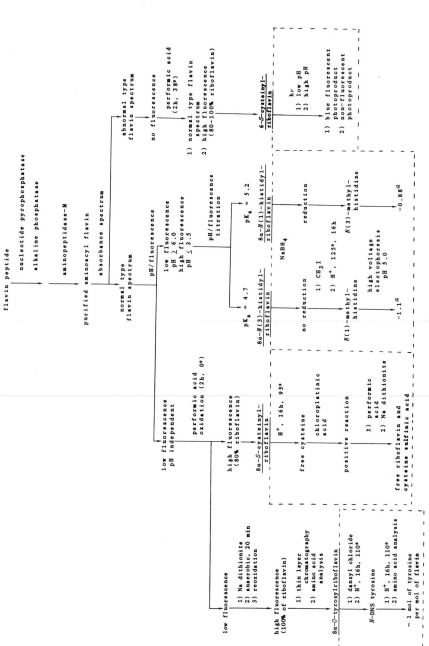

FIG. 2. Flow chart for identification of covalently bound flavins. a, Mobilities relative to FMN.

with dashed lines are additional tests which may be applied to the pure aminoacylflavin, since use of a single criterion may easily lead to erroneous identification. As already noted, for unambiguous identification the unknown aminoacylflavin should be compared, side by side, with synthetic standards with regard to R_f values in TLC systems, mobility on high voltage electrophoresis, or elution profile for HPLC columns.

Qualitative determination of the covalently bound flavin content of a pure protein presupposes knowledge of the structure involved. Provided that no other chromophore is present in the enzyme, which is seldom the case, the easiest test is calculation of the flavin content from the absorbance at 450 nm. With 6-S-cysteinylflavin this method is ruled out because of the anomalous absorption spectrum.

A more sensitive procedure for histidylflavins is based on the difference in fluorescence of the aminoacyl flavin between pH 3.2 and 7. Details of the assay were described by one of us in earlier volumes of this series.[45,46] A rather satisfactory method for the quantitative analysis of the tyrosylflavin is based on the reductive cleavage of the flavin from the *denatured* protein and measurement of its fluorescence.[47] (The flavin is not cleared by dithionite from the native protein.) Assay of 8α-S- and 6-S-cysteinylflavin by fluorescence measurements is more difficult because of the incomplete (60 to 80%) fluorescence yield after performic acid oxidation. For details of this method the original papers quoted in Table I may be consulted.

Synthesis

The chemical synthesis of 8α-substituted flavins is given in previous articles in this series;[46,48] that of 8α-O-tyrosylriboflavin in the paper by McIntire *et al.*,[47] and the synthesis of 6-S-cysteinylriboflavin in the article of Ghisla *et al.*[38]

Biosynthesis

Despite the importance of this group of flavoproteins and major efforts in several laboratories, very little is known about the mechanism by which FMN (riboflavin 5'-phosphate) or FAD is inserted into covalent linkage during the assembly of these enzymes. Even the fundamental question whether the attachment of the flavin to the protein is an enzymatic process

[45] T. P. Singer, J. Salach, P. Hemmerich, and A. Ehrenberg, this series, Vol. 18B, p. 416.
[46] T. P. Singer and D. E. Edmondson, this series, Vol. 66, p. 253.
[47] W. McIntire, D. E. Edmondson, D. J. Hopper, and T. P. Singer, *Biochemistry* **20**, 3068 (1981).
[48] D. E. Edmondson, W. C. Kenney, and T. P. Singer, this series, Vol. 53, p. 449.

or whether the structure of the assembled polypeptide chain predisposes it to facile binding of the flavin in covalent linkage[49] has not been resolved. Studies in Decker's laboratory[50,51] over the past decade on the biosynthesis of 6-hydroxy-D-nicotine oxidase a bacteria have demonstrated that the binding of the FAD is cotranslational; a polypeptide chain of at least 20,000 daltons attached to tRNA must be present before radioactive FAD becomes incorporated in the protein. No apoenzyme was detected in riboflavin-requiring mutants. These findings may either be interpreted to indicate that the critical histidine residue does not appear until the polypeptide chain reaches 20,000 dalton dimensions, after which spontaneous binding of the FAD occurs, or that the specificity of a putative flavinating enzyme which links FAD to the apoprotein requires the amino acid environment created when this chain length appears. No flavin (riboflavin, FMN, or FAD) activated in the 8α position has been found in preparations actively biosynthesizing enzymes which contain covalently bound flavin, in this or other laboratories.

[49] C. Walsh, *Acc. Chem. Res.* **13**, 148 (1981).
[50] H. H. Hamm and K. Decker, *Eur. J. Biochem.* **92**, 449 (1978).
[51] H. H. Hamm and K. Decker, *Eur. J. Biochem.* **104**, 391 (1980).

[38] Diphthamide in Elongation Factor 2: ADP-Ribosylation, Purification, and Properties

By JAMES W. BODLEY, PATRICIA C. DUNLOP, and BRIAN G. VANNESS

Diphthamide is a posttranslational derivative of histidine which occurs in a single location in protein synthesis elongation factor 2 (EF-2).[1] The occurrence of diphthamide was first brought to light through the study of a second post translational modification reaction, the ADP-ribosylation of EF-2 by diphtheria toxin (for reviews, see refs. 2 and 3).

In 1959 diphtheria toxin was found to kill susceptible eukaryotic cells by interfering with the process of protein synthesis.[4] In 1968 the toxin was shown to inactivate EF-2 by catalyzing the transfer of ADP-ribose from the coenzyme NAD^+ to the protein.[5] In 1974 preliminary evidence was

[1] B. G. Van Ness, J. B. Howard, and J. W. Bodley, *J. Biol. Chem.* **255**, 10710 (1980).
[2] A. M. Pappenheimer, Jr., *Annu. Rev. Biochem.* **46**, 69 (1977).
[3] O. Hayaishi and K. Ueda, *Annu. Rev. Biochem.* **46**, 95 (1977).
[4] N. Strauss and E. D. Hendee, *J. Exp. Med.* **127**, 144 (1959).
[5] T. Honjo, Y. Nishizuka, O. Hayaishi, and I. Kato, *J. Biol. Chem.* **243**, 3553 (1968).

obtained that ADP-ribose is attached to EF-2 by the toxin via an unusual amino acid designated amino acid X.[6] Several years later we began a systematic study[7] of this unusual amino acid which we named diphthamide.[8] Here we describe methods for the purification and study of diphthamide. These methods were developed through the study of EF-2 and diphthamide in the yeast *Saccharomyces cerevisiae* but have been applied, with some modifications, to other organisms.

ADP-Ribosylation of EF-2 by Diphtheria Toxin

The purification and study of EF-2 and diphthamide revolve around the specificity and nature of the diphtheria toxin reaction. With the aid of toxin and radioactive NAD^+ (either [*adenine*-2,8-^{3}H]NAD^+ or [^{32}P]NAD^+) it is possible to specifically radiolabel EF-2 either in crude extracts or in purified protein preparations. The reaction is specific in that only a single eukaryotic polypeptide is labeled and there are no toxin-specific ADP-ribose acceptors in eubacterial extracts. The toxin reaction also provides a direct measure of the quantity of EF-2 because the reaction is both irreversible (at neutral pH and in the absence of nicotinamide) and stoichiometric (one equivalent of ADP-ribose is incorporated per single peptide chain of the monomeric protein; M_r, ~100,000).

Several steps must be taken to achieve stoichiometric radiolabeling. First, ribosomes interfere with the reaction and should be removed from crude extracts. This may be accomplished either by centrifugation or protamine sulfate precipitation. Second, endogenous NAD^+ should be removed either by extensive dialysis or more conveniently by charcoal adsorption. Finally, the overall protein concentration in the toxin reaction mixture should be kept low, perhaps to minimize the effects of NAD^+ degrading or binding proteins. Generally, we have observed quantitative ADP-ribosylation in reactions where the total protein concentration is less than 5 mg/ml.

Care must also be taken to avoid the generation of proteolytic artifacts, a particular concern with extracts of *S. cerevisiae* which contain a number of active proteases. This problem can be minimized by conducting the reaction with a relatively high toxin concentration for a short labeling period at low temperature. Proteolytic artefacts can also be introduced during the preparation of samples for sodium dodecyl sulfate gel electrophoresis. We have found that this problem can be avoided by terminating the ADP-ribosylation reaction with the addition of 2 volumes

[6] E. A. Robinson, O. Hendriksen, and E. S. Maxwell, *J. Biol. Chem.* **249,** 5088 (1974).
[7] B. G. Van Ness, J. B. Howard, and J. W. Bodley, *J. Biol. Chem.* **253,** 8687 (1978).
[8] B. G. Van Ness, J. B. Howard, and J. W. Bodley, *J. Biol. Chem.* **255,** 10717 (1980).

of ice cold absolute ethanol. The resulting protein suspension is allowed to stand 5 min on ice and the alcohol is evaporated by heating at 90°. These samples can then be analyzed by standard electrophoretic techniques. Gill and Dinius[9] have noted that potential interference from the formation of poly(ADP-ribose) with mammalian extracts can be minimized by the addition of histamine or 1-methylimidazole to the toxin reaction mixture. We have not found poly(ADP-ribose) synthesis to be significant under our conditions of assay with yeast extracts.

Procedure

The ADP-ribosylation reaction is performed in a volume of 50 μl which contains 50 mM Tris–HCl, pH 7.4, 50 mM dithiothreitol, 25 μg of bovine serum albumin, 2–4 μM [*adenine*-2,8-^{3}H]NAD (New England Nuclear Corp.), 0.3–1.5 μg diphtheria toxin (Connaught Laboratories, Toronto, Canada), and typically 0.1–5 pmol of EF-2. For routine assays the radiolabeled NAD$^+$ is diluted with unlabeled NAD$^+$ to a specific activity 1000 mCi/mmol. The assay solutions are incubated for 15 min at 37° and the reaction is terminated by the addition of 50 μl 20% (w/v) cold trichloroacetic acid. The precipitate is collected on a Millipore filter (type HAWP 0.45 μm), washed with 10% (w/v) trichloroacetic acid, and subjected to liquid scintillation counting either after solubilization of the filter in Aquasol II (New England Nuclear Corp.) made 5% (w/v) with water, or without solubilization using a toluene-based fluor containing 10% (v/v) Beckman Biosolve BBS-3. For rapid stoichiometric labeling, the reaction may be performed at 0° with a 10-fold increase in diphtheria toxin concentration (15 μg/assay), using undiluted [^{3}H]NAD (specific activity 25 Ci/mmol). Under these conditions the reaction is usually complete within 5 min.

To prepare [^{3}H]ADP-ribosyl EF-2 in micromolar amounts the procedure is modified by adding diphtheria toxin ($\sim$2 μg/ml of protein solution), a 2-fold excess of [^{3}H]NAD$^+$ (specific activity, 5–10 Ci/mol) over the amount of ADP-ribose acceptor, and 50 mM dithiothreitol directly to partially purified EF-2. The rate of the reaction is followed by processing small aliquots as described above. When the reaction is complete (within 1 hr), 100% (w/v) trichloroacetic acid is added to a final concentration of 10%. The precipitate is collected by centrifugation, washed with ethanol, ethanol : ether (1 : 1,v/v), and ether, and dried under vacuum at room temperature.

[9] B. M. Gill and L. L. Dinius, *J. Biol. Chem.* **248**, 654 (1973).

Purification of EF-2

The preparation of large quantities of ADP-ribosyl EF-2 is facilitated by the partial purification of the protein prior to its modification by diphtheria toxin. When performed on a small scale (with <100 g wet weight of yeast), we have found that a single chromatography step is sufficient to purify the protein approximately 10- to 15-fold and to remove interfering substances so that EF-2 can be ADP-ribosylated without dilution. At this point the protein may be further purified to near homogeneity by additional chromatographic steps or used without further purification as a starting material for the preparation of pure ADP-ribosyl peptides. When performed on a large scale (with one to three 1 pound bricks of commercial pressed bakers' yeast), we have found that it is usually necessary to repeat the first chromatography step in order to achieve complete ADP-ribosylation. In these large scale preparations we have observed that it is difficult to avoid partial proteolysis of EF-2. The resulting material is suitable for the preparation of the tryptic ADP-ribosylpeptide but not for the preparation of native EF-2. The large scale procedure describe below is suitable for the subsequent preparation of the tryptic ADP-ribosylpeptide.[8] With some modifications it may also be applied to the preparation of the corresponding peptide from bovine liver or wheat germ.[10]

Procedure

Commercial bakers' yeast (3 pounds, Red Star Yeast Company) is suspended in 1.5 l of buffer containing 50 mM Tris–HCl, pH 7.4, 10 mM ethylenediaminetetraacetate, 5 mM 2-mercaptoethanol, and 100 μM phenylmethylsulfonyl fluoride. All steps are conducted at 0–4°. To break the cells, an aliquot of the suspension (~400 ml) is placed in a 2.5 liter glass bottle (such as that used to ship concentrated acids) containing 1 kg of 0.5-mm-diameter glass beads (A. H. Thomas Co.) and the bottle is shaken for a total of ~6 min at approximately 2 cycles/sec. The bottle may be shaken by hand or with the aid of a paint shaker (model 5066, Red Devil Inc.). At the conclusion of each 6-min cycle the broken cell suspension is decanted and an additional aliquot of the whole cell suspension is added to the same bottle. Cell breakage is usually 70–90% complete and may be followed qualitatively by microscopic examination. After the entire suspension has been shaken, the glass beads are rinsed with three 100-ml portions of the original buffer and the rinsing solution is added to the broken cell suspension.

[10] B. A. Brown and J. W. Bodley, *FEBS Lett.* **103,** 253 (1979).

Cell debris and unbroken cells are removed from the suspension by centrifugation at 23,000 g for 30 min. A 2% solution of protamine sulfate (adjusted to pH 7.4) is added to the supernatant from the centrifugation (125 ml protamine sulfate solution/liter of supernatant) and the resulting suspension is centrifuged as before. After centrifugation the fluid is decanted and its conductivity is adjusted to ⩽12 mS by adding solid KCl. The clear solution is then loaded onto a DEAE-Sephadex A-50 (Pharmacia) column (volume ~1.8 l) which has been equilibrated with chromatography buffer (50 mM Tris–HCl, pH 7.4, 0.1 mM ethylenediaminotetraacetate and 5 mM 2-mercaptoethanol) containing 0.14 M KCl. The column is washed with ~3 liters of the same buffer. EF-2 is eluted from the column with chromatography buffer containing 0.35 M KCl and fractions are collected. The fractions containing EF-2 are combined and dialyzed against chromatography buffer until a conductivity of ⩽13 mS is achieved. The chromatography is repeated as above but, after loading, the column is eluted with a 3-liter linear salt gradient (0.14 to 0.40 M KCl) in chromatography buffer. Fractions are collected and those which contain EF-2 are combined and ADP-ribosylated for the subsequent preparation of ADP-ribosyl peptides. Typical preparations yield specific activities of 0.7–1.3 nmol ADP-ribose/mg protein.

Purification of Diphthamide-Containing Peptides

ADP-ribosylation of yeast EF-2 from the preceding preparation, with diphtheria toxin and labeled NAD[+], followed by complete digestion with trypsin yields a single radiolabeled peptide.[7] A single labeled peptide is also obtained from the EF-2 of rat liver[6] and bovine liver and wheat germ[10] and all of these peptides are similar in sequence.

The tryptic peptide containing diphthamide may be purified to homogeniety in two ways. In the first method[6,8] the digest containing the labeled ADP-ribosyl-peptide is chromatographed on DEAE-Sephadex and the elements of ADP are removed by treating the partially purified peptide with a combination of phosphodiesterase and alkaline phosphatase. This modification of the peptide reduces its negative charge so that it is no longer retained by DEAE-Sephadex and rechromatography on this resin yields pure peptide. We have found this method most generally useful and it is described below. Its primary disadvantage is that the radioactive marker is lost. Alternatively, pure peptide can be obtained by chromatography of the partially purified material on a dihydroxyboryl-substituted resin.[7] This procedure takes advantage of the vicinal hydroxyls on the ribose(s) of the modified peptide and in principle is applicable whether or not the elements of ADP are present. The primary disadvantage of this method is that it will not remove contaminants which bind to dihydroxyboryl-substituted resins. We have also found that even small amounts of Tris, carried over from earlier steps, can prevent the retention of ribosyl peptides by these resins.

Procedure

The [³H]ADP-ribosyl EF-2 preparation from the preceding step is denatured by adding trichloroacetic acid to a final concentration of 10% (w/v). This and subsequent steps are conducted at room temperature unless otherwise indicated. The precipitated protein is collected by centrifugation and washed with ethanol, ethanol : ether (1 : 1, v/v), and ether. The precipitate is dried under vacuum and suspended in 10 mM Tris–HCl, pH 8.0, to a concentration of ~100 mg protein/ml. The pH of the suspension is adjusted to 8.0 by the addition of Tris base. N-Tosyl-L-phenylalanine (TPCK)-trypsin (Sigma) is added to a final concentration of 1 mg/ml and the suspension is incubated for 8 hr at 37° with periodic mixing. The trypsin addition and incubation are repeated. The suspension is diluted by the addition of 3 volumes of 10 mM Tris–HCl, pH 8.0, insoluble material is removed by centrifugation and the pH of the resulting solution is adjusted to 7.4 by the addition of HCl.

The tryptic digest is loaded onto a DEAE-Sephadex A-25 (Pharmacia) column (volume 100 ml) which is equilibrated with 10 mM Tris–HCl, pH 7.4, 5 mM NaCl. The column is washed with 2 volumes of this buffer and eluted with a 1 liter linear gradient from 5 to 250 mM NaCl in 10 mM Tris–HCl, pH 7.4. Fractions are collected and analyzed for radioactivity. The [³H]ADP-ribosyl peptide elutes midway through the gradient and these fractions are combined and MgCl₂ is added to a final concentration of 5 mM. Phosphodiesterase (8 units/μmol peptide, Worthington) and alkaline phosphatase (10 units/μmol peptide, Worthington) are added and the solution is incubated for 2 hr at 37°. The ribosyl peptide is desalted on a Sephadex G-15 column equilibrated and eluted with 10 mM Tris–HCl, pH 7.4, and applied to a second DEAE-Sephadex column equilibrated and eluted with 10 mM Tris–HCl, pH 7.4. The ribosyl peptide passes directly through the second DEAE-Sephadex column. The fractions containing the peptide are combined and desalted (to remove Tris) on a Sephadex G-15 column equilibrated and eluted with 10 mM ammonium acetate, pH 8.0.

Purification of Ribosyldiphthamide

Ribosyldiphthamide is obtained from the tryptic ribosyl peptide by sequential enzymatic digestion. Complete hydrolysis of the tryptic peptide with thermolysin yields the sequence apparently common to all

EF-2's,[10] isoleucyl(ribosyl)diphthamoylarginine. The C- and N-terminal residues are removed with carboxypeptidase B and aminopeptidase M, respectively. Ribosyldiphthamide is separated from other digestion products by chromatography on a dihydroxyboryl-substituted resin. If a large quantity of the amino acid is required it is frequently desirable to combine several tryptic peptide preparations prior to this step. In our experience six pounds of yeast yields ∼1 mg of ribosyldiphthamide suitable for structural studies.

Procedure

The tryptic peptide from the preceding step is incubated with thermolysin (25 μg enzyme/μmol peptide, Calbiochem-Behring Corp.) at 37° for 14 hr. The digestion is terminated by placing the tube in a boiling water bath for 10 min and any insoluble material is removed by centrifugation. Carboxy peptidase B (25 units enzyme/μmol peptide, Worthington) is added and the solution is incubated for 12 hr at 37°. The reaction is terminated as above and $MgCl_2$ is added to a final concentration of 15 mM. Aminopeptidase M (2.5 units enzyme/μmol peptide, Boehringer Mannheim Corp.) is added and the solution is incubated for 12 hr at 37°. The aminopeptidase addition and incubation is repeated two times and the reaction is terminated as above.

The digestion products are loaded on to a 0.8 × 12 cm column of dihydroxyboryl-substituted (aminoethyl)cellulose (acetylated) column (BioRad Laboratories) equilibrated with 50 mM ammonium acetate, pH 8.0. The column is washed with approximately 5 volumes of starting buffer and ribosyldiphthamide is eluted with 20 mM acetic acid, pH 3.4.

Criteria of Purity

The preparation of pure ribosyldiphthamide requires a degree of purification, relative to the total quantity of starting material, on the order of one million-fold. Thus the purity of the final product is a significant concern. We attempted unsuccessfully to obtain diphthamide with ADP-ribose, and hence the identifying radiolabel, intact. Apparently the bulky group interfers with the enzymatic cleavage of the adjacent peptide bonds. Once the label is removed the amino acid may be detected by amino acid analysis but, because it only reveals amino acids, this analysis is not an adequate criterion of purity for the preparation of material for structural studies. We have found proton NMR in ^2H_2O to be the most useful analytical method for this purpose. Proton NMR reveals all organic material, is nondestructive, and the solvent can be easily removed. While it is relatively insensitive, the total product produced by the procedure described here is easily sufficient for analysis by proton NMR with most instrumentation.

Ribosyldiphthamide produces 3 characteristic proton NMR resonances which are useful in detecting the compound and assessing its purity during preparation. These are the singlets at $\delta 7.6$ from the single aromatic imidazole proton and $\delta 3.3$ from the 9 identical protons of the trimethylammonio group and the doublet at $\delta 6.3$ from the glycosidic proton. It should be noted that especially the first 2 resonances are sensitive to the pD of the solvent,[1] a variable which can be minimized by acidification with ^2HCl. The spectrum of the pure compound shows a total of $\sim$24 nonexchangeable protons.

In our experience two types of contaminants have been particularly troublesome. First, if inorganic ions have not been completely removed broad resonances are seen. The resolution of the doublet at $\delta 6.3$ is diagnostic of this situation which can usually be corrected by rerunning the final purification step or by desalting the preparation on Sephadex G-10. The second is caused by the incomplete removal of acetic acid and Tris. Acetic acid is seen as a resonance at $\sim$$\delta 2.0$ and can usually be removed by repeated lyophilization from acidic solution. The single resonance from Tris, $\sim$$\delta 3.7$, is pH dependent and can be easily confused with the trimethylammonio resonance at $\delta 3.3$. Tris is difficult to remove because it binds to dihydroxyboryl-substituted resins and can not be readily separated from ribosyldiphthamide by gel permeation chromatography. Tris should be completely removed before digestion of the tryptic peptide and not be further employed in the preparation.

Analysis of Diphthine in Protein Hydrolyzates

The amino acid analyzer can be effectively employed for several purposes in the study of diphthamide and its derivatives.[8] It can, for example, be utilized to quantitatively determine the ribosyldiphthamide liberated during the digestion of the tryptic ribosyl peptide. It can also be utilized to detect the complete acid hydrolysis product of diphthamide, diphthine, in hydrolyzates of EF-2 and its peptides. However, since diphthamide is only one of approximately 1000 residues in pure EF-2, the detection of diphthine solely by the ninhydrin reaction is difficult with hydrolysates of the total protein.[11] Finally, we have developed a radiolabeling technique with which diphthine can be quantitatively determined in hydrolyzates of total cell protein by chromatography on the analyzer.[12] In this procedure,

[11] W. C. Merrick, W. M. Kemper, J. A. Kantor, and W. F. Anderson, *J. Biol. Chem.* **250**, 2620 (1975).

[12] P. C. Dunlop and J. W. Bodley, *J. Biol. Chem.* **258**, 4754 (1983).

described below, cells are grown on [β-³H]histidine and total protein hydrolyzates are chromatographed on a system optimized to resolve diphthine from histidine. The quantity of radioactivity recovered as diphthine (~0.03%) is a measure of the total protein histidine which was post translationally modified in the formation of diphthamide. This procedure has also been successfully applied to the quantitative determination of diphthine in other cell types.[13] Another application of this technique is described in Moehring and Moehring, this volume [39].

Procedure

Yeast[12] or other cell types[13] are grown on [β-³H]histidine (New England Nuclear Corp.) so as to achieve radiolabeling in excess of 10 μCi/mg of total cellular protein. We have found that it is usually necessary to purify the radioactive amino acid on the amino acid analyzer prior to its incorporation. The radioactive cells are washed by centrifugation from unlabeled medium and treated with 10% (w/v) trichloroacetic acid at 100° for 30 min. The protein precipitated by this treatment is washed with water and dried under vacuum. The dried protein ($\leq$ 4 mg containing $\geq$50 μCi) is transfered to a hydrolysis tube using 0.8 ml of 6 N HCl. The tube is sealed under vacuum and heated at 110° for 72–96 hr. The hydrolyzed samples are dried under vacuum and dissolved in 0.25 ml of 0.05 M citric acid just prior to analysis.

Analysis is conducted on a Beckman Model 120C amino acid analyzer equipped with a 9 × 500 mm column containing type AA-15 resin (22 μm particle size). The analyzer effluent should be collected directly at the base of the column into a fraction collector thus bypassing the ninhydrin detection system. The analyzer is eluted at a rate of 1.2 ml/min with the following three-buffer system: first buffer (0.2 N trisodium citrate, 0.1 N NaCl, titrated to pH 4.12 with conc. HCl); second buffer (0.35 N trisodium citrate, conc. HCl to pH 5.26); third buffer (0.2 N trisodium citrate, 0.4 N NaCl, conc. HCl to pH 6.40). The column was equilibrated with the first buffer and eluted with the second and third buffers at 70 and 120 min, respectively. With this system diphthine elutes with the second buffer at 109.5 min completely resolved from histidine which elutes with the third buffer at 157 min.

Remarks

Primarily on the basis of NMR analysis we proposed that diphthamide is 2-[3-carboxyamido-3-(trimethylammonio)propyl]histidine.[1] This struc-

---

[13] A. M. Pappenheimer, Jr., P. C. Dunlop, K. W. Adolph, and J. W. Bodley, *J. Bacteriol* **153**, 1342 (1983).

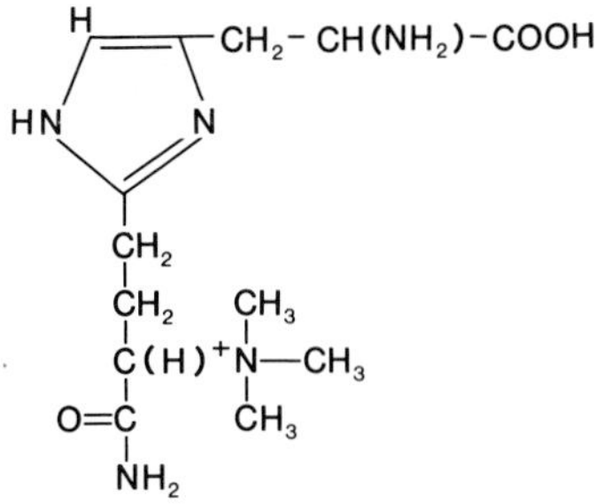

FIG. 1. Structure of diphthamide, 2-[3-carboxyamido-3-(trimethylammonio)propyl] histidine.

tural proposal has since gained support from biosynthetic labeling studies of the amino acid.[12] With growing yeast we found that [β-^{3}H]histidine, [α-^{3}H]methionine, and [*methyl*-^{3}H]methionine are incorporated into the amino acid in a ratio of $1:1:3$ as predicted by the proposed structure. Also as predicted, [^{35}S]methionine is not incorporated into diphthamide. In addition, preliminary accurate mass measurements of the molecular ion of ribosyldiphthamide by fast atom bombardment mass spectrometry conform to the proposed structure.[14] Finally, the proposed structure (Fig. 1) is in accord with genetic[15] and enzymatic (see Chapter 39) studies of Moehring and Moehring and co-workers.

Taken together our findings as well as those of Moehring and Moehring suggest that diphthamide occurs in the single toxin modification site of EF-2 and probably does not occur to any significant extent in other proteins. Moreover it would appear that diphthamide occurs in all EF-2s which are modified by diphtheria toxin, including those from archaebacteria.[13] But despite the unique nature and ubiquitous occurrence of this posttranslational modification in EF-2, its role in the function of the protein remains unknown. It is interesting to note that, in light of the low abundance and particular properties of diphthamide and its hydrolysis product diphthine, this interesting posttranslational modification would have been very difficult to discover without the aid of the posttranslational modification reaction catalyzed by diphtheria toxin.

[14] J. W. Bodley, R. Upham, F. W. Crow, K. A. Tomer, and M. L. Gross, unpublished observations.
[15] J. M. Moehring, T. J. Moehring, and D. E. Danley, *Proc. Natl. Acad. Sci. U.S.A.* **77,** 1010 (1980).

[39] Diphthamide: *In Vitro* Biosynthesis

By J. M. MOEHRING and T. J. MOEHRING

Elongation factor 2 (EF-2), the protein synthesis translocase enzyme, possesses a unique amino acid, a derivative of histidine with the proposed structure 2-[3-carboxyamido-3-(trimethylammonio)propyl]histidine.[1] Diphtheria toxin inactivates EF-2 by catalyzing the covalent attachment of the adenosine diphosphate ribose moiety of NAD^+ to the N-1 nitrogen of the histidine imidazole ring of this amino acid.[1,2] This amino acid has been given the trivial name diphthamide, and its acid hydrolysis product, from which the terminal amino group of the side chain has been cleaved, has been named diphthine.[3]

Diphthamide is synthesized by a complex posttranslational modification of a histidine residue of EF-2 believed to involve three enzymatic reactions.[1,4] We are able to carry out one of these steps, the trimethylation of the amino group of carbon 3 of the side chain, *in vitro*. By selection with diphtheria toxin we have isolated mutants of CHO-K1 Chinese hamster ovary cells possessing altered forms of EF-2 that are resistant to ADP-ribosylation by the toxin.[4,5] These mutants are of two types, based upon their behavior in dominance hybridization studies. In one type the mutation is codominant. Hybrids between wild-type and mutant cells produce both sensitive and resistant EF-2. These cells are mutant in a structural gene which codes for EF-2. We postulate that they are either lacking the histidine residue that would ordinarily be posttranslationally modified to make diphthamide, or that there may be other substitutions in amino acids near the critical histidine that interfere with the modification. In the other type the mutation is recessive. Hybrids between wild-type and mutant cells produce only sensitive EF-2. These mutants possess mutations that affect the posttranslational biosynthesis of diphthamide. We refer to these as modification, or MOD^-, mutants.

Complementation analysis has revealed that there are at least 3 distinct complementation groups of MOD^- mutants that can be isolated from CHO-K1 cells. The possibility that these groups reflect mutations in three

---

[1] B. G. van Ness, J. B. Howard, and J. W. Bodley, *J. Biol. Chem.* **255,** 10710 (1980).

[2] N. J. Oppenheimer and J. W. Bodley, *J. Biol. Chem.* **256,** 8579 (1981).

[3] B. G. van Ness, J. B. Howard, and J. W. Bodley, *J. Biol. Chem.* **255,** 10717 (1980).

[4] J. M. Moehring, T. J. Moehring, and D. E. Danley, *Proc. Natl. Acad. Sci. U.S.A.* **77,** 1010 (1980).

[5] T. J. Moehring, D. E. Danley, and J. M. Moehring, *Somatic Cell Genet.* **5,** 469 (1979).

distinct enzymes required to complete the side chain of diphthamide correlates well with the postulate that the reaction requires three enzymatic steps.[1] We believe that the diphthamide residue in the EF-2 of cells of these 3 complementation groups lacks different portions of the 4-carbon side chain. When exposed to normal modification enzymes in a cell–cell hybrid, the modification of the diphthamide is completed. In cells of two of the complementation groups the conditions of the living cell hybrid have, thus far, been required to complete the modification, but for EF-2 from cells of the remaining group we have been able to establish conditions to complete the modification *in vitro*.[6]

The diphthamide in EF-2 from this group, which we refer to as complementation group 1 (CG-1), apparently lacks the three methyl residues of the trimethylamino group on the side chain. We postulate a mutation in a methyl transferase enzyme required to transfer one or more of these methyl residues from *S*-adenosylmethionine (AdoMet) to the side chain. By *in vitro* modification we can demonstrate the conversion of the EF-2 from a form resistant to diphtheria toxin-catalyzed ADP-ribosylation to a toxin-sensitive form. We can also label the diphthamide residue with radioactive methyl groups from Ado[^{14}C]Met. A crude cell extract of CG-1 cells is prepared to provide the substrate EF-2 containing the incomplete diphthamide residue. To provide the donor methyl transferase enzyme essentially any other mammalian cell will do, as long as it is not CG-1. It is most convenient, however, to use extracts from other mutants that produce only non-ADPribosylatable EF-2 because no allowance for a background amount of native normal EF-2 will have to be made. Any ADP-ribosylatable EF-2 generated in the course of the reaction will be from the *in vitro* modification of CG-1 EF-2. We prefer to use, as donor, extract prepared from a fully resistant structural gene mutant, but extracts from cells of MOD⁻ CG-2 or CG-3 will also suffice.

Preparation of Cell Extract

Materials

Cultured cells, harvested, washed and pelleted. At least 5×10^8 of the CG-1 type and an equal number from another complementation group

Lysing buffer: 10 mM Tris–HCl, pH 7.5; 10 mM KCl; 1.5 mM Mg acetate · $4H_2O$; 6 mM 2-mercaptoethanol

1.25 M sucrose in H_2O

Tight Dounce homogenizer

[6] T. J. Moehring, D. E. Danley, and J. M. Moehring, *Mol. Cell. Biol.* **4,** in press (1984).

Add lysing buffer to cells, 1.6 ml buffer per each 10^8 cells, and suspend the cells, dispersing all clumps. Allow cells to swell at least 10 min in this hypotonic buffer and monitor swelling by observing wet mounts microscopically. Transfer mixture to chilled Dounce homogenizer in an ice bath and grind 10 strokes. Add 1.25 M sucrose to give a final concentration of 0.25 M sucrose (1:5 dilution), and continue homogenizing in the ice bath until all, or nearly all, cells are disrupted, as determined by observation of wet mounts.

Transfer homogenate to appropriate centrifuge tube and centrifuge at 100,000 g for 1 hr. Transfer supernatant to dialysis tubing and dialyze for 24 hr against three changes of complete lysing buffer (containing 0.25 M sucrose), each change 100× the volume of the cell extract. Determine the protein content of the extract by Lowry or equivalent method. We generally obtain 4–5 mg protein/ml. Aliquot the dialyzed extract and store under liquid nitrogen, or at $-70°$.

The Modification Reaction Mixture

Materials

TK buffer: 10 mM Tris–HCl, pH 7.5; 10 mM KCl

TK-Mg^{2+} solution: TK buffer + 4 mM Mg acetate · 4H$_2$O

ES (energy-generating system): 200 mM Tris–HCl, pH 7.5; 560 mM KCl; 36 mM 2-mercaptoethanol; 10 mM ATP; 1 mM GTP; 6 mM CTP; 100 mM creatine-PO$_4$; 1.6 mg creatine phosphokinase/ml (all from Calbiochem)

S-Adenosyl-L-methionine (chloride salt, Sigma), 1 mM in 0.01 M H$_2$SO$_4$, or

S-Adenosyl-L-[methyl-^{14}C]methionine, 50–60 mCi/mmol (Amersham or ICN)

Waterbath at 30°

Reactions mixtures are assembled in tubes in an ice bath and the reaction is run at 30° in a waterbath. Substrate CG-1 extract is included at 1 mg protein/ml and methyltransferase donor extract at from 0.5–1.5 mg protein/ml. If incorporation of labeled methyl groups into diphthamide is to be measured, each extract to be used in the mixture must also be incubated alone in an identical reagent mixture, as all extracts incorporate methyl groups from AdoMet into a number of proteins other than EF-2 in the course of the incubation. This background must be subtracted to determine methyl groups incorporated specifically into diphthamide in the two-extract mixture. Total volume of extracts added may comprise up to 60% of the final volume of the reaction mixture. TK-Mg^{2+}, ES, and 30 mM Tris–HCl are added so that each makes up 10% of the final volume.

AdoMet (^{14}C-labeled or unlabeled, as desired) is added to a final concentration of 0.02 mM. Final volume adjustments are made with TK buffer.

Samples are withdrawn at desired times and precipitated with 10% cold trichloroacetic acid (TCA) immediately, if incorporation of labeled methyl groups is to be followed. If conversion of EF-2 to ADP-ribosylatability is to be determined, samples are stored on ice until the end of the incubation. TCA precipitates are collected on Whatman GF-C filters, extensively washed with 5% TCA, dried, and counted in Econofluor (New England Nuclear) in a scintillation spectrometer. In Fig. 1 a typical incorporation of ^{14}CH$_3$ into combined and single-cell extracts is shown. Specific incorporation of methyl residues into diphthamide in combined extracts occurs at a faster rate, and is complete by 120 min. Incorporation of methyl into unmixed extracts, presumably the result of exchange reactions catalyzed by endogenous methyl transferases, is slower and does not reach a plateau in the incubation periods used. When the radioactively labeled proteins from these reaction mixtures are fractionated by polyacrylamide gel electrophoresis and the labeled protein bands detected by fluorography (Fig. 2) the same pattern of minor labeled bands is seen in mixed or single-cell extracts, but an additional strongly labeled band that comigrates with [^{14}C]ADP-ribose-labeled EF-2 from wild-type cells (M_r = 94,000) is seen in the mixed extracts.

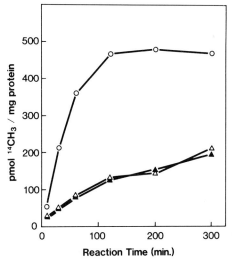

FIG. 1. Incorporation of ^{14}CH$_3$ into mixed and unmixed cell extracts. (○) Mixed extracts from CG-1 and structural gene mutant cells (picomoles ^{14}CH$_3$ per ml CG-1 protein). (△) Extract from CG-1 cells. (▲) Extract from gene mutant cells.

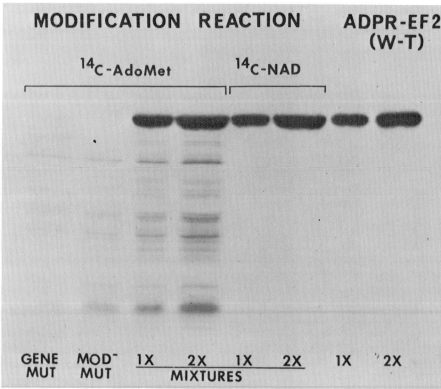

FIG. 2. Comparison of radioactively labeled proteins in cell extracts from modification mixtures and wild-type CHO-K1 cells. Proteins were concentrated in a collodion bag apparatus (Schleicher and Schaell, Keene, NH) and separated by sodium dodecyl sulfate–polyacrylamide slab gel electrophoresis (7.5% polyacrylamide) [U. K. Laemmli, *Nature (London)* **227,** 680 (1970)]. Radioactive proteins were detected by fluorography [J. P. Chamberlain, *Anal. Biochem.* **98,** 132 (1979)]. The dried gel was exposed to Kodak X-Omat AR-5 film for 7 days at $-70°$. From the left, extracts run in the first 4 lanes were incubated in the modification reaction mixture with Ado[^{14}C]Met. Extracts in the next two lanes were modified with unlabeled AdoMet and then ADP-ribosylated with [^{14}C]NAD. The last two lanes are extracts of wild-type CHO-K1 cells ADP-ribosylated with [^{14}C]NAD. Total protein run per lane was 90 μg for the gene mutant and MOD⁻ mutant alone, 175 μg for the 1× mixtures, 350 μg for the 2× mixtures, and the wild-type extracts 80 and 160 μg, respectively.

ADP-ribosylation of in Vitro Modified EF-2

Materials

Modified extract samples from previous step, containing 1 mg CG-1 protein/ml

EF-2 buffer: 20 mM Tris–HCl, pH 7.5; 0.1 mM EDTA; 1 mM
   dithiothreitol
Reaction mix: 50 mM Tris–HCl, pH 8.2; 12.5 mM dithiothreitol; 7.5
   mM Mg acetate · 4H$_2$O; 250 μM CaCl$_2$; 50 mM thymidine; 4.2 μM
   nicotinamide[U-^{14}C]adenine dinucleotide (100–250 mCi/mmol,
   Amersham)
Diphtheria toxin, 1000 μg/ml in EF-2 buffer (nicked) (obtainable from
   Connaught Medical Research Laboratories, Toronto, Ontario,
   Canada)
Waterbath at 25°

Reaction mixtures are assembled in an ice bath. The reaction mix
comprises 40% of the final mixture; extract samples comprise 20%. Diphtheria toxin is added to 200 μg/ml. Final volume is adjusted with EF-2
buffer.

The samples are incubated for 20 min at 25°. They are then precipitated with 10% cold TCA and subsequently collected and counted as were
the ^{14}CH$_3$ samples from the previous step. Conversion of CG-1 EF-2 to
ADP-ribosylatability proceeds more slowly than does methylation, and is
usually not complete in less then 6 hr (see Fig. 3). This implies that there
is a modification step, in addition to formation of the trimethylamino
group, that is necessary for the conversion of CG-1 EF-2 to an ADP-
ribosylatable state. We believe this may involve the addition of the termi-

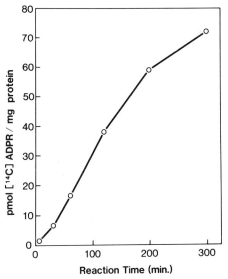

F$_{IG}$. 3. Diphtheria toxin-catalyzed ADP-ribosylation of *in vitro* modified CG-1 EF-2.

MODIFIED PROGRAM FOR AMINO ACID ANALYSIS OF
PROTEIN HYDROLYZATE

Buffer	pH	Na citrate (N)	NaCl (N)	Buffer change time[a] (min)
A	4.12	0.2	0.1	60
B	5.00	0.38	0.02	100
C	6.40	0.2	0.8	—

[a] Column equilibrated with buffer A; buffer change times measured from time of sample injection.

nal amino group of the diphthamide side chain.[6] Analysis of ADP-ribosylated extract mixtures by polyacrylamide gel electrophoresis shows that the same 94,000 molecular weight protein is labeled as was labeled by the $^{14}CH_3$ residues (see Fig. 2). Double-label experiments were carried out,

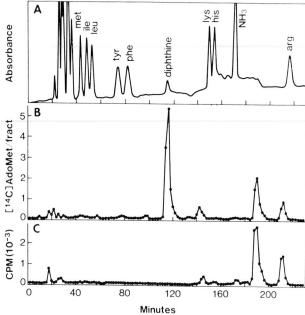

Fig. 4. Demonstration of $^{14}CH_3$-labeled diphthamide by amino acid analysis of protein hydrolyzate. (A) Ninhydrin elution profile, showing position of yeast diphthamide. (B) Radioactive profile of hydrolysate of mixed, *in vitro* modified extracts from CG-1 and gene mutant cells, reacted with Ado[^{14}C]Met. (C) Radioactive profile of hydrolyzate of extracts from CG-1 and gene mutant cells incubated separately in reaction mixtures containing Ado[^{14}C]Met, then mixed immediately prior to hydrolysis.

using Ado[^{3}H]Met in the modification reaction and [^{14}C]NAD in the ADP-ribosylation reaction. When these mixtures were fractionated by gel electrophoresis and the labeled band cut out and counted, the molar ratio of ^{3}H to ^{14}C was 3.1 : 1, which correlates with the fact that diphthamide has 3 methyl residues and binds 1 molecule of ADP-ribose.[1-3]

Demonstration of $^{14}CH_3$-Labeled Diphthamide by Amino Acid Analysis

Analysis of the amino acids from the modification mixture demonstrates that diphthamide is the primary amino acid to receive the $^{14}CH_3$ label. Modification mixtures containing 2 mg of CG-1 protein, either alone or mixed with 2 mg of structural gene mutant protein, were incubated for 180 min and precipitated with 10% TCA. Precipitates were centrifuged at 9000 g and washed 2× with distilled water. Precipitates were transferred to siliconized hydrolysis vials in 1 ml of 6 N HCl, sealed under vacuum, and hydrolyzed at 110° for 60 hr. Samples were evaporated to dryness on a vacuum centrifuge (Savant), rehydrated in H_2O, and raised to pH 11 with 5 N KOH to remove ammonia, dried again, then rehydrated and reduced to pH 1 with 6 N HCl. A final drying preceded rehydration in 0.2 N sodium citrate buffer, pH 2.2. Samples were analyzed on a Beckman Model 121 amino acid analyzer using a single 50-cm-long column (AA-15 resin) and a modified three-buffer methodology (see the table). (We are indebted to J. W. Bodley for making this buffer program available to us, and for kindly providing the sample of authentic yeast diphthamide with which we verified the position of our methyl-labeled diphthamide.) Analyzer effluent was collected and $^{14}CH_3$ counts were determined by scintillation counting in Aquasol 2 (New England Nuclear). The ninhydrin and radioactive profiles are shown in Fig. 4.

Section VI

Protein Glycosylations; ADP-Ribosylation

[40] ADP-Ribosylarginine

By NORMAN J. OPPENHEIMER

A growing number of enzymes and toxins have been found that utilize NAD^+ as a donor for the ADP-ribosylation of proteins where arginine serves as the putative acceptor.[1,2] To date these reactions represent unique examples for the glycosidation of proteins through formation of a glycosyl-guanidine linkage. This chapter will cover the chemical properties of glycosylguanidines and present the methodology for their enzymatic synthesis. The first section will focus on the acceptor specificity of guanidine-dependent ADP-ribosyltransferases and the chemistry of glycosylguanidines. The second section will review the literature regarding chemical synthesis and present procedures used in this laboratory for the enzymatic preparation of ADP-ribosylguanidine derivatives. An important aspect of this chapter is to point out the near total absence of modern analytical data on this important class of sugar derivatives.

Chemical and Enzymatic Properties

As shown in Scheme 1, the arginine-dependent ADP-ribosyltransferases catalyze two reactions: reaction I, the slow hydrolysis of NAD^+ to

$$NAD^+ + H_2O \longrightarrow ADP ribose + Nic \qquad I$$

$$NAD^+ + ARG \longrightarrow ADP ribosyl\text{-}ARG + Nic \qquad II$$

SCHEME 1

ADP-ribose and nicotinamide; and reaction II, the ADP-ribosylation of guanidino compounds.[3,4] The hydrolytic reaction is specific for water, and there is no evidence for methanolysis even in the presence of 7 *M* methanol.[5] This is in contrast to the facile methanolysis catalyzed by NAD^+ nucleosidases with transglycosidation activity.[6,7]

---

[1] P. H. Pekala and B. M. Anderson, *in* "The Pyridine Nucleotide Coenzymes" (J. Everse, B. M. Anderson, and K.-S. You, eds.), Chapter 10. Academic Press, New York, 1982.

[2] J. Moss and M. Vaughan, *Annu. Rev. Biochem.* **48,** 581 (1979).

[3] J. Moss, V. C. Manganiello, and M. Vaughan, *Proc. Natl. Acad. Sci. U.S.A.* **73,** 4424 (1976).

[4] J. Moss and M. Vaughan, *J. Biol. Chem.* **252,** 2455 (1977).

[5] N. J. Oppenheimer, *J. Biol. Chem.* **253,** 4907 (1978).

[6] M. Pascal and F. Schuber, *FEBS Lett.* **66,** 107 (1976).

[7] N. J. Oppenheimer *FEBS Lett.* **94,** 368 (1978).

The acceptor specificity for a number of guanidine-dependent ADP-ribosyltransferases has been studied by Moss and co-workers.[8,9] These studies demonstrate that the *in vitro* reactions catalyzed by the transferases are relatively nonselective. The general sequence of reactivity, agmatine > arginine methyl ester > arginine > guanidinopropionic acid indicates that the presence of a negative charge decreases to some extent the ability of the guanidino group to serve as an acceptor. No detailed kinetic data are available; thus it is not known whether this effect represents an alteration in K_m or V_{max}. Furthermore guanidine itself can also act as an acceptor.[9] Therefore the specificity shown *in vivo* by the toxins for the ADP-ribosylation of selected proteins probably reflects a recognition of larger aspects of the acceptor's tertiary structure.

The stereochemistry of the reaction has been determined by ^{1}H NMR for a number of these enzymes.[5,8–10] In all cases the formation of the ribosylguanidine linkage occurs with inversion at the ribose C-1' and yields an α-configuration as shown in Scheme 2. The ribosylguanidine

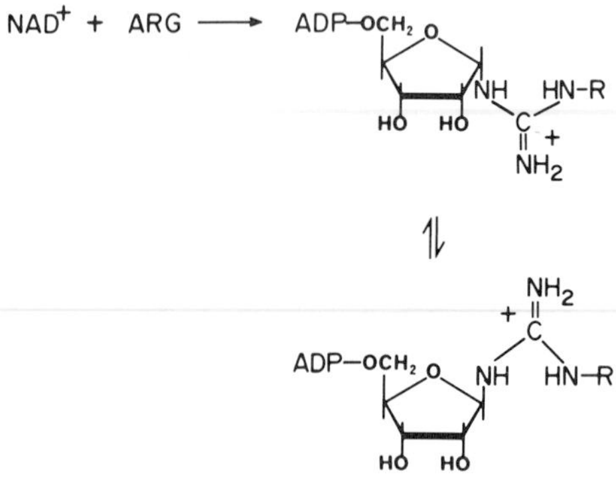

SCHEME 2

linkage is configurationally unstable and has been shown to mutarotate to a 1 : 1 mixture of α- and β-anomers.[5] The $t_{1/2}$ for anomerization under "physiological" conditions (37°, pH 7.4) is estimated to be between 3 and 6 hr, however, definitive studies, e.g., pH profiles, testing for general acid catalysis, etc., have not been conducted. Mutarotation can be readily

[8] J. Moss, S. Garrison, N. J. Oppenheimer, and S. H. Richardson, *J. Biol. Chem.* **254,** 6270 (1979).

[9] J. Moss, S. Garrison, N. J. Oppenheimer, and S. H. Richardson, *in* "Symposium on Cholera, Karatsu, 1978" (K. Takeya and Y. Zinnaka, eds.), p. 274. Fuji Printing Co., Tokyo, 1978.

[10] J. Moss, S. J. Stanley, and N. J. Oppenheimer, *J. Biol. Chem.* **254,** 8891 (1979).

monitored by ^{1}H NMR since the anomeric proton resonances are well resolved, α-C-1'H (5.336 ppm) and β-C-1'H (5.157 ppm).

Note that if the ADP-ribosylation of arginine represents a reversible mechanism for regulation of cellular functions, then the anomerization of the ribosyl-guanidine linkage could have a profound influence on their enzymatic degradation, i.e., this would require either a multiplicity of hydrolytic enzymes or a nonselective hydrolase to remove the ADP-ribose moiety. Evidence is thus far lacking, however, for either the enzymatic hydrolysis of the ribosyl arginine bond or the direct reversibility of ADP-ribosylation. These topics will clearly require further investigation.

The chemical stability of glycosylguanidines is reportedly similar to that of uredyl and thiouredyl glycosides.[11] They are quite stable to both acid and base and are resistant to attack by nucleophiles.[12] For example, hydroxylamine or dilute base causes only a slow release of the ADP-ribose moiety from cholera toxin-treated proteins whereas ADP-ribose hydroxamate is rapidly generated when glutamate serves as the acceptor.[13]

Synthesis of Glycosylguanidines

Chemical Synthesis. Glycosylguanidines are reportedly prepared by the direct reaction of guanidine with reducing sugars although the microcrystalline products have an empirical formula indicating a 3 : 2 ratio of sugar to guanidine.[14] Subsequent work cautions that the strongly basic nature of guanidine causes alkaline degradation of sugars analogous to that of sodium or potassium hydroxide.[15] The only characterized chemical synthesis is shown in Scheme 3. In this method the guanidine group is generated by attack of a primary amine on an isothiourea intermedi-

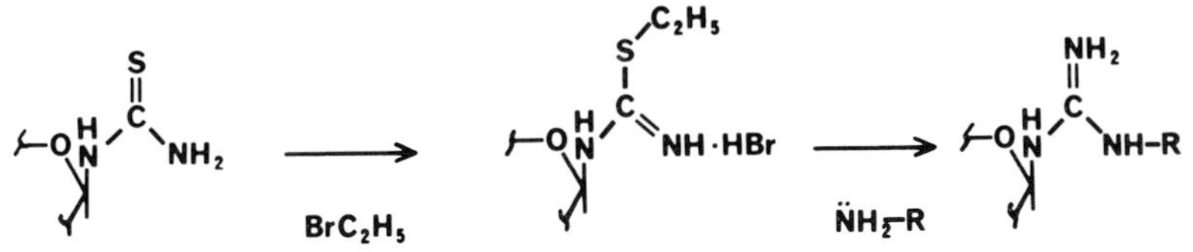

SCHEME 3

[11] H. Paulsen and K.-W. Pflughaupt, *in* "The Carbohydrates" (W. Pigman and D. Horton, eds.), p. 898. Academic Press, New York, 1980.

[12] C. G. Goff, *J. Biol. Chem.* **249**, 6181 (1974).

[13] O. Hayaishi and K. Ueda, *Annu. Rev. Biochem.* **46**, 95 (1977).

[14] R. S. Morrell and A. E. Bellars, *J. Chem. Soc.* **91**, 1010 (1907).

[15] E. J. Witzemann, *J. Am. Chem. Soc.* **46**, 790 (1924).

ate.[16,17] It is interesting to note that the direct reaction of glycosylamines with cyanamide, which should generate a glycosylguanidine, instead leads to formation of oxazolines.[18]

The reaction of proteins with hexose-isothiourea derivatives has been reported[17] although the conjugates have not been characterized either chemically or biologically. Based on these studies the potential exists for the preparation of ADP-ribosylisothioureas that could serve as chemical ADP-ribosylating agents, forming ADP-ribosylguanidino linkages to proteins through reaction with either the terminal amine or lysine residues.

Enzymatic Synthesis of ADP-ribosylguanidines[5]

The following is a generalized procedure for the preparation of micromolar amounts of ADP-ribosylguanidines. Adjustments in the incubation time will have to be made depending upon the specific guanidine acceptor being used. Choleragen (the intact toxin), 200 μg (Schwarz/Mann) is incubated at 37° in 2 ml of a reaction mixture consisting of 0.4 M potassium phosphate (pH 7.2), 80 mM guanidino derivative (note that there are few data regarding the K_m for guanidino compounds; therefore this value may have to be obtained for the most efficient utilization of the acceptor), 20 mM dithiothreitol (required to dissociate the enzymatically active A subunit of the toxin), and 6 mM NAD$^+$ (12 μmol). The reaction is monitored by measuring the decrease in the NAD$^+$ concentration using the convenient cyanide assay.[19] Aliquots are taken from the reaction mixture with time and added to 1 M KCN in 0.1 M potassium carbonate, pH 10. The amount of NAD-CN adduct formed which is obtained by measuring the UV absorption at 325 nm for the cyanide adduct (λ_{max} = 325 nm, ε = 5900) allows a direct quantitation of the remaining NAD$^+$. Using the incubation conditions described above for choleragen, less than 5% NAD$^+$ will remain after about 12 hr for guanidine hydrochloride and about 6 hr for L-arginine. Purification can be conducted either by ion-exchange chromatography on an anion-exchange column such as DEAE-cellulose or DEAE-Sephadex or by preparative C-18 reverse-phase HPLC. The optimal concentration of buffer and the optional use of 1–3% methanol will depend upon the guanidine derivative being used and must be determined experimentally. Under some conditions and for some derivatives it is possible to separate the α- and β-forms by HPLC, e.g., ADP-ribosyl-

[16] F. Micheel, W. Berlenbach, and K. Weichbrodt, *Chem. Ber.* **85**, 189 (1952).

[17] F. Micheel and A. Heesing, *Justus Leibigs Ann. Chem.* **604**, 34 (1957).

[18] R. M. Davidson, S. A. Margolis, E. White V, B. Coxon, and N. J. Oppenheimer, *Carbohydr. Res.* **111**, C16 (1983).

[19] S. P. Colowick, N. O. Kaplan, and M. M. Ciotti, *J. Biol. Chem.* **191**, 447 (1951).

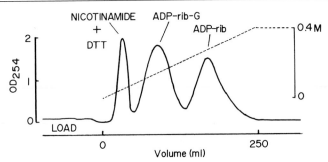

FIG. 1. Elution profile of ADP-ribosylguanidine from DEAE-cellulose (HCO₃⁻). The 2 ml reaction mixture is diluted to 50 ml with water and washed onto the column (1.5 by 15 cm) and a linear 0–0.4 M ammonium bicarbonate gradient (total volume 250 ml) is applied.

guanidine separates easily, whereas no separation is observed for the L-arginine derivative.

For ion-exchange chromatography to be effective the ADP-ribosylated acceptor must have a net charge of either -1 or >-2 (e.g., ADP-ribose has a net charge of -2). The elution profile for ADP-ribosylguanidine from DEAE-cellulose is shown in Fig. 1. Alternatively HPLC or cation exchange chromatography (if the net charge is $+1$ or greater) should be employed.

The pure α-anomer can be obtained by conducting the reaction as quickly as possible, e.g., using increased enzyme concentration, and by purifying the product at as high a pH as feasible (e.g., pH 8–10) and at 4°.

[41] Use of Guanylhydrazones as Substrates for Guanidine-Specific Mono-ADP-Ribosyltransferases

By GOPALAN SOMAN, JOYCE F. MILLER, and DONALD J. GRAVES

ADP-ribosyltransferases have been described that catalyze the mono-ADP-ribosylation of proteins and guanidino-containing compounds.[1] Because the enzyme is a part of certain bacterial toxins [e.g., choleragen[2] and a heat-labile enterotoxin from *Escherichia coli*[3]] and normal function-

[1] J. Moss and M. Vaughan, *Ann. Rev. Biochem.* **48**, 581 (1979).
[2] J. Moss and M. Vaughan, *J. Biol. Chem.* **252**, 2455 (1977).
[3] J. Moss and S. H. Richardson, *J. Clin. Invest.* **62**, 281 (1978).

ing cells and tissues,[4–8] there is considerable interest in evaluating factors that influence its activity and in defining what are the natural substrates for the enzyme. We describe herein new and easy procedures for measuring ADP-ribosyltransferase activity. No radioisotopes are required, and the assay can be done by following the disappearance of p-nitrobenzylidine aminoguanidine (NBAG) after ADP-ribosylation by high-performance liquid chromatography (HPLC) or by following product formation by spectrophotometry.[9] NBAG can be utilized to measure ADP-ribosyltransferase activity with purified enzymes or in crude systems, e.g., homogenates of turkey erythrocytes and rabbit skeletal muscle.

Materials and Methods

We synthesized p-nitrobenzylidine aminoguanidine by a general procedure for the synthesis of guanylhydrazones.[10] Aminoguanidine bicarbonate (6.8 g) in water was acidified to pH 5.6 with 1 N HCl. The solution (100 ml) was warmed to 60°C, and an equal molar amount of p-nitrobenzaldehyde (7.8 g) was added in portions with stirring. Ethanol was added to keep the p-nitrobenzaldehyde in solution. The final concentration of ethanol was 50%. The solution was stirred at room temperature for 16 hr. The solid was collected by the use of a rotary evaporator and recrystallized from ethanol. Alternatively, p-nitrobenzylidine aminoguanidine can be made by the procedure of Nishimura et al.[11] β-NAD, dithioerythritol (DTE), and choleragen were purchased from Sigma. Methylglyoxal bis-(guanylhydrazone) dihydrochloride monohydrate (MGBG) was obtained from Aldrich Chemical Co. The chemical structures of NBAG and MGBG are shown in Fig. 1 Cholera toxin was activated by incubation with DTE.[12] ADP-ribosyltransferase was partially purified from turkey erythrocytes according to Moss and Stanley.[5] The eluate from the phenyl-Sepharose chromatographic step was used as the source of enzyme for

[4] J. Moss and M. Vaughan, *Proc. Natl. Acad. Sci. U.S.A.* **75**, 3621 (1978).
[5] J. Moss and S. J. Stanley, *J. Biol. Chem.* **256**, 7830 (1981).
[6] M. J. S. Dewolf, P. Vitti, F. S. Ambesi-Impiombato, and L. D. Kohn, *J. Biol. Chem.* **256**, 12287 (1981).
[7] M. R. Hammerman, V. A. Hansen, and J. J. Morrissey, *J. Biol. Chem.* **257**, 12380 (1982).
[8] S. Filetti and B. Rapport, *J. Clin. Invest.* **68**, 461 (1981).
[9] G. Soman, K. B. Tomer, and D. J. Graves, *Anal. Biochem.* **134**, 101 (1983).
[10] F. Baiocchi, C. C. Cheng, W. J. Haggerty, Jr., L. R. Lewis, T. K. Liao, W. H. Nyberg, D. E. O'Brien, and E. G. Podrebarrag, *J. Med. Chem.* **6**, 431 (1963).
[11] T. Nishimura, C. Yamagaki, H. Toku, S. Yoshii, K. Hasegawa, and M. Saito, *Chem. Pharm. Bull.* **22**, 2444 (1974).
[12] P. A. Watkins, J. Moss, and M. Vaughan, *J. Biol. Chem.* **255**, 3959 (1980).

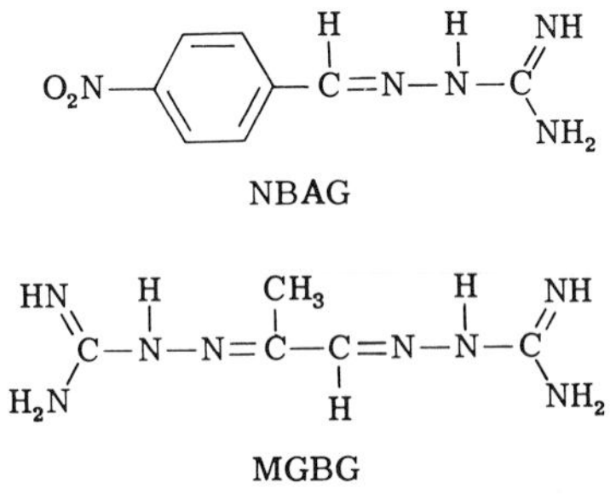

FIG. 1. Structure of guanylhydrazones.

activity measurements. Rabbit skeletal muscle (100 g) was ground in a meat grinder and homogenized in a Waring blender for 1 min in 250 ml of 50 mM phosphate buffer, pH 7.0.

Assay of ADP-ribosyltransferase by HPLC

The reaction system is prepared by adding 40 μl of 25 mM NAD, 10 μl of 160 mM DTE, 20 μl of activated choleragen (0.5 mg/ml), 10 μl of 4 mM p-nitropbenzylidine aminoguanidine, 10 μl of 0.2 M Tris-acetate, pH 7.4 unless otherwise stated, and 10 μl of H$_2$O to make a final volume of 100 μl. Higher concentrations of NBAG (final concentrations up to 2.0 mM) can be used with no inhibitory effects. After 30–60 min of incubation at 30°, the reaction is stopped by adding 100 μl of 10% trichloroacetate (TCA). The TCA filtrate is analyzed on a HPLC component system consisting of a Beckman Model 100 A pump, Beckman Model 210 injector, Hitachi (40–100) variable-wavelength spectrophotometer, and a Beckman recorder. The column used is ultrasphere ODS (25 cm × 4.5 mm i.d.) 5 μm particle size. Chromatography is done isocratically, and the running solvent is 30% methanol in 0.2 M triethylammonium acetate at pH 5.5. A flow rate of 1.25 ml/min with a pressure of 3300 psi is used, and the absorbance of the elutate is recorded at 315 nm. Figure 2 represents a typical HPLC elution pattern of the reaction mixture. ADP-ribosyltransferase activity can be quantitated by measuring the decrease of the area of the NBAG peak (A) or by the area of the product peaks (B and C). Soman et al.[9] found that the calculated area of the elution peak of NBAG correlates with the amount of NBAG injected. The decrease of the NBAG peak is nearly linear with time over an extended period of reaction, up to 90 min and the reaction is proportional to enzyme concentration (30 to 100 μg/ml) of choleragen.

The products of the reaction, represented by peaks B and C of Fig. 2, were isolated by preparative HPLC on a 25 cm × 10 mm i.d. column and

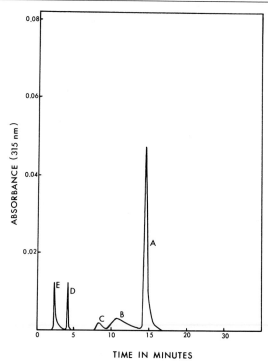

TIME IN MINUTES

FIG. 2. High-performance liquid chromatography of a sample of the cholera toxin reaction system with NBAG. The reaction system consisting of 10 mM NAD, 36 mM DTE, 0.4 mM NBAG, 50 μg/ml cholera toxin in 25 mM Tris-acetate, pH 7.4, was incubated for 30 min at 30° and analyzed as described under Materials and Methods; 10–20 μl of TCA filtrate was injected into the column. (A) NBAG, (B) and (C) ADP-ribosylated products, (D) DTE, and (E) NAD and nicotinamide.

characterized. Figure 3B gives the absorption spectrum of peak B of Fig. 2 in comparison with the starting material (Fig. 3A). Peak C has the identical spectral characteristics of peak B. The absorption spectrum of peaks B and C shows an intense absorption maximum due to the NBAG component and a new absorption at 259 nm corresponding to the ADP-ribose component. The spectrum of the product is equivalent to a 1:1 mixture of NBAG and ADP-ribose. On the basis of this result along with information derived from fast atom-bombardment mass spectrometry, we concluded that the product is mono-ADP-ribosylated NBAG.[9] The two products were found to interconvert[9] and likely are anomeric forms similar to those described earlier for the ADP-ribosylation of L-arginine.[13]

[13] N. J. Oppenheimer, *J. Biol. Chem.* **253**, 4407 (1978).

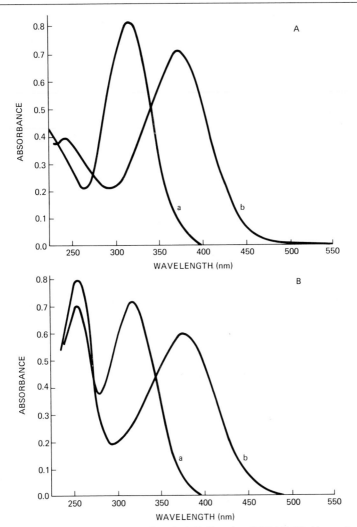

FIG. 3. Absorption spectra of NBAG and ADP-ribosylated NBAG (A) Absorption spectra of NBAG (a) in 0.1 N HCl, (b) in 0.1 N NaOH. (B) Absorption spectra of ADP-ribosylated NBAG peak B of Fig. 1 (a) in 0.1 N HCl, (b) in 0.1 N NaOH. The ADP-ribosylated products were collected, and spectra recorded.

Assay of ADP-ribosyltransferase Activity by Spectrophotometry

One hundred microliters of activated choleragen (50 μg) is incubated with 50 μl of 0.4 M Tris-acetate, pH 7.4, 50 μl H_2O, and 200 μl of 160 mM DTE for 30 min at 30°. H_2O (200 μl) and 400 μl of 25 mM NAD are added, and the reaction is initiated by adding 20 μl of 20 mM NBAG preincubated for 10 min with 36 mM DTE. After some initial unsteadiness (2–3 min), a steady linear phase is achieved.[9]

The reference cuvette contained everything except cholera toxin. Temperature equilibration is attained by pumping liquid from a thermostated bath to the reaction chamber. Absorbance increase at 370 nm is monitored continuously. Quantitation of the spectral increase to the amount of NBAG reacted is attained by analyzing the reaction system at different intervals by HPLC. The measurement of product formation at 370 nm is based upon the fact that the observed products, peaks B and C in the HPLC system, have a lower pK_a value (pK_a 7.5) compared with the substrate, which has a pK_a of 8.2.[9] Because of this and the fact that the deprotonated forms have a higher absorbance at 370 nm, product formation measured at pH 7.4 causes partial deprotonation of the product and, hence, an increase in absorbance. The time course of the reaction is linear with time and choleragen concentration.[9] The absorbance increase at 370 nm can be correlated with the amount of NBAG reacted by analyzing the reaction system at intervals by HPLC. The increase in molar absorptivity accompanying ADP-ribosylation of NBAG was calculated to be 2×10^3. Under the reaction conditions specified, a difference of 0.1 absorbance units at 370 nm corresponds to the formation of 5×10^{-2} μmol of product per ml of reaction. With 0.40 mM substrate, full scale deflection (0.1 units) represents 12.5% conversion of substrate to product. Thus, it seems, that the assay can be used effectively to measure less than 10% conversion of substrate to product.

Applications

Both assay procedures described can be utilized to evaluate kinetic parameters of the ADP-ribosylation reaction with NBAG. A K_m value of 1.1 mM was reported earlier for ADP-ribosylation.[9] We reported earlier that ADP-ribosylation of NBAG can be detected in homogenates of turkey erythrocytes and in partially purified preparations by utilizing the HPLC or spectrophotometric procedure.[9] Because of the turbidity of the homogenates, the amount of protein needs to be limited in the spectrophotometric assay, making it somewhat less useful than the HPLC system for evaluating ADP-ribosyltransferase activity in crude systems. We found that incubation of 500 μl of rabbit skeletal muscle homogenate with 50 μl of 25 mM NAD, 100 μl of 10 mM NBAG, 250 μl of 0.1 M phosphate, pH 7.0, and 100 μl of H_2O at 30° led to the formation of mono-ADP-ribosylated NBAG. A rate of 90 nmol of product/hr/g of muscle was estimated from the product peaks found by HPLC.[9] Because the assay is sensitive (4% product formation was detected), it seems that this system

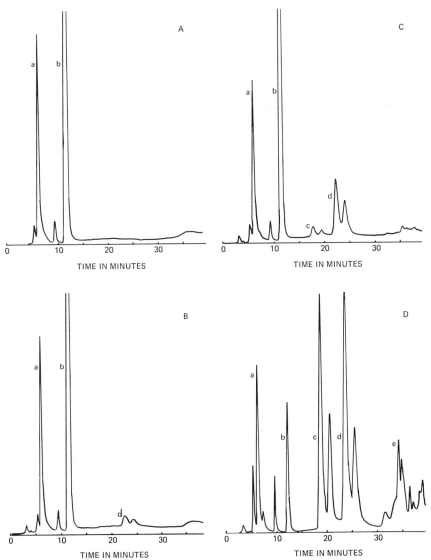

FIG. 4. (A–D) HPLC elution profile of the reaction system with MGBG and cholera toxin at different time intervals. The reaction system contained 10 mM NAD, 2 mM MGBG, 20 mM DTE, 25 mM Tris-acetate pH 7.4, and 200 μg/ml activated cholera toxin. At indicated time points, 25 μl of the reaction system was transferred to 25 μl 10% TCA. The TCA sample is filtered and centrifuged and 20 μl of TCA filtrate is injected into an ultrasphere ODS column (20 cm × 4.5 mm) equilibrated with 0.2 M TEAA, 0.1 M sodium perchlorate (pH 6.0). Chromatography was performed under the following conditions. A flow rate of 1.5 ml/min (initial pressure 2500 psi) was used. Elution was performed with the equilibrating buffer for 5 min. After that, a linear gradient of 0–15% methanol in the same buffer was used for 20 min. Elution was continued with 15% methanol in 0.2 M triethylammonium acetate (TEAA) in perchlorate for a period of 20 min. The absorbance of the eluent was recorded at 283 nm. Peak a is NAD, b is MABG, c, d, and e are ADP-ribosylated products.

could be utilized to evaluate guanidine specific ADP ribosyltransferase activity in other animal tissues.

Presently it is uncertain what structural and chemical features of the substrate influence ADP-ribosylation reactions. From the experiments of Moss et al.[4,14] and Mekalanos et al.[15] and ours[9,16] the results suggest that guanidine specific ADP-ribosylatransferase can react with a variety of guanidine derivatives. Chemical modification of arginyl groups in proteins and arginine[17] and the guanylhydrazone, NBAG[18] suggest that the deprotonated form of the guanidino group is the chemically reacting species. It is possible that features that influence chemical reactions with arginine might also influence ADP-ribosylation. A variety of guanylhydrazones with varying pK_a values can be synthesized to evaluate how important nucleophilicity is for ADP-ribosylation reactions. The guanylhydrazones, NBAG and MGBG, were found to be good substrates for reactions with cholera toxin.[9,16] Both of these compounds have pK_a values much lower than guanidine and are better substrates. MGBG which has two pK_a values of 7.5 and 9.2[19] can be mono ADP-ribosylated on both guanidino groups. Traces from the HPLC column at different times of reaction are shown in Fig. 4. The dual peaks, c and d, likely represent anomeric forms. It seems there is a preference for reaction at one end of MGBG as product d is formed before product c. At later stages of the reaction other product peaks (e) appear. Spectral analysis of the products isolated from the HPLC column showed that c and d are mono-ADP-ribosylated and fraction e contains di-ADP-ribosylated material.[9] Thus even a bulky ADP ribose constituent on the product does not block further ADP-ribosylation. Guanylhydrazones have use in the study of ADP-ribosylation reactions. They are chromophoric and reactive and can provide a sensitive nonradioactive assay for ADP-ribosylation. Studies with them could provide insight on how the enzymatic reaction occurs with the natural protein substrates.

Acknowledgments

This work was supported by Research Grant GM-09587 and GM-07187 from the National Institutes of Health, United States Public Health Service. This is Journal Paper J-11027 of the Iowa Agriculture and Home Economics Experiment Station, Ames, IA, Project 2120.

[14] J. Moss and M. Vaughan, in "Biology and Medicine" (O. Hayaishi and K. Ueda, eds.), p. 624. Academic Press, New York, 1982.
[15] J. J. Mekalanos, R. J. Collier, and W. R. Romig, J. Biol. Chem. **256,** 5849 (1979).
[16] D. J. Graves, G. Soman, and J. F. Miller, Fed. Proc., Fed. Am. Soc. Exp. Biol. **42,** 1808 Abstr. (1983).
[17] L. Pathy and J. Thesz, Eur. J. Biochem. **105,** 387 (1980).
[18] M. O. Hurst and D. J. Graves, unpublished results.
[19] M. G. Rosenblum and T. L. Loo, J. Chromatogr. **183,** 363 (1980).

[42] Toxin ADP-Ribosyltransferases That Act on Adenylate Cyclase Systems

By JOEL MOSS and MARTHA VAUGHAN

Effects of Choleragen (Cholera Toxin), *Escherichia coli* Heat-Labile
Enterotoxin and *Bordetella pertussis* Toxin on Adenylate
Cyclase Systems

cAMP plays a key role in control of metabolism and function of many cells. The activity of adenylate cyclase, the enzyme responsible for cAMP synthesis, is regulated by both inhibitory (e.g., opioid, muscarinic, and α-adrenergic agonists) and stimulatory (e.g., peptide hormones, β-adrenergic agonists) molecules.[1-3] Three bacterial toxins are known to exert their effects on cells through activation of the adenylate cyclase system. Choleragen and *E. coli* heat-labile enterotoxin, toxins involved in the pathogenesis of cholera and "traveler's diarrhea," respectively, enhance cyclase activity by acting on the stimulatory arm of the system.[4] Pertussis toxin, which may play a role in whooping cough, appears to increase cyclase activity by preventing the action of inhibitory ligands.[5]

An adenylate cyclase system is generally believed to consist of at least five components.[1-3] The catalytic unit (C) is responsible for the conversion of ATP to cAMP. Stimulatory and inhibitory ligands act by binding to agonist-specific cell surface receptors. Stimulatory (R_s) and inhibitory (R_i) receptors are coupled to the catalytic unit through different GTP-binding regulatory proteins termed G_s and G_i, respectively. Demonstration of hormonal stimulation or inhibition of adenylate cyclase activity requires agonist, its specific receptor, the appropriate GTP-binding protein (G_s or G_i) and GTP. In the current view, the catalytically active form of C is a C · G_s · GTP complex.[6] Hydrolysis of the bound GTP by a specific GTPase produces an inactive C · G_s · GDP complex.[6] In addition to increasing adenylate cyclase activity, β-adrenergic agonists such as isoproterenol have two other effects on membrane systems that have been related to

[1] E. M. Ross and A. G. Gilman, *Annu. Rev. Biochem.* **49,** 533 (1980).
[2] M. Rodbell, *Nature (London)* **284,** 17 (1980).
[3] D. M. F. Cooper, *Curr. Top. Membr. Transp.* **18,** 67 (1983).
[4] J. Moss and M. Vaughan, *Mol. Cell. Biochem.* **37,** 75 (1981).
[5] T. Murayama and M. Ui, *J. Biol. Chem.* **258,** 3319 (1983).
[6] D. Cassel, H. Levkovitz, and Z. Selinger, *J. Cyclic Nucleotide Res.* **3,** 393 (1977).

METHODS IN ENZYMOLOGY, VOL. 106   ISBN 0-12-182006-8

their action on the cyclase.[7,8] In turkey erythrocyte membranes, isoproterenol stimulates a specific low-K_m GTPase[7] and enhances the release of guanine nucleotide bound in the presence of the β-adrenergic agonist.[8] It was proposed that isoproterenol activates adenylate cyclase by accelerating the release of GDP from G_s, thereby freeing the nucleotide site on G_s to bind GTP.[9] By promoting the formation of $G_s \cdot$ GTP, isoproterenol might accelerate hydrolysis of GTP, i.e., apparently increase GTPase activity.

Choleragen and *E. coli* heat-labile enterotoxin appear to exert their effects on cells by enhancing the stimulatory arm of the cyclase system. The A subunits of the two toxins possess ADP-ribosyltransferase activity and catalyze the transfer of ADP-ribose from NAD to a subunit (42 and/or 47,000 daltons) of G_s.[10–12] ADP-ribosylation of G_s increases its sensitivity to stimulation by GTP.[13] Other observations suggest that the toxin acts to preserve the cyclase as the active C $\cdot$ G_s $\cdot$ GTP complex.[9] Cassel and Selinger[9] demonstrated that a low K_m GTPase believed to be associated with the cyclase system was inhibited by incubation of erythrocyte membranes with toxin. This would prevent formation of the inactive GDP $\cdot$ G_s species. In addition, choleragen promotes the release of GDP from G_s, permitting the protein to bind GTP and form the active GTP $\cdot$ G_s complex.[14]

Inhibitory ligands such as opiates and muscarinic or α-adrenergic agonists decrease adenylate cyclase activity via specific receptors coupled to G_i.[3] Inhibition is dependent on GTP at concentrations somewhat higher than those necessary for G_s action. In addition to promoting the hormonal inhibition of C through G_i, binding of GTP by G_i decreases the affinity of R_i for ligands.[3] The turn-off mechanism for the hormone- and GTP-dependent inhibition may be hydrolysis of GTP to GDP; Koski and Klee[15] observed that opiates stimulated a low K_m GTPase in NG108-15 (neuroblastoma × glioma hybrid) cells. They suggested that opiates might inhibit cyclase by facilitating the degradation of GTP bound to the C $\cdot$ G_s $\cdot$ GTP complex.

[7] D. Cassel and Z. Selinger, *Biochim. Biophys. Acta* **452**, 538 (1976).

[8] D. Cassel and Z. Selinger, *J. Cyclic Nucleotide Res.* **3**, 11 (1977).

[9] D. Cassel and Z. Selinger, *Proc. Natl. Acad. Sci. U.S.A.* **74**, 3307 (1977).

[10] D. Cassel and T. Pfeuffer, *Proc. Natl. Acad. Sci. U.S.A.* **75**, 2669 (1978).

[11] D. M. Gill and R. Meren, *Proc. Natl. Acad. Sci. U.S.A.* **75**, 3050 (1978).

[12] G. L. Johnson, H. R. Kaslow, and H. R. Bourne, *J. Biol. Chem.* **253**, 7120 (1978).

[13] S. Nakaya, J. Moss, and M. Vaughan, *Biochemistry* **19**, 4871 (1980).

[14] D. L. Burns, J. Moss, and M. Vaughan, *J. Biol. Chem.* **257**, 32 (1982).

[15] G. Koski and W. A. Klee, *Proc. Natl. Acad. Sci. U.S.A.* **78**, 4185 (1981).

Pertussis toxin appears to act by catalyzing the transfer of ADP-ribose from NAD to a subunit of G_i (41,000 daltons), thereby altering its activity.[16,17] ADP-ribosylation of G_i has different effects on cyclase activity depending on the cell or tissue system. Pertussis toxin blocks the inhibition of cyclase by opiates and muscarinic or α-adrenergic agonists, thus permitting the unopposed action of stimulatory ligands.[18] It also appears to enhance the action of stimulatory agonists on C.[17] In addition to effects on cyclase activity, pertussis toxin inhibits the opiate (enkephalin)-stimulated GTPase,[19] reduces the binding of inhibitory agonist to membranes,[20] and decreases the effects of guanyl nucleotides on inhibitory ligand binding.[18,20] All of these effects are compatible with the hypothesis that ADP-ribosylation of G_i by pertussis toxin uncouples it from the inhibitory receptor and thereby decreases the affinity of R_i for ligand.

Demonstration of Enzymatic Activity of Choleragen and E. coli Heat-Labile Enterotoxin

Choleragen and *E. coli* toxin are very similar in structure. Each is made up of one A and five B subunits.[21] The B subunits are responsible for toxin binding to the cell surface and the A subunit contains the ADP-ribosyltransferase. The holotoxin structure is necessary for toxin action on intact cells. For demonstration of enzymatic activity in broken cell systems or in the absence of cellular components, only the A subunit is required, but it must be "activated" to release the functional A_1 peptide.[22,23] In most commercial preparations of choleragen, the A subunit has been proteolytically nicked so that the A_1 and A_2 peptides are linked only by a single disulfide bond, reduction of which releases active A_1. The A subunit in *E. coli* toxin is apparently a single-peptide chain and to observe maximal catalytic activity must be proteolytically nicked (e.g., with trypsin) as well as reduced with thiol.[24]

[16] T. Katada and M. Ui, *Proc. Natl. Acad. Sci. U.S.A.* **79**, 3129 (1982).

[17] T. Katada and M. Ui, *J. Biol. Chem.* **257**, 7210 (1982).

[18] H. Kurose, T. Katada, T. Amano, and M. Ui, *J. Biol. Chem.* **258**, 4870 (1983).

[19] D. L. Burns, E. L. Hewlett, J. Moss, and M. Vaughan, *J. Biol. Chem.* **258**, 1435 (1983).

[20] J. A. Hsia, J. Moss, E. L. Hewlett, and M. Vaughan, *J. Biol. Chem.* **259**, 1086 (1984).

[21] J. Moss and M. Vaughan, *Annu. Rev. Biochem.* **48**, 581 (1979).

[22] J. Moss, V. C. Manganiello, and M. Vaughan, *Proc. Natl. Acad. Sci. U.S.A.* **73**, 4424 (1976).

[23] J. J. Mekalanos, R. J. Collier, and W. R. Romig, *J. Biol. Chem.* **254**, 5855 (1979).

[24] J. Moss, J. C. Osborne, Jr., P. H. Fishman, S. Nakaya, and D. C. Robertson, *J. Biol. Chem.* **256**, 12861 (1981).

Activated choleragen is prepared by incubation of the toxin (0.4 mg/ml) with 50 mM glycine · HCl (pH 8.0), 20 mM dithiothreitol, and ovalbumin (0.4 mg/ml) for 10 min at 30°.[25]

E. coli toxin was activated by incubation of the toxin (5.8 μg) with trypsin (0.5 μg), 50 mM potassium phosphate, pH 7.0, and 20 mM dithiothreitol (total volume 50 μl) for 1 hr at 30°. Soybean trypsin inhibitor (0.2 mg) and assay components were then added (final volume 0.3 ml).[24]

Assays for Choleragen and E. coli Heat-Labile Enterotoxin

Toxin action on cells involves multiple steps from initial binding to the membrane through activation of adenylate cyclase and finally alterations in metabolism or function (e.g., increased fluid secretion from intestinal cells). Assays based on cellular responses are beyond the scope of this chapter. Other assays for choleragen and E. coli toxin are based on their ability to activate adenylate cyclase, catalyze the ADP-ribosylation of G_s, purified proteins, or low-molecular-weight guanidino compounds and hydrolyze NAD to ADP-ribose and nicotinamide.[22,25–28] Activation of adenylate cyclase and ADP-ribosylation of G_s are not precise measures of toxin activity, since cellular preparations may contain inhibitory factors. Assay of NAD glycohydrolase or ADP-ribosyltransferase activity with pure substrates in the absence of cellular components, however, is quantitative and susceptible to standard kinetic analysis.

Assay 1. ADP-ribosylation of Protein or Agmatine

Transfer of ADP-ribose is measured directly using [^{32}P]NAD or [adenine-U-^{14}C]NAD as an ADP-ribose donor and protein or agmatine (or other guanindino compound) as an ADP-ribose acceptor.[25]

$$[^{32}P]NAD + acceptor \rightarrow [^{32}P]ADP\text{-ribose acceptor} + nicotinamide + H^+$$

Assays (0.3 ml total volume) containing 400 mM potassium phosphate (pH 7), 20 mM dithiothreitol, 2 mM NAD with 40,000 cpm of [^{32}P]NAD or [adenine-U-^{14}C]NAD, and protein such as histone, polyarginine, or lysozyme (5 mg/ml) are initiated with activated choleragen (50 μg) and incubated for 1 hr at 30°. Cold 20% trichloroacetic acid (0.1 ml) is added, and

[25] J. Moss and M. Vaughan, Proc. Natl. Acad. Sci. U.S.A. **75**, 3621 (1978).
[26] J. Moss and M. Vaughan, J. Biol. Chem. **252**, 2455 (1977).
[27] J. Moss and S. H. Richardson, J. Clin. Invest. **62**, 281 (1978).
[28] J. Moss, S. Garrison, N. J. Oppenheimer, and S. H. Richardson, J. Biol. Chem. **254**, 6270 (1979).

after 30 min at 0° the samples are transferred to 0.45-μm Millipore filters. Assay tubes are washed three times with 2 ml of cold 5% trichloroacetic acid. Dried filters are dissolved in 2 ml of Piersolve for radioassay.

Formation of [*adenine*-U-^{14}C]ADP-ribose agmatine is assayed as described above except that 75 mM agmatine replaces the ADP-ribose acceptor protein. At the end of the incubation, two 0.1-ml samples are transferred to columns (0.5 × 2 cm) of AG 1-X2 previously washed with 2 ml of 20 mM Tris(Cl$^-$), pH 7.0. [*adenine*-U-^{14}C]ADP-ribose agmatine is eluted with five 1-ml portions of Tris(Cl$^-$), pH 7, for radioassay. Preparation of the Ag 1-X2 is described in Assay 3.

Assay 2. ADP-ribosylation of [N-^{125}I]Guanyltyramine

Synthesis of ADP-ribosyl-[N-^{125}I]guanyltyramine from [N-^{125}I]guanyltyramine and NAD is directly measured.[29] The radioactive substrate is prepared by iodination of guanyltyramine [trivial name for 1-(p-hydroxyphenyl)-2-guanidinoethane] by the Iodogen method.[30]

NAD + [^{125}I]guanyltyramine → ADP-ribose-[^{125}I]guanyltyramine + nicotinamide + H$^+$

Assays (total volume 70 μl) containing 140 mM sodium phosphate (pH 7), 28 mM dithiothreitol, 0.006% Triton X-100 (w/v), 6 mM NAD, [^{125}I]GT (~50,000 cpm), and activated choleragen (0.7 ng) are incubated for 60 min at 30°. After addition of 7.5 ml of 0.01 M sodium borate, pH 8.5, the solution is filtered through two stacked 2.4-cm DEAE paper disks (Whatman DE 81), which are then washed with borate buffer (100 ml) and transferred to vials for radioassay. Iodination of guanyltyramine markedly increases its ability to serve as an ADP-ribose acceptor in the choleragen-catalyzed reaction; the K_m for [^{125}I]GT is 44 μM, whereas the K_i for GT is apparently at least two orders of magnitude higher. The K_m for NAD in this reaction is ~3.6 mM, similar to that observed in the NAD glycohydrolase reaction.[22]

Assay 3. Release of [carbonyl-^{14}C]Nicotinamide

Transfer of ADP-ribose from [*carbonyl*-^{14}C]NAD to acceptor (water, guanidino compound, or protein) is accompanied by the release of [*carbonyl*-^{14}C]nicotinamide.

$$\left.\begin{array}{l} [carbonyl\text{-}^{14}\text{C}]\text{NAD} \\ + \text{ acceptor} \end{array}\right\} \rightarrow \left\{\begin{array}{l} [carbonyl\text{-}^{14}\text{C}]\text{nicotinamide} \\ + \text{ ADP-ribose acceptor} + \text{H}^+ \end{array}\right.$$

[29] J. J. Mekalanos, R. J. Collier, and W. R. Romig, *J. Biol. Chem.* **254**, 5849 (1979).
[30] P. J. Fraker and J. C. Speck, Jr., *Biochem. Biophys. Res. Commun.* **80**, 849 (1978).

Assays (total volume 0.3 ml) containing activated choleragen (50 μg), 400 mM potassium phosphate, pH 7.0, 2 mM [carbonyl-^{14}C]NAD (40,000 cpm), and 20 mM dithiothreitol are incubated for 90 min at 30°. Two 0.1-ml samples from each assay are transferred to columns (0.5 × 4 cm) of AG 1-X2 resin (Bio-Rad) (prepared as described below) in Pasteur pipets. [carbonyl-^{14}C]Nicotinamide is eluted with five 1-ml portions of 20 mM Tris(Cl⁻), pH 7, for radioassay. Recovery of [carbonyl-^{14}C]nicotinamide (0.17 μM to 1.25 mM) is ~97%.[22]

As described, this assay measures the NAD glycohydrolase activity of toxin. Release of [carbonyl-^{14}C]nicotinamide is accelerated when an ADP-ribose acceptor is present. Guanidino compounds (e.g., 75 mM agmatine) or proteins such as histone, lysozyme, or polyarginine (5 mg/ml) may be used.[25] In these instances, the assay measures the sum of NAD glycohydrolase and ADP-ribosyltransferase activities. NAD glycohydrolase activity is enhanced by increasing the potassium phosphate concentration from 50 to 400 mM; less activity is observed in glycine (Cl⁻) or Tris(Cl⁻).

AG 1-X2 (Bio-Rad) is prepared by washing four times with an equal volume of water before pouring columns and columns are washed twice with 1 ml of water before application of sample. Resin from Sigma requires more extensive washing with eight volumes of 0.5 N NaOH, water until neutral, eight volumes of 0.5 N HCl, and finally by water until neutral.

Assays for Pertussis Toxin

Pertussis toxin (islet-activating protein), an oligomeric protein of M_r ~ 117,000 produced by *Bordetella pertussis,* consists of five subunits named S$_1$ (28,000), S$_2$ (23,000), S$_3$ (22,000), S$_4$ (11,700), and S$_5$ (9300). In the holotoxin, the enzymatically active subunit, S$_1$, is associated with a binding complex consisting of (S$_2$S$_4$)S$_5$(S$_2$S$_3$).[31]

Most proteins are not demonstrably ADP-ribosylated by pertussis toxin, and no simple model ADP-ribose acceptors (such as the guanidino compounds that serve as substrates for choleragen and *E. coli* toxin) have been found.[32] Pertussis toxin rather specifically ADP-ribosylates a 41,000-dalton membrane protein that is in all probability a component of the

[31] M. Tamura, K. Nogimori, S. Murai, M. Yajima, K. Ito, T. Katada, M. Ui, and S. Ishii, *Biochemistry* **21,** 5516 (1982).

[32] J. Moss, S. J. Stanley, D. L. Burns, J. Hsia, D. Yost, G. A. Myers, and E. L. Hewlett, *J. Biol. Chem.* **258,** 11879 (1983).

inhibitory guanyl nucleotide-binding protein (G_i) of adenylate cyclase.[16,17,33,34]

$$[^{32}P]NAD + G_i \rightarrow [^{32}P]ADP\text{-ribose-}G_i + \text{nicotinamide} + H^+$$

Assay 1: [³²P]ADP-ribosylation of Membrane Protein

The source of G_i can be a crude particulate fraction of cells, a membrane preparation, or solubilized and chromatographically purified G_i. [³²P]ADP-ribosylation of G_i in a particulate fraction from NG108-15 (neuroblastoma × glioma hybrid) cells is carried out in a total volume of 0.13 ml containing 65 μg of membrane, ~2 μg of toxin, 25 mM glycine/40 mM potassium phosphate, pH 7.5, 15 μM [³²P]NAD (~20 μCi/ml), 0.4 mM ATP, 0.4 mM GTP, 15 mM thymidine, 10 mM dithiothreitol, and 0.1 mg/ml ovalbumin in 0.13 ml. After incubation for 30 min at 37°, 2 ml of 10% trichloroacetic acid was added. Precipitated proteins were collected by centrifugation (18,000 g), solubilized in 1% sodium dodecyl sulfate, and separated by polyacrylamide gel electrophoresis (in sodium dodecyl sulfate). Radiolabeling of the 41,000-dalton peptide was quantified by densitometry after autoradiography of dried gels.[19]

Assay 2: Release of [carbonyl-¹⁴C]Nicotinamide from [carbonyl-¹⁴C]NAD

In the absence of a specific ADP-ribose acceptor, pertussis toxin acts as an NAD glycohydrolase and catalyzes the hydrolysis of NAD.[32]

$$[carbonyl\text{-}^{14}C]NAD + H_2O \rightarrow ADP\text{-ribose} + [carbonyl\text{-}^{14}C]\text{nicotinamide} + H^+$$

The isolated S_1 subunit is catalytically active, but the native holotoxin is relatively inactive. Enzymatic activity is demonstrable following incubation of holotoxin with high concentrations of thiol. NAD glycohydrolase activity of the holotoxin (10 μg) was assayed in a total volume of 0.3 ml containing potassium phosphate (pH 7.5), 250 mM dithiothreitol, 1 mg/ml ovalbumin, and 32.4 μM [carbonyl-¹⁴C]NAD. Assays were initiated with toxin and incubated for 6 hr at 30°. Two 0.1-ml samples were transferred to columns (0.5 × 2 cm) of AG 1-X2 (200–400 mesh, chloride form, Bio-Rad) previously washed twice with 1 ml of H_2O (0.5 × 2 cm). [carbonyl-¹⁴C]Nicotinamide was eluted with five 1-ml portions of 20 mM Tris(Cl⁻)

[33] D. R. Manning and A. G. Gilman, *J. Biol. Chem.* **258,** 7059 (1983).
[34] G. M. Bokoch, T. Katada, J. K. Northup, E. L. Hewlett, and A. G. Gilman, *J. Biol. Chem.* **258,** 2072 (1983).

and collected for radioassay. Under these conditions, release of [*carbonyl*-^{14}C]nicotinamide was ~1 nmol/min/mg of toxin.

With holotoxin, >100 mM dithiothreitol enhanced enzymatic activity ~20-fold; activity was less with >250 mM. The K_m for NAD with the dithiothreitol-activated toxin was ~25 μM and did not appear to vary with the extent of toxin activation.

[43] Coliphage-Induced ADP-Ribosylation of *Escherichia coli* RNA Polymerase

By C. G. Goff

Specific ADP-ribosylation of *Escherichia coli* RNA polymerase (EC 2.7.7.6) is induced by infection of the cell with any of the serologically related T-even coliphages (T2, T4, T6). Most studies of this system have utilized phage T4, and *E. coli* B or K12. A great deal is known about the enzymology of this reaction, the structure of the product, and the genetics of the T4-coded enzymes responsible, although the exact physiological functions of the ADP-ribosylation remain uncertain.

Another review of this subject has recently been written by Skorko.[1]

The bacteriophage T4 genome encodes two distinct (NAD$^+$) : protein ADP-ribosyltransferases, the *alt* and *mod* enzymes, as shown by Horvitz.[2] Surprisingly, neither enzyme is essential for T4 growth, at least in laboratory strains of *E. coli*.[3]

The T4 *alt* Gene and Enzyme

The *alt* gene maps between T4 genes 30 (DNA ligase) and 54 (a viron baseplate protein).[4] *Alt* is expressed primarily late in T4 infection, producing a 79,000-dalton precursor protein.[5] The precursor is proteolytically cleaved to the active *alt* ADP-ribosyl transferase (size estimates range from 61,000 to 70,000 daltons) during assembly of mature T4 phage parti-

[1] R. Skorko, *in* "ADP-Ribosylation Reactions" (O. Hayaishi and K. Ueda, eds.). Academic Press, New York, 1982.

[2] H. R. Horvitz, *J. Mol. Biol.* **90**, 739 (1974).

[3] C. G. Goff and J. Setzer, *J. Virol.* **33**, 547 (1980).

[4] C. G. Goff, *J. Virol.* **29**, 1232 (1979).

[5] H. R. Horvitz, *J. Mol. Biol.* **90**, 727 (1974).

cles.[2,6-8] Twenty-five to fifty molecules of the enzyme are packaged into each virion[4] (Fig. 1) and are injected into *E. coli* with the T4 DNA and other "internal proteins" upon infection.[9] The enzyme has been purified essentially to homogeneity from disrupted phage particles.[8] The *alt* enzyme functions as an ADP-ribosyltransferase immediately upon injection (within 1 min), prior to any T4-specific gene expression.[10] The fact that active *alt* enzyme preexists in the infecting phage explains why blockage of protein synthesis during infection has no effect on *alt* function.[11]

The T4 *mod* Gene and Enzyme

The *mod* gene maps almost half the T4 genome away from *alt*, in a large "nonessential" region between genes 39 and 56.[2] *Mod* activity is expressed early in T4 infection, beginning about 2 min after infection and reaching maximum rates by 3 to 4 min; the activity declines thereafter.[10-12] There is no evidence for a *mod* precursor. This enzyme has been purified about 100-fold from extracts of T4-infected cells; it has a molecular weight of approximately 26,000, as determined by gel filtration.[10]

Polypeptide Substrate Specificities of the *alt* and *mod* Enzymes

The effects of the *alt* and *mod* enzymes were discovered by researchers curious about the changes in patterns of messenger RNA synthesis which occur after T4 phage infection (see recent reviews by Rabussay and Geiduschek[13] and by Rabussay[14]). Changes in the structure of *E. coli* RNA polymerase, the enzyme known to be responsible for mRNA synthesis throughout T4 infection,[15] were therefore sought and studied as possible causes of transcription specificity changes. Although the fundamental 400,000-dalton "core" RNA polymerase (β, $\beta'\alpha_2$ subunit structure; see Chamberlin[16]) is conserved, several new T4-coded polypeptides

[6] A. Coppo, A. Manzi, J. Pulitzer, and H. Takahishi, *J. Mol. Biol.* **76,** 61 (1973).

[7] U. Laemmli and M. Favre, *J. Mol. Biol.* **80,** 575 (1973).

[8] H. Rohrer, W. Zillig, and R. Mailhammer, *Eur. J. Biochem.* **60,** 227 (1975).

[9] K. Abremski, personal communication (1979).

[10] R. Skorko, W. Zillig, H. Rohrer, H. Fujiki, and R. Mailhammer, *Eur. J. Biochem.* **79,** 55 (1977).

[11] W. Seifert, P. Qasba, G. Walter, P. Palm, M. Schachner, and W. Zillig, *Eur. J. Biochem.* **9,** 319 (1969).

[12] C. G. Goff, *J. Biol. Chem.* **249,** 6181 (1974).

[13] D. Rabussay and E. P. Geiduschek, *Compr. Virol.* **8,** 1 (1977).

[14] D. Rabussay, *Am. Soc. Microbiol. News* **48,** 398 (1982).

[15] R. Haselkorn, M. Vogel, and R. D. Brown, *Nature (London)* **221,** 836 (1969).

[16] M. Chamberlin, *in* "The Enzymes" (P. D. Boyer, ed.), 3rd ed., Vol. 15, p. 87. Academic Press, New York, 1982.

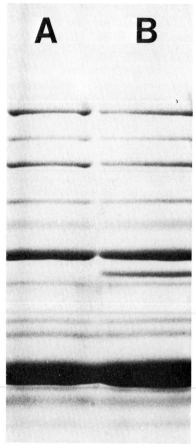

FIG. 1. Scoring T4 strains for *alt* phenotype on polyacrylamide gels.[4] About 2×10^{10} T4 phage particles were purified from a clarified lysate by pelleting at 100,000 g for 30 min. The phage pellet was resuspended and boiled 2 min in 2% SDS, 5% 2-mercaptoethanol, 10 mM Tris–Cl, pH 7.5, and subjected to electrophoresis on a 10 to 20% acrylamide gradient slab gel in the presence of SDS. After electrophoresis, the gel was stained in Coomassie brilliant blue R-250, then destained. This figure is a photograph of the stained gel. Electrophoresis was from top to bottom. The uppermost band is P34, a tail-fiber protein (molecular weight 145,000); the heavy band near the bottom is P23*, the processed form of the major capsid protein (molecular weight 47,000). The processed *alt* protein exhibits mobility corresponding to about 61,000 molecular weight. (A) T4 *ac-q* (*Alt⁻*); (B) T4D (*Alt⁺*).

are added, and the *E. coli*-coded σ subunit is released.[13] In addition, and of prime interest to us here, the two preexisting 40,000-dalton RNA polymerase α polypeptides become more negatively charged soon after infection.[11,17–19] This change in net charge we now know to be the consequence of ADP-ribosylation by first (upon T4 infection) the *alt* enzyme (resulting in "altered" α) and second (2 to 4 min later) the *mod* enzyme (resulting in "modified" α).

Initial characterization of "altered" and "modified" α was done on the *in vivo* products; they were differentiated only by their sequential appearance after T4 infection, and by the fact that α "alteration," unlike "modification," still occurs if T4 protein synthesis is blocked by chloramphenicol (as discussed above). With the development of *in vitro* assays for *alt* and *mod* enzymes, much more precise understanding became possible, including effects on other proteins besides α. What follows is a summary based on our current knowledge.

alt Substrate Specificity Is Relatively Low

Alt rapidly adds ADP-ribose to a fraction (maximum 50%) of the RNA polymerase α subunits in the cell; at low multiplicity of infection, only a small percentage of α is affected.[5,20] Approximately half of the ADP-ribose residues on "altered" α is attached to the guanido group of arginine residue 265 in the α polypeptide, at the sequence -Leu-Thr-Val-Arg-Ser-.[12,21] (The underlined sequence was isolated as an ADP-ribosylated Pronase peptide.) Another one-third of the ADP-ribose residues is attached to the guanido group of arginine residue 191 or 195 in "altered" α at the sequence -Ala-Arg-Val-Glu-Gln-Arg-Thr-; we do not know which of the two arginines in the purified Pronase peptide (underlined) is ADP-ribosylated.[12,22] The remaining ADP-ribose is on arginine-containing peptides which may be derived from the amino acid 191–195 region.[22] Thus there are at least two distinct sites of ADP-ribosylation on "altered" α. "Altered" α can be resolved into two electrophoretically distinct forms by SDS–polyacrylamide gel electrophoresis, one of which moves identically to "modified" α[2,3] (Fig. 2). The ratio of the two forms varies in

[17] G. Walter, W. Seifert, and W. Zillig, *Biochem. Biophys. Res. Commun.* **30,** 240 (1968).

[18] E. K. F. Bautz and J. J. Dunn, *Biochem. Biophys. Res. Commun.* **34,** 230 (1969).

[19] C. G. Goff and K. Weber, *Cold Spring Harbor Symp. Quant. Biol.* **35,** 101 (1970).

[20] D. Rabussay, R. Mailhammer, and W. Zillig, *in* "Metabolic Interconversion of Enzymes" (O. Wielard, E. Helmrich, and H. Holzer, eds.), p. 213. Springer-Verlag, Berlin and New York, 1972.

[21] Yu. A. Ovchinnikov, V. M. Lipkin, N. N. Modyanov, and O. Yu. Chertov, *FEBS Lett.* **76,** 108 (1977).

[22] C. G. Goff, unpublished results (1975).

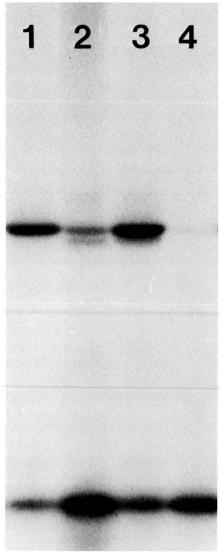

FIG. 2. Patterns of ^{32}P incorporated into the RNA polymerase α polypeptides after infection by various T4 strains. One-milliliter cultures of *E. coli* B/r growing at 30° in low phosphate medium were labeled for one generation with ^{32}P$_i$ (specific activity 1000 Ci/mol) and then infected (multiplicity of infection, 8) at a cell density of 4 × 10^8/ml with T4 phage (strains as indicated below). After 10 min the cells were pelleted by centrifugation, mixed with 25 mg unlabeled carrier cells, and lysed by repeated freezing and thawing with lysozyme. Nucleic acids were digested with DNase I and RNase A, and cell debris was removed by centrifugation (7000 rpm, 10 min). RNA polymerase was precipitated from the crude

different experiments. The form unique to "altered" α may be doubly ADP-ribosylated, or may reflect single ADP-ribosylation at a different site. In any case, it appears that the fraction of α molecules which carry ADP-ribose after *alt* enzyme activity is randomly distributed on RNA polymerase molecules, rather than occurring in pairs (with both α polypeptides of an RNA polymerase subpopulation being ADP-ribosylated).[20]

Many other proteins besides α can be ADP-ribosylated *in vitro* by *alt*, including egg white lysozyme, histone Fl, bovine serum albumin, the *alt* protein itself, other RNA polymerase subunits, and poly(L-arginine); the efficiency of ADP-ribosylation is roughly correlated with arginine content of the protein.[8] *Alt* also ADP-ribosylates the other RNA polymerase subunits *in vivo*, albeit at a low level relative to α.[5] There are conflicting claims about the stability of ADP-ribose attached to protein by *alt* enzyme *in vivo;* some reports suggest that the ADP-ribose is removed later during infection[2] while others find it stable.[3]

mod Substrate Specificity Is Very High

The *mod* enzyme rapidly adds a single ADP-ribose to essentially every RNA polymerase α polypeptide in the cell, at arginine #265 (one of the sites also affected by *alt*).[12] No other amino acid in RNA polymerase is affected by *mod*. However, one other protein, 13,000 daltons in molecular weight, is ADP-ribosylated *in vivo* by *mod*.[10] It is not known whether this is a T4-coded or *E. coli*-coded protein. *In vitro,* partially purified *mod* will inefficiently ADP-ribosylate poly(L-arginine), but no other proteins tested except α.[10] Surprisingly, the *mod* enzyme utilizes both α polypeptides within the RNA polymerase complex as readily as isolated α, while the otherwise less specific *alt* enzyme is apparently unable to ADP-ribosylate

---

extract with rabbit antiserum made against the purified enzyme. The precipitate was washed in 1% Triton X-100, 1 M NaCl, 10 mM Tris–Cl pH 7.5, and then dissolved in 2% SDS, 5% 2-mercaptoethanol, and 10 mM Tris–Cl pH 7.5. After heating at 95° for 2 min, samples containing 5000 dpm were subjected to electrophoresis on a 10% acrylamide slab gel in the presence of SDS. The gel was dried and exposed for 3 days to Kodak BB-1 X-ray film. More details of these procedures are presented elsewhere.[3,5] In the autoradiographs shown here, electrophoresis was from top to bottom. The band near the bottom, at the tracking dye front, probably contains oligonucleotides released from the active site of the enzyme by SDS denaturation. The upper bands are due to ADP-ribose covalently bound to the α polypeptides; the narrow, intense form of the band in sample 1 comigrates with "modified" α in a stained gel. The four tracks represent cells infected with (1) T4 *ac-q* (*alt⁻ mod⁺*), (2) T4 *del*(39-56)10 (*alt⁺ mod⁻*), (3) T4D (*alt⁺ mod⁺*), (4) T4(A⁻ × M⁻)2 (*alt⁻ mod⁻*). Uninfected cells, like sample (4), show no incorporation of ^{32}P into α (data not shown).

one of the two α polypeptides in the complex, perhaps due to steric constraints (*alt* is a larger molecule than *mod*).[23]

In cells infected by wild type T4, "alteration" of α precedes "modification." However, "altered" α is not an obligatory precursor of "modified" α, as proven by the existance of viable Alt⁻ Mod⁺ and Alt⁺ Mod⁻ mutants, which ADP-ribosylate α independently.[2] *Mod* simply completes the ADP-ribosylation of α begun by *alt*. The ADP-ribose added to α by the *mod* enzyme is stable *in vivo* throughout the rest of T4 infection.[2,3]

Evidence That *alt* and *mod* Are Both (NAD⁺):Protein Arginyl ADP-ribosyltransferases

Both "modified" and "altered" RNA polymerase α polypeptides have been shown to contain covalently bound phosphorus, and 5'-AMP which is released by venom phosphodiesterase.[8,19] In both "modified" and "altered" α the molar stoichiometry of adenine to α polypeptide is approximately 1:1, judged by UV absorption spectra.[12,24] Therefore, "modified" and "altered" α contain mono(ADP-ribose) rather than the poly(ADP-ribose) often found in eukaryotic cells (see other chapters in this volume).

An *in vitro* α "modifying" system utilizing extracts of T4-infected cells (3 min postinfection) has been shown by Goff to transfer radioactivity specifically into α from NAD⁺; label was incorporated quantitatively from the adenine ring, from both phosphates, and from ribose of the NMN moiety, but not from the nicotinamide ring of NAD⁺.[12] Moreover, hydrolysis of "modified" α at neutral pH releases authentic ADP-ribose.[10] Therefore "modified" α contains adenosine diphosphoribose.

The ³²P "modified" α produced with [³²P]NAD⁺ in this *in vitro* system could be digested with Pronase to produce ³²P-labeled peptides electrophoretically and chromatographically identical to ³²P-labeled Pronase peptides from "modified" α labeled *in vivo*. These *in vivo* labeled peptides have been shown (by leucine aminopeptidase digestion) to contain ³²P covalently attached to arginine.[12] The "modified" arginine, residue 265, is neither N-terminal nor C-terminal in the α polypeptide; therefore it has neither a free amino group nor a free carboxyl group, and must carry the ADP-ribose on its guanido side chain.

Similar experiments utilizing purified *alt* enzyme, NAD⁺, and egg white lysozyme as a protein substrate have shown that label from the

[23] W. Zillig, R. Mailhammer, R. Skorko, and H. Rohrer, *Curr. Top. Cell. Regul.* **12**, 263 (1977).
[24] W. Seifert, D. Rabussay, and W. Zillig, *FEBS Lett.* **16**, 175 (1971).

adenosine ring of NAD^+ is covalently transferred to the protein while label from the nicotinamide ring is concurrently liberated.[8] The purified *alt* enzyme transfers label in NAD^+ efficiently to poly(L-arginine).[8] Further evidence that *alt*, like *mod*, ADP-ribosylates arginine comes from *in vivo* data discussed above; *in vivo* ^{32}P-labeled "altered" α can be digested with Pronase to produce ^{32}P-labeled peptides, all of which contain arginine, and some of which are identical to the ^{32}P-labeled Pronase peptides from "modified" α.[12,22]

The Arginine-ADP-ribose Linkage in "Modified" and "Altered" α Is Probably a Glycosidic Bond between a Guanido Nitrogen and the Ribose 1-Carbon

Because 5'-AMP is released from "modified" α by venom phosphodiesterase,[8,12] the arginine–ADP-ribose bond can only involve linkage through the phosphates or the terminal ribose of adenosine diphosphoribose. Attempts to identify the specific linkage have involved chemical stability studies and analysis of the degradation products of ADP-ribosylated protein. The rate of release of [^{32}P]- and [^{3}H]adenine from both "altered" and "modified" proteins has been measured in acid, alkali, and hydroxylamine.[8,12] The data are consistent for both labels and for both "altered" and "modified" proteins. Half-lives of 16 to 18 min in 0.25 N HCl at 66°, 3 to 6 min in 0.25 N NaOH at 66°, and 7 min in 0.4 N NH_4OH–HCl at 66°, rule out most bonds other than an N-glycosidic linkage. The ADP-ribose substitution is more stable to acid and to hydroxylamine than to alkali, which is opposite to expectations for a phosphoamide bond.[25] The substitution is much less stable in acid than expected for a carbon-to-phosphorus bond (e.g., a phosphonic acid).[26] More direct evidence for ADP-ribosyl linkage through the ribose 1-carbon is the release of ADP-ribosyloxime from "altered" α by hydrolysis in the presence of NH_4OH–HCl, and release of authentic ADP-ribose by hydrolysis at pH 8.[10] Acid hydrolysis of "modified" peptides apparently releases authentic arginine, as judged by ion-exchange chromatography and by two-dimensional thin-layer chromatography of the dansylated material.[12] However, others have detected an unusual basic amino acid, perhaps ornithine, after combined proteolytic and acid hydrolysis.[10]

There have been no studies on the stereospecificity of the T4-induced ADP-ribosylation reaction. However, a number of other NAD^+ : arginine

[25] Z. A. Shalbarova, *Prog. Nucleic Acids Res.* **10**, 145 (1970).
[26] J. S. Kittridge, E. Roberts, and D. G. Simonsen, *Biochemistry* **1**, 624 (1962).

mono(ADP ribosyl)transferases, for example cholera toxin,[27] avian erythrocyte ADP-ribosyltransferase,[28] *E. coli* LT toxin,[29] and diphtheria toxin[30] have all been shown by NMR spectroscopy to cause inversion of the β-NAD$^+$ substrate, producing an α-anomeric ADP-ribose–L-arginine bond (see other chapters in this volume). Thus, it seems likely that the *alt* and *mod* enzymes also invert the configuration of this bond.

Assay Conditions and Kinetic Parameters for *alt* and *mod* Enzymes

Both these enzymes have been assayed *in vitro* under similar conditions in Zillig's laboratory.[8,10] The standard reaction buffer contains (in a reaction volume of 10 to 100 μl): 0.05 M Tris-acetate, pH 7.5, 0.022 M NH$_4$Cl, 0.001 M EDTA, 0.01 M 2-mercaptoethanol, and (for *alt* only), 10% glycerol. Substrates used are [*adenine*-^{3}H]NAD$^+$, 0.25 mM and 25–200 Ci/mol; and acceptor protein, 15–20 μg. Egg white lysozyme is routinely used as the acceptor for *alt;* normal *E. coli* RNA polymerase, for *mod*. Approximately 1–2 μg of purified *alt* enzyme, or 1–100 μg of partially purified *mod* enzyme, give easily measurable levels of incorporation. The reactions are incubated 30–60 min at 15 to 20°, then stopped by precipitation with 5% trichloroacetic acid. Precipitates are collected on nitrocellulose filters, washed, and counted.

Both *alt* and *mod* enzymes are maximally active at 15 to 20° and function over a broad pH range (6 to 9) with an optimum at pH 7.5. Both enzymes are essentially indifferent to divalent cations, but are inhibited by increased ionic strength (their activities are reduced about 50% by 0.1 M NH$_4$Cl). Both enzymes remain active for hours in assays at 15°. They do however differ in some kinetic parameters. *Alt* assays yield a nonlinear Lineweaver–Burk plot of activity as a function of NAD$^+$ concentration; from a linear part of the plot at high NAD$^+$ concentrations, the K_m for NAD$^+$ appears to be about 5×10^{-5} M at 15°. *Mod* exhibits normal kinetics, with a K_m for NAD$^+$ of 1.43×10^{-5} M at 20°. *Alt* is not inhibited by nicotinamide at pH 7.5; in contrast, *mod* exhibits a K_i for nicotinamide of 4×10^{-3} M in the standard assay conditions. Purified *alt* enzyme has a specific activity of 14 mU/mg; the purest *mod* preparation reported has a specific activity of 0.11 mU/mg (one unit incorporates 1 μmol of ADP-ribose per minute into protein).

[27] N. J. Oppenheimer, *J. Biol. Chem.* **253,** 4907 (1978).

[28] J. Moss, S. J. Stanley, and N. J. Oppenheimer, *J. Biol. Chem.* **254,** 8891 (1979).

[29] J. Moss, S. Garrison, N. J. Oppenheimer, and S. H. Richardson, *J. Biol. Chem.* **254,** 6270 (1979).

[30] N. J. Oppenheimer and J. W. Bodley, *J. Biol. Chem.* **256,** 8579 (1981).

Purification of *Alt* Enzyme[8]

The *alt* enzyme can be purified from previously purified T4 phage particles. Ten milliliters phage at 4×10^{12} infective particles/ml was dialyzed against assay buffer salts without substrates or glycerol (see section above) then mixed with one volume of glycerol and incubated 30 min at 37°. The partially lysed phage were digested 30 min at room temperature with pancreatic DNase (10 μg/ml), then diluted dropwise into 2.5 liters of water with stirring. The viscous solution of lysed phage was brought to $1\times$ assay buffer salts, digested 30 min more with 500 μg DNase, then concentrated to 300 ml with a hollow-fiber ultrafiltration system. The enzyme was precipitated by $(NH_4)_2SO_4$ added to 60% saturation. The precipitate was dissolved in 15 ml of 0.05 M Tris-acetate pH 7.5, 1.2 M NH_4Cl, 1 mM EDTA, 0.01 M 2-mercaptoethanol, and fractionated on a BioGel A-1.5M column (3 $\times$ 90 cm). The variable amount (10 to 100%) of enzyme activity which eluted as soluble protein (rather than bound to ruptured phage particles) was reprecipitated with $(NH_4)_2SO_4$, dissolved and dialyzed against assay buffer without substrates but with 10% glycerol, then passed over a DEAE-cellulose column in the same buffer. Essentially pure *alt* protein (as judged by SDS–acrylamide gel electrophoresis) eluted immediately in the flowthrough of the column. The enzyme was stored at −70°. No data on yield were reported.

Purification of *Mod* Enzyme[10]

E. coli B/r (200 liters) was grown to 8×10^8 cells/ml in a rich broth,[12] then infected with an *Alt* T4 mutant at a multiplicity of 8 phage per cell. The culture was chilled to 6° at 3–5 min postinfection, cell metabolism was blocked with 1.5 mM NaN_3, and the cells were rapidly harvested. The cells (400 g) were frozen, thawed in 1.5 volume assay buffer salts without substrate or glycerol (see above), and broken in a Gaulin press. Polymin P (polyethylene imine) was added to the lysate (to 0.7% by weight). The precipitate was collected by centrifugation and the *mod* protein eluted from the pellet by resuspension in assay buffer containing 0.25 M NH_4Cl. The precipitate was reextracted two times with the same buffer and the pooled supernatants were brought to 60% saturation with $(NH_4)_2SO_4$. The precipitate was dissolved and dialyzed against assay buffer salts, then applied to a DEAE-cellulose column in the same buffer. Apparently *mod* enzyme is unstable at low ionic strength, so this dialysis and column loading should not be prolonged. Protein was eluted by a gradient of 0.022–0.6 M NH_4Cl in assay buffer salts and the activity eluting between 0.1 and 0.4 M NH_4Cl was again precipitated, with 50%

(NH$_4$)$_2$SO$_4$. This precipitate was dissolved and dialyzed against assay buffer salts containing 0.5 M NH$_4$Cl, then passed over a Sephadex G-100 column (3 × 70 cm) in the same buffer. The low-molecular-weight fraction, containing the enzyme activity, was again concentrated by (NH$_4$)$_2$SO$_4$ precipitation, then dissolved and dialyzed against assay buffer salts containing 0.5 M NH$_4$Cl and 40% glycerol. This material was fractionated on a BioGel P-30 column (1 × 100 cm) in the same buffer. The enzyme from this column, purified about 85-fold from the crude extract, would be suitable for many experiments. The authors further purified *mod* protein by velocity gradient centrifugation; this step resulted in a 1.4-fold increase in specific activity but a 67% loss of enzyme. The overall yield of activity at this point was 4.1%. A further 2-fold purification, with concomitant 50% loss of activity, could be achieved by affinity chromatography on 5'-AMP-Sepharose. The purest material reported contains a number of polypeptides, with molecular weights of 18,000 to 28,000.

It is worth noting that an active crude extract of cells grown, infected, and harvested as discussed above can be made on a small scale using a rapid lysis method.[12] This extract will modify microgram amounts of RNA polymerase and, combined with an antibody precipitation step to partially purify the modified RNA polymerase, permits analysis of *mod* activity in a short time.

Physiological Function of *alt* and *mod* Enzymes

Alt and *mod* enzymatic function have been defined by biochemical studies of the reaction products, and by genetics of *alt* and *mod* mutants. However, the *alt* and *mod* mutants were not selected, but rather discovered in existing phage strains. Both mutants are viable and show no obvious defective growth phenotype, at least under laboratory conditions, in normal laboratory host strains.[2] So the discovery of *alt* and *mod* mutants did little to explain the *in vivo* function of the ADP-ribosylation. Since the two enzymes show overlapping specificity, it was possible to argue that either *alt* or *mod* function alone would carry out the same critical function (i.e., redundancy would mask the defect of either mutant). More recently however, Goff and Setzer constructed a double mutant, *alt⁻mod⁻*; it too grows normally, ruling out an essential role for the ADP-ribosylation.[3] There are many possible nonessential functions for *alt* and *mod;* for example they might confer a broader host range or a slight growth advantage on wild type T4.[2,3] No quantitative *in vivo* studies have yet been reported, but preliminary searches have not revealed *E. coli* mutants which specifically restrict growth of *alt⁻mod⁻* double mutant phage.[31]

[31] C. G. Goff, R. S. Israel, and A. Kessler, unpublished (1979).

A number of studies have examined the *in vitro* transcription properties of ADP-ribosylated RNA polymerase in search of physiological functions. A coupled transcription–translation system was used to show that "modified" RNA polymerase is less efficient *in vitro* than normal RNA polymerase at producing functional mRNA from *E. coli* DNA template.[32] Under the same conditions, RNA polymerase "modification" did not significantly affect the level of expression from T4 DNA, implicating the "modification" in shutoff of host cell gene expression after infection.[32] This assay system is complex, and the effect of added proteins (both bacterial and phage-coded) complicates interpretation of the results. Preliminary *in vivo* studies have shown no striking delay in the shutoff of *E. coli* stable RNA transcription after infection by *alt⁻mod⁻* mutant T4.[31]

Other workers have examined synthesis of defined RNA molecules from specific T4 promoters, in a purified *in vitro* transcription system. By reconstructing active RNA polymerase from carefully isolated subunits of the normal and "modified" enzyme, Goldfarb and Palm have shown that ADP-ribosylation of α reduces the transcription activity of RNA polymerase on T4 DNA about 4-fold. Apparently this is because the enzyme containing "modified" α fails to initiate transcription at certain promoters on T4 DNA.[33,34] Those promoters affected *in vitro* by "modification" include several responsible for expression of a class of T4 early genes which stop being transcribed *in vivo* at the time when ADP-ribosylation of α is completed.[33] This result argues that ADP-ribosylation is one of the switching mechanisms which direct the complex developmental program of T4 gene expression. It is interesting that the effect of "modification" defined by these *in vitro* studies is most clearly seen in transcription assays containing both 0.25 N KCl (which inhibits "modified" RNA polymerase) and 1% Triton X-405 (which relieves the salt inhibition, perhaps by releasing a T4-coded protein from the RNA polymerase σ subunit[35]). It seems very unlikely that α "modification" has a function independent of the other changes occurring in T4 infected cells.

Despite these *in vitro* results, Brody has recently found that the switch-off of early T4 genes mentioned above still occurs normally *in vivo* after infection by an *alt⁻mod⁻* mutant T4.[36] Thus, we still do not have a complete understanding of why T4 ADP-ribosylates *E. coli* RNA polymerase.

[32] R. Mailhammer, H.-L. Yang, G. Reiness, and G. Zubay, *Proc. Natl. Acad. Sci. U.S.A.* **72**, 4928 (1975).
[33] A. Goldfarb, *Proc. Natl. Acad. Sci. U.S.A.* **78**, 3454 (1981).
[34] A. Goldfarb and P. Palm, *Nucleic Acids Res.* **9**, 4863 (1981).
[35] J. C. Rhoton and A. Stevens, *Biochemistry* **14**, 5074 (1975).
[36] E. Brody, personal communication (1983).

[44] NAD: Arginine Mono-ADP-Ribosyltransferases from Animal Cells

By JOEL MOSS and MARTHA VAUGHAN

Animal tissues contain two types of enzymes that catalyze ADP-ribose transfer, mono-ADP-ribosyltransferases and poly(ADP-ribose) synthetases; the latter group is discussed in other chapters in this volume. The types of reactions catalyzed by mono-ADP-ribosyltransferases have not been fully identified. One subset of the transferases has, however, been investigated in detail. These enzymes, purified to homogeneity from turkey erythrocytes, are referred to as transferase A and transferase B (in order of discovery). They catalyze the transfer of ADP-ribose from NAD to cytosolic and membrane proteins as well as various purified proteins [Reaction (1)] and to low-molecular-weight guanidino compounds such as arginine [Reaction (2)].[1,2] In the absence of such acceptors, the transferases exhibit NAD glycohydrolase activity, i.e., they catalyze the hydrolysis of NAD [Reaction (3)], albeit at a much slower rate. The physiological substrates (ADP-ribose acceptors) for these enzymes are unknown.

$$\text{NAD} + \text{protein} \rightarrow \text{ADP-ribose protein} + \text{nicotinamide} + \text{H}^+ \qquad (1)$$
$$\text{NAD} + \text{guanidino-R} \rightarrow \text{ADP-ribose-guanidino-R} + \text{nicotinamide} + \text{H}^+ \qquad (2)$$
$$\text{(arginine)} \qquad\qquad \text{(ADP-ribose-arginine)}$$
$$\text{NAD} + \text{H}_2\text{O} \rightarrow \text{ADP-ribose} + \text{nicotinamide} + \text{H}^+ \qquad (3)$$

Assays

Enzymatic activity can be assayed by measuring the amount of either ADP-ribosylated acceptor or concomitantly released nicotinamide. Assay conditions for transferases A and B differ as noted.

Assay 1. Release of [carbonyl-^{14}C]nicotinamide

$$[carbonyl\text{-}^{14}\text{C}]\text{NAD} + \text{acceptor} \rightarrow \text{ADP-ribose acceptor}$$
$$+ [carbonyl\text{-}^{14}\text{C}]\text{nicotinamide} + \text{H}^+$$

Assay determines total [carbonyl-^{14}C]nicotinamide release accompanying ADP-ribose transfer to protein acceptors, low-molecular-weight guanidino compounds, or water. The standard assay for transferase A con-

[1] J. Moss and M. Vaughan, *Proc. Natl. Acad. Sci. U.S.A.* **75,** 3621 (1978).
[2] D. A. Yost and J. Moss, *J. Biol. Chem.* **258,** 4926 (1983).

METHODS IN ENZYMOLOGY, VOL. 106   ISBN 0-12-182006-8

tained 50 mM potassium phosphate (pH 7.0), 0.3 mg ovalbumin, 30 μM [carbonyl-^{14}C]NAD (40,000 cpm), and 75 mM arginine methyl ester in a total volume of 0.3 ml.[1] For transferase B,[2] assays contained 50 mM sodium phosphate, pH 7.1, 56 μM [carbonyl-^{14}C]NAD (45,000 cpm), ovalbumin (1 mg/ml), and 75 mM arginine methyl ester. Assays were initiated with transferase (~1 ng) and incubated for 30 to 60 min at 30° (transferase A) or 37° (transferase B). Duplicate 0.1-ml samples were then transferred to columns (0.5 × 2 cm) of AG 1-X2 (100–200 mesh) previously washed twice with 1.2 ml of water. [carbonyl-^{14}C]Nicotinamide was eluted with four 1.2-ml portions of water and collected for radioassay. This assay is similar to that described for choleragen[3] (this volume [42]).

Assay 2. Formation of [adenine-U-^{14}C]ADP-ribose protein or [^{32}P]ADP-ribose protein

$$\begin{bmatrix} [adenine\text{-}U\text{-}^{14}C]NAD \\ or \\ [^{32}P]NAD \end{bmatrix} + \text{protein} \rightarrow \begin{bmatrix} [adenine\text{-}U\text{-}^{14}C]ADP\text{-ribose protein} \\ or \\ [^{32}P]ADP\text{-ribose protein} \\ + \text{ nicotinamide} + H^+ \end{bmatrix}$$

The principal advantage of the assay is that it directly measures the formation of an ADP-ribosylated product. Assays (total volume, 0.3 ml) containing 50 mM potassium phosphate (pH 7), ovalbumin (5 mg/ml) or other protein acceptor, 30 μM [adenine-U-^{14}C] or [^{32}P]NAD (~40,000 cpm), and 200 mM NaCl [or other activator, such as histone, 100 μg/ml, or 3% 3-[(3-cholamidopropyl)dimethylammonio]-1-propane sulfonate (CHAPS)] were initiated with transferase A (1–6 ng) and incubated for 30 min at 30°; the reaction was terminated by addition of 0.1 ml of cold 20% trichloroacetic acid. When histone was used as the ADP-ribose acceptor, the final concentration of trichloroacetic acid was 20%. After 30 min at 0°, precipitated proteins were collected on 0.45-μm Millipore filters. Filters were washed three times with 2 ml of cold 5% trichloroacetic acid. Dried filters were transferred to vials and dissolved in 2 ml of Piersolve for radioassay.[1,4]

Assay 3. Formation of [adenine-U-^{14}C]ADP-ribose agmatine

[adenine-U-^{14}C]NAD + agmatine → [adenine-U-^{14}C]ADP-ribose agmatine
+ nicotinamide + H[+]

As it measures directly the ADP-ribosylated product, this assay is especially useful in crude systems where NAD glycohydrolases and phospho-

[3] J. Moss, V. C. Manganiello, and M. Vaughan, *Proc. Natl. Acad. Sci. U.S.A.* **73,** 4424 (1976).
[4] J. Moss, S. J. Stanley, and J. C. Osborne, Jr., *J. Biol. Chem.* **256,** 11452 (1981).

diesterases may catalyze agmatine-independent degradation of [*carbonyl*-^{14}C]NAD to release [*carbonyl*-^{14}C]nicotinamide. Assays (total volume 0.3 ml) containing 20 mM potassium phosphate (pH 7.0), 6 mM agmatine, and 30 μM [*adenine*-U-^{14}C]NAD (40,000 cpm) were initiated with transferase A (1 ng) and incubated for 30 min at 30°.[5] Two 0.1-ml samples were then transferred to columns of AG 1-X2 prepared and eluted as in Assay 1.

Assay 4. Formation of ADP-ribose-[^{125}I]guanyltyramine

$$\text{NAD} + [^{125}\text{I}]\text{guanyltyramine} \rightarrow \text{ADP-ribose-}[^{125}\text{I}]\text{guanyltyramine} + \text{nicotinamide} + \text{H}^+$$

The assay was developed for use with choleragen,[6] a bacterial NAD: arginine ADP-ribosyltransferase (see this volume [42]). The principal advantages of the assay are the relatively low cost of [^{125}I]guanyltyramine and the possibility of using high concentrations of NAD, which minimizes problems (particularly in the [*carbonyl*-^{14}C]nicotinamide release assay) caused by NAD glycohydrolases in some impure enzyme preparations. The primary disadvantage of the assay, as noted below, is that [^{125}I] guanyltyramine is metabolized by some tissue preparations to give products that may cochromatograph with ADP-ribosyl[^{125}I]guanyltyramine.[6] Assays (0.1 ml total volume) containing 50 mM potassium phosphate (pH 7.0), ovalbumin (1 mg/ml), 0.2–1.0 nM [^{125}I]guanyltyramine (100,000 cpm) with and without 3 mM NAD were initiated with ADP-ribosyltransferase (2 ng) and incubated for 60 min at 30°.[7] After addition of 1 ml of water, the contents of each assay tube were transferred to a column (1 ml) of DEAE-Sephadex A-25 followed by two 10-ml portions of water. ADP-ribosyl [^{125}I]guanyltyramine was eluted with 10 ml of 0.5 N HCl into a vial for radioassay.

The product synthesized by the purified transferase in the presence of NAD is believed to be ADP-ribosyl[^{125}I]guanyltyramine. Iodination of guanyltyramine apparently results in both mono- and diiodinated products; two ADP-ribosylated [^{125}I]guanyltyramine products can be resolved by appropriate chromatographic procedures.[7] With crude enzyme systems, various [^{125}I]guanyltyramine degradation products that chromatograph with the ADP-ribosylated product are formed in NAD-dependent reactions. In such cases, the product must be identified further by other means, such as high-performance liquid chromatography.

---

[5] J. Moss and S. J. Stanley, *Proc. Natl. Acad. Sci. U.S.A.* **78**, 4809 (1981).

[6] J. J. Mekalanos, R. J. Collier, and W. R. Romig, *J. Biol. Chem.* **254**, 5849 (1979).

[7] P. A. Watkins, D. A. Yost, A. W. Chang, J. J. Mekalanos, and J. Moss, *Fed. Proc., Fed. Am. Soc. Exp. Biol.* **42**, 2046 (1983).

ADP-ribosyltransferase A from Turkey Erythrocytes

Purification

Turkey blood (5 liters) was shipped to the laboratory on ice (from Pel-Freez). Erythrocytes were pelleted by centrifugation at 25,000 $g \times 30$ min, washed with iced physiological saline, and stored at $-20°$. For enzyme preparation, cells were thawed at 4°, suspended in an equal volume of 50 mM potassium phosphate (pH 7.0), and homogenized in a Waring blender. The homogenate was centrifuged (25,000 g, 30 min). The supernatant (5 liters), after bringing the salt concentration to 1.2 M NaCl in 50 mM potassium phosphate (pH 7.0), was mixed with phenyl-Sepharose (1 liter).[8] After the gel had settled, the supernatant was discarded. The gel was washed five times with 4 liters of 50 mM potassium phosphate (pH 7.0)/1 M NaCl, then eluted three times with 7.5 liters of 50 mM potassium phosphate (pH 7.0)/50% propylene glycol. The eluate was applied to (carboxymethyl)cellulose (4 liters) equilibrated with 50 mM potassium phosphate (pH 7.0)/0.1 M NaCl. The gel was then washed with 12 liters of 50 mM potassium phosphate (pH 7.0)/0.1 M NaCl/25% propylene glycol and eluted with 8.2 liters of 50 mM potassium phosphate (pH 7.0)/1 M NaCl/25% propylene glycol. The enzyme was concentrated by binding to phenyl-Sepharose (100 ml) and eluting with six portions (total 770 ml) of 50 mM potassium phosphate (pH 7.0)/50% propylene glycol. The eluate was diluted with an equal volume of 50 mM potassium phosphate and applied to NAD-agarose (50 ml). The gel was washed with 200 ml of 25% propylene glycol/50 mM potassium phosphate and eluted with 300 ml of 50% propylene glycol/50 mM potassium phosphate. The eluate was mixed with 300 ml of 50 mM potassium phosphate, pH 7.0, and applied to 50 ml of concanavalin A-agarose column which was washed with 50 mM potassium phosphate, pH 7.0/20% propylene glycol (100 ml) and eluted with 50 mM potassium phosphate, pH 7.0/20% propylene glycol/0.5 M NaCl/0.3 M methyl glucoside. The enzyme was concentrated on a phenyl-Sepharose column as described above. Results of a representative purification are shown in Table I.

Properties of Transferase A

Purified transferase A exhibited one major band on sodium dodecyl sulfate–polyacrylamide gel electrophoresis. The enzyme chromatographed as a single peak on Ultrogel AcA 54 in the presence of 50 mM potassium phosphate, pH 7.0/30% glycerol/1 M NaCl.[8] Activity cochro-

[8] J. Moss, S. J. Stanley, and P. A. Watkins, *J. Biol. Chem.* **255,** 5838 (1980).

TABLE I
PURIFICATION OF ADP-RIBOSYLTRANSFERASE A FROM TURKEY ERYTHROCYTES[a]

Step	Protein	Units (μmol min^{-1})	Specific activity (μmol min^{-1} mg^{-1})	Yield (%)	Fold purification
1. Supernatant	578,000	376	0.00065	100[b]	1
2. Phenyl-Sepharose	5,920	368	0.0622	97.8	95.
3. (Carboxymethyl)-cellulose	92.9	203	2.18	53.9	3,360
4. NAD-agarose	6.60	123	18.6	32.6	28,500
5. Concanavalin A-agarose	0.195	68.9	353	18.3	544,000

[a] Modified from Moss *et al.*[8]
[b] Supernatant contains both transferases, A and B. The fraction of total activity attributable transferase B is, however, relatively small, since B is inhibited by the salt present in t transferase A assay.

matographed with protein and specific activity was constant across the protein peak.

In dilute aqueous solutions of transferase A, activity declined rapidly at 0–4° but was stable for several months in the presence of 50% propylene glycol. The stability of the transferase at 30° was increased by histone,[9] lysolecithin,[10] and several nonionic or zwitterionic detergents.[10]

Activators and Their Effects on Physical Properties

Transferase A activity, assayed with relatively low concentrations of agmatine (20 mM) as ADP-ribose acceptor, was increased >10-fold by certain inorganic salts.[4] Chaotropic salts were the more potent; their order of effectiveness corresponded to the Hofmeister species (SCN$^-$ > Br$^-$ > Cl$^-$ > F$^-$ > PO$_4^{3-}$). NaCl was maximally effective at ~250 mM and NaSCN at ~100 mM. In the presence of 250 mM NaCl, the K_m for arginine methyl ester was 1.3 mM and Lineweaver–Burk plots were linear. In the absence of salt, plots were nonlinear, consistent with substrate activation, with limiting slopes corresponding to K_ms of 3.8 and 50 mM. The effect of salt on ADP-ribosylation of proteins was variable; depending on the protein, NaCl increased, decreased, or had no effect on the rate. High concentrations of NaCl increased NAD glycohydrolase activity to a greater extent than transferase activity (with ovalbumin as ADP-ribose acceptor).[4]

In the absence of salt, with low concentrations of ADP-ribose acceptor, histones increased activity.[5,9] Identical maximal activities were at-

[9] J. Moss, S. J. Stanley, and J. C. Osborne, Jr., *J. Biol. Chem.* **257,** 1660 (1982).
[10] J. Moss, J. C. Osborne, Jr., and S. J. Stanley, Biochemistry (in press).

:ained with optimal concentrations of salt or histone or both together.[5] Histone activation (like salt activation) was rapid and reversible[9]; it was maximal with ~10 μg/ml, a concentration approximately two orders of magnitude lower than that required for histones to serve effectively as ADP-ribose acceptors under standard assay conditions in the presence of agmatine.[5]

Transferase A activity was also increased by certain phospholipids. Lysophosphatidylcholine containing a C_{18} or C_{16} saturated fatty acid was most effective; the order was $C_{18} \cong C_{16} > C_{14} > C_{12} > C_{10} \cong C_8$.[10] Phosphatidylcholine, lysophospholipids not containing the choline moiety, and choline were inactive.[10] Maximal activation by lysolecithin was less than that induced by histone or salt. Addition of lysolecithin did not decrease activation by histone or salt.[10]

Several nonionic detergents, such as Triton X-100, and the zwitterionic detergent CHAPS also enhanced activity but to a lesser extent than did salt, histones, or lysolecithin. Some of these detergents in high concentrations were inhibitory. In the presence of histone or salt plus detergent, activity approximated that with histone or salt alone.[10,11] In the presence of lysolecithin plus detergent, however, activity was equal to that with detergent alone, perhaps due to detergent-induced disruption of lysolecithin micelles.

Histone or NaCl in suitable concentrations converted the transferase from an inactive, high-molecular-weight form to an active protomeric species.[4,9] The inactive form eluted in the void volume of a Sephadex G-200 column[4] and sedimented rapidly in a propylene glycol gradient.[9] In the presence of histone, transferase A sedimented more slowly.[9] In the presence of salt, the enzyme chromatographed on Sephadex G-200 with a K_{av} similar to that of chymotrypsinogen ($M_r \cong 25,300$).[4] As the molecular weight of the transferase estimated by sodium dodecyl sulfate–gel electrophoresis is ~28,000,[8] it appears that the catalytically active form of the enzyme observed in the presence of histone or salt is a protomeric species.

Substrate Specificity

Transferase A can use several purified proteins, polyarginine, and low-molecular-weight guanidino compounds, such as arginine and agmatine, as ADP-ribose acceptors.[12,13] The enzyme ADP-ribosylated both soluble and particulate proteins from a number of tissues.[13] The ability of a protein to serve as an acceptor was dependent on assay conditions (salt,

[11] J. Moss and S. J. Stanley, unpublished observations.
[12] J. Moss, S. J. Stanley, and N. J. Oppenheimer, *J. Biol. Chem.* **254**, 8891 (1979).
[13] P. A. Watkins and J. Moss, *Arch. Biochem. Biophys.* **216**, 74 (1982).

nucleotide concentrations).[1,13] High concentrations of arginine methyl ester inhibited ADP-ribosylation of protein.[1] Low-molecular-weight guanidino compounds, with a positive charge in the vicinity of the guanidino group, e.g., agmatine, arginine methyl ester, were more effective acceptors than those with a negative charge, such as D- or L-arginine and guanidinopropionate.[12] NAD was preferred over NADP as a donor of ADP-ribose and phospho-ADP-ribose, respectively.[12]

Apparent K_m values for NAD and low-molecular-weight guanidino compounds such as arginine methyl ester were influenced by activation (i.e., NaCl) and the concentration of the other substrate. With 100 μM NAD and 200 mM NaCl, the K_m for arginine methyl ester was 1.3 mM. In the presence of a high concentration of ADP-ribose acceptor and 200 mM NaCl, the K_m for NAD was ~30 μM; it appeared to be <10 μM in the absence of acceptor (unpublished observations).

Purification of ADP-Ribosyltransferase B from Turkey Erythrocytes[2]

Turkey erythrocytes (1.5 liters) prepared as described for purification of transferase A were homogenized in 3 liters of 200 mM sodium phosphate (pH 7.1) in a Waring blender for 15 sec. After centrifugation of the homogenate (10 min, 25,000 g), the supernatant was adjusted to 1.2 M NaCl and mixed with phenyl-Sepharose (300 ml) for 24 hr at 4°. The phenyl-Sepharose on a fritted glass funnel was then washed twice with 4 liters of 1.2 M NaCl. The combined washes were stirred with 100 ml of

TABLE II

PURIFICATION OF ADP-RIBOSYLTRANSFERASE B FROM TURKEY ERYTHROCYTES[a]

Step	Protein (mg)	Units (μmol min^{-1})	Specific activity (μmol min^{-1} mg^{-1})	Yield (%)	Fold purification
1. Supernatant	277,000	[b]	—	—	—
2. Phenyl-Sepharose	200,000	55	0.0003	100	1
3. Concanavalin A-agarose	62.5	46	0.74	84	3,750
4. (Carboxymethyl)-cellulose	9.0	34	3.8	60	19,300
5. Procion-Red-agarose	1.0	30	32	55	160,000
6. Ultrogel AcA 54	0.6	27	53	49	268,000

[a] Data from Yost and Moss.[2]
[b] Total amount of NAD : arginine ADP-ribosyltransferase activity was 170 units (determined by Assay 1); activity is predominantly ADP-ribosyltransferase A.

TABLE III
CHARACTERISTICS OF ADP-RIBOSYLTRANSFERASES A AND B

Property	Transferase A	Transferase B
Molecular weight[a]	28,000	32,000
K_m—NAD	15 μM	36 μM
K_m—arginine methyl ester	1.3 mM	3 mM
V_{max} (μmol min^{-1} mg^{-1})[b]	350	60
Effect of histones		
On size	Decrease	None
On activity	Increase	None
Effect of NaCl		
On size	Decrease	None
On activity	Increase	Decrease

[a] Molecular weight of protomer. In the absence of NaCl or histone, transferase A forms large aggregates.

[b] With arginine methyl ester as ADP-ribose acceptor; assays for transferases A and B as noted in the text.

concanavalin A-Sepharose for 24 hr at 4°. The gel was washed twice with 2 liters of 50 mM sodium phosphate, pH 7.1/1 M NaCl and twice with 2 liters of 50 mM sodium phosphate, pH 7.1. The enzyme was eluted with six 200-ml portions of 50 mM sodium phosphate, pH 5.8/0.4 M α-methyl mannoside and applied to (carboxymethyl)cellulose (~1 liter resin) equilibrated with 50 mM sodium phosphate, pH 5.8. The gel was washed with 4 liters of 50 mM sodium phosphate, pH 5.8. The enzyme was eluted with 50 mM sodium phosphate, pH 8.1, and applied to Procion Red-agarose (20 ml). The gel was washed with 50 mM sodium phosphate (pH 7.1)/75 mM NaCl and eluted with 50 mM sodium phosphate, pH 7.1/1 M NaCl. Fractions containing transferase activity were concentrated on an Amicon YM-10 ultrafiltration membrane unit to a volume of 2 ml and applied to a column (1.5 × 90 cm) of Ultrogel AcA 54, which was eluted with sodium phosphate (pH 7.1)/100 mM NaCl. Active fractions were pooled and concentrated as described above. Results of a representative purification are shown in Table II.

The purified transferase B exhibited one band on sodium dodecyl sulfate–polyacrylamide gel electrophoresis (M_r = 32,000). It was considerably more stable than transferase A and could be stored for several months at 0–4° in 50 mM sodium phosphate, pH 7.1/100 mM NaCl without loss of activity. The properties of transferases A and B are compared in Table III.

[45] Overview of Poly(ADP-ribosyl)ation

By Mark E. Smulson and Takashi Sugimura

Historically, the ADP-ribosylation reaction was first ascertained by experiments in 1965 which demonstrated the formation of poly(ADP-ribose) from the substrate NAD^+ as catalyzed by chromatin-associated enzyme(s). It was shortly revealed that the formation of poly(ADP-ribose) by chromatin enzyme(s) was actually initiated by ADP-ribosylation on chromatin proteins.

The following chapters of this volume encompass the scope of recent advancements in very basic methodology in this area.

The ADP-ribosylation reaction is catalyzed by enzymes which transfer the ADP-ribose residue of NAD^+ to various acceptor proteins (Fig. 1). The enzymatic reactions for mono(ADP-ribosyl)ation and for poly(ADP-ribosyl)ation differ in many aspects. The DNA requirement for enzymatic reaction and formation of the unique polymer, poly(ADP-ribose), is the most fascinating characteristic of the poly(ADP-ribosyl)ation reaction, catalyzed by poly(ADP-ribose) polymerase or synthetase.

Since NAD^+ was noticed as a cofactor of fermentation in 1904 by Harden and Young, it has been established to be a cofactor for many dehydrogenases, such as alcohol dehydrogenase, lactate dehydrogenase and malate dehydrogenase. Around 1965, two new reactions were noted to utilize NAD^+. NAD^+ was shown to be used in the DNA ligase reaction in prokaryotes, where the nucleotide is cleaved at the pyrophosphate bond to yield NMN and a $5'$-AMP-DNA intermediate; $5'$-AMP is subsequently released when DNA is ligated. The other is the ADP-ribosylation reaction described here. The enzyme involved in poly(ADP-ribosyl)ation has been purified from various tissues including calf and pig thymus, rat liver, Ehrlich ascites tumor cells, and HeLa cells.

The poly(ADP-ribosyl)ation reaction has been shown to correlate with the number of strand breaks present in the DNA which is required by the polymerase for activity. However, DNA does not serve as a template as it does in the cases of DNA polymerases and RNA polymerases, but rather appears to activate the enzyme at the nicked site of DNA.

Poly(ADP-ribose) attached to acceptors is the product of poly(ADP-ribosyl)ation reaction. These acceptors include histones, nonhistone nuclear proteins and poly(ADP-ribose) polymerase itself, and methods for the characterization of these acceptors are provided in the following chapters. Poly(ADP-ribose) is formed by a successive ADP-ribosylation reac-

$$\text{Acceptor} + n \cdot \text{NAD}^+ \xrightarrow[\substack{\text{poly(ADP-ribose)}\\ \text{polymerase}}]{\text{DNA}} \text{Acceptor-(ADP-ribose)}_n + n \cdot \text{nicotinamide}$$

$$+ n \cdot \text{H}^+$$

$$\text{Acceptor} + \text{NAD}^+ \xleftarrow[]{\substack{\text{ADP-ribosyl}\\ \text{transferase}}} \text{Acceptor-(ADP-ribose)} + \text{nicotinamide}$$

$$+ \text{H}^+$$

FIG. 1. Poly(ADP-ribosyl)ation (top) and mono(ADP-ribosyl)ation (bottom) reactions.

tion using ADP-ribose residues as acceptors. Both a linear form and a branched form of poly(ADP-ribose) have been established (Fig. 2). ADP-ribose residues are linked through ribose–ribose bonds in the linear chain and by ribose–ribose–ribose bonds at branch points. The chemistry of bonds between riboses has been clarified recently as D-ribofuranosyl-O-α-(1→2)-D-ribofuranoside bond, using ^{13}C NMR, ^{1}H NMR, mass spectrometry and gas–liquid chromatography.

Poly(ADP-ribose) is hydrolyzed by two major classes of enzymes: (1) poly(ADP-ribose) glycohydrolase, which hydrolyzes the ribose–ribose bond or the ribose–ribose–ribose bond in the branched form of the polymer; (2) phosphodiesterase (or pyrophosphatase), which hydrolyzes the pyrophosphate bonds. However, poly(ADP-ribose) glycohydrolase plays the more important role *in vivo* for hydrolysis of poly(ADP-ribose). A third enzyme, ADP-ribosyl histone splitting enzyme, cleaves the bond between ADP-ribose and histones. Accordingly, when poly(ADP-ribose) glycohydrolase and ADP-ribosyl histone splitting enzyme function in concert, poly(ADP-ribosyl) histone can be converted to unmodified histone, which may again serve as acceptor for poly(ADP-ribose) polymerase (Fig. 3).

Poly(ADP-ribose) was initially discovered and characterized by the use of an *in vitro* reaction with isolated nuclei and NAD$^+$. Methods for

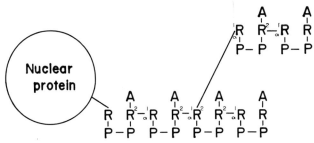

FIG. 2. A branching form of poly(ADP-ribose). A, Adenine; R, ribose; P, phosphate.

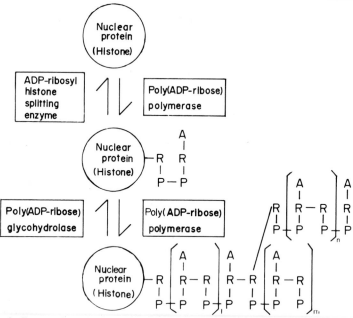

FIG. 3. Enzymatic poly(ADP-ribosyl)ation and hydrolysis reactions. A, Adenine; R, ribose; P, phosphate. l, m, n represent the numbers of repeating monomer units.

quantitating the small quantities of poly(ADP-ribose) present *in vivo* have been recently developed, and various approaches are discussed in the following chapters. These methods will be very useful in studying the turnover of poly(ADP-ribose) and biological functions of poly(ADP-ribosyl)ation, e.g., mutation and cell transformation. A number of review articles on ADP-ribosylation reactions have been published,[1-9] and should be useful to researchers interested in this field.

[1] T. Sugimura, *Prog. Nucleic Acid Res. Mol. Biol.* **13**, 127 (1973).

[2] H. Hilz and P. R. Stone, *Rev. Physiol., Biochem. Pharmacol.* **76**, 1 (1976).

[3] O. Hayaishi and K. Ueda, *Annu. Rev. Biochem.* **46**, 95 (1977).

[4] M. R. Purnell, P. R. Stone, and W. J. D. Whish, *Biochem. Soc. Trans.* **8**, 215 (1980).

[5] T. Sugimura, M. Miwa, H. Saito, Y. Kanai, M. Ikejima, M. Terada, M. Yamada, and T. Utakuji, *Adv. Enzyme Reg.* **18**, 195 (1980).

[6] M. E. Smulson and T. Sugimura, eds., "Novel ADP-Ribosylations of Regulatory Enzymes and Proteins." Elsevier/North-Holland, New York, 1979.

[7] P. Mandel, H. Okazaki, and C. Niedergang, *Prog. Nucleic Acid Res. Mol. Biol.* **27**, 1 (1982).

[8] O. Hayaishi and K. Ueda, eds., "ADP-Ribosylation Reactions." Academic Press, New York, 1982.

[9] M. Miwa, O. Hayaishi, S. Shall, M. Smulson, and T. Sugimura, eds., "ADP-Ribosylation, DNA Repair, and Cancer." Jpn. Sci. Soc. Press, Tokyo, 1983.

[46] Structure of Poly(ADP-ribose)

By Masanao Miwa and Takashi Sugimura

Preparation and Purification of Poly(ADP-ribose)

In this method a calf thymus nuclear preparation was used as a source of poly(ADP-Rib) [poly(ADP-ribose)] polymerase.

Preparation of Crude Poly(ADP-Rib) Polymerase. Fifty grams of fresh or frozen calf thymus was minced or cracked to small pieces, mixed with 100 ml of 0.88 M sucrose in 3 mM MgCl$_2$ and homogenized in a Waring blender operated at 50 V for four 5-min periods with 1-min intervals for cooling. Then another 100 ml of 0.88 M sucrose in 3 mM MgCl$_2$ was added and the homogenate was filtered through 4 layers of gauze. The filtrates from 250 g of calf thymus were pooled and centrifuged for 10 min at 10,000 g at 4°, and the resulting pellet was used as crude poly(ADP-Rib) polymerase.

Incubation for Synthesis of Poly(ADP-Rib). The incubation mixture consisted of 20 ml of 1 M Tris–HCl buffer, pH 8.0, 6 ml of 1 M MgCl$_2$, 4 ml of 0.1 M nonlabeled NAD$^+$ or [*adenosine*-^{14}C]NAD$^+$ (5 × 10^3 cpm/μmol), crude poly(ADP-Rib) polymerase preparation, and 0.88 M sucrose to give a final volume of 200 ml in the homogenizing vessel of a Waring blender. The mixture was homogenized and incubated for 15 min at 37°. Then a mixture of 400 ml of ethanol and 25 ml of 5 M sodium acetate buffer, pH 5.0, was added.

Purification of Poly(ADP-Rib)

Pronase Digestion. The precipitate, containing formed poly(ADP-Rib) with ethanol, was washed with ethanol and ether. The ether was removed by air drying and the pellet was suspended in 300 ml of 50 mM Tris–HCl buffer, pH 8.0, and clumps of pellet were broken up by homogenization in a Potter-Elvehjem type glass-Teflon homogenizer. Then 200 mg of Pronase E and 0.02% NaN$_3$ were added and the mixture was incubated overnight at 37°. After incubation for 1 hr, 200 μg of pancreatic DNase I and MgCl$_2$ to a final concentration of 5 mM were added to reduce the viscosity of the mixture.

First Phenol Extraction. The incubation mixture was mixed with 0.8 volume of water-saturated phenol and shaken for 20 min at room temperature. The aqueous layer obtained by centrifugation was mixed with 2 volumes of ethanol and 0.125 volume of 5 M sodium acetate buffer, pH 5.0, and stored overnight at 4°.

METHODS IN ENZYMOLOGY, VOL. 106

Nuclease Digestion. Poly(ADP-Rib) and nucleic acids were collected by centrifugation and washed first with ethanol and then with ether. The pellet was dissolved in 40 ml of 50 mM Tris–HCl buffer, pH 8.0, and digested by adding 2 mg of pancreatic DNase I, 2 mg of pancreatic RNase, and 0.5 mg of nuclease P$_1$ with MgCl$_2$ to a final concentration of 5 mM. The incubation was performed overnight at 37° in the presence of 1% (v/v) toluene. During the incubation, the mixture was maintained at about pH 8 by addition of 2 N NaOH. Then 20 mg of Pronase E was added and incubation was continued for 30 min at 37°.

Second Phenol Extraction. After treatment with nuclease, the solution was mixed with 0.8 volume of water-saturated phenol and shaken for 20 min at room temperature. The aqueous layer was collected after centrifugation, and after removal of phenol with ether, it was dialyzed overnight against 0.1 M sodium phosphate buffer, pH 6.8.

Hydroxylapatite Column Chromatography. The dialyzed poly(ADP-Rib) fraction was next applied to a 1.5 × 12 cm hydroxylapatite column (BioGel HTP, Bio-Rad Laboratories). In preparation of the column, hydroxylapatite was suspended in buffer, and fine particles were carefully removed by decantation. Furthermore, when packing the column, the flow of effluent was temporarily stopped to allow bigger particles to settle to the bottom of the column.

After the sample had been applied to the column, the column was washed with about 3 column volumes of 0.1 M sodium phosphate buffer, pH 6.8, and elution of poly(ADP-Rib) was carried out with a gradient formed with 200 ml of 0.1 M sodium phosphate buffer, pH 6.8, and 200 ml of 0.5 M sodium phosphate buffer, pH 6.8, at a flow rate of 30 ml/hr.

Figure 1 shows the profile on hydroxylapatite column chromatography of a large scale preparation of poly(ADP-Rib) synthesized by incubation of the calf thymus nuclei from 1250 g of thymus.

Poly(ADP-Rib) was eluted in 3 fractions with absorbance at 260 nm and radioactivity when labeled NAD$^+$ was added: Fraction I, the breakthrough fraction; Fraction II, eluted with 0.15–0.25 M sodium phosphate buffer, pH 6.8, just before the main peak, and Fraction III, the main peak eluted with 0.25–0.40 M sodium phosphate buffer, pH 6.8. The fraction eluted with 0.40–0.50 M sodium phosphate buffer, pH 6.8, was termed Fraction IV.

The sizes of poly(ADP-Rib) in Fractions I, II, and III were less than 10, 10–25, and more than 20 repetitions of ADP-Rib residues, respectively. The average size of poly(ADP-Rib) in Fraction IV was a little larger than that in Fraction III. Fraction I was contaminated with oligonucleotides and required further purification.

Fractions II, III, and IV were dialyzed against distilled water and

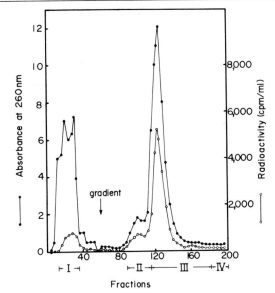

FIG. 1. Hydroxylapatite column chromatography of poly(ADP-Rib).

concentrated in a rotary evaporator. Poly(ADP-Rib) was precipitated by addition of 2 volumes of ethanol and 0.125 volume of 5 M sodium acetate buffer, pH 5.0. The precipitated poly(ADP-Rib) was recovered by centrifugation, washed with ethanol and ether, and stored at $-20°$. Usually 20 mg of poly(ADP-Rib) was obtained.

Properties of Poly(ADP-Rib)

Since Fraction III of poly(ADP-Rib) is most easily obtained in a pure state, its properties are described below. The purity of Fraction III of poly(ADP-Rib) was confirmed in three ways. First, the content of organic phosphate was measured, and assuming that the structure of poly(ADP-Rib) is (ADP-Rib)n, the dry weight of poly(ADP-Rib) was then calculated. This calculated value corresponded to more than 98% of the dry weight of Fraction III of poly(ADP-Rib). Second the dry weight of poly(ADP-Rib) corresponded with the radioactivity calculated from the specific radioactivity of the substrate NAD^+. Third, when Fraction III of poly(ADP-Rib) was hydrolyzed by snake venom phosphodiesterase and analyzed by Dowex 1 column chromatography,[1] no ultraviolet absorbing materials were eluted in the position of CMP and dCMP, indicating that there was

[1] S. Fujimura and T. Sugimura, this series, Vol. 18, Part B, p. 223.

no significant contamination with RNA or DNA. Therefore, the purity of the preparation was more than 98%.

The specific radioactivity, radioactivity/A_{260} unit, of Fraction III was 22% higher than the expected specific radioactivity of ADP-Rib, due to the hypochromicity of poly(ADP-Rib), resulting from its secondary structure.

Elemental Analysis. The contents of C, H, N, (O), P, and Na were 30.07, 4.48, 10.62, (35.13), 10.20, and 9.50%, respectively. The percentage of O was calculated by subtracting the percentages of other elements. If poly(ADP-Rib) is a polymer of repeating units of ADP-Rib as dibasic sodium salt, the expected values for C, H, N, O, P, and Na with 1 molecule of H_2O per ADP-Rib unit should be 29.86, 3.50, 11.61, 37.13, 10.27, and 7.63%, respectively. The observed values agree reasonably with the expected values for $(C_{15}H_{19}N_5O_{13}P_2Na_2 \cdot H_2O)n$.

Ultraviolet Absorption Spectrum. The ultraviolet absorption spectrum of Fraction III was similar to that of poly(A), with a peak in the ultraviolet region at 258 nm at pH 7.0 and a molar absorbance, $E(2P)$, of 13,000 at pH 7.0. The A_{280}/A_{260} ratio of Fraction III was 0.26 in 0.1 M sodium phosphate buffer, pH 6.8.

When heated from 25 to 98°, Fraction III of poly(ADP-Rib) showed about 12% increase of A_{260}. On digestion with snake venom phosphodiesterase, Fraction III of poly(ADP-Rib) showed about 15% increase of A_{260}. When mixed with poly(U), Fraction III of poly(ADP-Rib) did not show hypochromicity. Poly(ADP-Rib) was not adsorbed to a poly(U) Sepharose column under conditions in which poly(A) bound to the column.

Melting Point. The melting (decomposition) point of Fraction III of poly(ADP-Rib) was at 219–220°.

Amino Acid Analysis. Fraction III of poly(ADP-Rib) was hydrolyzed in 6 N HCl for 24 hr at 110°, and subjected to amino acid analysis. Glycine was the main amino acid, found at 0.59 mol/mol of adenine residue. This glycine was probably formed by decomposition of the adenine ring, because poly(A) also yielded 0.58 mol of glycine per adenine residue under the same conditions. Therefore, the purified poly(ADP-Rib) does not seem to have any amino acid attached to a terminus.

Size of Poly(ADP-Rib). Poly(ADP-Rib) in Fraction III showed a very broad size distribution; the biggest molecule was estimated by gel filtration on BioGel A-50m and gel electrophoresis in 20% polyacrylamide to be more than 4.5×10^5 daltons.

Electron Micrograph. Fraction III of poly(ADP-Rib) had a branching structure (Fig. 2).[2] Assuming that an ADP-Rib residue is twice the length

[2] K. Hayashi, M. Tanaka, T. Shimada, M. Miwa, and T. Sugimura, *Biochem. Biophys. Res. Commun.* **112,** 102 (1983).

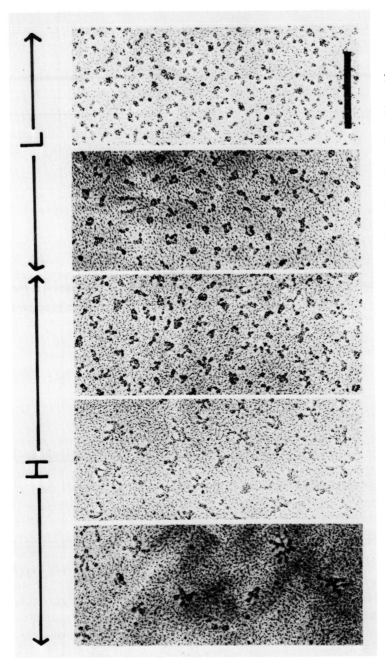

FIG. 2. Electron microgram of H and L fractions of poly(ADP-Rib). Bar indicates 1 μm. From Hayashi et al.[2]

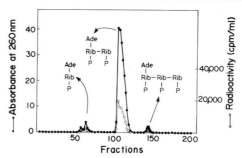

FIG. 3. DEAE-Sephadex A-25 column chromatography of hydrolysis products of poly([14C]ADP-Rib) treated by snake venom phosphodiesterase.

of a mononucleotide residue in DNA, the biggest molecule of poly(ADP-ribose) was calculated to have 300 residues of ADP-Rib and about 10 branched points.

Solubility and Buoyant Density. A higher molecular weight fraction (H fraction) and a lower molecular weight fraction (L fraction) were separated from Fraction III by sucrose density gradient centrifugation. In 2 M NaCl the L fraction was soluble, whereas the H fraction was gradually precipitated. On $CsSO_4$ density gradient equilibrium centrifugation, the L fraction formed a band with a peak at a buoyant density of 1.57 g/cm^3, whereas the H fraction was recovered from the bottom of the tube, presumably because it was precipitated at high salt concentration.

Preparation of the Monomer Unit and the Branched Portion of Poly(ADP-Rib)

Preparation of the Monomer Unit of Poly(ADP-Rib), Ado(P)-Rib-P. About 150 mg of fraction III (or II) of poly(ADP-Rib) was dissolved in 2.4 ml of distilled water and mixed with 24.6 units of snake venom phosphodiesterase, which had been purified by the method of Oka *et al.*,[3] and final concentrations of 1 mM $MgCl_2$ and at 0.02% NaN_3. The mixture was incubated at 37° overnight. The process of hydrolysis was followed by subjecting small samples to thin-layer chromatography in a mixture of isobutyric acid : concentrated ammonia : water (66 : 1 : 33, v/v). The reaction was considered to be complete when ultraviolet absorbing material disappeared from the origin and hydrolytic products appeared in positions corresponding to those of Ado(P)-Rib-P and 5′-AMP.

The incubation mixture was mixed with 0.55 g of urea/ml to give a final concentration of 7 M and applied to a 0.5 × 76 cm DEAE-Sephadex A-25 column equilibrated with 50 mM Tris–HCl buffer, pH 7.5, in 7 M urea.

[3] J. Oka, K. Ueda, and O. Hayaishi, *Biochem. Biophys. Res. Commun.* **80,** 841 (1978).

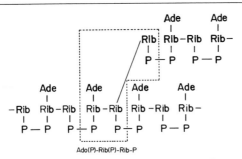

FIG. 4. Branching structure of poly(ADP-Rib). Branched portion, Ado(P)-Rib(P)-Rib-P, is enclosed with broken lines. Ade, Adenine; Rib, ribose; P, phosphate.

Material was eluted with 400 ml of a linear gradient of 0–0.35 M NaCl in 50 mM Tris–HCl buffer, pH 7.5, in 7 M urea at a flow rate of 10 ml/hr. Three peaks of material were separated (Fig. 3). The first peak corresponded to 5′-AMP. A very minor peak was seen just before 5′-AMP, which would be that of Ado-Rib-P and Ado(P)-Rib derived from the termini of poly(ADP-Rib) by partial digestion with an unidentified pyrophosphatase.[4] The major peak was that of Ado(P)-Rib-P, the monomer unit of poly(ADP-Rib). The peak appearing after that of Ado(P)-Rib-P was identified as that of Ado(P)-Rib(P)-Rib-P, the branched portion of poly(ADP-Rib)[5] (Fig. 4). The percentage of Ado(P)-Rib(P)-Rib-P in the total hydrolyzate of poly(ADP-Rib) varied in different fractions, being about 2, 1, and 0.5% in Fractions III, II, and I, respectively.

The fractions of Ado(P)-Rib-P and Ado(P)-Rib(P)-Rib-P were each pooled, diluted 3-fold with distilled water, and applied to a DE 52 column equilibrated with 20 mM triethylammonium bicarbonate (TEAB) buffer, pH 9.0. Ado(P)-Rib-P and Ado(P)-Rib(P)-Rib-P were both adsorbed to the column; the columns were washed with 10 column volumes of 20 mM TEAB to remove urea, and then the adsorbed Ado(P)-Rib-P or Ado(P)-Rib(P)-Rib-P was eluted with 2 M TEAB. The TEAB was evaporated in a rotary evaporator to obtain desalted preparations of Ado(P)-Rib-P and Ado(P)-Rib(P)-Rib-P.

Structure and Properties of Ado(P)-Rib-P and Ado(P)-Rib(P)-Rib-P

Structure of Ado(P)-Rib-P. Ado(P)-Rib-P shows the same λ_{max} and λ_{min} at pH 2, 7 and 12 as 5′-AMP. The chemical linkage of the ribose–

[4] M. Tanaka, M. Miwa, K. Hayashi, K. Kubota, T. Matsushima, and T. Sugimura, *Biochemistry* **16**, 1485 (1977).
[5] M. Miwa, M. Ishihara, S. Takishima, N. Takasuka, M. Maeda, Z. Yamaizumi, T. Sugimura, S. Yokoyama, and T. Miayazawa, *J. Biol. Chem.* **256**, 2916 (1981).

ribose bond in Ado(P)-Rib-P was shown to be 1″→2′ by thin-layer and gas–liquid chromatographies. The ribose–ribose bond of Ado(P)-Rib-P was not hydrolyzed by poly(ADP-Rib) glycohydrolase.

The structure of Ado(P)-Rib-P was studied by ^{13}C NMR spectrometry.[6] The ^{13}C NMR spectra of Ado(P)-Rib-P, poly(ADP-Rib), ADP-Rib, and poly(A) are shown in Fig. 5. The spectrum of Ado(P)-Rib-P is very similar to that of poly(ADP-Rib). The ^{13}C NMR peaks of ADP-Rib were assigned using data on 5′-AMP. There are α- and β-anomers of ribofuranose attached to the ADP moiety of ADP-Rib, and also two anomeric forms of ribofuranose in ribose 5-phosphate. Ribose in aqueous solution is known to exist in four tautomeric forms: 56% as β-ribopyranose, 20% as α-ribopyranose, 18% as β-ribofuranose, and 6% as α-ribofuranose. The ^{13}C resonance peaks of the α- and β-ribofuranoses were identified by their relative intensities. Downfield displacement by 2.5 ppm was observed in C-5′ due to the presence of a phosphate group.

The ^{13}C resonances of 10 peaks for two sets of ribose moieties are observed and they were assigned by taking account of ^{31}P–^{13}C spin–spin couplings, temperature-dependent shift of the resonance due to base–base stackings, and data on ADP-Rib, 5′-AMP, methyl-α-D- and methyl-β-D-ribofuranoses.[6]

The downfield displacement of the C-2′ of Ado(P)-Rib-P by about 5 ppm compared to the resonance of this carbon in 5′-AMP and ADP-Rib, indicates that the glycosidic bond is formed at this carbon atom. This finding is consistent with the observation that the ribose–ribose bond is 1″→2′. A ribofuranose type sugar structure was indicated for the ribose moiety in adenosine. The binding of the ribose moiety to the adenosine moiety was indicated by the similarity in the ^{13}C resonance of the sugars to the resonances of 5′-AMP and ribose 5-phosphate. Two lines of evidence indicate the α-anomeric configuration of the ribose–ribose linkages. First, the ^{13}C resonances of C-1″, C-2″, C-3″, C-4″, and also C-5″ (considering the downfield displacement effect of 5″-phosphate) were very similar to those of methyl-α-D-ribofuranoside, but not methyl-β-D-ribofuranoside. Here methyl-β-D-ribofuranoside was proved to be a good reference compound in the case of ribostamycin and hexa-N-acetyl neomycin B. Second, the ^{13}C resonance of C-1″ of Ado(P)-Rib-P was found to be shifted downfield by 5.4–5.7 ppm from the corresponding ^{13}C resonance of the α-anomer of ADP-Rib and by 0.6–0.9 ppm from that of the β-anomer of ADP-Rib. A downfield displacement of about 5–7 ppm has generally been observed in the case of glycosidic bond formation or

[6] M. Miwa, H. Saitô, H. Sakura, N. Saikawa, F. Watanabe, T. Matsushima, and T. Sugimura, *Nucleic Acids Res.* **4**, 3997 (1977).

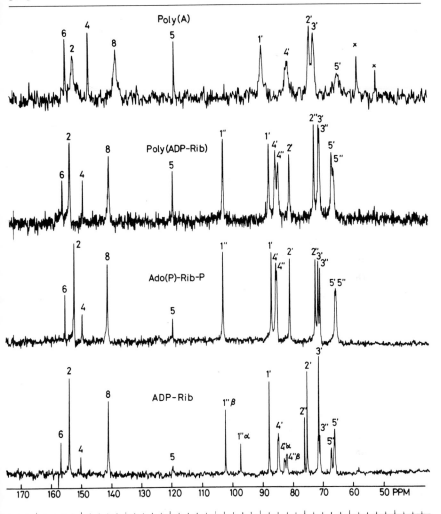

FIG. 5. ^{13}C NMR spectra of Ado(P)-Rib-P, poly(ADP-Rib), ADP-Rib, and poly(A). From Miwa *et al.*[6]

O-isopropylation of pentoses and hexoses. The downfield displacement by 5.4–5.7 ppm of the ^{13}C resonance of C-1″ of Ado(P)-Rib-P could be explained by glycosidic bond formation to α-linked ribofuranoside of Ado(P)-Rib-P. The above data confirmed that Ado(P)-Rib-P has an α-(1″→2′) ribofuranosyl ribofuranoside bond.

Structure and Properties of Ado(P)-Rib(P)-Rib-P. The structure of Ado(P)-Rib(P)-Rib-P was determined to be 2′-[1″-ribosyl-2″-(1‴-ribosyl)] adenosine-5′,5″,5‴-tris(phosphate) or O-α-D-ribofuranosyl-(1‴→2″)-O-α-D-ribofuranosyl-(1″→2′)-adenosine-5′,5″,5‴-tris(phosphate).

The ultraviolet spectrum was similar to that of 5′-AMP and Ado(P)-Rib-P. The ribose–ribose–ribose bond of Ado(P)-Rib(P)-Rib-P was also not hydrolyzed by poly(ADP-Rib) glycohydrolase.

[47] Evidence for Poly(ADP-ribosyl) Derivatives of Carboxylates in Histone

By Kunihiro Ueda and Osamu Hayaishi

Poly(ADP-ribosyl)ation of histone has been studied most extensively among various poly(ADP-ribosyl)ation reactions,[1] and evidence has been obtained that a majority of ADP-ribosyl groups are attached to histone (H2B and H1) through an ester bond between the terminal ribose residue and the carboxyl group of either an internal glutamic acid residue[2–5] or a COOH-terminal residue.[5] Details, however, remain to be clarified, including the location and the steric conformation of the bond on the ribose as well as the nature of bonds that are not esterlike.[6] This chapter describes the methods of preparation of various ADP-ribosyl derivatives of histone and summarizes evidence for their being carboxyl esters. Possible heterogeneity of ADP-ribosyl histone linkages[7] is also discussed.

Preparation of ADP-Ribosylated Histone Derivatives

Nuclei are isolated from rat liver by the sucrose-calcium procedure of Chauveau *et al.*,[8] as modified by Busch[9] and Ogata *et al.*[4] Chromatin is

[1] O. Hayaishi and K. Ueda, *in* "ADP-ribosylation Reactions: Biology and Medicine" (O. Hayaishi and K. Ueda, eds.), p. 3. Academic Press, New York, 1982.
[2] P. T. Riquelme, L. O. Burzio, and S. S. Koide, *J. Biol. Chem.* **254**, 3018 (1979).
[3] L. O. Burzio, P. T. Riquelme, and S. S. Koide, *J. Biol. Chem.* **254**, 3029 (1979).
[4] N. Ogata, K. Ueda, and O. Hayaishi, *J. Biol. Chem.* **255**, 7610 (1980).
[5] N. Ogata, K. Ueda, H. Kagamiyama, and O. Hayaishi, *J. Biol. Chem.* **255**, 7616 (1980).
[6] P. Adamietz and H. Hilz, *Hoppe-Seyler's Z. Physiol. Chem.* **357**, 527 (1976).
[7] L. O. Burzio, *in* "ADP-ribosylation Reactions: Biology and Medicine" (O. Hayaishi and K. Ueda, eds.), p. 103. Academic Press, New York, 1982.
[8] J. Chauveau, Y. Moulé, and C. Rouiller, *Exp. Cell Res.* **11**, 317 (1956).
[9] H. Busch, this series, Vol. 12, Part A, p. 421.

prepared from the isolated nuclei by the method of Ueda et al.[10] except that all solutions are supplemented with protease inhibitors (0.2 mM phenylmethylsulfonyl fluoride and 5 mM NaHSO$_3$). Poly(ADP-ribosyl) histones are prepared by either of the following two methods; method 1 employs chromatin as the source of enzyme and acceptor, and isolates, first, modified and unmodified histones together, the separation of which is effected later by enzymatic digestion and purification of peptides; method 2, in contrast, utilizes nuclei, and separates modified histones from unmodified ones at the beginning by the use of a borate gel which interacts specifically with the cis-diol portion of ADP-ribose,[11] followed by enzymatic digestion and peptide purification as in method 1. Generally speaking, method 1 yields histones of better purity, while method 2 is suitable for preparation on a larger scale.

Method 1[4]

The reaction mixture containing 100 mM Tris–HCl (pH 7.5), 10 mM MgCl$_2$, 1 mM dithiothreitol, 5 mM NaHSO$_3$, 1 mM [ribose(in NMN)-^{14}C]NAD[12] (4.2 × 10^8 cpm, 2.5 Ci/mol), and chromatin (about 1 g of protein; prewarmed to 15°) in a total volume of 80 ml is incubated for 15 min at 37°. The reaction is terminated by the addition of 250 ml of ice-cold 150 mM NaCl containing 5 mM NaHSO$_3$. The mixture is centrifuged at 13,000 g for 10 min. The precipitate is washed with the same solution until no ^{14}C is detectable in the supernatant fraction. The final precipitate is mixed with 440 ml of 5% HClO$_4$, and the mixture is centrifuged at 13,000 g for 10 min. The HClO$_4$ extraction is repeated three more times. From the supernatant fractions pooled is histone H1 precipitated with 18% (w/v) CCl$_3$COOH[5,13] (yield = 230 mg of protein, 2.3 × 10^6 cpm). The HClO$_4$ residue is suspended in 590 ml of 80% ethanol, followed by centrifugation as above. The extraction with ethanol is repeated once. The supernatant fractions pooled are mixed with five volumes of acetone containing 0.2% HCl. The precipitate collected by centrifugation (13,000 g, 10 min) contains histones H2A, H3, H4,[13] and protein A24.[14] The residue after 80% ethanol extraction is resuspended in 590 ml of 250 mM HCl, and the mixture centrifuged as above. The HCl extraction is repeated four more

[10] K. Ueda, R. H. Reeder, T. Honjo, Y. Nishizuka, and O. Hayaishi, *Biochem. Biophys. Res. Commun.* **31,** 379 (1968).

[11] H. Okayama, K. Ueda, and O. Hayaishi, *Proc. Natl. Acad. Sci. U.S.A.* **75,** 1111 (1978).

[12] N. Ogata, K. Ueda, and O. Hayaishi, *Anal. Biochem.* **115,** 58 (1981).

[13] E. W. Johns, *Methods Cell Biol.* **16,** 183 (1977).

[14] I. L. Goldknopf and H. Busch, *in* "The Cell Nucleus" (H. Busch, ed.), p. 149. Academic Press, New York, 1978.

times. Histone H2B is precipitated from the supernatant fractions with five volumes of acetone[4,13] (yield = 241 mg of protein, 6.6 × 10⁶ cpm).

ADP-ribosyl Histone H1 and Its Fragments.[5] The 5% HClO₄ extract precipitated by 18% CCl₃COOH contains almost exclusively histone H1, as judged by SDS–polyacrylamide gel electrophoresis. This preparation comprises several subfractions of H1, all accompanied by ¹⁴C, and consists of ADP-ribosylated and non-ADP-ribosylated H1s in a ratio of about 1 : 4. The average length of ADP-ribose chains attached to the histone is 1.1.

The precipitated histone is dissolved in 90 ml of 50% CH₃COOH, and mixed with 1.4 ml of 0.1 M N-bromosuccinimide (NBS) freshly prepared in 50% CH₃COOH. The mixture is incubated for 2 hr at 25°, and, after the addition of another 1.4 ml of 0.1 M NBS, incubated further for 2 hr. The mixture is diluted fourfold with distilled water, and then lyophilized. The lyophilized material is dissolved in 9 ml of 20 mM HCl, and the solution subjected to filtration through a Sephadex G-100 column (2.4 × 143 cm) using 20 mM HCl as an eluent. Large and small fragments, referred to as NBS_C (COOH-terminal) and NBS_N (NH₂-terminal), respectively, are recovered by lyophilization.

The NBS_N fragment (54 mg of protein, 7.5 × 10⁵ cpm) is dissolved in a solution (1.4 ml) containing 100 mM Tris–HCl (pH 7.4), 5 mM MgCl₂, 5 mM CaCl₂, and 50 μg of snake venom phosphodiesterase. The mixture is incubated for 1 hr at 22°, and, after the addition of 200 μg of trypsin, for additional 2 hr. The digest is purified by sequential chromatographies on columns of Sephadex G-25 (1.9 × 198 cm; in 20 mM HCl), Dowex 1-formate (1 × 19 cm; eluted with a linear gradient of 0–70 mM HCOOH), and DEAE-Sephadex A-25 [0.7 × 85 cm; eluted with a linear gradient of 0–70 mM NaCl in 5 mM Na phosphate (pH 7.1)]; each time the main ¹⁴C fractions are collected and concentrated by lyophilization. The final eluate contains pure phospho[¹⁴C]ribosyl pentadecapeptide (NH₂-terminal residues 1–15) of histone H1 (¹⁴C yield from NBS_N = 17%). The peptide (1.8 × 10⁵ cpm) is dissolved in a solution (60 μl) containing 100 mM K phosphate (pH 7.0), 5 mM CaCl₂, and 6 μg of Pronase, and the mixture incubated for 2 hr at 25°. Chromatography of the digest on a Dowex 1-formate column (0.7 × 26 cm; eluted with linear gradients of 0–10 mM and then 0.5–1.5 M HCOOH) yields two major ¹⁴C peaks. The first peak contains phospho[¹⁴C]ribosyl octapeptide (Ala-Ala-Pro-Ala-Pro-Ala-Glu-Lys, residues 8–15), while the second peak contains phospho-[¹⁴C]ribosyl peptapeptide (AcSer-Glu-Thr-Ala-Pro, residues 1–5). Further digestion of the second peak material (5.5 × 10⁴ cpm) with carboxypeptidase Y (125 μg) in 33 μl of 50 mM K phosphate (pH 6.0) for 23 hr at 20°, and chromatography of the digest on a Dowex 1-formate column (0.3 ×

5 cm; eluted with a linear gradient of 0–2 M HCOOH) yield phospho[^{14}C]-ribosyl tripeptide (AcSer-Glu-Thr, NH$_2$-terminal residues 1–3) (^{14}C yield from NBS$_N$ = 2.3%). On the other hand, the digestion of the first peak material (1.7 × 10^4 cpm) with proteinase K (1.4 μg) in 17 μl of 50 mM Tris–HCl (pH 7.5) for 1 hr at 25°, followed by similar Dowex 1 column chromatography with a gradient of 0–5 mM HCOOH, yields phospho-[^{14}C]ribosyl pentapeptide (Ala-Pro-Ala-Glu-Lys, residues 11–15) (^{14}C yield from NBS$_N$ = 6.0%).

The NBS$_C$ fragment (184 mg of protein, 7.8 × 10^5 cpm) is dissolved in a mixture (4.1 ml) containing 100 mM Na citrate (pH 5.7), 10 mM KCN, 1 mM EDTA, and 5.8 mg of papain. After incubation for 1 hr at 25°, the mixture is applied to a phosphocellulose column (1.2 × 30 cm), and the column is eluted with a linear gradient (20–200 mM) of NaCl in 20 mM ammonium acetate (pH 4.0). The main ^{14}C peak eluting around 290 mM NaCl is concentrated by lyophilization, and the concentrated material is filtered through a BioGel P-2 column (1.6 × 85 cm) in 20 mM HCl. The major ^{14}C fraction (4.6 mg of protein, 1.0 × 10^5 cpm) is lyophilized, and the lyophilized material is dissolved in a solution (113 μl) containing 200 mM Tris–HCl (pH 7.5), 5 mM MgCl$_2$, 0.1 μg of snake venom phospho-diesterase, and 15 μg of alkaline phosphatase. After incubation for 30 min at 25°, the mixture is applied to a phosphocellulose column (0.6 × 9.3 cm). The column is eluted with the same NaCl gradient as used for the papain digest. The ^{14}C peak material is further purified by passing through a BioGel P-2 column (1 × 96 cm) in 20 mM HCl. This eluate contains pure [^{14}C]ribosyl tetrapeptide (Ala-Lys-Lys-Lys) originating from the COOH-terminal residues 210–213 (the residue numbers are tentatively assigned by comparison with the reported sequences of rabbit thymus histone H 1 s[15]) (^{14}C yield from NBS$_C$ = 6.0%).

Another ADP-ribosyl peptide is purified from the papain digest of NBS$_C$; the fraction eluting from the first phosphocellulose column at about 270 mM NaCl slightly ahead of the COOH-terminal peptide peak (1.4 × 10^5 cpm) is chromatographed successively on three phosphocellu-lose columns. The first column (0.3 × 14 cm) is eluted with a linear gradient (220–360 mM) of NaCl, the second (0.6 × 75 cm) is, after wash-ing with 20 mM NaCl, eluted with 400 mM NaCl, and the third (0.6 × 75 cm) is eluted with 350 mM NaCl; all NaCl solutions are supplemented with 50 mM HCOOH, and each time the ^{14}C main peak is collected and desalted using Sephadex G-10 gel filtration (or diluted with 50 mM HCOOH) prior to application to the next column. The final eluate is

[15] C. von Holt, W. N. Strickland, W. F. Brandt, and M. S. Strickland, *FEBS Lett.* **100,** 201 (1979).

concentrated by lyophilization, and the concentrated material is filtered through a BioGel P-4 column (1.0×99 cm) using 20 mM HCl as an eluent. The ^{14}C peak material is concentrated by lyophilization, and subjected to cellulose thin-layer chromatography in a solvent system of n-butanol/pyridine/CH$_3$COOH/H$_2$O (30 : 20 : 6 : 24). The area containing ^{14}C is eluted with 0.5 M HCl. This eluate contains ADP-[^{14}C]ribosyl peptide which has alanine at the NH$_2$-terminus and the amino acid composition of 5 Lys, 4 Ala, 1 Thr, 1 Gly, 1 Glu, and 1 Pro, and thus, most probably, originates from the residues 112–124 (Ala-Ala-Thr-Gly-Glu-Ala-Lys-Lys-Lys-Pro-Lys-Lys-Ala) (^{14}C yield from NBS$_C$ = 0.4%).[16]

ADP-ribosyl Histone H2B and Its Fragments.[4] The crude histone H2B preparation, that is the 250 mM HCl extract precipitated by acetone, is, in 100-mg portions, dissolved in 2 ml of 8 M urea containing 1% 2-mercaptoethanol, and the solution passed through a BioGel P-60 column (1.9×190 cm) in 20 mM HCl containing 100 mM NaCl. The prominent ^{14}C peak material is desalted by use of Sephadex G-25 in 20 mM HCl, lyophilized, and redissolved in 1 ml of 50 mM Na acetate (pH 5.1) containing 5 mM NaHSO$_4$. Upon filtration of this solution through a Sephadex G-150 column (1.9×97 cm) in the same buffer, a single ^{14}C peak is obtained. This peak material exhibits a single protein band upon SDS–polyacrylamide gel electrophoresis, and contains ADP-[^{14}C]ribosylated and non-ADP-ribosylated histone H2Bs in a ratio of about 1 : 12 (yield = 201 mg of protein, 5.9×10^6 cpm). Average length of ADP-ribose chains attached to the histone is 1.0.

A portion (100 mg, 2.7×10^6 cpm) of the ADP-ribosyl histone H2B is incubated for 25 min at 25° in a solution (16 ml) containing 100 mM Tris–HCl (pH 7.5), 5 mM each of MgCl$_2$, CaCl$_2$, and AMP, 2.6 mg of trypsin [treated with L-(1-tosylamido-2-phenylethyl)chloromethyl ketone], and 0.12 unit of snake venom phosphodiesterase. The reaction is terminated by the addition of 0.8 ml of 2 M HCOOH. Chromatography of the digest on a Sephadex G-25 column (2.6×89 cm) in 20 mM HCl gives a prominent ^{14}C peak. The peak fraction is lyophilized, redissolved in 10 ml of distilled water, and applied to a Dowex 1-formate column (0.7×13.5 cm). The column is eluted with a linear gradient (0–2 mM) of HCOOH. The ^{14}C peak material eluting around 1 mM HCOOH is lyophilized, redissolved in 0.2 ml of 20 mM HCl, and filtered through a BioGel P-2 column (1×49 cm) in 20 mM HCl. The ^{14}C peak fraction contains pure phospho[^{14}C]ribosyl pentapeptide (Pro-Glu-Pro-Ala-Lys) of the NH$_2$-terminal residues 1–5 (^{14}C yield from ADP-ribosyl histone H2B = 42%).

[16] N. Ogata, K. Ueda, and O. Hayaishi, *in* "Novel ADP-Ribosylations of Regulatory Enzymes and Proteins" (M. Smulson and T. Sugimura, eds.), p. 333. Elsevier/North-Holland, New York, 1980.

Further digestion of the pentapeptide (7.6×10^4 cpm) is performed by incubations with carboxypeptidase B (6.3 μg) in 60 μl of 20 mM Tris–HCl (pH 7.5) for 1 hr, and then, after adjusting pH to about 3.5 with 3 μl of 2 M HCOOH, with acid protease B (63 μg) for 36 hr at 25°. A 59-μl portion of the digestion mixture, supplemented with 200 μl of 20 mM HCl, is loaded on a BioGel P-2 column (1 $\times$ 99 cm). Elution of the column with 20 mM HCl yields two major ^{14}C peaks. The two peaks contain identical tripeptide (Pro-Glu-Pro) originating from the same NH$_2$-terminal residues 1–3, but associated with different modification groups, i.e., phospho[^{14}C]ribosyl with the earlier peak and [^{14}C]ribosyl with the latter (^{14}C yield from ADP-ribosyl histone H2B = 22 and 18%, respectively). The latter material is produced from the former presumably by a phosphatase activity detected in the acid protease B preparation used.

ADP-ribosyl Histone H2A. The crude histone H2A preparation, that is the 80% ethanol extract precipitated by acetone–HCl (about 200 mg of protein, 1.2×10^6 cpm), is dissolved in 10 ml of 8 M urea containing 1% 2-mercaptoethanol and 10 mM HCl, and the solution applied to a BioGel P-60 column (3.2 $\times$ 90 cm). Among several ^{14}C peaks eluted by 20 mM HCl containing 100 mM NaCl, ADP-ribosyl histone H2A is enriched in the second peak. This peak material is precipitated by 18% (w/v) CCl$_3$COOH, and the precipitate, washed with an acetone–HCl (1 mM) mixture, is dissolved in 820 μl of 2 M NaCl containing 6 M guanidine–HCl. The solution is applied to a Sephadex G-100 column (5 $\times$ 49 cm), and the column eluted with 50 mM Na acetate (pH 5.4) containing 50 mM NaHSO$_3$. Two prominent ^{14}C peaks are obtained. The first (minor) peak contains primarily ADP-[^{14}C]ribosyl histone H3, whereas the second (major) peak contains almost exclusively ADP-[^{14}C]ribosyl histone H2A; both peaks are associated with unmodified respective histones (yield of ADP-ribosyl histone H2A = 11 mg of protein, 1.5×10^5 cpm).

Method 2[11]

The reaction mixture containing 100 mM Tris–HCl (pH 7.5), 10 mM MgCl$_2$, 1 mM dithiothreitol, 0.4 mM phenylmethylsulfonyl fluoride, 5 mM NaHSO$_3$, 0.1 mM [*ribose*(*in NMN*)-^{14}C]NAD,[12] (1.0×10^8 cpm, 2.5 Ci/mol), and isolated nuclei (about 2 g of protein) in a total volume of 200 ml is incubated for 10 min at 20°. The reaction is terminated by the addition of 22 ml of 2.5 M HCl. After stirring in an ice bath for 1 hr, the mixture is centrifuged at 12,000 g for 20 min. The supernatant fraction is then made 20% with respect to CCl$_3$COOH, and the precipitate formed is collected by centrifugation as above. The precipitate is washed three times with ethyl ether and dissolved in 40 ml of 0.1 mM K phosphate (pH 6.0) containing 6 M guanidine–HCl, followed by dialysis against 10 mM

CH_3COOH and lyophilization. The lyophilized material is dissolved in 50 mM morpholine–HCl (pH 8.2) containing 6 M guanidine–HCl, and applied to a dihydroxyboryl BioGel P-60 column (1.2 × 22 cm) equilibrated with the same buffer. A bulk of unmodified histone and some nonhistone proteins passes through the column, whereas a majority (>80%) of ADP-[^{14}C]ribosyl histone is retained. After washing the column with the equilibration buffer until the A_{280} of the eluate decreases to <0.02, the column is eluted with 150 mM K phosphate (pH 6.5) containing 6 M guanidine–HCl. This eluate contains a whole population of ADP-ribosyl histones and nonhistone proteins, but essentially none of unmodified protein. The inclusion of guanidine–HCl in buffers is indispensable for solubilization of proteins without aggregates and also inhibition of nonspecific interactions between proteins and the gel. Dihydroxyboryl BioGel P-60 is prepared in the laboratory, or replaced by commercially available boronate cellulose[17] or other borate-bound resins with a pore size sufficient for histone penetration. The borate-gel eluate is dialyzed against 5 liter of 10 mM CH_3COOH, and lyophilized.

The lyophilized material is dissolved in 3 ml of 20 mM K phosphate (pH 6.0) containing 7 M urea, and applied to a CM-cellulose column (1.2 × 5 cm). The column is equilibrated and washed with the same buffer, and eluted with a linear gradient of 0–0.4 M KCl in this buffer. Among several ^{14}C peaks, the major one eluting at approximately 0.15 M KCl and the following, minor, one eluting at about 0.25 M KCl are separately collected, dialyzed against 10 mM CH_3COOH and lyophilized. These peaks contain almost exclusively ADP-[^{14}C]ribosyl histone H2B and H1, respectively. Average lengths of ADP-ribose chains attached to H2B and H1 are 1.0 and 1.5, respectively.

Further purification of these histones is carried out by the procedures described in method 1. If the preparation is to be used for fragmentation and peptide analysis, this purification is not necessary. Enzymatic digestion of the ADP-ribosyl histones and subsequent purification of peptides are carried out as in method 1. In many chromatographic steps, the size of column may be reduced compared with that of method 1, because a large amount of unmodified proteins has been excluded at a borate-gel step.

Evidence for Poly(ADP-ribosyl) Carboxylates

Carboxyl Groups in Isolated Peptides. All of the smallest fragments of histone with ADP-ribosylation sites so far isolated have a glutamic acid

---

[17] P. Adamietz, K. Klapproth, and H. Hilz, *Biochem. Biophys. Res. Commun.* **91**, 1232 (1979).

residue or a COOH-terminus[4,5,16]:

H1	AcSer-Glu-Thr	(residues 1–3)
	Ala-Pro-Ala-Glu-Lys	(residues 11–15)
	Ala-Ala-Thr-Gly-Glu-Ala-Lys-Lys-Lys-Pro-Lys-Lys-Ala	
		(residues 112–124)
	Ala-Lys-Lys-Lys-COOH	(residues 210–213)
H2B	Pro-Glu-Pro	(residues 1–3)

The results obtained by Riquelme *et al.*[2] and Burzio *et al.*[3] supported this view.

Further digestion of these fragments with various proteolytic enzymes is very difficult, if not impossible, as far as the modification group remains attached; this is, probably, due to a steric hindrance caused by a bulky attachment, although a cluster of unfavored amino acids (acetylserine, proline, lysine) may also contribute to the resistance.

Sensitivity of ADP-ribosyl Histone Bond to Alkali and Neutral NH$_2$OH. As early as in 1969,[18] the linkage between ADP-ribose and histone was shown to be stable in acid, but very labile in alkali and neutral NH$_2$OH. This property has thereafter been confirmed with various preparations of ADP-ribosyl histone.[7] Figure 1 shows the stabilities of purified ADP-ribosyl histone H2B and its fragment, phosphoribosyl pentapeptide. ADP-ribose bound to the histone decreases, in 1 hr at 25°, to 95, 42, and 17% in 0.1 M HCl, 0.1 M Na carbonate (pH 9.6), and 0.1 M NaOH, respectively (Fig. 1A). Hydrolysis of phosphoribosyl pentapeptide proceeds at about the same rate as the intact ADP-ribosyl histone molecule (Fig. 1B). The effects of temperature on hydrolysis at pH 9.6 of these two derivatives are also similar (Fig. 1C). The hydrolysis of ADP-ribosyl histone H2B by 0.1 M NaOH proceeds biphasically; the half-lives of the fast and slow phases are about 2.5 and 50 min, respectively. The ADP-ribosyl histone bond is cleaved also by neutral NH$_2$OH, and the cleavage proceeds in two phases with the half lives of about 2 and 150 min at 25° (Fig. 1B). Essentially the same properties are observed with ADP-ribosyl histone H1 and its NBS fragments; the half-lives of ADP-ribosyl histone H1 in 0.1 M NaOH and neutral NH$_2$OH are, under both conditions, 2 min (fast phase) and 80 min (slow phase).[5]

The faster phases of hydrolysis by alkali and NH$_2$OH are comparable with those of carboxylic acid esters such as aminoacyl-tRNA or 1-(indole-3-acetyl)-β-D-glucose. The nature of the slower phases is not known at present; all other known N- or O-glycosides are much more stable under these conditions.

[18] Y. Nishizuka, K. Ueda, K. Yoshihara, H. Yamamura, M. Takeda, and O. Hayaishi, *Cold Spring Harbor Symp. Quant. Biol.* **34**, 781 (1969).

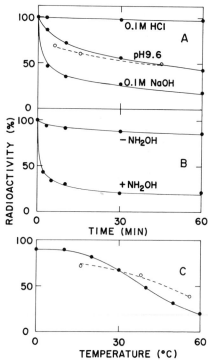

FIG. 1. Stabilities of ADP-ribosyl histone H2B and phosphoribosyl pentapeptide.[4] (A) ADP-[^{14}C]ribosyl histone H2B was incubated with either 0.1 M HCl, 0.1 M Na carbonate (pH 9.6), or 0.1 M NaOH at 25°, and, at intervals, the mixture is examined for acid (20% CCl$_3$COOH)-insoluble ^{14}C (●). Phospho[^{14}C]ribosyl pentapeptide (Pro-Glu-Pro-Ala-Lys) of H2B is incubated in 0.1 M Na carbonate (pH 9.6) at 25° for various time lengths, and the mixture examined for ^{14}C remaining on the peptide using paper chromatography (○). (B) ADP-[^{14}C]ribosyl histone H2B was incubated in either 2 M NH$_2$OH (pH 7.0) or 0.1 M K phosphate (pH 7.0) at 25°, and acid-insoluble ^{14}C determined (●). (C) ADP-[^{14}C]ribosyl histone H2B (●) or phospho[^{14}C]ribosyl pentapeptide (○) was incubated in 0.1 M Na carbonate (pH 9.6) for 5 min, at various temperatures, and acid-insoluble ^{14}C or the peptide-bound ^{14}C, respectively, is determined as described in A.

Sensitivity of ADP-ribosyl NBS$_C$ Bond to Trypsin.[5] The linkage between ADP-ribose and NBS$_C$ (the COOH-terminal half of histone H1 bisected by NBS) is very sensitive to trypsin; the enzyme releases almost instantaneously >70% of ADP-ribose from NBS$_C$, and also >85% of ribose from ribosyl tetrapeptide (Ala-Lys-Lys-Lys-COOH). These hydrolysis are completely inhibited by a specific inhibitor, N-α-p-tosyl-L-lysine chloromethyl ketone. These results, together with the known esterase activity of trypsin acting on α-carboxyl esters of lysine derivatives, support the view that ADP-ribose is linked to the α-carboxyl group of the terminal lysine residue via an ester bond.

Net Charges of Ribosyl Peptides.[4,5] The net electric charge of ribosyl tripeptide (Pro-Glu-Pro; NH$_2$-terminal residues 1–3) of histone H2B, as estimated from the electrophoretic mobility at pH 6.5,[19] is 0, whereas that of ribosyl pentapeptide (Pro-Glu-Pro-Ala-Lys; residues 1–5) is +1. These values are consistent with the view that the γ-carboxyl group of glutamic acid residue is blocked by a neutral substituent. This result also excludes the possibility of ADP-ribosylation of the α-imino group of NH$_2$-terminal proline residue, or the net charge should be −2.

A similar analysis of two ribosyl peptides (AcSer-Glu-Thr-Ala-Pro, residues 1–5, and Ala-Ala-Pro-Ala-Pro-Ala-Glu-Lys, residues 8–15) of histone H1 indicates their net charges to be −1 and +1, respectively. These values also coincide with the view that the γ-carboxyl group of glutamic acid residues is blocked.

Infrared Spectrum. More direct evidence for ADP-ribosyl carboxylates is obtained by infrared (IR) spectral analysis of isolated fragments in comparison with synthetic derivatives. The difference spectrum of FT-(Fourier transformation)-IR of ADP-ribosyl H2B pentapeptide (Pro-Glu-Pro-Ala-Lys) and ADP exhibits a peak ($\nu_{C=O}$) at 1730 cm^{-1}, which coincides with that of chemically synthesized α- or β-ribosyl tripeptide (Pro-Glu-Pro). This result is suggestive of a γ-carboxyl ester of the glutamic acid residue.

Miscellaneous. Lysine is found in three of the five fragments with ADP-ribosylation sites so far isolated (see above). The possibility of ADP-ribosylation on any of these lysine residues through an aldimine bond is excluded by the effect of NaBH$_4$ treatment prior to alkaline hydrolysis; the treatment of phosphoribosyl pentapeptide (residues 1–5) of histone H2B or ribosyl tetrapeptide (residues 210–213) of H1 does not stabilize the bond against alkali.

The NH$_2$-terminal fragment of histone H2B has an N-blocked serine residue. Quantification of phosphate in phosphoribosyl pentapeptide (residues 1–5) reveals no extra phosphate with this fragment. This result excludes the possibility of ADP-ribosylation on phosphoserine as proposed by Ord and Stocken.[20]

The proline in the NH$_2$-terminal pentapeptide (residue 1–5) of ADP-ribosylated histone H2B is reactive with dansyl chloride.[3] This reactivity, however, does not necessarily indicate the presence of free α-imino group and thus a carboxyl ester in the original peptide, because dansylation is

[19] R. E. Offord, *Nature (London)* **211**, 591 (1966).
[20] M. G. Ord and L. A. Stocken, *Biochem. J.* **161**, 583 (1977).

carried out at pH 10, and such alkaline conditions should hydrolyze rapidly the ADP-ribosyl peptide bond and expose the α-imino group.

Discussion

All lines of evidence presented above are suggestive of an ester bond, ADP-ribosyl carboxylate, as the major ADP-ribosyl histone linkage. This bond is unstable in both alkali and neutral NH_2OH. Studies of chemical stabilities, however, suggest the presence of another type of bond which is relatively, but not completely, stable under these conditions. Furthermore, Adamietz and Hilz[6] proposed a third type of bond which is resistant to both alkali and NH_2OH. Such heterogeneity has been analyzed with intact ADP-ribosyl histones and their large fragments such as NBS_N and NBS_C, but not yet with smaller fragments.

As for the molecular basis of the heterogeneity in chemical stabilities, three cases appear to be possible; first, the same type of ADP-ribosyl carboxylate has different stabilities in different circumstances; second, ADP-ribosylation takes place also at sites other than carboxyl groups of histone; and, third, a carboxylate is bound to several different sites on ribose. The first possibility is partly supported by our finding that an apparently identical bond exhibits different stabilities at different stages of fragmentation; for example, phosphoribosyl tripeptide (residues 1–3) of histone H2B appears to be less stable than its parent fragment, phosphoribosyl pentapeptide (residues 1–5). Burzio et al.[3] also noticed a difference in stabilities between acetylated and nonacetylated oligo(ADP-ribosyl) histone H1.

The second possibility is supported by the recent finding of arginine-dependent ADP-ribosyltransferase. This enzyme was initially discovered by Moss et al.[21] in the cytoplasm of avian erythrocytes, and thereafter found in the cytoplasm,[22] membrane,[23] or nucleus[24] of many other animal cells. By the action of this enzyme, a single ADP-ribose residue is transferred to guanido nitrogen of various acceptors including histone. The resultant N-glycosides are stable in both alkali and neutral NH_2OH. Recently, Shimoyama et al.[24] reported that this type of enzyme is fairly active in hen liver nuclei under certain conditions, and that the ADP-ribose attached to histone may be elongated in vitro into a polymer by the

[21] J. Moss, S. J. Stanley, and P. A. Watkins, J. Biol. Chem. **255,** 5838 (1980).
[22] J. Moss and S. J. Stanley, J. Biol. Chem. **256,** 7830 (1981).
[23] Z. K. Beckner and M. Blecher, Biochem. Biophys. Acta **673,** 477 (1981).
[24] M. Shimoyama, Y. Tanigawa, A. Kitamura, K. Kawakami, and H. Nomura, Int. Congr. Biochem., 12th, Abstracts, p. 180 (1982).

action of poly(ADP-ribose) synthetase. Another possibility that γ-carboxyl (glutamyl) and α-carboxyl (COOH-terminal) esters have somewhat different stabilities remains to be investigated.

As for the third possibility, the hydroxyl group of ribose engaged in the ester formation has not been identified yet with ADP-ribosyl histones. Until recently, based on the fact that ADP-ribose is originally linked to nicotinamide in NAD through the hemiacetal group at position 1″ and that ADP-ribose polymerization (including branching) is effected by bonding the 1″-hemiacetal group to the 2′ (or 2″)-hydroxyl group of a preattached ADP-ribose unit,[1] it had been tacitly assumed that the same 1″-hemiacetal group is used for ADP-ribosylation of protein. A preliminary determination of periodate consumption by phosphoribosyl peptides of histone H1 and H2B appeared to support this view.[5] However, our recent studies on ADP-ribosyl protein lyase[25] (ADP-ribosyl histone splitting enzyme[26]) suggest a possibility that the bond may be at position 2″ or 3″; the product of ADP-ribosyl histone H2B split by this enzyme is not ADP-ribose, but its dehydrated derivative, ADP-3″-deoxypent-2″-enofuranose.[25,27] It appears possible that a carboxylate migrates on the ribose during storage or manipulation of ADP-ribosyl histone (or peptides) and that these different structures have different stabilities. In this context, it is noteworthy that ADP-ribosyl histone H2B (or its pentapeptide) molecules resistant to the lyase are less sensitive to neutral NH$_2$OH.[25]

[25] J. Oka, K. Ueda, O. Hayaishi, H. Komura, and K. Nakanishi, *J. Biol. Chem.* **259**, 986 (1984).
[26] H. Okayama, M. Honda, and O. Hayaishi, *Proc. Natl. Acad. Sci. U.S.A.* **75**, 2254 (1978).
[27] H. Komura, T. Iwashita, H. Naoki, K. Nakanishi, J. Oka, K. Ueda, and O. Hayaishi, *J. Am. Chem. Soc.* **105**, 5164 (1983).

[48] Purification and Characterization of (ADP-Ribosyl)$_n$ Proteins

By PETER ADAMIETZ and HELMUTH HILZ

Posttranslational modification of proteins by transfer of ADPR (adenosine 5′-diphosphoribose) groups is a general phenomenon seen in most eukaryotic cells. It is also used as a tool of various bacterial toxins to modify metabolic functions of the host (for reviews, see refs. 1–3). The existence *in vivo* of different types of (ADPR)$_n$ protein conjugates, their independent changes under varying metabolic or growth conditions and their uneven subcellular distribution[4] indicate that the process of ADP-ribosylation serves multiple functions. To unravel the precise functions, it appears indispensable to analyze the modification of specific acceptor proteins. Since it was shown that ADP-ribosylation *in vivo* of defined acceptor proteins in normally proliferating and in alkylated tissues differs markedly from *in vitro* analyses, including permeabilized cells[5–7] special emphasis is laid on the analysis of endogenous (ADPR)$_n$ protein conjugates from living cells. This chapter describes the detection, purification, and analysis of ADP-ribosyl polypeptides from eukaryotic cells. No consideration will be given to ADPR acceptor proteins modified by bacterial toxins and phage products.

Purification of (ADPR)$_n$ Protein Conjugates

Principle

Since the half-life of poly(ADPR) residues, at least in repair-induced cells, is shorter than 1 min,[8] it appears essential to "freeze" the *in vivo* status by immediate homogenization in an acidic milieu. This can be done by mixing with trichloroacetic acid (TCA) when the total fraction of cellular (ADPR)$_n$ protein conjugates is to be isolated. Use of perchloric acid (PCA) instead of TCA allows the separate analysis of histone H1 and HMG protein conjugates because of their solubility in 5% PCA. Nuclear (ADPR)$_n$ protein conjugates can be purified from "citric acid nuclei," while the purification from other compartments usually requires isolation of the subcellular fraction under nondenaturing conditions. The acid-insoluble fractions are brought into solution with the aid of guanidine hydrochloride and chromatographed on aminophenyl boronate matrices. While

[1] H. Hilz and P. R. Stone, *Rev. Physiol., Biochem. Pharmacol.* **76**, 1 (1976).
[2] O. Hayaishi and K. Ueda, *Annu. Rev. Biochem.* **46**, 95 (1977).
[3] P. Mandel, H. Okazaki, and C. Niedergang, *Prog. Nucleic Acid Res. Mol. Biol.* **27**, 1 (1982).
[4] P. Adamietz, K. Wielckens, R. Bredehorst, H. Lengyel, and H. Hilz, *Biochem. Biophys. Res. Commun.* **101**, 96 (1981).
[5] P. Adamietz, R. Bredehorst, and H. Hilz, *Eur. J. Biochem.* **91**, 317 (1978).
[6] N. T. Man and S. Shall, *Eur. J. Biochem.* **126**, 83 (1982).
[7] A. Kreimeyer, K. Wielckens, P. Adamietz, and H. Hilz, *J. Biol. Chem.* **259**, 890 (1984).
[8] K. Wielckens, A. Schmidt, E. George, R. Bredehorst, and H. Hilz, *J. Biol. Chem.* **257**, 12872 (1982).

ADPR protein conjugates are covalently linked to the resin at slightly alkaline conditions, DNA, RNA, and most proteins can be washed out. Conjugates are subsequently released at acidic pH values and analyzed. It is essential to avoid exposure to alkaline pH values at elevated temperature as far as possible since the NH_2OH-sensitive conjugates are quite unstable under these conditions. Chromatography on boronate matrices which must be carried out at pH 8.2–8.3, should therefore not exceed 1–2 hr at 4°.

Preparation of Acid-Insoluble Fractions

Cells, freeze-clamped tissues, or subcellular fractions are homogenized with cold 20% TCA (15 ml/g tissues), centrifuged at low speed (2000 g, 10 min) to avoid the formation of a gum-like pellet. The sediment is carefully resuspended and washed 4 times with 20% TCA (5 ml/g tissue), twice with ethanol and twice with ether to obtain a dry powder. The powder can be stored in the cold (−20°) at least for 4 weeks without degradation of conjugates. For separation of conjugates on aminophenyl boronate matrices, the dry powder or the ethanol-washed precipitate is dissolved in 6 M guanidine hydrochloride–5 mM mercaptoethanol. DNA which at higher concentrations interferes with binding of the (ADPR)$_n$ protein conjugates to the boronic acid matrix, is removed by ultracentrifugation (5 hr at 210,000 g and at 4°). The DNA-free supernatant contains at least 95% of both (ADPR)$_n$ residues and protein as revealed by in vitro analysis.

The same procedure is applied, when the perchloric acid-insoluble fraction is to be prepared. The acid-soluble supernatant containing the histones and HMG protein conjugates must be extensively dialyzed at slightly acidic pH values before application to boronate columns in order to remove free ribonucleotides.

Preparation of "Citric Acid Nuclei"

The choice of citric acid as isolation medium offers two important advantages over the application of conventional neutral buffers. First, under these conditions adverse enzymic activities that could degrade poly(ADPR) like phosphodiesterase I and poly(ADPR) glycohydrolase are eliminated or inhibited. Second, losses of poly(ADPR) protein conjugates from the nuclei are kept at a minimum since solubilization of nonhistone proteins is observed at more alkaline pH values while histones are extracted at more acidic conditions.[9] The effects of proteolytic enzymes

[9] C. W. Taylor, L. C. Yeoman, and H. Busch, *Exp. Cell Res.* **82,** 215 (1973).

still active at pH 3.5 are counteracted by the addition of PMSF. The following isolation procedure was found to give optimal results with respect to total yield of nuclear conjugates. It refers to AH 7974 hepatoma cells grown for 5 to 7 days in the peritoneum of rats ($\female$ Wistar, 220 g). The procedure may be applied to all types of cells growing in culture or in the ascites fluid of rats or mice.

Cells (10^9) ($\approx$20 mg DNA) are freed from ascites fluid by low-speed centrifugation (5 min 1500 g) and cautiously resuspended in 25 ml of 25 mM citric acid/0.5 mM phenylmethylsulfonyl fluoride (PMSF), giving a final pH of 3.5. After 5 min in the hypotonic medium NP 40 is added to a final concentration of 0.5%, and plasma membranes are disrupted by homogenization in a glass-Teflon homogenizer (Braun, Melsungen). Release of nuclei is controlled microscopically. Usually, 3–5 strokes are sufficient. The suspension is diluted with an equal volume of 0.9 M sucrose/25 mM sodium citrate pH 3.5/0.2 mM PMSF, transferred to a suitable centrifuge tube, and underlayered with 20 ml of 0.9 M sucrose/25 mM sodium citrate pH 3.5/0.2 mM PMSF. After centrifugation for 7 min at 3000 g in a swingout rotor the nuclei are found at the bottom of the tube. All steps are done in the cold. The entire procedure may be performed within 20 min.

Rat *liver nuclei* are released with the aid of a motor-driven glass-Teflon homogenizer (Braun, Melsungen) in the presence of 0.5 M sucrose/25 mM citric acid/0.5 mM PMSF. The homogenate is filtered through gauze and made 1.6 M with respect to sucrose. The nuclei are isolated by sedimentation through a 2.2 M sucrose layer during 30 min centrifugation at 40,000 g and at 4° (SW 27 rotor, Beckmann Instruments, München). This procedure takes about 45 min.

In both cases time-consuming washings of nuclei are omitted. Rather, the nuclei are treated immediately after separation with perchloric or trichloroacetic acid to minimize degradative processes. The yield of nuclei has been found to be quite high, mainly due to the low shearing forces needed for disruption of plasma membranes in the presence of detergent.

When nuclei have to be isolated under conditions that preserve the activity of poly(ADPR) polymerase rather than the endogenous conjugates, the preceding isolation technique may be applied with a neutral buffer system. The Hewish/Bourgoyne procedure[10] specifically developed to prevent enzyming degradation of DNA during preparation has been used successfully for *in vitro* formation of (ADP-ribosyl)$_n$ conjugates.[11] In this case washing of nuclei with buffers containing 1% Triton X-100 is of

---

[10] D. R. Hewish and L. A. Burgoyne, *Biochem. Biophys. Res. Commun.* **52**, 504 (1973).
[11] P. Adamietz, K. Klapproth, and H. Hilz, *Biochem. Biophys. Res. Commun.* **91**, 1232 (1979).

advantage since it removes contaminating cytoplasmic phosphodiesterase and protease that both degrade poly(ADPR) protein conjugates.

Preparation of mitochondria and of submitochondrial particles may be performed according to refs. 12 and 13. These fractions can be extracted with chloroform/methanol (2 : 1, v : v) to remove lipid, or precipitated immediately with TCA (PCA). Endogenous ADPR conjugates are then isolated as outlined below.

Chromatography of (ADPR)$_n$ Protein Conjugates on Aminophenyl Boronate Matrices

Aminophenyl boronate cellulose (E. Merck, Darmstadt) is washed on a column (10 × 2.5 cm) first with 6 M guanidine hydrochloride/0.2 M sodium phosphate pH 5.5 to remove impurities. The column is then equilibrated with 6 M guanidine hydrochloride/50 mM morpholine, pH 8.2, at 4°. Samples (50 mg protein) dissolved in 25 ml cold 6 M guanidine hydrochloride are adjusted to pH 8.2 by the addition of 6 M guanidine hydrochloride/1 M morpholine pH 9.0 immediately before starting the chromatographic separation. The sample is applied to the column during a period of 30 min. Removal of nonbinding contaminants is achieved by washing the column with 6 M guanidine hydrochloride/50 mM morpholine pH 8.2 at an elevated flow rate (130 ml/hr). The washing procedure takes about 20 min. The protein fraction that is bound specifically to the column is then eluted with a buffer containing 6 M guanidine hydrochloride/0.2 M sodium phosphate pH 5.5 at a flow rate of 40 ml/hr.

Absence of contaminating proteins may be checked by rechromatography of an aliquot of the eluate in which case all the protein should be bound again. On the other hand, no binding of proteins occurs when the sample is subjected to alkaline treatment prior to rechromatography in order to release the (ADPR)$_n$ residues from their acceptors. In this case only the free (ADPR)$_n$ residues are rebound by the boronate matrix.

Application of boronate chromatography to the purification of (ADPR)$_n$ protein conjugates synthesized *in vitro* by incubating nuclei from Ehrlich ascites tumor (EAT) cells with [^{3}H]NAD has revealed a yield of 90% with respect to (ADPR)$_n$ residues (cf. ref. 11).

It should be mentioned that the procedure is not specific for (ADP-ribosyl)$_n$ proteins. Glycoconjugates with *cis*-diol-containing prosthetic groups may also be bound. However, when applied to isolated nuclei of EAT cells, no such glycoconjugates were detected.[11] It should also be noted that at least some mono(ADP-ribosyl) proteins have a relatively low

[12] E. Kun, *Biochemistry* **15**, 2328 (1976).
[13] P. L. Pederson and C. A. Schnaitman, *J. Biol. Chem.* **244**, 5065 (1969).

affinity to the boronate matrix. They may therefore be lost or their content reduced by this procedure. Further, it appears that a small subfraction (5%) of the ADPR conjugates from EAT cell nuclei did not bind to the column, suggesting that structural features interfered with the boronic ester formation.[11] That such environmental effects exist may be seen from the inability of phosphoribosyl-AMP to bind to the boronate matrix[14] although it contains a free *cis*-diol group.

Isolation of Histone H1 (ADPR)$_n$ Conjugates[15]

Histone H1 conjugates so far represent the only example for the isolation of pure conjugates of a single polypeptide, free from the unmodified protein. Histone H1 conjugates, including the so-called "histone H1 dimer" are soluble in cold 5% perchloric acid.[16] Extraction from tissues, cells, or nuclei is therefore easily accomplished by homogenization with 5% perchloric acid which eliminates most proteins and the nucleic acids. Nucleotides and contaminants like the HMG proteins are separated by chromatography on cation-exchange columns: The combined perchloric acid extracts and washings are made 1 M with respect to guanidine hydrochloride, buffered to pH 4 by the addition of solid sodium acetate, and dialyzed against 6 M urea/0.35 M guanidine hydrochloride/0.1 M phosphate pH 6.8 (dialysis buffer) for 6 hr. The solution is then applied to a cation exchange column (Bio-Rex 70, Bio-Rad, München) previously equilibrated with the dialysis buffer, and washed thoroughly with dialysis buffer to remove contaminants. Histone H1 and its conjugates are finally eluted by 1.08 M guanidine hydrochloride/6 M urea/0.1 M phosphate pH 6.8, the eluate extensively dialyzed against 5 mM HCl and lyophilized.

Analysis of (ADPR)$_n$ Protein Conjugates

Electrophoretic Separation

Histone H1 (ADPR)$_n$ conjugates are well separated on 15% acetic acid/urea gels according to Traub and Boeckmann.[17] Therefore, samples containing about 10 μg protein are applied to flat gels and run for at least 5 hr at 180 V. A number of different protein bands can usually be visualized by staining with 0.1% Amido Black (Fig. 1). They migrate with (slightly)

[14] T. Minaga, J. McLick, N. Pattabiraman, and E. Kun, *J. Biol. Chem.* **257**, 11942 (1982).
[15] H.-C. Braeuer, P. Adamietz, U. Nellessen, and H. Hilz, *Eur. J. Biochem.* **114**, 63 (1981).
[16] P. R. Stone, W. S. Lorimer, and W. R. Kidwell, *Eur. J. Biochem.* **81**, 9 (1977).
[17] P. Traub and G. Boeckmann, *Hoppe-Seyler's Z. Physiol. Chem.* **359**, 571 (1978).

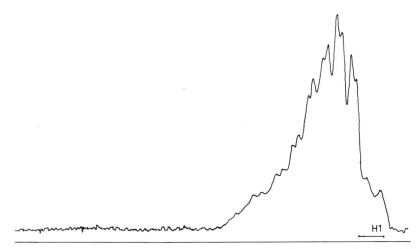

Fig. 1. (ADPR)$_n$ histone H1 conjugates from Ehrlich ascites tumor cells. Scan of an electropherogram after staining with Coomassie Blue (from Braeuer et al.[15]).

lower mobilities as compared to unmodified histone H1 depending on the number and/or length of the (ADPR)$_n$ residues attached to the histone H1 molecule.[15]

The denatured (ADPR)$_n$ protein conjugates of the perchloric acid-insoluble fraction of nuclei or mitochondria usually exhibit an extreme tendency to aggregate. Analysis is therefore performed in the presence of 0.1% sodium dodecyl sulfate and 6 M urea. Since the purified (ADPR)$_n$ protein conjugates are obtained from boronate cellulose chromatography in a rather dilute solution concentration by lyophilization or ultrafiltration (YM 10 filters, Amicon, Witten) is obligatory. Exchange of guanidine hydrochloride by sodium dodecyl sulfate for lyophilization is achieved by a two-step dialysis using 6 M urea as an intermediate component to avoid precipitation. All operations are performed at low temperature (4°), under slightly acidic conditions (pH 6.0–6.5), and in the presence of reducing agents [5 mM mercaptoethanol or 1 mM dithiothreitol (DTT)]. Optimal results with respect to simultaneous resolution of high-molecular-weight nonhistone proteins and low-molecular-weight histones are obtained by applying the discontinuous system[18] with the inclusion of 6 M urea to 7.5–15% exponential polyacrylamide gradient gels. Since the nuclear (ADPR)$_n$ protein conjugates are sensitive toward alkaline conditions, a continuous buffer system consisting of 30 mM Tris/phosphate pH 6.0/0.1% dodecyl sulfate/6 M urea can also be applied. Proteins are stained by any one of

[18] U. K. Laemmli, Nature (London) 227, 680 (1970).

the conventional procedures [e.g., 0.2% Coomassie blue dissolved in methanol/acetic acid/water (5 : 1 : 5, v : v)].

Detection of $(ADPR)_n$ Protein Conjugates on Polyacrylamide Gels

Analysis of radioactively labeled $(ADPR)_n$ protein conjugates on polyacrylamide gels may be performed by different methods: If higher amounts of radioactive label are available visualization of labeled bands by fluorography is convenient. Therefore the gel has to be impregnated with a special scintillator before exposure to X-ray film as first described by Laskey and Bonner.[19] Comparison of the band pattern with that obtained by protein staining (Fig. 2) allows the detection of $(ADPR)_n$ residues in individual protein bands in a semiquantitative manner. If less radioactive material is available, the gel is sliced (1–2 mm slices) and radioactivity determined by liquid scintillation counting, which provides a quantitative survey of the distribution of $(ADPR)_n$ residues over the gel. The method is much more sensitive when the oligo- and poly(ADPR) residues contained in the gel slices are extracted by acid hydrolysis (5% TCA, 1 hr at 95°) prior to liquid scintillation counting.

Blotting

The position of protein-bound poly(ADPR) on polyacrylamide gels may also be determined by a blotting procedure using anti-poly(ADPR) antibodies. The sequence of reactions as outlined in the scheme below is conveniently performed on a solid surface such as nitrocellulose to which the poly(ADPR) protein conjugates can be transferred by electroblotting (cf. ref. 20). Satisfying results were obtained by applying an electric field of 8 V/cm for 3 hr at 20° using a solution of 15.6 mM Tris/120 mM glycine/20% methanol as transfer buffer. Depending on the properties of the poly(ADPR) antibodies poly(ADPR) amounts of down to 2 fmol/mm² have so far been detected with the following procedure.

Wash the nitrocellulose sheet at least 6 times for 5 min with 2.5% bovine serum albumin (BSA)/2.5% ovalbumin in the cold. Incubate with diluted anti-poly(ADPR) antiserum, developed in rabbits, for 16 hr at 4°. Wash as above. Incubate with anti-rabbit IgG from goat (Miles), diluted 1 : 40 for 3 hr at 4°. Wash again as above. Incubate with peroxidase–antiperoxidase complex from rabbit (Miles) diluted 1 : 20 for 3 hr at 4°, and wash. Incubate with 0.05%, 3,3′-diaminobenzidine/0.01% hydrogen

[19] W. M. Bonner and R. A. Laskey, *Eur. J. Biochem.* **46,** 83 (1974).
[20] H. Towbin, T. Staehlin, and J. Gordon, *Proc. Natl. Acad. Sci. U.S.A.* **76,** 4350 (1979).

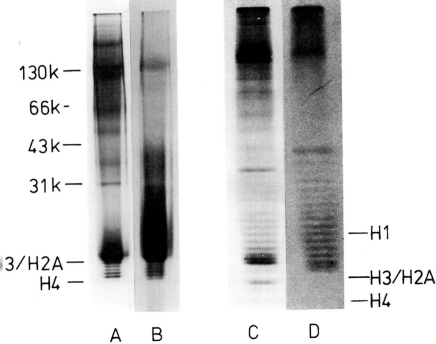

FIG. 2. Electrophoretic separation of [³H]poly(ADPR) protein conjugates isolated from EAT cell nuclei. EAT cell nuclei were incubated with [³H]NAD, and the perchloric acid insoluble poly(ADPR) conjugates were isolated by boronate chromatography.[11] The purified fraction was separated by electrophoresis either on 7.5–15% exponential gradient gels using the discontinuous buffer system of Lämmli[18] (A + B) or on 7% acetic acid urea gels according to[17] (C + D). The gels were stained with Coomassie Blue (A + C) and subjected to fluorography[19] (B + D). β-Galactosidase (130 kilodaltons), bovine serum albumin (66 kilodaltons), ovalbumin (45 kilodaltons), DNase I (31 kilodaltons), and unmodified core histones were used as marker proteins.

peroxide for 5 min at 4°. All solutions are buffered to pH 7.5 with 10 mM Tris–HCl, and they contain 0.9% sodium chloride and 0.02% sodium azide.

An example of this procedure is given in Fig. 3.

Radioimmunoassay of Nonlabeled Conjugates on Gels

Unlabeled mono- and poly(ADPR) conjugates on polyacrylamide gels can be determined by immunological methods, using antibodies to 2′-(5″-phosphoribosyl)-5′-AMP (PR-AMP) [the specific poly(ADPR) breakdown

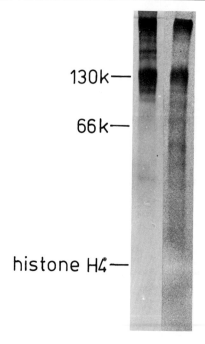

130k—

66k—

histone H4—

A B

Fɪɢ. 3. Detection of protein bound poly(ADPR) on SDS gels with anti-poly(ADPR) antibodies. ADP-ribosylated nonhistone proteins were synthesized in isolated EAT cell nuclei by incubation with 1 mM NAD, freed from histones by precipitation with 0.2 M H₂SO₄, and purified by boronate chromatography. When separated by SDS gel electrophoresis using the Tris/phosphate buffer system at pH 6.0 the protein components of the conjugates could be stained by Coomassie Blue (A), while the poly(ADPR) moieties were visualized with the help of anti-poly(ADPR) antibodies (B). Therefore the conjugates had to be transferred to nitrocellulose that was treated subsequently as described in the text. The positions of β-galactosidase (130 kilodaltons), bovine serum albumin (66 kilodaltons), and histone H4 are indicated.

product][21,22] and to 5′-AMP, respectively.[23] Thus, when gel slices are incubated with 0.3 M NaOH at 56° for 60 min, the neutralized extracts can be quantified for mono(ADPR) equivalents with anti-5′-AMP antibodies as described elsewhere in this volume. Determination of poly(ADPR) is based on the chemical conversion of the polymer to PR-AMP by treat-

[21] K. Wielckens, R. Bredehorst, P. Adamietz, and H. Hilz, *Eur. J. Biochem.* **117,** 69 (1981).
[22] H. Sakura, M. Miwa, Y. Kanai, T. Matsushima, and T. Sugimura, *Nucleic Acids Res.* **5,** 4025 (1978).
[23] R. Bredehorst, K. Wielckens, and H. Hilz, *Eur. J. Biochem.* **92,** 129 (1978).

ment with Mg^{2+}/NaOH.[24] In this case, the gel must be processed in the following way: Fixation of the conjugates with ethanol/water/acetic acid (3 : 1 : 8, v : v : v) to remove urea and dodecyl sulfate is followed by treatment with 6 mM $MgCl_2$ for 2 hr. After slicing, the individual gel sections are further incubated with 300–500 μl 6 mM $MgCl_2$ for several hours, or over night. Finally, a small volume of concentrated sodium hydroxide is added to reach a concentration of 0.33 M. Subsequent incubation for 6 hr at 56° leads to the degradation of mono(ADPR) and poly(ADPR) residues to 5'-AMP and PR-AMP, respectively. Quantitation by radioimmunoassay is then performed on neutralized aliquots.

Pure or crude (ADPR)$_n$ protein conjugates can be characterized for their content in monomeric and polymeric ADPR residues, for hydroxylamine-sensitive and hydroxylamine-resistant linkages, and for chain length distribution. In these cases, the modifying groups must be released from the acceptor proteins either by treatment with alkali (0.3 M NaOH, 30 min, 56°) or by NH_2OH (0.5 M, 60 min, 37°). In the former case, the ribosyl phosphate group at the acceptor site is lost and free 5'-AMP (from mono(ADPR) residues) or free oligomer/polymer is formed. Hydroxylamine treatment releases ADPR and poly(ADPR) with an intact reducing ribose. Quantitation of the released groups is then carried out as described elsewhere in this volume. Distribution of poly(ADPR) chain lengths may be performed by passing the released polymers over a hydroxylapatite column[25] and analyzing the individual fractions for PR-AMP and 5'-AMP after treatment with phosphodiesterase I.

An alternative procedure for the purification of poly(ADPR) protein conjugates was recently described by Smulson et al.[26] They used immune affinity columns to extract ADP-ribosylated proteins from nuclear preparations. However, the specificity of the procedure has not yet been characterized.

Acknowledgment

This work was supported by the Deutsche Forschungsgemeinschaft.

[24] P. Adamietz and R. Bredehorst, Anal. Biochem. 112, 314 (1981).
[25] T. Sugimura, N. Yoshimura, M. Miwa, H. Nagai, and M. Nagao, Arch. Biochem. Biophys. 147, 660 (1971).
[26] M. Wong, Y. Kanai, M. Miwa, M. Bustin, and M. Smulson, Proc. Natl. Acad. Sci. U.S.A. 80, 205 (1983).

[49] Quantification of Protein-Bound ADP-Ribosyl and (ADP-Ribosyl)$_n$ Residues

By KLAUS WIELCKENS, REINHARD BREDEHORST, and HELMUTH HILZ

Proteins modified by single ADPR groups [mono(ADPR)conjugates] and by polymeric ADPR residues [poly(ADPR)conjugates] are found in many eukaryotic cells (cf. ref. 1). These ADPR protein conjugates can be further divided in those that are sensitive to neutral 0.4 M hydroxylamine (NH$_2$OH-sensitive conjugates) and others that are hydroxylamine resistant.[2] The different types show independent variations[1] and uneven subcellular distribution.[3] Poly(ADP-ribosyl)ation appears to be restricted to the cell nucleus, and histone and nonhistone proteins can serve as acceptors *in vitro* and *in vivo*.[4-6] Although the precise function of poly(ADP-ribosyl)ation is not known, available evidence links the process to DNA repair, and cell differentiation (cf. ref. 7). Little is known about the function(s) associated with mono(ADP-ribosyl)ation.

The patterns of ADP-ribosylation obtained *in vitro* differ markedly from the situation found in intact tissues (cf. refs. 5,8). Analyses performed by NAD incorporation into nuclear preparations or permeabilized cells are therefore of limited value compared to the quantitation of endogenous conjugates. However, amounts of protein-bound (ADPR)$_n$ residues are usually rather low. In unstimulated cells, amounts of mono(ADPR) groups linked to proteins are about 100 times lower than NAD, and polymeric ADPR residues can be lower than NAD by a factor of 10^4 (cf. refs. 1,7). Even in cells treated with an alkylating agent, the maximal amounts reached remain far below normal amounts of cellular NAD.[1] Reliable quantitation of endogenous (ADPR)$_n$ residues therefore requires

[1] H. Hilz, K. Wielckens, and R. Bredehorst, *in* "ADP-Ribosylation Reactions" (O. Hayaishi and K. Ueda, eds.), p. 305. Academic Press, New York, 1982.

[2] P. Adamietz and H. Hilz, *Hoppe-Seyler's Z. Physiol. Chem.* **357**, 527 (1976).

[3] P. Adamietz, K. Wielckens, R. Bredehorst, and H. Hilz, *Biochem. Biophys. Res. Commun.* **101**, 96 (1981).

[4] K. Ueda, A. Omachi, M. Kawaichi, and O. Hayaishi, *Proc. Natl. Acad. Sci. U.S.A.* **72**, 205 (1975).

[5] P. Adamietz, R. Bredehorst, and H. Hilz, *Eur. J. Biochem.* **91**, 317 (1978).

[6] P. Adamietz, *in* "ADP-Ribosylation Reactions" (O. Hayaishi and K. Ueda, eds.), p. 77. Academic Press, New York, 1982.

[7] K. Wielckens, R. Bredehorst, P. Adamietz, and H. Hilz, *Adv. Enzyme Regul.* **20**, 23 (1982).

[8] A. Kreimeyer, K. Wielckens, P. Adamietz, and H. Hilz, submitted for publication.

METHODS IN ENZYMOLOGY, VOL. 106

highly sensitive tests and in the case of poly(ADPR) an additional purification/concentration step.

Here we describe procedures for the determination of mono(ADPR) and poly(ADPR) residues in protein conjugates which are based on radioimmunoassays. They also allow discrimination between hydroxylamine-sensitive and -resistant conjugates. Quantitation of poly(ADPR) using different methodologies is also described elsewhere in this volume.

Determination of Protein-bound Mono(ADPR) Residues

Principle

Protein-bound mono(ADPR) groups are released from the TCA-insoluble tissue fraction either by neutral NH_2OH (hydroxylamine-sensitive ADPR protein conjugates) and subsequently converted to 5'-AMP by treatment with alkali, or by NaOH, which converts hydroxylamine-sensitive plus hydroxylamine-resistant ADPR protein conjugates directly to 5'-AMP.[9] Neither (2')3'-AMP (from the alkaline hydrolysis of RNA) nor poly(ADPR) (which is alkali stable) interferes even in crude extracts, if the subsequent quantification of 5'-AMP exhibits high specificity. Here, a radioimmunoassay is described that is based on highly specific anti-5'-AMP antibodies. It allows the determination of ADPR residues in TCA-insoluble tissue fractions (Fig. 1).

Preparation of TCA-Insoluble Fraction

Freeze-clamped tissues or cells were homogenized with cold TCA (5 ml/g tissue) to give a final concentration of 20% w/w using a tissue blender (Ultraturrax; Jahnke und Kunkel, Freiburg), or an all-glass homogenizer. The precipitate is sedimented (10 min at 4° and 3000 g), and the pellet is washed four times by resuspending with cold 20% TCA (5 ml/g tissue), twice with ethanol, and twice with ether. The powder may be stored at −20° when moisture is excluded.

Note: High centrifugational forces tend to convert the TCA insoluble fraction to a gum-like pellet, that is very difficult to resuspend. If histone H1 conjugates are to be analyzed separately, precipitation with 5% perchloric acid is indicated since it avoids the formation of sticky precipitates.

The dry powder is suspended in water (20–100 mg/ml), sonicated, and aliquots are withdrawn for DNA and protein determinations, and for the quantification of ADPR residues.

[9] R. Bredehorst, K. Wielckens, A. Gartemann, H. Lengyel, K. Klapproth, and H. Hilz, Eur. J. Biochem. 92, 129 (1978).

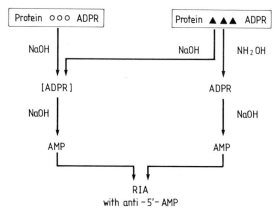

FIG. 1. Quantitation of total and of hydroxylamine-sensitive mono(ADPR) protein conjugates.

Hydroxylamine-Sensitive ADPR Protein Conjugates

This subfraction is defined as those conjugates that release their ADPR residues when treated with 0.5 M NH$_2$OH at 37° during 60 min.

Aliquots of 300 μl of the resuspended TCA-insoluble fraction are incubated with 100 μl freshly prepared 2 M NH$_2$OH pH 7.5 for 60 min at 37°. Four hundred microliters of 1 M HClO$_4$ is then added and the samples are centrifuged after 15 min in ice (Eppendorf minifuge, 2 min at 9000 g). Of the supernatant 600 μl is mixed with 300 μl 0.68 M K$_3$PO$_4$, and the KClO$_4$ formed is removed by centrifugation. The ADPR groups released by the NH$_2$OH treatment are hydrolyzed by incubation of 700 μl of the supernatant with 100 μl 8 M NaOH for 60 min at 56°. Samples are then neutralized with 100 μl 8 M HCl and 150 μl 0.4 M Tris–HCl pH 7.5, and serial dilution with decreasing amounts is evaluated by radioimmunoassay. To correct for nonspecific 5'-AMP contamination control extractions are performed by incubation of the resuspended TCA-insoluble material with 100 μl 2 M NaCl/0.4 M Tris–HCl at 0° for 5 min instead of NH$_2$OH.

If (ADPR)$_n$ proteins soluble in 5% perchloric acid (like histone H1 or HMG protein conjugates) are to be included, the procedure is modified. Four hundred microliters of 40% trichloroacetic acid instead of perchloric acid is added to the NH$_2$OH incubation mixture. After centrifugation (2 min, 9000 g), 600 μl of the supernatant is removed and extracted five times with five volumes water-saturated ether. The volume is then brought to 900 μl by the addition of water, and 700 μl of the mixture is analyzed as described for the standard procedure. Control extractions with NaCl–Tris are performed as above.

Alkali-Sensitive ("Total") ADPR Protein Conjugates

This fraction represents the sum of the hydroxylamine-sensitive and the hydroxylamine-resistant conjugates.

Aliquots (300 μl) of the resuspended TCA-insoluble fraction are incubated with 100 μl 4 M NaOH for 60 min at 56°. Samples are cooled, neutralized with 200 μl 2 M HCl, and freed from protein by the addition of 200 μl 2 M HClO$_4$. After 15 min on ice samples are centrifuged (2 min, 9000 g). Of the supernatant 600 μl is mixed with 300 μl 0.68 M K$_3$PO$_4$ and freed from KClO$_4$ by centrifugation. To 700 μl supernatant, 200 μl 3.8 M NaCl and 150 μl 0.4 M Tris–HCl pH 7.5 are added to adjust the salt concentration to standard conditions. Serial dilutions are then analyzed for ADPR equivalents in the radioimmunoassay. Control extractions with NaCl are performed as described above (hydroxylamine-sensitive fraction). To include histone H1 conjugates, perchloric acid is substituted by trichloroacetic acid followed by the removal of the acid by ether extraction (see preceding section).

Radioimmunoassay for ADPR (or 5'-AMP)

Of an ADPR standard solution (240 pmol) 200 μl is incubated with 40 μl 6.0 M NaOH at 56° for 60 min, cooled, neutralized with 40 μl 6 M HCl and 20 μl 1 M Tris–HCl pH 7.5. Aliquots corresponding to 0.5–80 pmol ADPR are used for calibration of the radioimmunoassay.

One hundred microliters containing alkali-treated sample or ADPR standard is mixed with 100 μl 5'-[^{3}H]AMP (5 × 10^4 dpm, 1.4 pmol), 50 μl anti 5'-AMP antiserum (0.1 nmol binding sites), and 50 μl buffer pH 7.5 (30 μmol sodium phosphate, 6 μmol sodium pyrophosphate, 350 μg human γ-globulin). After incubation for 16–18 hr in ice, the bound tracer is separated by precipitation with 1.7 ml cold 3.2 M ammonium sulfate solution. After 10 min in ice, the precipitate is sedimented 3000 g, 20 min, 4°), dissolved in 1 ml 1 mM NaOH, and counted in 15 ml Biofluor (NEN, Dreieichenhain). The values are corrected for a small unspecific tracer binding to normal rabbit serum. Processing of the data by computer-aided spline approximation for curve fitting helps in the analysis of large numbers of samples.

Appendix I

Antigens and Anti 5'-AMP Antiserum. When application to crude tissue fractions is intended, antisera must effectively discriminate 5'-AMP from 3'(2')-AMP. This cannot be achieved by antisera raised against periodate-oxidized 5'-AMP haptens. 5'-AMP analogs linked to antigenic proteins via N^6-carboxymethyl groups, however, gave rise to excellent specificities.[10]

Preparation of N^6-Substituted 5'-AMP Hapten. N^6-Carboxymethyl-AMP was synthesized according to Lindberg and Mosbach,[11] using 4 mmol 5'-AMP as the starting material. A simple alternative converts commercially available N^6-carboxymethyl-NAD (Sigma Chemicals, München) to N^6-carboxymethyl-AMP by treatment with snake venom phosphodiesterase. The use of ^{3}H-labeled AMP or cm^6-AMP facilitates the analysis of the reaction.

Coupling of the Hapten to Serum Albumin.[12] Bovine serum albumin (62.5 mg) was dissolved in 3 ml water, and the pH value adjusted to 5.5. N-Ethyl-N-(3-dimethylaminopropyl)carbodiimide (125 μmol) was added and the total volume brought to 5.0 ml. cm^6 5'-AMP solution (1.25 ml 50 mM) was added in 20-μl aliquots over a period of 90 min. The mixture was allowed to stand in the dark at room temperature for 21 hr with continuous adjustment at the pH to 5.5. Low-molecular-weight components were then removed by dialysis against 0.15 M NaCl. Calculation of the epitope density revealed an average of 10 mol cm^6 5'-AMP per mol serum albumin.

A significant increase in antibody affinity was achieved when cm^{6}ADP-ribose was coupled to serum albumin instead of cm^6 5'-AMP.[12] Due to conversion *in situ* to a cm^6 5'-AMP conjugate by intracellular or pericellular phosphodiesterase this antigen induced 5'-AMP specific antibodies. Since cm^{6}ADP-ribose is easily prepared from commercially available cm^{6}NAD (Sigma) by treatment with NAD glycohydrolase,[12] production of 5'-AMP specific antibodies with the aid of cm^{6}ADP-ribose–albumin conjugates offers an alternative route to 5'-AMP-specific antibodies.

Immunization of Rabbits. Antigen (0.5 mg) is dissolved or suspended in 1 ml of 0.15 M NaCl and emulsified with an equal volume of complete Freund's adjuvant, using the double syringe method: The thick, stable water-in-oil emulsions (controls should be performed by phase-contrast microscopy) are injected intradermally at multiple sites of rabbits. At 1- to 2-month intervals the rabbits are given booster injections with 0.25 mg antigen in 1 ml 0.15 M NaCl emulsified with 1 ml complete Freund's adjuvant. From the tenth day after the booster injections rabbits are bled several times through the lateral ear veins, and sera were prepared by

[10] R. Bredehorst, M. Schlüter, and H. Hilz, *Biochem. Biophys. Acta* **652,** 16 (1981).
[11] M. Lindberg and K. Mosbach, *Eur. J. Biochem.* **53,** 481 (1975).
[12] R. Bredehorst, A. Ferro, and H. Hilz, *Eur. J. Biochem.* **82,** 105 (1978).

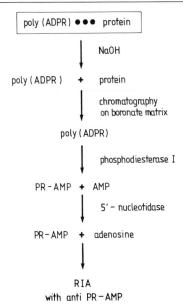

FIG. 2. Determination of protein-bound poly(ADPR) residues.

standard procedures. Optimal antisera were obtained after the third or fourth booster injection.

Determination of Protein-Bound Poly(ADPR)

Principle

Protein-bound oligo and poly(ADPR) chains are released by treatment with alkali, which leaves the polymer intact but converts mono(ADPR) to 5'-AMP (Fig. 2). Because of the low tissue concentrations of poly (ADPR), purification and concentration are required. This is achieved by covalent chromatography on boronate matrices which eliminates protein, DNA, and RNA split products. Highly efficient binding and salt-free release of the polymer is achieved by the use of Bio-Rex 70 aminophenyl boronate matrices. The released material is converted by phosphodiesterase I to PR-AMP which can be quantified by a specific radioimmunoassay.[13,14]

[13] K. Wielckens, R. Bredehorst, P. Adamietz, and H. Hilz, *Eur. J. Biochem.* **117**, 69 (1981).
[14] K. Wielckens, E. George, T. Pless, and H. Hilz, *J. Biol. Chem.* (in press).

Preparation of the TCA-Insoluble Fraction. This is done essentially as described in the preceding section on "Determination of Protein-Bound Mono(ADPR) Residues.

Release of Poly(ADPR) from Proteins. TCA-insoluble tissue fraction (20–50 mg) is dissolved in 2.5 ml 6 M guanidine–HCl/50 mM morpholine buffer pH 8.5 with the aid of an all-glass homogenizer, and the pH is adjusted to 8.5. Release is effected by treatment with NaOH (0.3 M, 1 hr at 56°). After neutralization with HCl and careful adjustment of the pH to 8.5, [^{3}H]poly(ADPR) marker (7000 dpm, sp act 51 Ci/mmol) is added to correct for losses during the subsequent processing. The sample is centrifuged (20 min at 3000 g) and the supernatant is used for chromatography on the boronate gel.

Boronate Chromatography. Before use, the boronate matrix is washed with 6 M guanidine–HCl/2 M morpholine buffer pH 8.5 until the effluent has reached the same pH value. Subsequently, the matrix is equilibrated with 6 M guanidine–HCl/50 mM morpholine buffer pH 8.5 to be ready for use.

Of the equilibrated boronate gel 300 mg, wet weight is mixed with each sample [2–3 ml supernatant containing the released poly(ADPR)]. The tubes are stoppered and rotated over night at room temperature. The content is then transferred to 2-ml plastic syringes plugged with silanized glass wool, and the solution is allowed to drain by gravity. Subsequently the syringes are centrifuged (5 min at 200 g), the matrix washed with 3 ml 6 M guanidine–HCl/50 mM morpholine buffer pH 8.5, and centrifuged again. To remove residual guanidine, the gels are washed with 5 ml 1.0 M NaCl/50 mM morpholine buffer pH 8.5 and freed from excess fluid by centrifugation. Poly(ADPR) is then eluted with 3 ml 1 mM sorbitol, the eluate lyophilized, and the lyophilisate dissolved in 600 μl water that has been boiled previously to eliminate traces of phosphatases. Two 50-μl aliquots are hydrolyzed in 1 ml 5% TCA (60 min 95°) and counted for ^{3}H to determine the yield (usually 80–90%).

Conversion of Polymer to PR-AMP. The remaining 400 μl of the eluate is incubated with 50 μl 0.5 M Tris–HCl buffer pH 7.5, 50 μl 0.1 M glucose 1-phosphate, 10 μl (2.4 mU) purified[15] snake venom phosphodiesterase, and 10 μl (1 U) 5′-nucleotidase (Sigma, grade IV) for 1 hr at 37°. (5′-Nucleotidase is included in order to remove contaminating 5′-AMP which at higher concentrations can interfere with the radioimmunoassay. This enzyme does not degrade PR-AMP.) Completion of poly(ADPR) and AMP degradations was controlled by incubating 10-μl aliquots of the reaction mixture with 2 μl 10 mM ADPR for the same time. Subsequent thin-layer chromatography on PEI-cellulose with water as the solvent

---

[15] W. Heyns and P. Dehoor, *Biochim. Biophys. Acta* **358**, 1 (1974).

showed a complete shift of ADPR to adenosine, when both enzymes were fully active.

After completion of hydrolysis, the enzymes are heat-inactivated (20 min, 95°), and dilutions (usually 1 : 2 to 1 : 16) are prepared for analysis by radioimmunoassay.

Radioimmunoassay of PR-AMP. One hundred microliters of sample or standard solutions (0.1–20 pmol) is mixed with 100 μl PR-[^{3}H]AMP (5 × 10^4 dpm, 0.5 pmol) in 0.45 M NaCl. Of an antiserum dilution (equivalent to bind 40% of the tracer made up with buffer A (0.3 M sodium phosphate, 0.06 M sodium pyrophosphate, and 3.5 mg/ml human immunoglobulin, pH 7.5) 100 μl is added. The tubes are incubated for 16–18 hr in ice, mixed with 1.7 ml 3.2 M ammonium sulfate, and left on ice for 10 min. The bound tracer is then separated by centrifugation, the pellets dissolved in 1.2 ml 1 mM NaOH, transferred to counting vials, and counted with a standard scintillation cocktail. The data are processed by a computer program using spline approximation for curve fitting.

PR-AMP standard solution is prepared from purified poly(ADPR) as described for the preparation of the antigen. Purity is controlled by HPLC.

Differentiation between Hydroxylamine-Sensitive and -Resistant Poly(ADP-ribose) Protein Conjugates

Acid-insoluble material (50 mg) is suspended in 5 ml freshly prepared 0.5 M NH$_2$OH pH 7.5, and the pH is adjusted if necessary. After incubation for 1 hr at 37°, 3.5 M perchloric acid is added to a final concentration of 0.5 M, and the sample is allowed to stand for 15 min on ice. The insoluble material is sedimented by centrifugation (15 min, 3000 g) and the pellet is washed twice with 2 ml 0.5 M perchloric acid. The combined supernatants (containing the hydroxylamine-sensitive fraction) are neutralized with 1 M K$_2$CO$_3$, and the KClO$_4$ formed is sedimented. The resulting supernatant can be analyzed for poly(ADPR) without purification on the boronate matrix. The HClO$_4$-insoluble material containing the hydroxylamine-resistant fraction is dissolved in 6 M guanidine–HCl/50 mM morpholine pH 8.5, purified by boronate chromatography and processed as described in the standard procedure. Control extractions are made with 0.5 M NaCl/0.1 M Tris–HCl pH 7.5 in order to correct for free chains.

This procedure relies on the solubility in 5% perchloric acid of free poly(ADPR) molecules with chain lengths up to at least 30 units,[16] while the intact protein conjugates are precipitated under these conditions.

[16] P. Adamietz, R. Bredehorst, and H. Hilz, *Biochem. Biophys. Res. Commun.* **81,** 1377 (1978).

Since it cannot be completely excluded that very long poly(ADPR) chains and branched structures of large sizes are insoluble in 5% perchloric, the hydroxylamine-sensitive poly(ADPR) fraction as obtained by this procedure may be an underestimate. Interference by perchloric acid-soluble conjugates of histone H1 or HMG proteins (usually present in very small amounts) can be corrected by control extraction with NaCl instead of NH_2OH.

Quantification of Poly(ADPR) Residues in Polyacrylamide Gels

Direct determination of poly(ADPR) in polyacrylamide gel slices is hampered by the slow and size-dependent diffusion of $(ADPR)_n$ residues into the extraction fluid, and by the inability of phosphodiesterase to penetrate the gel slices in reasonable time periods. These difficulties can be overcome when degradation of poly(ADPR) to PR-AMP is accomplished by treatment with Mg^{2+} and alkali instead of phosphodiesterase digestion,[17] although only 40% of the theoretical PR-AMP equivalents are detected by the radioimmunoassay. The relatively low yield of the PR-AMP in this reaction can be corrected by comparing with a standard poly(ADP-ribose) solution. In most cases, relative amounts may suffice anyway.

Slices from stained or instained gels containing the separated conjugates are incubated first with 6 mM MgCl$_2$ for 1 hr at room temperature to allow saturation of poly(ADPR) with Mg^{2+} ions. (If alkali is added together with MgCl$_2$, Mg^{2+} is precipitated before it reaches the polymer in the gel.) Subsequently, the slices were heated with NaOH (0.33 M final concentration) for 6 hr at 56°. Aliquots of the supernatants are neutralized with HCl and Tris–HCl buffer pH 7.5, and incubated for 1 hr at 37° with glucose 1-phosphate and 5′-nucleotidase as described for the standard procedure. After heat inactivation of the enzyme, the samples are analyzed for PR-AMP by radioimmunoassay.

Appendix II

Production of Anti-PR-AMP Antibodies. Poly(ADPR) is synthesized from NAD with Ehrlich ascites tumor cell nuclei or rat liver nuclei as an enzyme source.[13] Nuclei (200 mg/ml) are incubated at 25° with 4 mM NAD, 50 mM Tris–HCl pH 8.2, 40 mM KCl, 5 mM MgCl$_2$, and 100 μg DNase/ml. After 1 hr the reaction is stopped by the addition of cold TCA (final concentration: 20%). The acid-insoluble material is washed 3 times

---

[17] P. Adamietz and R. Bredehorst, *Anal. Biochem.* **112,** 314 (1980).

with cold 20% TCA and twice with ethanol. The precipitate is then dissolved in 6 M guanidine–HCl 15 mM morpholine pH 8.5. NaOH is added to a final concentration of 0.3 M, and the sample is incubated for 1 hr at 56° to release poly(ADPR) from proteins. After neutralization with concentrated HCl the pH is carefully adjusted to pH 8.5 using 2 M morpholine, and the solution is applied to a Bio-Rex 70 aminophenyl boronate column (1 × 2 cm). After extensive washing with 6 M guanidine–HCl/50 mM morpholine buffer pH 8.5 and 1 M NaCl, poly(ADPR) is eluted with 1 mM sorbitol concentrated by rotary evaporation, and digested with purified (by chromatography on Blue Sepharose[15]) snake venom phosphodiesterase (0.1 U/ml) in 50 mM Tris–HCl buffer pH 7.5 at 37° for 1 hr. The reaction is stopped by heating (95°, 45 min). One gram nuclei usually yields 1.2 μmol PR-AMP.

PR-AMP is isolated by preparative paper chromatography (Schleicher and Schüll, 2043b, HCl-washed). The chromatograms are developed in isobutyric acid/25% NH$_3$/water (60 : 1 : 33) and the PR-AMP band (R_f = 0.25) is cut into small pieces, extracted by water, and concentrated by rotary evaporation.

The *antigen* is prepared by treatment of 1.5 μmol PR-AMP in 0.5 ml H$_2$O with 0.5 ml 0.1 M NaIO$_4$/20 mM phosphate buffer pH 7.2 for 1 hr at room temperature. Excess NaIO$_4$ is destroyed by the addition of 20 μl 1 M glycerol. Bovine serum albumin (4 mg) (Hoechst, Frankfurt) in 100 μl 150 mM K$_2$CO$_3$ pH 9.5 is added and the pH adjusted to 9.5 with additional K$_2$CO$_3$. After standing for 1 hr at room temperature NaBH$_4$ is added to give a concentration of 50 mM, and the mixture is incubated at 25° for 12 hr (cf. ref. 18). The PR-AMP-serum albumin conjugate formed is then freed from low-molecular-weight compounds by dialysis for 24 hr. The epitope density of the antigen usually obtained is 4–6 mol PR-AMP/mol serum albumin.

Immunization of rabbits is done by injection of 1 mg antigen dissolved in 0.5 ml isotonic saline and emulsified with 0.5 ml complete Freund's adjuvant. Three booster injections (0.5 mg antigen each) are given at 1 month intervals. Sera are prepared by standard procedures and stored at −20°.

Preparation of Highly Labeled PR-[³H]AMP. Since the sensitivity of a radioimmunoassay depends to a high degree on the specific radioactivity of the tracer, it is necessary to prepare highly labeled PR-AMP. This is accomplished by synthesizing highly labeled poly(ADPR) directly from NMN and [³H]ATP of high specific radioactivity in isolated nuclei under conditions of fully activated poly(ADPR) polymerase.

[18] B. Erlanger and S. Beiser, *Proc. Natl. Acad. Sci. U.S.A.* **52,** 68 (1964).

Two millicuries [2,8,5'-³H]ATP (NEN, 40–60 Ci/mmol) is concentrated by rotary evaporation and mixed with 25 mg nuclei (EAT cells) in a final volume of 0.2 ml containing 50 mM Tris–HCl pH 8.2, 5 mM MgCl$_2$, 40 mM KCl, 1 mM dithiothreitol, 50 mU NAD pyrophosphorylase (Boehringer), 1.5 mM phosphoenolpyruvate, 1 U pyruvate kinase (Boehringer), 1.2 U myokinase (Boehringer), 170 μM NMN, and 50 U DNase I. After 20 min at 25° the reaction is stopped with TCA (20% final concentration). Isolation of [³H]poly(ADPR) and preparation of PR-[³H]AMP is then performed as described above under Production of Anti-PR-AMP Antibodies. The overall yield of PR-[³H]AMP with respect to [³H]ATP is 30–40%.

Preparation of Bio-Rex 70-Aminophenyl Boronate Matrix. Most of the commercially available boronate matrices are not suitable for the isolation of poly(ADPR) since they contain positively charged groups, which will retain the polyanionic poly(ADPR) chains during elution of the polymer with water or 1 mM sorbitol. Elution by lowering pH in the presence of high salt concentrations which could overcome this difficulty would introduce at the same time the new problem of desalting the samples prior to the radioimmunoassay. A boronate matrix specifically suited for the purification and concentration of poly(ADPR) can be easily synthesized from aminophenylboronic acid and the cation exchange resin Bio-Rex 70. This boronate matrix binds poly(ADPR) at alkaline pH in the presence of high salt concentrations and releases it without retardation when the high salt buffer is replaced by water or 1 mM sorbitol—due to labilization of the cyclic boronate ester bonds and the repulsion of the polymer by the residual cationic groups of the matrix.

Fifty grams of Bio-Rex 70 (100–200 mesh) is suspended in water and allowed to swell overnight. Then, fines are removed by decantation. Five grams of aminophenylboronic acid hydrochloride is dissolved in 50 ml water and mixed with the Bio-Rex 70 slurry. After adjustment of the pH value to 5.0, 5 g N-ethyl-N'-(3-dimethylaminopropyl)carbodiimide is added and the mixture is rotated for 18 hr at room temperature. After extensive washing with water and finally with 6 M guanidine hydrochloride 50 mM morpholine ethanesulfonate buffer pH 6.0, the boronate gel is stored in the same buffer at 4°.

Acknowledgment

Our work was supported by the Deutsche Forschungsgemeinschaft.

[50] Determination of *in Vivo* Levels of Polymeric and Monomeric ADP-Ribose by Fluorescence Methods

By Myron K. Jacobson, D. Michael Payne,
Rafael Alvarez-Gonzalez, Hector Juarez-Salinas,
James L. Sims, and Elaine L. Jacobson

The determination of *in vivo* levels of polymeric and monomeric ADP-ribose presents difficult analytical problems with respect to both selectivity and sensitivity. For example, to quantify poly(ADP-ribose) levels, picomole amounts of adenine-containing compounds derived from poly(ADP-ribose) must be measured in samples containing micromole amounts of adenine ring present in both RNA and DNA. With regard to quantification of monomeric ADP-ribose covalently bound to protein, foremost among possible interfering substances are NAD^+ and NADH, which are present in at least 100-fold excess and often more. Even a small fraction of these nucleotides trapped in acid insoluble fractions can lead to a large overestimation of levels of protein-bound ADP-ribose. It is, therefore, crucial that methods designed to measure *in vivo* levels of ADP-ribose residues be sensitive and that the selectivity be carefully demonstrated.

The overall analytical approach of the two methods is shown in Fig. 1. Two key features that are common to both methods are crucial in providing the necessary selectivity and sensitivity. The first of these features is the utilization of immobilized boronate resins to selectively and quantitatively adsorb polymeric or monomeric ADP-ribose from cell or tissue extracts. The second feature common to both methods is conversion of adenine-containing compounds to highly fluorescent $1,N^6$-etheno derivatives which can be quantified at the picomole level. For polymeric ADP-ribose, the adenine-containing compounds are formed by the enzymatic hydrolysis of the polymer to generate unique adenosine derivatives from all internal residues. As shown in Fig. 2, the nucleosides ribosyladenosine (RAdo) and diribosyladenosine (R_2Ado) are generated from linear and branched residues, respectively. For monomeric ADP-ribose, the method involves chemical release of intact ADP-ribose residues from protein and quantification following conversion to the $1,N^6$-etheno(ADP-ribose).

Preparation of Acid Insoluble Fractions

Cultured Cells. For monolayer cell cultures, the medium is decanted and the monolayer is washed once with phosphate-buffered physiological

METHODS IN ENZYMOLOGY, VOL. 106

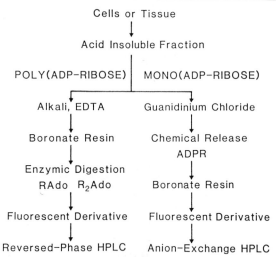

Fig. 1. Overall analytical approach for quantification of *in vivo* levels of polymeric and monomeric ADP-ribose.

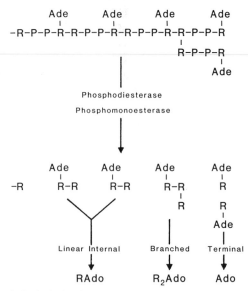

Fig. 2. Enzymatic hydrolysis of poly(ADP-ribose) to generate unique adenosine derivatives from internal residues.

saline. Ice cold 20% (w/v) trichloroacetic acid is added, and the dish is placed on ice for 15 min. Acid insoluble material is scraped from the dish with a rubber policeman, transferred to a centrifuge tube, and collected by centrifugation. The pellet is washed once with ice cold 20% trichloroacetic acid and twice with diethyl ether, dried, and stored at −20° until analysis.

Animal Tissues. Prior to treatment with trichloroacetic acid, tissues are rapidly excised and immediately dropped into liquid nitrogen. The desired amount of frozen tissue is placed in a mortar containing liquid nitrogen, pulverized to a fine powder with a pestle and transferred to a centrifuge tube. After the liquid nitrogen has evaporated, 5 volumes of ice cold 20% (w/v) trichloroacetic acid are added and the tissue is homogenized for two 30-sec intervals with a Polytron tissue homogenizer at a setting of 6. The tube is immersed in an ice bath for cooling during the homogenization. Acid insoluble material is collected by centrifugation, washed once with trichloroacetic acid as above, washed twice with 5 volumes of cold 95% ethanol and twice with 5 volumes of diethyl ether. Removal of residual ether by gentle vacuum results in a fine powder which is stored at −20° until analysis.

Measurement of Polymeric ADP-Ribose

Introduction. The following procedure is designed for acid-insoluble fractions as described above. The assay described is designed for up to 10^8 tissue culture cells or for up to 1.3 g (wet weight) of tissue. For larger amounts, scale up accordingly.

Specialized Reagents. Dihydroxyboryl Bio-Rex 70 (DHB Bio-Rex). This resin is synthesized as described by Wielckens *et al.* (this volume [49]) and stored in 50 mM MOPS, 6 M guanidinium chloride, pH 6.0, and is stable for several months. The required amount of resin is washed by a batch procedure on the day of assay. The following procedure is for up to 1 ml of packed resin; scale up accordingly for larger amounts. Wash once with 5 ml of 250 mM ammonium acetate, pH 9.0 containing 6 M guanidinium chloride (Buffer A), three times with 5 ml of distilled H$_2$O, once with 5 ml of Buffer A, and then resuspend in an equal volume of Buffer A. The resin is now ready for use in step 2 below.

Matrix Gel PBA-60. This resin is commercially available from Amicon Corporation. Prepare 0.5-ml columns in 1-ml plastic syringes containing a small plug of glass wool. Wash each column with 5 ml of 250 mM ammonium acetate, pH 4.5 and then with 10 ml of 250 mM ammonium acetate, pH 9.0 (Buffer B). The columns are now ready for use in step 5 below.

These columns can be recycled and used many times. Following use, they are stored tightly capped in 50 mM MOPS, pH 7.0 in the dark at 4°.

Venom Phosphodiesterase, Alkaline Phosphomonoesterase Mixture. Step 3 below utilizes a mixture of these enzymes to digest poly(ADP-ribose) to nucleosides as shown in Fig. 2. This mixture must be preincubated and dialyzed to remove nucleosides from RNA and DNA which are present as contaminants in the alkaline phosphatase preparation. A 10-ml solution containing 10 mM MOPS, 50 mM MgCl, pH 7.4, 1000 units of bacterial alkaline phosphatase (Sigma, Type III-S), and 100 units of snake venom phosphodiesterase (Worthington VPH) is incubated for 2 hr at 37°. This solution is then dialyzed in the cold against 1 liter of 10 mM MOPS, 50 mM MgCl$_2$, pH 7.4 for 24 hr with buffer changes at 8 hr intervals. The dialysate is diluted to 40 ml with the same buffer and stored frozen in 1-ml aliquots.

Step 1. Dissolution of Sample. Steps 1 and 2 are designed to be carried out in a 50-ml polypropylene screw cap conical centrifuge tube. To each acid insoluble pellet add 2.0 ml of 1 M KOH, 50 mM EDTA and mix by vortexing.[1] Incubate at 60° for 1 hr. Occasional vortexing during the incubation aids in the complete dissolution of the pellet.[2] In cases where trichloroacetic acid has not been thoroughly removed from the pellet, it may be necessary to utilize a glass homogenizer to disperse the pellet such that complete dissolution is possible. Following this incubation, a small aliquot can be removed from each sample for determination of DNA content if desired.

Step 2. Binding and Elution from Boronate Resin. Add 2.7 ml of Buffer A to each sample and mix well. Adjust the pH to 9.0 ± 0.2 by addition of approximately 100 μl of conc. HCl.[3] Add 0.2 ml of a 50% (v/v) suspension of DHB Bio-Rex and incubate at room temperature for 2 hr with gyrorotary shaking sufficient to keep the resin suspended. Transfer the resin suspension to 1-ml plastic syringes containing a small plug of glass wool and wash the resin with 5 ml of Buffer A.[3a] Wash the resin with

---

[1] With each set of samples to be analyzed, a sample containing exogenously added [adenine-^{14}C]poly(ADP-ribose) is included as a recovery control.

[2] The treatment with KOH, EDTA not only provides a convenient method for dissolution of the sample, but it also results in the release of polymer from protein and the conversion of monomers of ADP-ribose to 5'-AMP. Both the presence of protein bound to the polymer and an excess of monomers of ADP-ribose can potentially interfere with binding and/or release of the polymer from DHB Bio-Rex (step 2).

[3] The efficiency of binding of poly(ADP-ribose) to the resin can be conveniently determined by removing one aliquot from the recovery control prior to addition of the resin and a second aliquot after incubation and centrifugation.

[3a] Alternatively, 0.8 × 4 cm polypropylene Econo-Columns (Bio-Rad Laboratories) can be used in place of 1-ml syringes for affinity chromatography.

10 ml of 1.0 M ammonium acetate buffer, pH 9.0. Elute with 4.0 ml of 10 mM HCl and collect the eluates in 15-ml screw cap centrifuge tubes containing 500 μl of Buffer B and 500 μl of 0.5 M MgCl$_2$.[4]

Step 3. Enzymatic Digestion to Nucleosides. Add 100 μl of venom phosphodiesterase, alkaline phosphomonoesterase mixture and incubate at 37° for 3 hr.

Step 4. Formation of Fluorescent Derivatives. To the above, add 555 μl of 2.5 M ammonium acetate, pH 4.5 and 85 μl of 7 M chloroacetaldehyde. Incubate at 60° for 4 hr with the tubes tightly capped. This treatment results in the quantitative conversion of adenosine, ribosyladenosine, and diribosyladenosine to their respective 1,N^6-etheno (ε) derivatives.

Step 5. Preparation for Analysis of Fluorescence. Dilute to 10 ml with Buffer B and adjust the pH to 9.0 ± 0.2 with conc. NH$_4$OH (approximately 200–250 μl). Centrifuge at low speed to remove any insoluble material. Apply to a 0.5-ml column of Matrex Gel PBA-60. Wash the column with 10 ml of Buffer B. Elute the nucleosides with 4.0 ml of 200 mM sodium citrate buffer, pH 4.0. The samples are now ready for analysis by HPLC.

Step 6. Analysis by High-Performance Liquid Chromatography. The fluorescent nucleosides derived from poly(ADP-ribose) are separated on a Beckman–Altex Ultrasphere-ODS reversed-phase column (250 × 4.6 mm i.d.). All samples are injected in a volume of 2.0 ml in 200 mM sodium citrate buffer, pH 4.5. The column is eluted isocratically at 1.0 ml min^{-1} and room temperature with a mixture of 7 mM ammonium formate, pH 5.8 and 100% methanol. The relative amounts of buffer and methanol (v/v) are varied in the range of 6 to 10% methanol according to the age of the column, so that retention times remain fairly constant. Fluorescence monitoring is done with a Varian Fluorichrom filter fluorimeter equipped with a deuterium light source, a Varian 220-I interference filter for excitation, and a Varian 3-75 emission filter.

Standardization of the Assay. Figure 3 shows an analysis of a sample derived from chicken pancreas, in which fluorescent peaks of εAdo, εRAdo, and εR$_2$Ado are observed. It should be pointed out that while εRAdo and εR$_2$Ado are diagnostic for poly(ADP-ribose), εAdo is not since it can be derived from other cellular molecules including RNA. Small amounts of adenosine also contaminate the phosphodiesterase, phosphomonoesterase preparation. ε-Deoxyadenosine, which can be present as a result of residual DNA contamination, chromatographs

---

[4] An aliquot from the recovery control should be counted to determine the efficiency of elution, which is routinely greater than 95%.

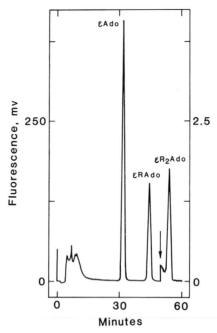

Fig. 3. HPLC analysis of a sample derived from chicken pancreas. The tissue sample was processed and analyzed as described in the text. The pen deflection at 50 min (arrow) resulted from a 100-fold increase in the sensitivity of the fluorimeter. Numbers on the right-hand ordinate represent fluorescence after the change in sensitivity.

slightly earlier than εRAdo, but its presence is eliminated completely in step 5 since deoxynucleosides do not bind to Matrex Gel PBA-60. However, the reader is advised that if the Matrex Gel PBA-60 step is eliminated, significant amounts of ε-deoxyadenosine are observed.

Chicken pancreas has a relatively high content of poly(ADP-ribose), so the presence of branched residues is readily detected. However, since branched residues comprise less than 2 mol% of total internal residues, εR$_2$Ado is not detected in many routine measurements.

The relationship between fluorescence peak height and quantity of εRAdo and εR$_2$Ado is linear from 0.5 to 100 pmol. A routine working range is from 0.5 to 5 pmol. εRAdo can be isolated in relatively large quantities by applying the procedure described here to samples of permeable cells that have been incubated *in vitro* with NAD$^+$.[5] εRAdo can be standardized by UV absorption assuming a molar extinction coefficient of

[5] R. Alvarez-Gonzalez, H. Juarez-Salinas, E. L. Jacobson, and M. K. Jacobson, *Anal. Biochem.* **135**, 69 (1983).

6.0×10^3 at pH 7.0 and 275 nm[6] or from radioactivity if radiolabeled NAD^+ of known specific activity is used (the endogenous polymeric ADP-ribose is negligible when cells have been incubated *in vitro* with NAD^+). εR_2Ado gives essentially the same fluorescence quantum yield as $\varepsilon RAdo$ and thus may be standardized relative to $\varepsilon RAdo$.

Validation of the Assay. In view of the very low levels of polymeric ADP-ribose *in vivo,* controls that demonstrate the selectivity of the method are very important. These have been described in detail for this method elsewhere[7] but will be briefly discussed here. As with any sensitive fluorescent method, a demonstration that the reagents used do not contain interfering fluorescent contaminants is essential. This has not proven to be a general problem for this assay; however, we have infrequently observed fluorescent impurities in some lots of guanidine. Care must also be taken to prevent and detect bacterial growth in the reagents since this is a common source of fluorescent materials. The omission of venom phosphodiesterase treatment results in the complete absence of fluorescent material at the elution positions of $\varepsilon RAdo$ and εR_2Ado, while addition of DNase or RNase treatments has no effect. Omission of chloroacetaldehyde treatment also results in the absence of fluorescent material, which rules out the possibility of interference by endogenous fluorescent compounds derived from cell extracts. Treatment with adenosine deaminase prior to treatment with chloroacetaldehyde also results in an absence of fluorescence at the $\varepsilon RAdo$ and εR_2Ado positions, since the N^6 amino group is required for the formation of the etheno derivatives. The identity of $\varepsilon RAdo$ and εR_2Ado has also been further established by limited acid hydrolysis of εR_2Ado to $\varepsilon RAdo$ and εAdo and of $\varepsilon RAdo$ to εAdo.[8]

Recovery controls are routinely performed using exogenously added [*adenine*-^{14}C]poly(ADP-ribose) which can be synthesized *in vitro* using nucleotide permeable cells and purified using DHB Bio-Rex chromatography.[5] The use of [*adenine*-^{3}H]poly(ADP-ribose) is not recommended because some of the ^{3}H is lost during the alkaline incubation required for dissolution of the acid insoluble pellet. The most crucial steps with regard to selective losses are the binding to and elution from the DHB Bio-Rex resin (step 2). We routinely observe that binding is greater than 90% and elution is quantitative. It is important to note that the binding is independent of the chain length of the polymer. The overall recoveries through the procedure are routinely greater than 80%.

[6] J. A. Secrist, J. R. Barrio, N. J. Leonard, and G. Weber, *Biochemistry* **11**, 3499 (1972).
[7] J. L. Sims, H. Juarez-Salinas, and M. K. Jacobson, *Anal. Biochem.* **106**, 296 (1980).
[8] H. Juarez-Salinas, H. Mendoza-Alvarez, V. Levi, M. K. Jacobson, and E. L. Jacobson, *Anal. Biochem.* **131**, 410 (1983).

Measurement of Monomeric ADP-Ribose

Introduction. A real difficulty with regard to the quantification of monomeric ADP-ribose residues covalently bound to proteins is our limited knowledge of the chemical nature of the linkage(s) that exist(s) *in vivo*. Enzymes from eukaryotic sources have been purified that can catalyze the covalent attachment of single ADP-ribosyl residues to acceptor proteins via N-glycosylic linkages to the guanidino group of arginine residues.[9] Additionally, ADP-ribosylated histones have been isolated following *in vitro* incubations with radiolabeled NAD$^+$. In these cases, the ADP-ribosyl residues were attached to proteins via ester linkages to the γ-carboxyl group of glutamic acid residues and to the α-carboxyl group of terminal lysine residues.[10] The wide difference in chemical stability of N-glycosylic and carboxylate ester linkages is such that ADP-ribose covalently bound via these types of linkages can be differentially released from proteins. Thus the following procedure allows the differential quantification of protein-bound ADP-ribose with the chemical stability characteristic of these two types of linkages. However, since the chemical identities of the linkages between proteins and ADP-ribose *in vivo* have not been rigorously demonstrated, we refer to these two classes as "N-glycosylic-like" and "carboxylate ester-like" only for the sake of simplicity.

The following procedure is designed for acid insoluble material derived from up to 2.5×10^7 tissue culture cells or from up to 0.3 g (wet weight) of tissue. For larger amounts, scale up accordingly. Except where noted otherwise, the pH of each solution is adjusted to the indicated value at 25°.

Specialized Reagents. Sephadex G-25 Superfine. Columns for centrifuge "desalting" are prepared as follows. Hydrate 30 g of Sephadex G-25 superfine (enough for 20 columns) in excess distilled water overnight, then filter the suspension. For each column, resuspend 6.25 g of moist Sephadex in 5 ml of 50 mM MOPS, 10 mM EDTA, 6 M guanidinium chloride, pH 4.0 (Buffer C). Transfer 10 ml of this suspension (equivalent to 5.0–5.5 ml packed bed volume) to a polypropylene Econo-Column (Bio-Rad Laboratories) and centrifuge at 600 g for 5 min, using a 15-ml polypropylene centrifuge tube for collecting the effluent. To completely equilibrate the Sephadex, add 2.5 ml of Buffer C, centrifuge as above, then add another 2.5 ml of Buffer C and store at 5° until ready for use. Prior to sample application (step 1, below), centrifuge at 600 g for 15 min.

---

[9] For review, see J. Moss and M. Vaughn, *in* "ADP-Ribosylation Reactions" (O. Hayaishi and K. Ueda, eds.), p. 637. Academic Press, New York, 1982.

[10] For review, see L. Burzio, *in* "ADP-Ribosylation Reactions" (O. Hayaishi and K. Ueda, eds.), p. 103. Academic Press, New York, 1982.

Dihydroxyboryl Bio-Rex 70 (DHB Bio-Rex). This resin is the same as that described in the previous section for measurement of polymeric ADP-ribose. The desired amount of resin is washed by a batch procedure on the day of use. The following washing procedure is for 1 ml of packed resin. Wash once with 10 ml of cold 100 mM ammonium acetate, 10 mM EDTA, 6 M guanidinium chloride, pH 9.0 (Buffer D), three times with 10 ml of cold distilled H_2O, and then resuspend in 8.2 ml of cold Buffer D. The resin is now ready for use in step 3 below.

Dihydroxyboryl Sepharose (DHB Sepharose). Wash 100 ml of packed Sepharose 4B with five 400-ml portions of 0.1 M NaCl in a Büchner funnel, with drying between washes. Then wash the gel with 1 liter of deionized water, resuspend in 100 ml of cold deionized water, and place in an ice bath. Suspend 25 g of finely divided cyanogen bromide in 50 ml of cold deionized water and add to the Sepharose suspension with constant stirring. Maintain the pH at 11 ± 0.2 by the dropwise addition of 4 M NaOH. When the pH is stable for 5 min, add 50 g of crushed ice and rapidly filter the gel, then wash twice with 400 ml of cold 0.1 M NaHCO$_3$, pH 9.0. Resuspend the gel in 100 ml of cold 0.1 M NaHCO$_3$, pH 9.0, containing 10 g of 6-aminohexanoic acid and stir at 4° for 18 hr. Filter the gel and wash with 5 liters of deionized water, then resuspend in 400 ml of deionized water containing 1 g of 1-ethyl-3-(3-dimethylaminopropyl)car-bodiimide hydrochloride. Adjust the pH to 5.0 and stir for 15 min. Dissolve 800 mg of *m*-aminophenylboronic acid hemisulfate in 10 ml of deionized water, adjust the pH to 5.0 with NaOH, and add to the Sepharose suspension. Stir at room temperature for 16 hr, then filter and wash with 1 liter of 0.5 M NaCl followed by 10 liters of deionized water. Resuspend the gel in 100 ml of deionized water and store at 4° until used.

Prepare 0.5-ml columns in 1-ml plastic syringes containing a small plug of glass wool.[3a] Wash each column once with 5 ml of 250 mM ammonium chloride, pH 9.0 (Buffer E), once with 5 ml of 10 mM H$_3$PO$_4$, 25 mM KCl, and once more with 5 ml of Buffer E. The columns are now ready for use in step 5 below.

Step 1. Dissolution of Sample and Removal of Noncovalently Bound Nucleotides.[11] Steps 1–3 are designed to be carried out in 15-ml polypropylene conical centrifuge tubes. Dissolve the acid insoluble material in

---

[11] Since even a small fraction of the total cellular NAD$^+$ or NADH trapped in acid insoluble pellets can lead to a significant overestimation of ADP-ribose covalently bound to protein, it is important to remove these compounds. We have found that this can be achieved by dialysis or standard gel filtration chromatography, as well as by the procedure described here. However, we have found the procedure described is more convenient, especially for the analysis of multiple samples.

2 ml of ice cold Buffer C.[12] In some cases, homogenization in the tube with a Teflon pestle is necessary to achieve complete dissolution. Noncovalently bound nucleotides are removed by the column centrifugation method.[12a] For each sample to be analyzed, apply a 500-μl aliquot of the dissolved sample to each of two 5-ml columns of Sephadex G-25 superfine and centrifuge at 600 g for 15 min, collecting the effluents in 15-ml polypropylene centrifuge tubes. For quantitative recovery of proteins, apply an additional 250 μl of Buffer C to each column and centrifuge as above, collecting the effluents in the same tubes.

Step 2. Chemical Release of ADP-ribose from Protein. Adjust the pH of each sample to 7.0 ± 0.1 by the addition of 5 to 10 μl of 5 M NH$_4$OH. To one sample add an equal volume of 50 mM MOPS, 10 mM EDTA, 6 M NH$_4$Cl, 6 M NH$_2$OH, pH 7.0 (neutral hydroxylamine), and to the second sample add an equal volume of 50 mM MOPS, 10 mM EDTA, 6 M NH$_4$Cl, pH 7.0 (neutral buffer). Cap the tubes tightly and incubate at 37° for 6 hr.

Step 3. Binding and Elution from Boronate Resin. Place the samples on ice and adjust the pH to 9.4 (5°) by addition of conc. NH$_4$OH (~70 μl).[13] To each sample add 4.5 ml of the suspension of washed DHB Bio-Rex (~0.5 ml of packed resin). Incubate at 0–5° for 30 min with gyrorotary shaking sufficient to keep the resin suspended. Centrifuge and discard the supernatant. Wash the resin once with 5 ml of cold Buffer D and twice with 5 ml of cold 1 M ammonium acetate, pH 9.0. Add 1.0 ml of 1 M HCl and warm the suspension to room temperature. Transfer the suspension to a 1-ml plastic syringe plugged with glass wool and collect the eluate.[3a] Wash the column with 4.0 ml of distilled water and continue collecting the eluate (the pH of the 5-ml eluate should be 4.0–4.5).[14]

Step 4. Formation of 1,N^6-Etheno(ADP-ribose) (εADP-ribose). To each sample add 555 μl of 2.5M ammonium acetate, pH 4.5 and 80 μl of 7 M chloroacetaldehyde. Incubate at 60° for 4 hr with the tubes tightly capped. Cool the samples to room temperature.

Step 5. Preparation for Analysis of Fluorescence. Adjust the pH of each sample to 9.0 ± 0.2 by addition of conc. NH$_4$OH (~200 μl) and apply

---

[12] With each set of samples to be analyzed, a sample containing exogenously added [*adenine*-^{14}C]mono(ADP-ribosylated) histones is included as a recovery control. Alternatively, [*adenine*-^{14}C]ADP-ribose can be added to one sample at the end of step 1.

[12a] E. Helmerhorst and G. B. Stokes, *Anal. Biochem.* **104,** 130 (1980).

[13] The efficiency of binding of mono(ADP-ribose) to the resin can be conveniently determined by removing one aliquot from the recovery control prior to addition of the resin and a second aliquot after incubation and centrifugation.

[14] An aliquot from the recovery control should be counted to determine the efficiency of elution.

to a 0.5 ml column of DHB-Sepharose. Wash with 10 ml of Buffer E and 0.5 ml of 10 mM H$_3$PO$_4$, 25 mM KCl. Elute with 2 ml of 10 mM H$_3$PO$_4$, 25 mM KCl.

Step 6. Analysis by High-Performance Liquid Chromatography. εADP-ribose is separated from residual interfering substances on a Whatman Partisil-10 SAX column (250 × 4.6 mm i.d.) preceded by a guard column (50 × 1.5 mm i.d.) containing the same material. Samples are routinely injected in a volume of 1.0 ml in 10 mM H$_3$PO$_4$, 25 mM KCl, pH 4.5, although up to 2.0 ml may be injected with no change in resolution. The column is eluted isocratically with 100 mM potassium phosphate, pH 4.7 at room temperature and a flow rate of 1.0 ml min^{-1}. Fluorescence monitoring is as described in the previous section.

Standardization of the Assay. Figure 4 shows a typical analysis of a sample derived from adult rat liver, in which fluorescent peaks of εADP-ribose are observed. The relationship between fluorescence peak height and quantity of εADP-ribose is linear from 1.0 to 100 pmol. εADP-ribose can be prepared from εNAD$^+$ by incubation in 0.1 *M* NaOH at 37° for

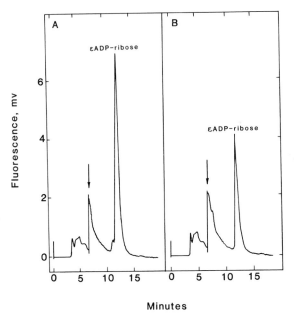

FIG. 4. HPLC analysis of a sample derived from adult rat liver. (A) Sample treated with neutral hydroxylamine. (B) Sample treated with neutral buffer. The tissue sample was processed and analyzed as described in the text. The pen deflection at 6.5 min (arrow) resulted from a 10-fold increase in the sensitivity of the fluorimeter. Numbers on the ordinate represent fluorescence after the change in sensitivity.

30 min, followed by DHB Sepharose chromatography as described above. Alternatively, NAD^+ or [*adenine*-^{14}C]NAD^+ can be subjected to alkaline hydrolysis, followed by derivatization and chromatography as described in steps 4 and 5 above. εADP-ribose can be standardized by UV absorption or specific radioactivity as described for εRAdo in the preceding section.

Validation of the Assay. The requirements for a rigorous demonstration of selectivity in the measurement of polymeric ADP-ribose also apply to the measurement of protein-bound monomeric ADP-ribose. Omission of extract from the assay results in the absence of fluorescent material eluting at the position of εADP-ribose, demonstrating that the reagents do not contain interfering compounds. Omission of chloroacetaldehyde treatment also results in the absence of a fluorescent peak, ruling out the possibility that the cell extracts contain endogenous fluorescent compounds which interfere with the analysis. Finally, addition of [*adenine*-^{14}C]ADP-ribosylated histones to a cell extract prior to analysis results in one peak of radioactivity which is coincident with the fluorescent peak and elutes at the position of εADP-ribose.

Since the N-glycosylic linkage between ADP-ribose and arginine is stable at pH 7.0 in the absence of hydroxylamine, the εADP-ribose observed from samples treated with neutral buffer only (Fig. 4B) is derived from ADP-ribose residues bound to proteins via another, less stable type(s) of linkage(s) (e.g., "carboxylate ester-like"). The εADP-ribose observed from samples treated with neutral hydroxylamine (Fig. 4A) is derived from a combination of ADP-ribose residues bound to proteins via linkages that require hydroxylamine for cleavage plus those residues bound via the "carboxylate ester-like" linkages. Therefore, the difference between the amounts of εADP-ribose observed in the presence and absence of hydroxylamine provides an estimate of the amount of ADP-ribose residues bound to proteins via a second class(es) of more stable linkage(s) (e.g., "N-glycosylic-like").

Recovery controls can be performed using either of two different exogenously added radiolabeled markers. The first is [*adenine*-^{14}C]ADP-ribosylated histones which can be synthesized *in vitro* using mono(ADP-ribosyl) transferase purified from turkey erythrocytes.[15] Alternatively, [*adenine*-^{14}C]ADP-ribose can be added to the extract prior to chemical release of protein-bound ADP-ribose (step 2). The overall recovery is routinely about 50%. The overall recoveries are the same for both markers in the presence of hydroxylamine and for radiolabeled ADP-ribose in the absence of hydroxylamine, confirming that losses are nonselective.

[15] J. Moss, D. Yost, and S. Stanley, *J. Biol. Chem.* **258**, 6466 (1983).

[51] Quantification of *in Vivo* Levels of Poly(ADP-ribose): Tritium Labeling Method and Radioimmunoassay

By MASANAO MIWA and TAKASHI SUGIMURA

Tritium Labeling Method

Principle. The specific degradation product of poly(ADP-ribose), 2′-(1″-ribosyl)-adenosine-5′,5″-bis(phosphate), also called 2′-(5″-phosphoribosyl)-5′AMP [Ado(P)-Rib-P], is dephosphorylated and oxidized in the *cis*-diol region by $NaIO_4$. Then the dialdehyde compound is labeled with $KB[^3H]H_4$ of high specific radioactivity (Fig. 1). This method is also applicable for quantification of the branched portion, namely 2′-[1″-ribosyl-2″-(1‴-ribosyl)]adenosine-5′,5″,5‴-tris(phosphate) [Ado(P)-Rib(P)-Rib-P].

Reagents

Poly([^{14}C]ADP-ribose) (poly([^{14}C]ADP-Rib)) (542 mCi/mmol)
KB^3H_4 (3 Ci/mmol)
Unlabeled Ado-Rib* (Ado-Rib pentalcohol) (Ado-Rib was oxidized with $NaIO_4$ and then reduced with KBH_4.)

Procedures. About 2×10^7 cultured cells or an equivalent amount of tissue was mixed with 10^4 cpm of poly([^{14}C]ADP-Rib) and treated with 10% trichloroacetic acid. The precipitate was collected by centrifugation and suspended in 50 mM Tris–HCl buffer, pH 8.0. Proteinase K (Merck) was added at a concentration of 1 mg/ml and the mixture was incubated overnight at 37° in the presence of 0.02% NaN_3. Aliquots were taken for DNA determination. The mixture was extracted with phenol and the aqueous layer was collected.

The aqueous layer was extracted with ether to remove traces of phenol, then heated for 2 min at 100° and rapidly cooled. Nuclease P_1 (Yamasa Shoyu Co., Japan) was added at a final concentration of 10 μg/ml and incubation was performed for 2 hr at 50°. The hydrolysis of denatured DNA and RNA was checked by cellulose thin-layer chromatography with a mixture of isobutyric acid : ammonia : H_2O (66 : 1 : 33, v/v) as developing solvent. When all ultraviolet absorbing materials had disappeared from the origin and moved to the positions of 5′-AMP and Ado(P)-Rib-P, 1 N NaOH was added to adjust the pH of the reaction mixture to 8. Then snake venom phosphodiesterase from *Crotalus adamanteus,* snake venom 5′-nucleotidase from *Crotalus atrox,* bovine serum albumin, and

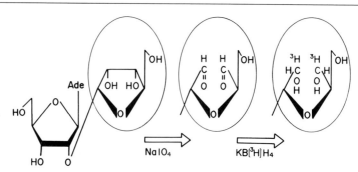

Fig. 1. Principle of tritium labeling.

$MgCl_2$ were added at final concentrations of 0.01 unit/ml, 0.01 unit/ml, 100 μg/ml, and 5 mM, respectively, and the mixture was incubated for 2 hr at 37°. By this procedure, all the nucleotides derived from DNA or RNA should be converted to nucleosides. The reaction was checked by PEI-cellulose thin-layer chromatography with water as developing solvent. The nucleosides all moved from the origin and therefore disappearance of ultraviolet absorbing materials from the origin indicated completion of the reaction. Poly(ADP-Rib) was hydrolyzed mainly to Ado(P)-Rib-P with small amounts of Ado(P)-Rib(P)-Rib-P and adenosine. This reaction was monitored by cellulose thin-layer chromatography with iso-butyric acid : ammonia : water (66 : 1 : 33, v/v) as developing solvent, by following the disappearance of the radioactivity of poly([14C]ADP-Rib) and appearance of radioactivity of [14C]Ado(P)-Rib-P.

After the reaction was completed, the mixture was diluted 5-fold, and applied to a DE 52 column (1 × 1 cm) equilibrated with 20 mM triethylammonium bicarbonate (TEAB) buffer, pH 7.5. The column was washed with 20 ml of 20 mM TEAB buffer, pH 7.5, and the adsorbed material containing Ado(P)-Rib-P (and also Ado(P)-Rib(P)-Rib-P) was eluted with 5 ml of 2 M TEAB buffer, pH 7.5. The eluate was evaporated in a rotary evaporator to remove TEAB.

The eluted Ado(P)-Rib-P (and also Ado(P)-Rib(P)-Rib-P[1]) was hydrolyzed in 1 ml of 60 mM sodium bicine buffer, pH 8.0, with 0.2 unit of alkaline phosphomonoesterase (Sigma) for 6 hr (or overnight) at 37°. The reaction was monitored by PEI-cellulose thin-layer chromatography with water as developing solvent. Dephosphorylation of [14C]Ado(P)-Rib-P to [14C]Ado-Rib was followed by the disappearance of the radioactivity from the origin and its appearance in the position of Ado-Rib. The reaction mixture was concentrated to about 200 μl in a rotary evaporator and subjected to HPLC on an octadecylsilicone column, μBondapak C-18

[1] M. Miwa and T. Sugimura, this volume [46].

column, (0.4 × 30 cm, Waters) using 5% methanol as solvent. The recovered [^{14}C]Ado-Rib was evaporated and dissolved in 200 μl of water.

Oxidation. The sample was mixed with 0.1 A_{260} unit of adenosine and adjusted to pH 8 with 1 N NaOH. Then 3 μl of freshly prepared 9 mM NaIO$_4$ (2 mg/ml) was added and the mixture was shaken for 3 hr at room temperature in the dark. The reaction was followed by subjecting 5 μl samples of the mixture to HPLC on a μBondapack column with 0.5% monoethanolamine phosphate buffer, pH 3.0, in 7% methanol as solvent. Adenosine disappeared with appearance of oxidized adenosine as a broad peak. When adenosine was oxidized, Ado-Rib should also be oxidized.

Tritium Labeling. The mixture was shaken with 3 μl of 0.1 M KB^{3}H$_4$ (3 Ci/mmol) at room temperature in the dark. The reaction was followed by injecting 5 μl samples of the mixture into a μBondapak HPLC column (0.5 × 30 cm, Waters) with 0.5% monoethanolamine phosphate buffer, pH 3.0, in 7% methanol as solvent. When oxidized adenosine was reduced to adenosine trialcohol, the retention time did not change from that of oxidized adenosine, but the peak became very sharp. In these conditions oxidized [^{14}C]Ado-Rib should be converted to [^{14}C,^{3}H]Ado-Rib pentalcohol, [^{14}C,^{3}H]Ado-Rib*.

When tritium labeling was complete, 500 μl of 1 N CH$_3$COOH was added and the mixture was stood for 1 hr, then evaporated to small volume for the next step. Then 0.1 A_{260} unit of authentic unlabeled Ado-Rib* in 200 μl of H$_2$O was added as a carrier and the sample was applied to a μBondapak HPLC column (0.4 × 30 cm, Waters) and eluted at a flow rate of 1 ml/min with 0.5% monoethanolamine phosphate buffer, pH 3.0, in 7% methanol (Fig. 2).[2] Ado-Rib* was eluted at about 14 min and the fractions containing Ado-Rib* were pooled, concentrated, and again subjected to chromatography. Ado-Rib* was purified further by HPLC on a LiChrosorb RP-18 column (0.5 × 25 cm, E. Merck, Darmstadt) with the same solvent. Fractions of 1 ml of the Ado-Rib* peak were collected and of these fractions was processed for a sample oxidizer (Packard). The ratio of ^{3}H/^{14}C of the peak fraction was calculated.

Calculation of the Ado-Rib Content. If x pmol of Ado-Rib is derived from endogenous poly(ADP-Rib), it should be converted to x pmol of [^{3}H]Ado-Rib* by the tritium labeling method. A pmol of [^{14}C]Ado-Rib derived from poly([^{14}C]ADP-Rib) should also be converted to A pmol of [^{14}C,^{3}H]Ado-Rib*. The ^{3}H/^{14}C ratio of [^{14}C,^{3}H]Ado-Rib* derived from authentic poly([^{14}C]ADP-Rib) can be determined by tritium labeling of authentic poly([^{14}C]ADP-Rib) alone assuming R_A (R_A = [^{3}H]$_A$/[^{14}C]$_A$).

When the sample contains $(x + A)$ pmol of Ado-Rib, it will be converted to $(x + A)$ pmol of [^{14}C,^{3}H]Ado-Rib*. Then the ratio of ([^{3}H]$_x$ +

[2] M. Kanai, M. Miwa, Y. Kuchino, and T. Sugimura, *J. Biol. Chem.* **257**, 6217 (1982).

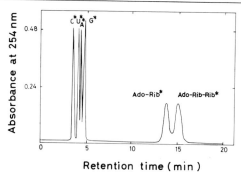

Fig. 2. High-performance liquid chromatogram of standard nucleoside trialcohols, Ado-Rib* and Ado-Rib-Rib*. C*, Cytocine trialcohol; U*, uridine trialcohol; A*, adenosine trialcohol; G*, guanosine trialcohol; Ado-Rib*, Ado-Rib pentalcohol; Ado-Rib-Rib*, Ado-Rib-Rib heptalcohol. From Kanai et al.[2]

$[^3H]_A/[^{14}C]_A = R_x$ can be determined by tritium labeling of a sample containing A pmol of authentic poly($[^{14}C]$ADP-Rib).

$$R_x/R_A - 1 = [^3H]_x/[^3H]_A$$

If tritium labeling is linearly proportional to the amount of Ado-Rib, $[^3H]_x/[^3H]_A$ should be equal to x/A. The linearity was confirmed with at least up to 200 pmol of authentic Ado-Rib. Within this range, the following equations should be valid: $R_x/R_A - 1 = x/A$ and then $x = A(R_x/R_A - 1)$.

For the determination of R_A value, poly($[^{14}C]$ADP-Rib) was digested with snake venom phosphodiesterase and with alkaline phosphomonoesterase, and then treated like other samples after the step of oxidation.

This tritium-labeling method is sensitive and also convenient, since it does not require a parallel experiment for calculation of recovery. If necessary, the recovery can easily be calculated from the recovery of ^{14}C radioactivity.

Comment. The amount of branched portion, Ado-Rib-Rib can be measured similarly.[1] KB^{3}H$_4$ was dissolved in 0.1 M KOH with added unlabeled KBH to give a final specific activity of 3 Ci/mmol. It was divided into small aliquots and stored at $-80°$. Since the specific activity of KB^{3}H$_4$ decreased considerably during storage, it was necessary to determine the R_A value for each measurement.

Radioimmunoassay

Radioimmunoassay of poly(ADP-Rib) is very simple and is an efficient method for detecting poly(ADP-Rib). This method is more sensitive for poly(ADP-Rib) of higher molecular weight than for that of lower molecular weight.[3,4] This assay can also be used to determine the amount of anti poly(ADP-Rib) antibody present in biological samples such as sera of patients with autoimmune diseases.[5]

Reagents

Poly([^{14}C]ADP-Rib), Fraction III (1.1×10^9 cpm/μmol ADP-Rib) prepared from rat liver nuclei incubated with [U-^{14}C]ATP plus NMN.

Anti poly(ADP-Rib) antiserum or IgG.

Unlabeled poly(ADP-Rib), Fractions III and II.

Procedure. The incubation mixture of 500 μl contained poly-([^{14}C]ADP-Rib) (3200 cpm, 2 ng), an adequate amount of anti-poly(ADP-Rib) IgG or antiserum (usually 10 μg of 50-fold diluted antiserum) which binds 70–80% of poly([^{14}C]ADP-Rib), 100 μg/ml bovine serum albumin, and phosphate buffered saline (PBS) with or without a competitor. The mixture was incubated for 60 min at 37° and then for 30 min at 0°.

The reaction mixture was then filtered through a Millipore filter (pore size 0.22 or 0.45 μm). The filter was washed 3 times with 5 ml of PBS and dried, and the radioactivity of the immune complex trapped on the filter was counted. This filter assay is based on the property of immunoglobulin to bind to a Millipore filter. Therefore, poly(ADP-Rib), which binds to the antibody, was trapped on the Millipore filter. Alternatively, the immune complex was collected by addition of 10 μl of nonimmunized rabbit serum and 500 μl of saturated ammonium sulfate solution and centrifugation of the mixture after standing it overnight at 4° and the radioactivity of the precipitate was measured. The assays by membrane filter binding and ammonium sulfate precipitation gave similar results.

Properties of Anti-Poly(ADP-Rib) Antibody. RNA and DNA[3] did not compete with the binding of the antibody to Fraction III of poly-([^{14}C]ADP-Rib). The antibody had higher affinity to poly(ADP-Rib) of larger molecular weight. Fraction III of poly(ADP-Rib) competed most with the binding of the antibody to poly([^{14}C]ADP-Rib), Fraction II less and Fraction I[4] far less.

When this antibody is used in the fluorescent antibody technique, in which the cell or tissue was fixed with cold ethanol or acetone, preincu-

[3] Y. Kanai, M. Miwa, T. Matsushima, and T. Sugimura, *Biochem. Biophys. Res. Commun.* **59**, 300 (1974).

[4] H. Sakura, M. Miwa, M. Tanaka, Y. Kanai, T. Shimada, T. Matsushima, and T. Sugimura, *Nucleic Acids Res.* **4**, 2903 (1977).

[5] Y. Kanai, Y. Kawaminami, M. Miwa, T. Matsushima, T. Sugimura, Y. Moroi, and R. Yokohari, *Nature (London)* **265**, 175 (1977).

bated with NAD$^+$, and treated with anti-poly(ADP-Rib) antibody and fluorescein isothiocyanate (FITC)-labeled anti-rabbit IgG[6] or treated with FITC-labeled anti-rabbit IgG,[7] a difference in the amounts of poly(ADP-Rib) formed could be observed even in single cells. This difference in fluorescence intensity reflected a difference in activity of poly(ADP-Rib) polymerase.

Radioimmunoassay of Anti-Poly(ADP-Rib) Antibody. The radioimmunoassay system described above can be used for measurement of the titer of anti-poly(ADP-Rib) antibody in biological specimens, especially human sera of patients with autoimmune diseases. For assay of antibody in serum, anti-poly(ADP-Rib) antibody or IgG is omitted from the incubation mixture, and 10 μl of human serum is added and the activity is assayed by the membrane filter binding method. This method is also useful for screening cell lines for production of monoclonal antibody against poly(ADP-Rib).

Comment. Anti-poly(ADP-Rib) antibody is also useful for isolating and characterizing poly(ADP-ribosyl)ated proteins.[8]

[6] K. Ikai, K. Ueda, and O. Hayaishi, *J. Histochem. Cytochem.* **28,** 670 (1980).
[7] Y. Kanai, S. Tanuma, and T. Sugimura, *Proc. Natl. Acad. Sci. U.S.A.* **78,** 2801 (1981).
[8] N. Malik, M. Miwa, T. Sugimura, P. Thraves, and M. Smulson, *Proc. Natl. Acad. Sci. U.S.A.* **80,** 2554 (1983).

[52] Poly(ADP-ribose) Synthetase

By KUNIHIRO UEDA, JINGYUAN ZHANG, and OSAMU HAYAISHI

Poly(ADP-ribose) synthetase is a nuclear enzyme that catalyzes the transfer of ADP-ribose moiety of NAD to an acceptor protein and then to this protein-bound ADP-ribose; the former reaction (*initiation*) produces an ADP-ribosyl carboxylate ester of protein, whereas the latter reaction (*elongation*) produces a long polymer of ADP-ribose linked by ribosyl–ribose glycosidic bonds. In the course of elongation, branches are occasionally introduced into the polymer chain also by this enzyme. Furthermore, the enzyme undergoes automodification, serving not only as a catalyst but also as an acceptor. These and other enzymological properties of poly(ADP-ribose) synthetase were recently reviewed.[1]

[1] K. Ueda, M. Kawaichi, and O. Hayaishi, *in* "ADP-ribosylation Reactions: Biology and Medicine" (O. Hayaishi and K. Ueda, eds.), p. 175. Academic Press, New York, 1982.

Poly(ADP-ribose) synthetase has already been purified to apparent homogeneity from various sources including rat liver,[2-4] calf thymus,[5-7] bovine thymus,[8] pig thymus,[9] Ehrlich ascites tumor cells,[10] HeLa cells,[11] lamb thymus,[12] mouse testicle,[13] and human tonsils.[14] The procedures used for these purifications are similar; the most favored recipe starts with salt extraction, followed by chromatographies on DNA-agarose (or DNA-cellulose) and hydroxyapatite. Gel filtration or affinity chromatography on Blue Sepharose is also optionally employed. Enzyme preparations from rat liver and calf thymus were described previously in this series.[15,16] We describe here only the modifications of the methods introduced thereafter, and present the method of preparation of a new form of enzyme, that is, an immobilized poly(ADP-ribose) synthetase.

Assay Methods

Principle. The assay is based on the quantification of [^{14}C]ADP-ribose incorporated into acid(20% CCl_3COOH)-insoluble material from [*adenine*-^{14}C]NAD.

Reagents

Tris–HCl buffer, 1 *M,* pH 8.0
[*adenine*-^{14}C]NAD, 2 m*M* (5000 cpm/nmol)
Calf thymus DNA (Sigma Chemical Co., St. Louis, Missouri; highly polymerized), 1 mg/ml; where indicated, sonicated with a Branson

[2] H. Okayama, C. M. Edson, M. Fukushima, K. Ueda, and O. Hayaishi, *J. Biol. Chem.* **252**, 7000 (1977).
[3] M. Kawaichi, K. Ueda, and O. Hayaishi, *J. Biol. Chem.* **255**, 816 (1980).
[4] M. Kawamura, Y. Tanigawa, A. Kitamura, Y. Miyake, and M. Shimoyama, *Biochim. Biophys. Acta* **652**, 121 (1981).
[5] P. Mandel, H. Okazaki, and C. Niedergang, *FEBS Lett.* **84**, 331 (1977).
[6] C. Niedergang, H. Okazaki, and P. Mandel, *Eur. J. Biochem.* **102**, 43 (1979).
[7] S. Ito, Y. Shizuta, and O. Hayaishi, *J. Biol. Chem.* **254**, 3647 (1979).
[8] K. Yoshihara, T. Hashida, Y. Tanaka, H. Ohgushi, H. Yoshihara, and T. Kamiya, *J. Biol. Chem.* **253**, 6458 (1978).
[9] C. Tsopanakis, E. Leeson, A. Tsopanakis, and S. Shall, *Eur. J. Biochem.* **90**, 337 (1978).
[10] T. Kristensen and J. Holtlund, *Eur. J. Biochem.* **88**, 495 (1978).
[11] D. B. Jump and M. Smulson, *Biochemistry* **19**, 1024 (1980).
[12] S. J. Petzold, B. A. Booth, G. A. Leimbach, and N. A. Berger, *Biochemistry* **20**, 7075 (1981).
[13] M. Agemori, H. Kagamiyama, M. Nishikimi, and Y. Shizuta, *Arch. Biochem. Biophys.* **215**, 621 (1982).
[14] S. G. Carter and N. A. Berger, *Biochemistry* **21**, 5475 (1982).
[15] H. Okayama, K. Ueda, and O. Hayaishi, this series, Vol. 66, p. 154.
[16] Y. Shizuta, S. Ito, K. Nakata, and O. Hayaishi, this series, Vol. 66, p. 159.

    sonifier for 10 sec (×10) at 0° (average size, 400–600 base pairs long)

Calf thymus whole histone (Sigma, type IIA), 1 mg/ml

Calf thymus histone H1 (prepared by the method of Johns[17]), 1 mg/ml

$MgCl_2$, 0.2 M

Dithiothreitol, 20 mM

CCl_3COOH, 100, 20, and 5% (w/v)

Heat(50°, 20 min)-inactivated chromatin,[18] 5 mg protein/ml

3-Aminobenzamide, 25 mM

KCl, 4 M

Procedure. For soluble enzyme systems, essentially the same procedure as described previously[16] is employed; briefly, the reaction mixture (100 μl) containing 10 μl each of Tris–HCl, [^{14}C]NAD, DNA, and histone (type IIA or H1), 5 μl each of $MgCl_2$ and dithiothreitol, and enzyme is incubated for 3 min at 25° (or 10 min at 37°[15]), then mixed with 4 ml of 20% CCl_3COOH, and the mixture is filtered through a Millipore filter. ^{14}C retained on the filter after washings with 5% CCl_3COOH is determined with a liquid scintillation technique.

    For immobilized enzyme systems, the same procedure, except for the use of sonicated DNA and gel-bound enzyme in a double-scaled reaction mixture (200 μl), is used when poly(ADP-ribosyl)ations of the endogenous (enzyme) and exogenous (histone) acceptors need not be distinguished. When the separation of the two types of acceptor is required, the reaction is terminated by the addition of 100 μl of 3-aminobenzamide,[19] and the mixture is filtered through a small polypropylene column (0.8 × 4 cm; Bio-Rad Lab., Richmond, CA). The column is washed 3 times with 2-ml portions of a mixture of KCl/Tris–HCl/3-aminobenzamide/H_2O (10:1:4:5). The filtrate pooled is mixed with 1.5 ml of 100% CCl_3COOH, while the gel retained in the column is suspended in 5 ml of 5% CCl_3COOH. Histone-bound and enzyme-bound ^{14}C's are determined in the filtrate and on the gel, respectively, using Millipore filters as described above.

Purification Procedure

    Enzyme preparation from rat liver is carried out exactly as described previously[15]; the procedure includes the steps of chromatin preparation,

---

[17] E. W. Johns, *Methods Cell Biol.* **16,** 183 (1977).

[18] K. Yoshihara, *Biochem. Biophys. Res. Commun.* **47,** 119 (1972).

[19] M. R. Purnell and W. J. D. Whish, *Biochem. J.* **185,** 775 (1980).

KCl extraction, hydroxyapatite column chromatography, $(NH_4)_2SO_4$ fractionation, Sephadex G-150 gel filtration, and phosphocellulose column chromatography. Purification from calf thymus is performed by a modification of the previous method,[16] namely, through the steps of NaCl extraction, DNA-agarose column chromatography, hydroxyapatite column chromatography, and Sephacryl S-200 gel filtration. The last step has been changed from the previous method that used two combined columns of Sephadex G-150 in buffer A (Tris–HCl (50 mM, pH 8.0)/NaCl (0.2 M)/glycerol (10%)/2-mercaptoethanol (10 mM))[16] to the present method that uses a single Sephacryl S-200 column (5 × 80 cm; Pharmacia Fine Chemicals AB, Uppsala, Sweden) in buffer A containing 1 M NaCl. The yield and purity of the final preparation are essentially identical in both methods.

Preparation of Immobilized Enzyme

Dry powder of CN-activated Sepharose 4B (Pharmacia), 750 mg, is suspended in 60 ml of 1 mM HCl, and allowed to swell for 30 min. All the following procedures are carried out at 0–4°. After removal of HCl, the gel is washed with 60 ml of H_2O on a small glass-filter funnel. The gel is treated with a solution (11 ml) of K phosphate (0.1 M, pH 8.0)/KCl (0.35 M)/glycerol (20%)/dithiothreitol (1 mM) for 10 min; this treatment abolishes about two-thirds of active groups, and leaves 11.5 μmol of active groups on the gel. The wet gel, packed by brief centrifugation, is suspended in 2.5 ml of a solution of poly(ADP-ribose) synthetase [purified from calf thymus; 636 μg of protein in K phosphate (0.1 M, pH 8.0)/KCl (0.35 M)/glycerol (20%)/dithiothreitol (0.2 mM)]. After stirring on a Vortex mixer for 3 min, the mixture is filtered through a glass-filter funnel, and the gel retained is washed with 10 ml of a solution of Tris–HCl (50 mM, pH 8.0)/KCl (0.35 M)/glycerol (20%)/dithiothreitol (4 mM). Approximately a half of enzyme protein is bound to the gel under these conditions; the other half is recovered, together with a half of original enzyme activity, in the filtrate. The gel is further washed with 300 ml of a solution of Tris–HCl (50 mM, pH 8.0)/KCl (2 M)/dithiothreitol (1 mM), followed by the same solution supplemented with 20% glycerol. The gel is finally suspended in the latter solution, 30 ml, and stored unfrozen at $-15°$. The enzyme activity is stable for at least 3 months under these conditions. Prior to the use in assays, the gel is washed with a solution of Tris–HCl (50 mM, pH 8.0)/KCl (2 M)/dithiothreitol (1 mM) and then with Tris–HCl (10 mM, pH 8.0)/dithiothreitol (1 mM). The gel suspension contains 53 mg of wet gel and 10.9 μg of enzyme protein per ml.

REQUIREMENTS FOR IMMOBILIZED POLY(ADP-RIBOSE)
SYNTHETASE ACTIVITY[a]

System	ADP-ribose incorporated		
	Total (pmol)	Gel-bound[b] (pmol)	Soluble[c] (pmol)
Complete	757	187	570
$-Mg^{2+}$	697	171	526
$-$Histone H1	465	437	28
$-Mg^{2+}$, histone H1	119	106	13
$-$Histone H1, DNA	27	27	0

[a] The reaction was carried out under the standard conditions except for the incubation time (8 min) and the omission(s) as indicated, using 3.3 μg of immobilized enzyme.
[b] Represents automodification of the enzyme.
[c] Includes both histone-bound (where histone is included) and free poly(ADP-ribose) released from acceptors during the incubation; the latter amounts to ~7% of total ADP-ribose incorporation.

Properties

Enzymatic and physicochemical properties of soluble enzymes were described previously.[15,16] Important findings made thereafter include that this enzyme not only elongates but also initiates poly(ADP-ribose) chains on various acceptors such as histone and the enzyme itself, that the enzyme activity is dependent on strand ends, and not intact chain, of DNA, and that the enzyme has an intrinsic NAD glycohydrolase activity. For more details, the readers are referred to our recent review.[1]

Immobilized enzyme has a specific activity of about one-fourth to one-tenth of that of the original enzyme; the value fluctuates among the preparations as well as with the assay conditions used. The table shows the requirements for the immobilized enzyme activity. DNA is almost absolutely required for the activity, as is for the activity of soluble enzymes. Histone H1, a potent activator of automodification of soluble enzymes,[3] inhibits markedly the automodification of the immobilized enzyme in the presence of Mg^{2+}, and slightly stimulates it in the absence of Mg^{2+}. The product polymer synthesized on the immobilized enzyme has branches. The intrinsic NAD glycohydrolase activity is higher in the immobilized enzyme compared with soluble enzymes.[20]

[20] J. Zhang, K. Ueda, and O. Hayaishi, in preparation.

[53] Poly(ADP-ribose) Glycohydrolase and
Phosphodiesterases in the Analysis of
Poly(ADP-ribosyl)ated Proteins

By WILLIAM R. KIDWELL and MICHAEL R. PURNELL

Modification of nuclear proteins by poly(ADP-ribose) involves the sequential addition of ADP-ribose moiety of NAD to form a linear or branched polymer. As yet the biological significance of this process is unknown although there are indications that it may be involved in chromatin rearrangement, recovery from DNA damage or in cellular differentiation. A number of reviews on the subject have been published.[1–6]

In this chapter we shall attempt to describe the uses to which poly(ADP-ribose) degrading enzymes have been applied. Poly(ADP-ribose) can be removed from proteins in two ways. The first involves scission of the 1'-2'' glycosidic linkage by poly(ADP-ribose) glycohydrolase to give mono-ADP-ribosylated proteins and free ADP-ribose. The other way is cleavage of the pyrophosphate linkage by phosphodiesterases to give one molecule of 5'-AMP from each nonreducing terminus and one molecule of 2'-(5''-phosphoribosyl)-5'-AMP (PR-AMP) from each internal residue (Fig. 1). The former mechanism is thought to account for the majority of poly(ADP-ribose) degradation *in vivo*.

Assay Conditions

Poly(ADP-ribose) Glycohydrolase

Poly(ADP-ribose) glycohydrolase(a)
0.15 *M* NaCl(b)
20 m*M* sodium phosphate pH 7.0(c)
10 m*M* 2-mercaptoethanol or dithiothreitol(d)
2 m*M* EDTA(e)
Poly(ADP-ribose) radioactively labeled(f)

[1] T. Sugimura, *Prog. Nucleic Acid Res. Mol. Biol.* **13**, 127 (1973).
[2] H. Hilz and P. R. Stone, *Rev. Physiol., Biochem. Pharmcol.* **76**, 1 (1976).
[3] O. Hayaishi and K. Ueda, *Annu. Rev. Biochem.* **46**, 95 (1977).
[4] M. R. Purnell, P. R. Stone, and W. J. D. Whish, *Biochem. Soc. Trans.* **8**, 215 (1980).
[5] P. Mandel, H. Okazaki, and C. Niedergang, *Prog. Nucleic Acid Res. Mol. Biol.* **27**, 1 (1982).
[6] O. Hayaishi and K. Ueda, eds., "ADP-Ribosylation Reactions," p. 279. Academic Press, New York, 1982.

1. Glycohydrolase and protein hydrolase[6a]

Protein-$(ADP-R)_n$ + Glycohydrolase ⟶ Protein-ADP-R + n-1 (ADP-R)

Protein-ADP-R + Protein hydrolase ⟶ Protein + ADP-R like Compound

2. Phosphodiesterase

Protein-$(ADP-R)_n$ + Phosphodiester- ⟶ Protein-R-P + n-1 (PR-AMP) + 5'-AM
ase

Fig. 1. Degradative pathways for poly(ADP-ribosyl) group.

Notes

(a) As yet no detailed procedure has been reported for the purification to homogeneity of this enzyme. The 200-fold purified preparation from calf thymus as described by Miwa et al.[7] is suitable for most applications.

(b) A high salt concentration prevents inhibition of the enzyme by any single-stranded DNA which might contaminate the preparation.[8]

(c) The enzyme has a fairly broad pH optimum. Tris–HCl should not be used because it has been reported to be inhibitory.[8]

(d) Thiol reagents stabilize the enzyme and also inhibit phosphodiesterases which cleave poly(ADP-ribose).

(e) Poly(ADP-ribose) glycohydrolase has no divalent metal ion requirement whereas phosphodiesterases do.

(f) The usual methods of determining poly(ADP-ribose) glycohydrolase activity are to measure conversion of acid-insoluble polymer to an acid-soluble form or by quantitating ADP-ribose production. If free polymer is used the average chain length should be as long as possible. If protein-bound poly(ADP-ribose) is used as substrate, it should be remembered that the enzyme acts exoglycosidically and it does not remove the last ADP-ribose molecule from protein.

In this laboratory we have modified the ion-exchange procedure of Burzio et al.[9] to separate liberated ADP-ribose from poly(ADP-ribose). After incubating the reaction mixture at 37° for a period of time, a 50-μl aliquot was added to 250 μl 6 M urea, 0.3 M acetic acid. This terminates the glycohydrolase reaction and allows the subsequent chromatographic separation of monomer and polymer to be carried out at room temperature. Of the mixture 250 μl is applied to a small column containing 0.5 ml

[7] M. Miwa, M. Tanaka, T. Matsushima, and T. Sugimura, J. Biol. Chem. 249, 3475 (1974).

[8] P. R. Stone, W. S. Lorimer, III, J. Ranchalis, M. Danley, and W. R. Kidwell, Nucleic Acids Res. 5, 173 (1978).

[9] L. O. Burzio, P. T. Riquelme, and S. S. Koide, Anal. Biochem. 66, 434 (1975).

(aminoethyl)cellulose which is previously equilibrated with 0.3 M acetic acid. Mono-ADP-ribose was eluted with three 500-μl aliquots of 6 M acetic acid. The eluant is mixed with a water-compatible scintillation fluid and the radioactivity determined.

Phosphodiesterase

Phosphodiesterase(a)
4 M urea(b)
20 mM 1-amino-2-methylpropanol–HCl pH 8.9(c)
2 mM NaCN(d)
Radioactive poly(ADP-ribose)(e)

Notes

(a) Phosphodiesterases which cleave the pyrophosphate linkage of poly(ADP-ribose) have been detected in a number of tissues including rat liver[10] and plant tissue culture cells.[11] Most workers however, use the enzymes from snake venom which are commercially available. It is important to recognize that commercial enzyme preparations have contaminants which may interfere with the analyses to be undertaken. These contaminants include 5'-nucleotidase,[12,13] phosphomonoesterase[12,13] proteolytic activity[14] and nucleotides (unpublished data). Affinity chromatography on NAD-agarose[12] or Blue Sepharose[13] can substantially reduce the levels of 5'-nucleotidase and phosphomonoesterase. In this laboratory we have found that these enzymes can also be removed by passage through an immobilized iminodiacetic acid affinity column by virtue of their zinc binding properties. Other approaches to prevent interference by these enzymes include the inclusion of competitive substrates such as 5'-AMP or glucose 1-phosphate to reduce dephosphorylation of 5'-AMP and PR-AMP.

(b) Urea (4 M) is included because of its solubilizing properties which is important when protein-bound poly(ADP-ribose) is being analyzed. A twofold activation of phosphodiesterase by 4 M urea has been reported.[15] In this laboratory we did not observe this but we did obtain a stimulation of phosphatase activity which may account for the earlier finding made

[10] M. Futai, D. Mizuno, and T. Sugimura, *Biochem. Biophys. Res. Commun.* **28,** 395 (1967).
[11] M. Miwa, M. Tanaka, H. Shinsi, M. Takeuchi, T. Matsushima, T. Sugimura, and S. Shall, *J. Biochem. (Tokyo)* **77,** 3 (1975).
[12] T. Tatsuki, S. Iwanaga, and T. Suzuki, *J. Biochem. (Tokyo)* **77,** 831 (1975).
[13] J. Oka, K. Ueda, and O. Hayashi, *Biochem. Biophys. Res. Commun.* **80,** 841 (1978).
[14] M. Gronow and M. R. Chapleo, *FEBS Lett.* **103,** 352 (1979).
[15] H. Hilz, U. Wiegers, and P. Adamietz, *Eur. J. Biochem.* **56,** 103 (1975).

with an unpurified commercial enzyme preparation and bis(nitrophenyl)-phosphate.[15]

(c) The enzyme has a pH optimum of 8.9–9.5.

(d) Cyanide was reported to be an extremely potent inhibitor of proteases from snake venom.[16] For many applications protease contamination is not important and safety considerations make the addition of cyanide undesirable unless proteins are to be characterized subsequent to phosphodiesterase digestion.

(e) Phosphodiesterase activity can be analyzed by the methods described above for poly(ADP-ribose) glycohydrolase.

Uses

Characterization of Poly(ADP-ribose). Following incubation of poly (ADP-ribose) synthetase with radioactive NAD and acid precipitation, the polynucleotide nature of the precipitated radioactivity can be demonstrated by its lability in hot acid (0.5 M HClO$_4$ at 70°). Some loss of acid precipitability occurs if the product is incubated with alkali (0.3 M NaOH at 56° for 30 min). The extent of loss is dependent on the chain length of the polymer because alkali treatment results in hydrolysis of the protein–poly(ADP-ribose) linkage while leaving the polymer intact. Mono-ADP-ribose is converted by alkali to 5'-AMP and an ADP-like compound. Mono ADP-ribose-derived radioactivity can be removed from poly(ADP-ribose) by chromatography on aminoethylcellulose.[17] Neutral hydroxylamine is known to cleave some polymer and monomer from protein.

For definitive characterization of poly(ADP-ribose), however, poly (ADP-ribose) glycohydrolase and snake venom phosphodiesterase digestions are the most commonly used methods. As stated previous, phosphodiesterase digestions can be complicated by contaminants such as alkaline phosphatase and nucleotidase. These can be alleviated by purification of the enzyme, inclusion of appropriate inhibitors or addition of alkaline phosphatase. The latter treatment dephosphorylates the venom phosphodiesterase products, 5'-AMP and PR-AMP to adenosine and ribosyladenosine, respectively, which can be resolved by thin-layer chromatography on PEI-cellulose [(polyethyleneimine)cellulose] using distilled water as solvent.[18] When trying to resolve 5'-AMP and PR-AMP, 5'-GMP is frequently employed as a marker for PR-AMP as they behave similarly in most chromatography systems. If the use of enzymes must be avoided,

[16] G. Pfleiderer and G. Sumyk, *Biochim. Biophys. Acta* **51,** 482 (1961).
[17] P. R. Stone, M. R. Purnell, and W. J. D. Whish, *Anal. Biochem.* **110,** 108 (1981).
[18] M. Miwa, N. Saikawa, Z. Yamaizuma, S. Nishimura, and T. Sugimura, *Proc. Natl. Acad. Sci. U.S.A.* **76,** 595 (1979).

0.33 M NaOH, 6 mM MgCl$_2$ at 56° for 6 hr has been reported to mimic the action of phosphodiesterase.[19] There are, however, some minor problems associated with this method, namely under the above conditions significant amounts of exchange occurs if poly([*ade*-³H]ADP-ribose) is used. This can be avoided by the use of some other form of polymer labelling, such as ³²P or ¹⁴C. A second problem is that the presence of high salt may interfere with subsequent analysis, especially thin-layer chromatography.

Poly(ADP-ribose) glycohydrolase digestion results in loss of acid insolubility of polymer-derived radioactivity. In interpreting the results from such a treatment, the following points should be borne in mind: (1) the enzyme will not remove the last poly(ADP-ribose) residue from protein; (2) the polymer may be protected by denatured protein or be buried in tightly bound complexes; (3) the enzyme will not hydrolyze polymer which does not have an intact 5'-AMP terminus. This situation arises if the polymer has been subjected to the previous action of phosphodiesterase. The first two considerations are eliminated if the polymer is removed from protein by alkali followed by neutralization.[8]

To date the sole means of identifying and quantifying branch points in poly(ADP-ribose) has been to resolve the phosphodiesterase derived product (see Ref. 3). Poly(ADP-ribose) glycohydrolase digests branched polymer to give the same product as linear polymer and consequently digestions with this enzyme give no information about branching.

Chain Length Determinations

An average chain length determination can be obtained by performing a snake venom phosphodiesterase digestion, resolving 5'-AMP and PR-AMP and dividing the total amount of adenine nucleotides by the amount of 5'-AMP.[20] This has since been revised to correct for the presence of branched poly(ADP-ribose) by subtracting the amount of branch points from the amount of 5'-AMP for the denominator.[21] Once again the major problem in this procedure is the presence of phosphatases in venom phosphodiesterase. This is particularly important because one of the phosphates in PR-AMP is the preferred substrate and the partially dephosphorylated PR-AMP coelutes with 5'-AMP in most ion-exchange resolution systems. A further problem is that if the washing procedure is incomplete during the preparation of poly(ADP-ribose) some NAD may be present

[19] P. Adamietz and R. Bredehorst, *Anal. Biochem.* **112,** 314 (1981).
[20] Y. Nishizuka, K. Ueda, K. Yoshihara, H. Yamamura, M. Takeda, and O. Hayaishi, *Cold Spring Harbor Symp. Quant. Biol.* **34,** 781 (1969).
[21] T. Sugimura and M. Miwa, *in* "ADP-Ribosylation Reactions" (O. Hayaishi and K. Ueda, eds.), p. 43. Academic Press, New York, 1982.

TABLE I
EFFECT OF DNA ON THE DEGRADATION OF
HISTONE H1–POLY(ADP-RIBOSE) COMPLEX BY
POLY(ADP-RIBOSE) GLYCOHYDROLASE[a]

Treatment	Mass ratio DNA/H1	Percentage poly(ADP-ribose) hydrolyzed
Control	0	15
+22.5 μg DNA	1.04	23
+45 μg DNA	2.09	30
+67.5 μg DNA	3.13	2.7[b]

[a] Adapted from Stone et al.[8]
[b] At high DNA–H1 ratios there is available DNA which binds to and inhibits glycohydrolase. This inhibition is reduced by increasing the ionic strength of the incubation buffer, i.e., to 0.15 M NaCl.

which upon phosphodiesterase digestion gives rise to 5′-AMP. This problem, and the presence of mono-ADP-ribosylated proteins, can be eliminated very simply by the use of aminoethylcellulose chromatography after an alkaline treatment as described by Stone et al.[17]

Other methods of chain length determination have been proposed including digestions by both glycohydrolase and phosphodiesterase,[22] hydroxyapatite chromatography,[23] gel electrophoresis[24,25] and ion-exchange chromatography.[25,26]

In this laboratory three independent methods were used to determine the chain length of poly(ADP-ribose) associated with the histone H1 "dimer" complex.[27] As can be seen from Table I, all three methods gave values which were in close agreement. It is important to remember that this relationship is true only in the absence of significant amounts of branched polymer but this is a minor component in most systems.

[22] M. Goebel, P. R. Stone, H. Lengyel, and H. Hilz, *Hoppe-Seyler's Z. Physiol. Chem.* **358,** 13 (1977).

[23] M. Tanaka, M. Miwa, K. Hayaishi, M. Miwa, T. Matsushima, and T. Sugimura, *Biochemistry* **16,** 1485 (1977).

[24] M. Tanaka, K. Hayashi, H. Sakura, M. Miwa, T. Mutsushima, and T. Sugimura, *Nucleic Acids Res.* **5,** 3183 (1978).

[25] P. Adamietz, R. Brederhorst, and H. Hilz, *Biochem. Biophys. Res. Commun.* **81,** 1377 (1978).

[26] M. Kawaichi, J. Oka, K. Ueda, and O. Hayaishi, *Biochem. Biophys. Res. Commun.* **101,** 672 (1981).

[27] P. R. Stone, W. S. Lorimer, III, and W. R. Kidwell, *Eur. J. Biochem.* **81,** 9 (1977).

TABLE II

CHAIN LENGTH DETERMINATION OF POLY (ADP-RIBOSE) ASSOCIATED
WITH HISTONE H1 "DIMER"[a]

Method	Average chain length
Phosphodiesterase digestion followed by thin-layer chromatography analysis of products	14.5–15.7
Exhaustive poly(ADP-ribose) glycohydrolase digestion in 0.1 M acetic acid followed by acid precipitation[b]	14.6–14.8
Hydroxyapatite chromatography of poly(ADP-ribose) following alkali treatment and proteolysis to remove proteins	Approximately 15

[a] Data adapted from Stone et al.[27]

[b] This method is based on the fact that poly(ADP-ribose glycohydrolase cleaves all the polymers to ADP-ribose plus mono-ADP-ribosyl protein. For each chain there is only one of the latter species. The chain length is readily determined by independently quantitating the amount of acid-soluble and acid-precipitable radioactivity derived from the labeled polymer in the polymer–protein complex.

Quantification of Poly(ADP-ribose) Levels in Vivo

The major difficulty in measuring poly(ADP-ribose) is the extremely low levels present in the cell. For this reason all quantitative analyses make use of either glycohydrolase or phosphodiesterase digestions to provide specificity. For example, in the radioimmunoassay based on anti-poly(ADP-ribose) antibodies developed in this laboratory,[28] poly(ADP-ribose) glycohydrolase is used to confirm that competing material is indeed poly(ADP-ribose).

Prospectives

The major problem confronting workers attempting to identify the proteins modified by poly(ADP-ribose) is the dramatic changes in physical properties caused by the attachment of variable length chains of highly charged polymer.[29] To some extent these problems can be alleviated by digestion of the preparation with either poly(ADP-ribose) glycohydrolase or phosphodiesterase. The use of glycohydrolase is complicated by its

[28] W. R. Kidwell and M. Mage, Biochemistry 15, 1213 (1976).
[29] P. Adamietz, K. Klapproth, and H. Hilz, Biochem. Biophys. Res. Commun. 91, 1232 (1979).

sensitivity to acceptor protein conformation. This is exemplified by its use in the digestion of the H1 "dimer." The data in Table II show that addition of DNA makes the attached poly(ADP-ribose) more accessible to glycohydrolase hydrolysis, presumably because the dimer binds to DNA and unfolds. For H1 dimer, a similar effect is observed by increasing the ionic strength of the incubation medium.[8]

The use of phosphodiesterase in the identification of acceptor proteins has been explored by Hilz's group.[29] They observed that if poly(ADP-ribosyl)ated proteins were incubated with phosphodiesterase they were no longer retained by affinity chromatography on a boronate column. This is somewhat puzzling because theoretically they should bind by virtue of the *cis*-diol of the ribose phosphate. Whether this is due to steric or ionic interference, a contaminated phosphodiesterase preparation, or lability of the linkage at high pH, is uncertain at the present time.

[54] Poly(ADP-ribosyl)ation of Nucleosomal Chromatin: Electrophoretic and Immunofractionation Methods

By MARK E. SMULSON

The nuclear enzyme poly(ADP-Rib) polymerase alters the structure of chromatin by the generation of chains of NAD-derived poly(ADP-Rib) which are covalently bound to nuclear proteins. It has been speculated that this natural biopolymer may well be a ubiquitous cross-linking agent in eukaryotic chromatin. It is possible that the poly(ADP-Rib) synthesizing and degrading activities of eukaryotic cell nuclei constitute a system for affecting a transient and localized condensation of chromatin by permitting a reversible interaction of adjacent or neighboring nucleosomes within chromatin. It is therefore of interest that the enzymatic activity for this system can be found associated directly with nucleosomal forms of chromatin. The polymerase has a strict requirement for DNA for activity. Accordingly, a limited population of oligonucleosomal particles contain all the requirements necessary for poly(ADP-ribosyl)ation, namely, the polymerase, nucleosomal histone (and nonhistone) acceptors, and DNA.

The technology in this chapter focuses upon methods (1) to study poly(ADP-ribosyl)ation at the nucleosomal level of chromatin; (2) to analyze the acceptors for the modification; and (3) to isolate selectively by immunofractionation from bulk polynucleosomes, the very limited domains of chromatin undergoing the reaction.

Nucleosomal Poly(ADP-ribosyl)ation

Preparation of Nucleosomes

Nuclei are washed with buffer containing 0.25 M sucrose, 1 mM CaCl$_2$, 5 mM Tris–HCl (pH 7.5), and 80 mM NaCl and centrifuged at 1000 g. In most experiments the inclusion of a protease inhibitor is recommended to be included in all buffers and enzyme assay solutions. The pellet is immediately suspended in the above buffer at 1 × 10^8 nuclei/ml.

Micrococcal nuclease is added (4 units/1 × 10^6 nuclei), and the nuclease incubated at 5° for 30 min. The reaction is terminated by the addition of a solution of EDTA to a final concentration of 1 mM.

The suspension is centrifuged at 3000 g; the nuclear pellet is suspended at 5° in 1 mM EDTA solution and allowed to lyse for 30 min.

The suspension is centrifuged at 6000 g.

Alternatively, the nuclei are washed and resuspended in digestion buffer as above, the nuclei are incubated at 37° for 5 min and then digested with micrococcal nuclease (30 units/1 × 10^8 nuclei) for 1 min at 37°.

The reaction is terminated and nuclei are lysed for 20 min as described above.

Isolated oligonucleosomes are separated on 10–30% (w/v) linear sucrose gradients containing 1 mM sodium phosphate (pH 6.8), 0.2 mM EDTA, and 80 mM NaCl. 80% glycerol (v/v) (0.4 ml) is present at the bottom of the gradient as a cushion. Ten to twenty A_{260} units (0.5 ml) of isolated chromatin is layered onto the gradients. The gradients are centrifuged in a Beckman SW40 rotor at 40,000 rpm for 4.5 hr at 4° and fractions collected.

Remarks. The above procedure has been used successfully with HeLa cell nuclei. We have noted only limited loss in nucleosomal bound poly (ADP-Rib) polymerase activity when the sucrose gradient fractions were frozen for 4–6 weeks. Two other points should be emphasized when considering enzymatic activity of poly(ADP-Rib) polymerase in isolated oligonucleosomal units. We have noted that nucleosomes released early during micrococcal nuclease digestion possess a high specific activity for polymerase compared to those nucleosomes released later.[1] However, if one incubated nuclei with radioactive NAD and subsequently digests the nuclei with micrococcal nuclease, much of the nuclear incorporation [i.e.,

[1] D. C. Jump, T. R. Butt, and M. E. Smulson, *Biochemistry* **18**, 983 (1979).

poly(ADP-ribosyl)ated domains of chromatin] is resistant to digestion. Second, we have noted a successive increase in specific activity for polymerase with increasing oligonucleosome chain length.[2-4] Highest activity has been noted in sucrose gradient fractions containing polynucleosomes with chains of 6–8.[2,3] Using antibody directed against poly(ADP-rib)polymerase, we have speculated that this may be due to primary binding sites of the enzyme in chromatin at average periodicities of 6–8 nucleosomes.[5]

Poly([^{32}P]ADP-ribosyl)ation of Oligonucleosomes

Chromatin fragments of different nucleosome chain size (~0.07 A_{260}) are incubated with a reaction mixture containing 50 mM Tris–HCl, pH 8., 2 mM MgCl$_2$, 1 mM DTT, and the desired concentration of NAD (0.02–0.3 ml total volume). Labeled NAD is included in the reaction, usually approximately 0.25 μCi [^{32}P]NAD.

Since chromatin fragments are fractionated from sucrose gradients, the contributions of salts from the gradient buffer in the test mixture are the following: 10 mM NaCl, 0.02 mM EDTA, 0.1 mM sodium phosphate, pH 6.8, and about 0.7–2% sucrose.

The reaction is carried out for 5 min at 20° and terminated by one of the following procedures: (1) addition of 10 mM nicotinamide followed by rapid chilling of the tubes on ice (for chromatin gel); (2) addition of 3 volumes of cold absolute ethanol (for analysis of protein on sodium dodecyl sulfate gels); and (3) addition of 20% trichloroacetic acid and 5 mM pyrophosphate (for enzyme assay).

A linear gradient of 3.5–8% of polyacrylamide (acrylamide/N,N-methylenebisacrylamide, 20 : 1 w/w) is established in slab glass plates.

The gel buffer contains 89 mM Tris base, 89 mM boric acid, and 2.5 mM EDTA at pH 8.3. Preelectrophoresis is performed for about 2 hr at 5° at 150 V/slab before the application of the samples.

^{32}P-Labeled poly(ADP-ribosyl)ated chromatin is applied to the gels and electrophoresis performed for 8–10 hr at 150 V.

The nucleoproteins are fixed on the gel with 10% acetic acid, 40% methanol stained with ethidium bromide, and the gel is processed for autoradiography.

Remarks. When oligonucleosomes of differing size are incubated with limiting substrate (25 nM [^{32}P]NAD), the labeled poly(ADP-Rib) is found

[2] T. R. Butt, J. F. Brothers, C. P. Giri, and M. Smulson, *Nucleic Acids Res.* **5**, 2775 (1978).
[3] D. B. Jump, T. R. Butt, and M. Smulson, *Proc. Natl. Acad. Sci. U.S.A.* **76**, 1628 (1979).
[4] D. B. Jump, T. R. Butt, and M. E. Smulson, *Biochemistry* **19**, 1031 (1980).
[5] N. Malik, M. Bustin, and M. Smulson, *Nucleic Acids Res.* **10**, 2939 (1982).

directly associated with the *same* sized nucleosome particles used for the assay (Fig. 1, lane 2). However, it is important to note that poly(ADP-ribosyl)ation is extremely sensitive to NAD concentration. As poly(ADP-Rib) chains become longer, proteins and chromatin particles tend to aggregate (see below), and this can lead to misinterpretation of data.

When oligonucleosomes are incubated under conditions optimal for poly(ADP-Rib) polymerase activity with an increasing range of NAD from 25 nM to 1 mM (Fig. 1), analysis of the products on native chromatin gels (Fig. 1A) indicates that NAD at these concentrations does not affect the overall mobility of the bulk of the oligonucleosomes in this gel system. However, autoradiography of the incorporated ADP-ribosylated particles reveals a progressive increase in the size of the modified products, which is directly related to substrate concentration (Fig. 1B). It was noted that when NAD concentrations were greater than 10 μM, the majority of the incorporated product migrates on chromatin gels to positions corresponding to much larger forms of chromatin than initially used for incubations. The aggregation can be most conveniently studied at 100 μM NAD. At higher concentrations (lane 6), the labeled reaction product is so large as to preclude its entrance into 3% polyacrylamide gels. The aggregation was noted also when analyzing these chromatin preparations by sedimentation velocity.[6,7]

It should also be noted that since oligonucleosomal particles are enzymatically active for polymerase, and in addition contain acceptor proteins and DNA, required for poly(ADP-ribosyl)ation, enzymatic activity may be analyzed directly on polyacrylamide gels of separated particles.[7,8]

In addition, methods have been provided for the removal of poly(ADP-Rib) polymerase from purified polynucleosomes, and its subsequent reconstitution to biologically relevant sites.[4,9] The location of polymerase on immunoblots of nucleosomes can also be detected with antibody directed against the polymerase.[5] This immunological method has also been useful recently in analysis of the poly(ADP-ribosyl)ation of histone H1 into a cross-linked complex, and in the natural occurrence of such a complex.[10]

[6] T. R. Butt and M. E. Smulson, *Biochemistry* **19,** 5235 (1980).
[7] T. R. Butt, B. DeCoste, D. B. Jump, N. Nolan, and M. E. Smulson, *Biochemistry* **19,** 5243 (1980).
[8] C. P. Giri, M. H. P. West, M. L. Ramirez, and M. Smulson, *Biochemistry* **17,** 3501 (1978).
[9] M. Wong, N. Malik, and M. Smulson, *Eur. J. Biochem.* **128,** 209 (1982).
[10] M. Wong, Y. Kanai, M. Miwa, M. Bustin, and M. Smulson, *Proc. Natl. Acad. Sci. U.S.A.* **80,** 205 (1983).

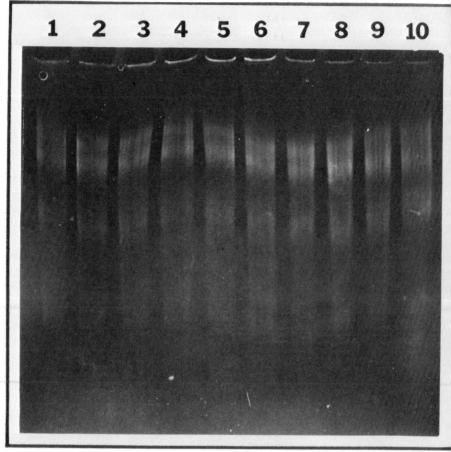

FIG. 1. NAD concentration-dependent complex formation of polynucleosomes, as analyzed by chromatin gel electrophoresis. An aliquot of chromatin (0.07 A_{260} unit) from oligonucleosomes was incubated with various concentrations of NAD at 20° for 5 min while
[^{32}P]NAD was maintained constant at 0.25 μCi per assay. The reaction was terminated by
placing the tubes on ice and by the addition of nicotinamide to a final concentration of 5 mM.
The samples were subjected to electrophoresis on 3–8% gradient polyacrylamide gels as
described in the text. (A) Ethidium bromide stain; (B) autoradiograph. (Lane 1) Chromatin
incubated with 25 mM NAD for 5 min at 5°. (Lanes 2–6) Chromatin incubated at 20° with 25
nM, 1 μM, and 1 mM NAD, respectively. In lanes 7–10, the samples were incubated with
100 μM NAD for 5 min and terminated with nicotinamide, and 10 μg of proteinase K was
added; the samples were incubated for an additional 1, 5, 10, and 20 min at 20°, respectively.
Taken from Butt and Smulson.[6]

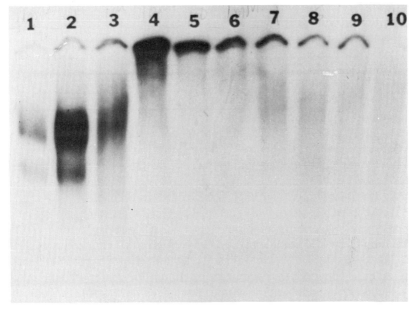

Fig. 1B.

Analysis of Labeled Proteins in Polyacrylamide Gels Containing Sodium
Dodecyl Sulfate

Analysis of Nucleosomal Acceptor Proteins

The poly(ADP-ribosyl)ation reaction is carried out on chromatin particles of various complexity. The total proteins are precipitated with 3
volumes of absolute ethanol at $-20°$ for 12 hr.

The precipitated material is centrifuged at 8000 g for 10 min, dissolved
in 10 mM Tris–HCl, pH 6.8, 1% sodium dodecyl sulfate, 6 M urea, 1%
2-mercaptoethanol, and 20% glycerol, and boiled for 2 min prior to electrophoresis.

Electrophoresis is performed in 10% polyacrylamide (acrylamide/
N,N-methylenebisacrylamide, 100 : 1 w/v) slab gels with 50 mM sodium
phosphate, pH 7.0, and 0.1% sodium dodecyl sulfate as the gel buffer.

Analysis of Acid-Soluble Proteins on Acetic Acid/Urea Gels

ADP-ribosylated proteins are extracted with 0.4 N H$_2$SO$_4$ for 30 min
and precipitated with 20% trichloroacetic acid. The precipitates are
washed with acidified acetone, dried, and dissolved in a solution containing 0.9 N acetic acid, 8 M urea, and 1% mercaptoethanol. (The proteins

extracted from chromatin samples incubated with concentrations of NAD higher than 10 μM are difficult to dissolve in sample buffer. Accordingly the samples are vortexed and left at room temperature for up to 2 hr to achieve total solubilization.)

Polyacrylamide (15%) gel electrophoresis is performed as described by Panyim and Chalkley.[11]

All the gels are fixed with 10% acetic acid, 40% methanol and stained with amido black. The gels are dried on a Bio-Rad Model 224 slab drier and exposed to X-ray film for an appropriate period.

Remarks. By far, the major *in vitro* acceptor for poly(ADP-ribosyl)ation is the polymerase itself undergoing automodification.[1,4] This protein migrates at 116,000 daltons on SDS gels, however, will aggregate considerably when incubated with high NAD concentrations.

In the case of histone H1, a cross-linked dimer complex has been characterized as one major product of poly(ADP-ribosyl)ation.[12,13] The formation of this H1 complex is favored by high NAD concentrations.[10,13] In general, one *must be cautious* of aggregation of poly(ADP-Rib) acceptors on gels, when NAD concentrations, 10 μM or greater are employed.

Recently, immunological methods have been developed, to study the *in vivo* occurrence of this cross-linked complex of histone H1 without the use of radioisotopes.[10]

Immunoaffinity Fractionation of the Poly(ADP-ribosyl)ated Domains of Chromatin

The poly(ADP-ribosyl)ation modification may only occur on those regions of chromatin possessing single-strand DNA breaks, accumulated during either DNA replication or repair. As such, the modification may exist on only a very small percentage of chromatin (less than 1%) which is undergoing these reactions, at any one time. Based upon a high turnover rate and resultant low level of poly(ADP-ribosyl)ation, a very sensitive method is required to purify those domains of chromatin undergoing this reaction. The recent availability of highly active antisera, directed specifically against the modified moiety of this reaction, poly(ADP-Rib) polymer, has allowed the development of an immunological affinity method to bind selectivity, those domains of chromatin undergoing this reaction.[14]

[11] S. Panyim and R. Chalkley, *Arch. Biochem. Biophys.* **130,** 337 (1969).
[12] P. R. Stone, W. S. Lorimer, and W. R. Kidwell, *Eur. J. Biochem.* **18,** 9 (1977).
[13] N. L. Nolan, T. R. Butt, M. Wong, A. Lambrianidou, and M. E. Smulson, *Eur. J. Biochem.* **113,** 15 (1980).
[14] N. Malik, M. Miwa, T. Sugimura, P. Thraves, and M. Smulson, *Proc. Natl. Acad. Sci. U.S.A.* **80,** 2554 (1983).

Coupling of Anti-Poly(ADP-Rib) IgG to Cyanogen Bromide-Activated Sepharose-4B

Cyanogen bromide-activated Sepharose 4B (3 g) is soaked for 2 hr in 1 mM HCl. The swollen and equilibrated Sepharose is washed with 1 mM HCl and is suspended in 15 ml of coupling buffer (0.1 M NaHCO$_3$ pH 8.5, 1 M NaCl) containing 12 mg of anti-poly(ADP-Rib) IgG. The gel suspension is mixed end-over-end for 2.5 hr.

The IgG coupled Sepharose is suspended in 1 M ethanolamine pH 8.0 for 2 hr and subsequently washed, alternately with low pH buffer (0.1 M acetate buffer containing 1 M NaCl) and high pH buffer (1% ammonium bicarbonate).

The washed gel is suspended and equilibrated in phosphate-buffered saline (PBS).

Preparation of HeLa Chromatin

HeLa nuclei were treated with micrococcal nuclease (30 units/1 × 10^8 nuclei) for 3 min at 37° and separated into discrete sized particles on 10–30% linear sucrose gradients by centrifugation as described above.

Incubation with [^{32}P]NAD

Nucleosomes (approximately 1.5 A_{260}) are suspended in 10 mM Tris–HCl, pH 8.0, containing 2 mM DTT and 5 mM Mg^{2+}. The reaction is initiated by the addition of 5 μCi of [^{32}P]NAD containing 10 μM nonradioactive NAD and incubated at room temperature for 5 min.

The reaction is terminated by the addition of nicotinamide to a final concentration of 20 mM. The labeled nucleosomal sample is passed through a small column of Sephadex G-25 to remove unincorporated NAD.

Chromatography on Anti-Poly(ADP-Rib) Columns

The in $vitro$ incubated oligomers (4–8 N; 4.95 × 10^6 cpm) in 1 ml are slowly applied to the column over a 45-min period (Fig. 2A). Washing is initiated with PBS; a peak of A_{260} nm material elutes between fractions 5 and 10. This represents greater than 90% of the original polynucleosome sample.

In addition to absorbance, acid-precipitable radioactivity, representing poly(ADP-ribosyl)ated nucleosomal acceptors, should be monitored in all fractions. Negligible ^{32}P incorporation is noted in the unbound fractions, where the majority of nucleosomes elute.

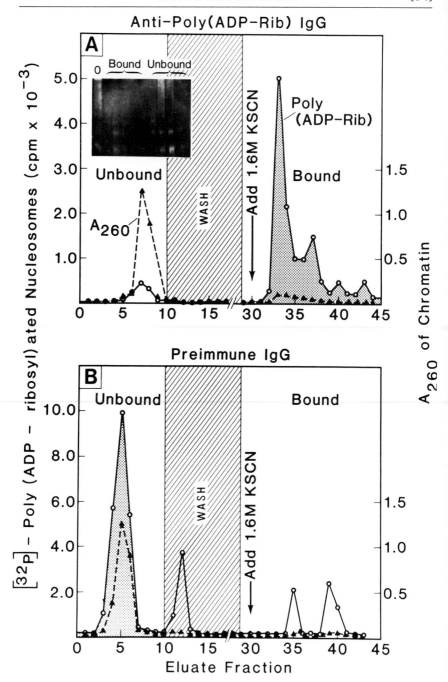

When no further ^{32}P-labeled (ADP-Rib)$_n$, nor A_{260} elutes, the column is washed with 20 ml of PBS. The bound material is released by the addition of 1.6 M KSCN (fraction 30). An almost immediate elution of acid-precipitable ^{32}P-labeled poly(ADP-ribosyl)ated material should be noted. Approximately 95% of the total nucleosomal radioactive incorporation is represented in five 1.0 ml fractions (fractions 32–37).

The presence of KSCN prevents reliable quantitation of A_{260} absorbance; however, when bound fractions are dialyzed, nucleoprotein structures and detectable A_{260} can be observed (insert, Fig. 2).

When a similarly prepared column containing preimmune IgG coupled to Sepharose is utilized (Fig. 2B), the majority of the acid-insoluble ^{32}P-labeled poly(ADP-ribosyl)ated material remains unbound to the column and elutes coincidental with the bulk of the nucleosomes, which passes directly through the column (fractions 3–7). Only negligible radioactivity is eluted with KSCN. In addition, neither A_{260} nor nucleoprotein structure is noted to be nonspecifically adsorbed by the Sepharose, and subsequently liberated by KSCN.

Remarks. Using the above procedure,[14] we showed that the eluted chromatin components contained poly(ADP-ribosyl)ated histones H1, H2A, H2B, H3, and H4 as well as automodified poly(ADP-Rib)polymerase (Fig. 3). Using [^{3}H]lysine and arginine-labeled chromatin, it was also shown that the poly(ADP-ribosyl)ated histones, attached to stretches of oligonucleosomes bound to the column possessed a 6-fold higher specific activity of modification compared to histones of unbound chromatin. This indicated that non-poly(ADP-ribosyl)ated nucleosomes, connected and proximal to the modified regions, were copurified by this procedure. This allowed the characterization of the oligonucleosomal DNA around poly(ADP-ribosyl)ated chromatin domains to be compared with the unbound, bulk chromatin. The data indicated that immunofractionated, poly(ADP-ribosyl)ated, oligonucleosomal DNA contain significant amounts of internal single-strand breaks compared with bulk chromatin.[14]

Fig. 2. Selective retention of modified nucleosomes domains by anti-poly(ADP-Rib) IgG-Sepharose. (A) Oligonucleosomes (4–8N) were poly(ADP-ribosyl)ated with [^{32}P]NAD as described in the text. Nucleosomes (1.0 ml, 1.5 A_{260} nm) were applied to a 6 ml bed volumn of anti-poly(ADP-Rib) IgG-Sepharose 4B over 45 min. The column was washed in 1 ml fractions with PBS. The bound material was eluted with 1.6 M KSCN at fraction 30. Absorbance at 260 mm (▲) was monitored on fractions prior to additions of KSCN. Acid-precipitable radioactivity (○) was determined. The inset represents a native 3% polyacrylamide electrophoretic separation of dialyzed and concentrated samples of bound (left) and unbound (right) chromatin, stained with ethidium bromide. (B) A similarly prepared poly(ADP-ribosyl)ated oligonucleosomal sample was analyzed on a preimmune IgG coupled Sepharose column. Taken from Malik *et al.*[14]

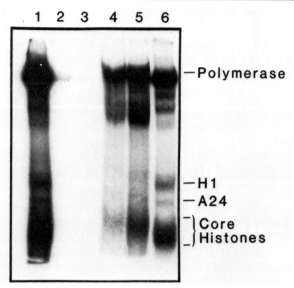

FIG. 3. Total nucleosomal poly(ADP-Rib) acceptors of chromatin purified by anti-poly (ADP-Rib). The experiment was performed essentially as described in Fig. 2 except that carrier-containing samples were ethanol precipitated and subjected to sodium dodecyl sulfate–polyacrylamide gel electrophoresis, and autoradiography. Lane 1, nucleosomes applied to the column; lanes 2, 3, unbound nucleosome; lanes 4–6, bound nucleosomes. Taken from Malik *et al.*[14]

The bound nucleoprotein complexes are found to be enzymatically active for poly(ADP-Rib) polymerase after elution from the antibody column. In contrast, the unbound nucleosomes, representing 90% of the unfractionated chromatin, are totally inactive in the poly(ADP-ribosyl)ation reaction.[14]

The immunofractionation method has been quite useful for other types of studies involving poly(ADP-ribosyl)ation. For example, poly(ADP-ribosyl)ated polyoma virus minichromosomes have been fractionated by this method.[15] It was noted that replicative intermediate minichromosomes were retained better on the antibody columns than were mature minichromosomes. In addition, the column has recently been utilized to selectively purify poly(ADP-ribosyl)ated histone H1.[16] Phosphorylated species of histone H1 cochromatographed with the poly(ADP-ribosyl)ated molecules, suggesting that common domains of chromatin may be mutually modified by several posttranslational systems.[16]

[15] A. Prieto-Soto, B. Gourlie, M. Miwa, V. Pigiet, T. Sugimura, N. Malik, and M. Smulson, *J. Virol.* **45,** 600 (1983).
[16] M. Wong, M. Miwa, and M. Smulson, *Biochemistry* **22,** 2384 (1983).

[55] Hydroxyproline Glycosides in the Plant Kingdom

By DEREK T. A. LAMPORT

Rodlike hydroxyproline-rich glycoproteins are indispensable scaffolding components of the eukaryotic extracellular matrix. While collagen is the major structural protein of animal matrix,[1] extensin is the analogous plant protein occurring in the young extracellular matrix or *primary cell wall*,[2] but *not* in the mature matrix or *secondary cell wall*. Differences between these matrix glycoproteins reflect the dichotomy between plants and animals. Thus vertebrate collagen typically contains very little carbohydrate, and that occurring as glycosylated hydroxylysine.[3] On the other hand extensin of higher plants contains at least 50% carbohydrate and this occurs mainly as short arabinooligosaccharide O-substituents of the hydroxyproline residues.[4] Even larger substituents occur in the arabinogalactan proteins (AGPs) which are also extracellular glycoproteins[5,6] but, unlike extensin, freely soluble and generally present at much lower levels. Such a remarkable dichotomy between "plant and animal hydroxyproline" is currently the subject of much experimentation and speculation. My object here is to (1) describe suitable methods for preparation and assay of hydroxyproline arabinooligosaccharides and other hydroxyproline glycosides, (2) describe their biosynthesis such as is known, (3) describe their structure (where worked out), incidentally indicating the diversity of hydroxyproline glycosylation in the plant kingdom, and (4) summarize current approaches toward understanding the role of glycosylated hydroxyproline residues in structural proteins.

Preparation and Assay of Hydroxyproline Glycosides

Hydroxyproline glycosides were discovered in hydroxyproline-rich glycopeptides obtained by enzymic degradation of cell walls from cell suspension cultures.[7] These glycopeptides contained carbohydrate which did not undergo β-elimination in dilute alkali, indicating a possible in-

[1] D. R. Eyre, *Science* **207**, 1315 (1980).
[2] D. T. A. Lamport and J. W. Catt, *Encycl. Plant Physiol., New Ser.* **13B**, 133 (1981).
[3] W. T. Butler and L. W. Cunningham, *J. Biol. Chem.* **241**, 3882 (1966).
[4] D. T. A. Lamport, *Nature (London)* **216**, 1322 (1967).
[5] A. E. Clarke, R. L. Anderson, and B. A. Stone, *Phytochemistry* **18**, 521 (1979).
[6] D. T. A. Lamport, *in* "The Biochemistry of Plants" (J. Preiss, ed.), Vol. 3, p. 501. Academic Press, New York, 1980.
[7] D. T. A. Lamport, *Biochemistry* **8**, 1155 (1969).

volvement of hydroxyproline residues in a glycopeptide linkage. More rigorous conditions of alkaline hydrolysis sufficient to cleave peptide bonds, released a series of low-molecular-weight compounds which behaved as free rather than peptidylhydroxyproline yet contained one or more arabinose residues. Since then it has become clear that other sugars besides arabinose may also be involved. There is good evidence that the highly soluble extracellular arabinogalactan proteins have a polysaccharide attached via O-galactosylhydroxyproline,[8,9] while in primitive green algae such as *Chlamydomonas* the substituents are heterooligosaccharides.[10]

A summary of our standard protocol (P04479) for release of any of these glycosides involves a somewhat empirical alkaline hydrolysis in 0.2 M Ba(OH)$_2$ (i.e., saturated at 25°) in a sealed vial at 105–110° as follows.

Hydroxyproline Glycoside Profile Determination

Weigh out material containing a minimum of 100 μg hydroxyproline into a 1-ml microvial.

Add 0.5 ml 0.2 M barium hydroxide solution.

Securely cap vial and place in block heater for 18 hr at 105–110°.

Cool, then add 4 μl conc. sulfuric acid and mix well.

Check that pH is neutral or slightly alkaline.

Centrifuge neutralized hydrolyzate for 15 min at 10K rpm.

Reduce volume of supernate and apply to appropriate column.

Either monitor column eluate continuously or collect discrete fractions for manual assay.

More concentrated alkali can be used for the hydrolysis[11] (e.g., 5 N NaOH at 100° for 24 hr) but the barium hydroxide hydrolysis is less prone to operator error, especially when freeze-drying the arabinosides which are extremely acid labile.

Choice of separation methodology depends to some extent on the amount of material involved, and also of course on available equipment. Three useful methods are as follows: (1) cation exchange chromatography on a highly cross-linked resin such as Technicon Chromobeads B (0.6 × 75 cm) H$^+$ form eluted with a 0–0.5 N HCl gradient;[12] (2) slow gel filtration on long columns (~200 cm) of BioGel P2 (−400 mesh), eluted with

[8] D. G. Pope, *Plant Physiol.* **59**, 894 (1977).
[9] G. B. Fincher, W. H. Sawyer, and B. A. Stone, *Biochem. J.* **139**, 535 (1974).
[10] D. H. Miller, D. T. A. Lamport, and M. Miller, *Science* **176**, 918 (1972).
[11] A. K. Allen, N. N. Desai, A. Neuberger, and J. M. Creeth, *Biochem. J.* **171**, 665 (1978).
[12] D. T. A. Lamport and D. H. Miller, *Plant Physiol.* **48**, 454 (1971).

water at ~200 cm head pressure;[13] and (3) high-voltage paper electrophoresis at pH 1.9 (8% acetic, 2% formic, 90% water, v/v/v), e.g., at 5 kV for 3 hr on 110 cm Whatman No. 4 paper.[4]

While methods 1 and 2 can be preparative, method 3 is only semipreparative. However all three depend on use of specific colorimetric determination of hydroxyproline (this may explain why HPLC and gas chromatographic methods have not been exploited) although the color yield may be dependent[13] or independent[12] of the sugar substituent, probably reflecting the initial oxidation step of the assay used. One should also note that alkaline hydrolysis racemizes the hydroxy-L-proline residues thereby yielding some allohydroxy-D-proline. Racemization changes the relative positions of hydroxyl and carboxyl groups from trans to cis with a concomitant change in properties sufficient to effect a resolution on cation-exchange chromatography.

Biosynthesis of Hydroxyproline Glycosides

To date only one paper deals with *in vitro* biosynthesis of hydroxyproline glycosides. Karr[14] isolated a particulate (37,000 g pellet) from sycamore cell suspension cultures. This crude preparation catalyzed the incorporation of arabinose from UDP-L-[^{14}C]arabinose into particulate material which, on base hydrolysis, released ^{14}C-labeled hydroxyproline arabinosides. Mg^{2+} and Mn^{2+} (1 mM) enhanced the arabinosyltransferase activity by about 40% at the pH optimum of 6.0. Triton X-100 solubilized only 10–20% of the particulate transferase activity which, quite remarkably, has not been further fractionated or characterized, and therefore remains in the glycosyltransferase (pentosyl subclass) limbo, EC 2.4.2.

This lack of progress is, at least in part, due to lack of readily available substrate. The endogenous substrate is naturally only available for assay of activity in the crude particulate (vesicle?) fractions. However, Karr also used a hydroxyproline peptidyl substrate obtained from the cell wall fraction by deglycosylation and tryptic digestion. These tryptides served as substrate for the particulate preparations provided that sonic disruption of the cells occurred in the presence of the peptide substrate, which presumably entered the vesicles as they "rehealed" during sonication. Recently however, precursor extensin has become readily available,[15] and can of course be dearabinosylated[16] by treatment at pH 1 for 1 hr at 100°,

[13] F. M. Klis and H. Eeltink, *Planta* **144**, 479 (1979).
[14] A. L. Karr, Jr., *Plant Physiol.* **50**, 275 (1972).
[15] J. J. Smith, E. P. Muldoon, and D. T. A. Lamport, *Phytochemistry* **23**, in press (1984).
[16] D. T. A. Lamport, L. Katona, and S. Roerig, *Biochem. J.* **133**, 125 (1973).

or completely deglycosylated by treatment for 1 hr at 0° in anhydrous HF containing 10% (v/v) methanol.[17,18] Thus substrate limitation for the study of solubilized arabinosyltransferases may no longer exist.

As three different arabinosidic linkages are involved we can expect to find at least three arabinosyltransferases.

The Structure of Hydroxyproline Glycosides

Complete characterization of hydroxyproline glycosides defines the sugar conformation, anomeric configuration, and linkage analysis, achieved so far only for the tri- and tetraarabinosides isolated from tobacco cell walls. Methylation analysis[19] defined the linkages and sugar conformation,[14,20] while the acid lability also indicated furanosidic linkages.[11] Unequivocal determination of the arabinosyl anomeric configuration is more difficult because specific β-L-arabinofuranosidases are, with one exception,[21] unknown. (On the other hand, fungal α-L-arabinosidases are common.) Thus some ambiguity arose when Knee[22] reported very slow cleavage of hydroxyproline tetraarabinofuranoside, (yielding 17% free arabinose) by α-L-arabinofuranosidase, which was at variance with optical rotation data (positive $[\alpha]$D) implicating β linkages.[11,23] Reconciliation of these data by a ^{13}C NMR approach[24] showed three β-linked arabinofuranosides, while the nonreducing terminal residue was α-linked! Thus the structures deduced are as follows:

α-L-Araf-(1–3)-β-L-Araf-(1–2)-β-L-Araf-(1–2)-β-L-Araf-(1–Hyp

β-L-Araf-(1–2)-β-L-Araf-(1–2)-β-L-Araf-(1–Hyp

Other hydroxyproline glycosides remain to be characterized. The algal heterooligosaccharides from *Chlamydomonas*[10] contain arabinose galactose and glucose up to ~10 residues total, while those from *Chloroccocum* contain arabinose, galactose, and an unknown 6-deoxyhexose.[25] The arabinogalactan polysaccharide of the hydroxyproline-rich AGPs consists of a partially characterized β1,3-linked galactopyranoside backbone with 1,6-linked galactose branches terminated by α 1,3-linked arabinofurano-

[17] A. Mort and D. T. A. Lamport, *Anal. Biochem.* **82,** 289 (1977).
[18] M. P. Sanger and D. T. A. Lamport, *Anal. Biochem.* **128,** 66 (1983).
[19] B. Lindberg and J. Lonngren, this series, Vol. 50, p. 3.
[20] Y. Akiyama and K. Kato, *Agric. Biol. Chem.* **40,** 2343 (1976).
[21] P. M. Dey, *Biochim. Biophys. Acta* **302,** 393 (1973).
[22] M. Knee, *Phytochemistry* **14,** 2181 (1975).
[23] Y. Akiyama and K. Kato, *Agric. Biol. Chem.* **41,** 79 (1977).
[24] Y. Akiyama, M. Mori, and K. Kato, *Agric. Biol. Chem.* **44,** 2487 (1980).
[25] D. H. Miller, *J. Phycol.* **14,** 189 (1978).

side residues.[5] Interestingly both the algal heterooligosaccharides and the higher plant AGPs share *O*-galactosylhydroxyproline as a glycopeptide linkage (in addition to *O*-arabinosylhydroxyproline) characterized in AGPs as 4-*O*-β-D-galactopyranosylhydroxy-L-proline.[9] The AGPs also contain small amounts of *O*-glucosylhydroxyproline.[26] Not surprisingly both galactosyl and glucosylhydroxyproline are considerably more stable to acid hydrolysis than the arabinosyl compounds. Ultimately we can expect completion of all these structures and confirmation by synthesis. Indeed Vercellotti has already synthesized some of the analogous model compounds.[27,28]

The Role of Glycosylated Hydroxyproline Residues

From their position outside the cell, their composition and structure, one can argue that glycosylated hydroxyproline-rich glycoproteins play a structural role in the extracellular matrix. But what is the structural role of the arabinosides? Current approaches range from purely physiological correlations and stochastic molecular model building, to CD spectral comparisons of native extensin precursor before and after deglycosylation, and electron microscopy.

Early work[12] showed that the hydroxyproline arabinoside profile was species specific. However it is now clear that the profile does show small but significant changes dependent on the growth state. For example there are *fewer* glycosylated hydroxyproline residues during the early growth of bean cultures,[13] while increased glycosylation occurs in diseased plants.[29] The "wrap-around" model provides a possible explanation: CPK models show that oligoarabinosides can "nest" along the groove of the extended extensin helix, and may therefore enhance structural stability through hydrogen bonding.[6] CD spectra support that possibility. Thus both extensin peptides[30] and similar algal proteins[31] display CD spectra typical of polyproline II type helices (3 residues/turn, pitch 9.4A). Indeed current work[32] shows that glycosylation actually *enhances* the typical polyproline II CD spectrum of extensin precursor, strongly supporting the thesis that arabinosides do indeed increase the structural stability of a rod-shaped

[26] U. V. Mani, Y. Akiyama, S. Mohrlok, and D. T. A. Lamport, *Plant Physiol.* **63,** Suppl., Abstr. 167 (1979).

[27] J. R. Vercellotti and E. K. Just, *Carbohydr. Res.* **5,** 102 (1967).

[28] P. D. Feil and J. R. Vercellotti, *Carbohydr. Res.* **31,** 311 (1973).

[29] M.-T. Esquerre-Tugaye and D. T. A. Lamport, *Plant Physiol.* **64,** 314 (1979).

[30] D. T. A. Lamport, *Recent Adv. Phytochem.* **11,** 79 (1977).

[31] R. B. Homer and K. Roberts, *Planta* **146,** 217 (1979).

[32] G. J. Van Holst and J. E. Varner, *Plant Physiol.,* in press (1984).

macromolecule recently visualized for the first time via electron micros-copy.[32]

The role of the larger oligo- and polysaccharide substituents of hy-droxyproline remains obscure. However, AGPs of the *Lilium* stigma act as adhesives for pollen,[33] while reassembly of the algal (notably *Chlamy-domonas*) cell wall into a crystal lattice depends on integrity of the sugar residues,[34] hinting that hydroxyproline glycoside profiles of higher plants may reflect more than structural stability of the polypeptide backbone.

Conclusion

Because both plant and animal extracellular matrices depend, at least initially, on rodlike hydroxyproline-rich macromolecules, one suspects an evolutionary connection despite the clear cut dichotomy between glyco-sylated extensin and nonglycosylated collagen. Recent speculations link the "sudden appearance" of the metazoa with the increased availability of oxygen which allowed its use for nonrespiratory purposes such as organic matrix formation.[35] There is in addition the recent discovery of exceedingly ancient blastula-like fossil algae such as *Eovolvox*[36] with a hydroxyproline-rich matrix presumably similar to the extant *Volvox*. If an evolutionary change from glycosylated to nonglycosylated hydroxypro-line occurred, relic species remain undiscovered. Thus we cannot answer the question of evolutionary affinities between collagen and extensin yet!

[33] F. Loewus and C. Labarca, *in* "Biogenesis of Plant Cell Wall Polysaccharides" (F. Loewus, ed.), p. 175 Academic Press, New York, 1973.

[34] J. W. Catt, G. J. Hills, and K. Roberts, *Planta* **138,** 91 (1978).

[35] K. M. Towe, *in* "Life in the Universe" (J. Billingham, ed.), p. 297. MIT Press, 1981.

[36] J. Kazmierczak, *Acta Palaeontol. Pol.* **26,** 299 (1981).

Author Index

Numbers in parentheses are footnote reference numbers and indicate that an author's work is referred to although the name is not cited in the text.

H

Haas, A. L., 240, 257, 258(68, 70, 71), 259(68), 260
Haas, G. M., 40
Habener, J. F., 220
Haberland, M. E., 135(6), 136
Hadley, M. E., 105, 110(21), 111(21)
Haenni, A. H., 160
Hagege, R., 59
Hagenmaier, H. E., 253
Haggerty, W. J., Jr., 404
Haimi, L., 223
Hallak, M. E., 225
Halter, J. B., 85
Hamada, H., 192
Hamelin, J., 221
Hamm, H. H., 378
Hammerling, U., 238, 240
Hammerman, M. R., 404
Hammons, G. T., 81
Hancock, W. S., 289
Hannappel, E., 166, 193, 194(8), 196(8), 197
Hanneken, A., 86
Hannum, C., 165(10), 166
Hansen, V. A., 404
Hantke, K., 42, 205, 206(1), 365
Harding, H. W., 253
Hardy, M. F., 287, 291(5)
Hardy, S. J. S., 267
Hargrove, J. L., 183
Harmon, W., 80
Harper, E., 62
Harris, G. I., 220
Harris, J. E., 24
Harris, S. A., 133
Harrison, J. H., 135(7), 136
Hartley, B. S., 358
Hartman, F. C., 136
Hartman, P. G., 255
Harvey, C., 194
Hase, T., 128(25), 129, 130
Hasegawa, K., 404
Haselkorn, R., 419
Hashida, T., 501
Hata, T., 340
Hatfield, D., 157, 158(3, 4), 160(4)
Hathaway, G. M., 119(20), 121
Haughland, R. B., 63
Hauschka, P. V., 48, 49(64), 50(64)

Hawke, D., 24
Hawkins, T. C., 239, 240, 241(18)
Hayaishi, K., 510
Hayaishi, O., 74, 119(23, 24), 121, 253, 378, 401, 440, 446, 450, 451(4), 452(4, 5), 454(4), 455(11, 12), 457, 458(4, 5), 459(4, 5), 461, 462, 472, 500, 501, 502(15, 16), 503(16), 504(1, 3, 15, 16), 505, 507, 509, 510
Hayashi, K., 444, 445, 447
Hayashi, R., 126, 340
Hayashi, S., 368, 369(27), 370, 371(27)
Hayashida, H., 221
Heath, H., 356
Heckman, J. E., 153
Heerma, W., 58
Heesing, A., 402
Held, W., 267
Helfman, P. M., 99, 102(4), 107, 110(4, 25), 113(4, 25), 114
Heller, E., 218
Heller, H., 240, 257, 258(69–72)
Helmerhorst, E., 492
Hemmerich, P., 369, 371(5), 377
Hempel, K., 282
Hempstead, J., 197
Hendee, E. D., 378
Henderson, L. E., 24, 26, 44, 209
Hendricks, S. B., 360, 362(6)
Hendriksen, O., 379, 382(6)
Henkart, M., 237
Hennighausen, L. G., 159
Hennrich, N., 62
Henschen, A., 126, 218
Hensen, E., 52, 53, 54
Hensens, O. D., 30
Herbert, E., 171
Heremans, J. F., 97
Herlihy, W. C., 33, 34(14), 38, 41
Hermann, J., 128(23), 129
Hermanson, G. T., 83
Hermodson, M. A., 23
Herndon, C. S., 136
Herp, A., 212
Hershey, J., 344
Hershey, N. D., 192
Hershko, A., 240, 257, 258(65, 67, 69–74), 259, 260
Herz, A., 193(15), 194
Hess, R. L., 79, 85

Subject Index

A

Abelson murine leukemia virus, transforming protein, fatty acid bound in, 209

S-Acetonyl CoA, synthesis, 181, 182

N^α-Acetylamino acid, release, from amino-terminal acetylated protein, 169, 170

N^2-Acetyllysine, formulation, 9

N^6-Acetyllysine, formulation, 9

O-Acetylserine (thiol)-lyase, spectral properties, 119

N^α-Acetyltransferase
distribution, 166
rat
distribution in various organs, 176, 177
rat pituitary, 170–179
activity in lobes, 177, 178
kinetic constants, 178, 179
properties, 172–175
subcellular localization, 175, 176
recognition site, 166–168

Actin
amino-terminal acetylation, 179–192
D. discoideum
NH₂-terminal processing, pathway, 180, 181
NH₂-terminal tryptic peptide, 185, 186
methylated lysines in, 288
newly synthesized, removal of acetyl methionine from amino-terminal, 185–191
nonacetylated, preparation, 182, 183
posttranslational NH₂-terminal acetylation, in reticulocyte lysate, 183–185
processed, NH₂-terminus, tryptic peptide analysis, 189–191
processing, assay improvement, 190–192

by NH₂-terminal peptide mapping, 185, 189–191
two-dimensional gel electrophoresis, 185–188
with two amino acids preceding NH₂-terminal aspartate or glutamate, processing, 192

Acyl residue, nomenclature, 7

5′-Adenosine monophosphate, antibody, in determination of ADP-ribosyl residues, 473–477

S-Adenosylmethionine: protein (arginine) N-methyltransferase, see Protein methylase I

S-Adenosylmethionine: protein-carboxyl O-methyltransferase, see Protein methylase II

S-Adenosylmethionine: protein (lysine) N-methyltransferase, see Protein methylase III

Adenylate cyclase system
ADP-ribosyltransferases acting on, 411–418
components, 411, 412
inhibitors, 412

ADP-ribose, see also Mono(ADP-ribose); Poly(ADP-ribose)
determination, in vivo, preparation of acid-insoluble fractions, 483–485
in vivo levels, determination by fluorescence methods, 483–494

ε-ADP-ribose
fluorescent peak, 493, 494
formation, 492

ADP-ribose-[¹²⁵I]guanyltyramine, formation, assay of ADP-ribosyltransferase, 432

[³²P]ADP-ribose protein, formation, assay of ADP-ribosyltransferase, 431

[adenine-U-¹⁴C]ADP-ribose protein, formation, assay of ADP-ribosyltransferase, 431

tyrosinase
S$^\beta$-(2-histidyl)cysteine in, 355–357
tripeptide Cys-Thr-His, 355–357
[*carbonyl*-^{14}C]Nicotinamide, release
assay of ADP-ribosyltransferase, 430, 431
in assay of toxin activity, 415, 416
from [*carbonyl*-^{14}C]NAD, in assay of
pertussis toxin, 417, 418
p-Nitrobenzylidine aminoguanidine
structure, 404, 405
substrate, for assay of ADP-ribosyltransferase, 404–410
Norleucine
ion-exchange chromatography, 18, 20
symbol, 6
systematic name, 6
Norvaline
symbol, 6
systematic name, 6
Nucleic acid, constitutents, nomenclature, 7
Nucleosome, poly(ADP-ribosyl)ation, 512–517

O

Oligonucleosome, poly(ADP-ribosyl)ation, 514–517
Ornithine, 10
ion-exchange chromatography, 19, 20
nomenclature, 6
Ornithine aminotransferase, spectral properties, 118
Ornithine decarboxylase
pyridoxal peptide, amino acid sequence, 128
spectral properties, 119
Osteocalcin, structure, determination, 48–50
Ovalbumin
amino-terminal acetylation, 165
enzymatic hydrolysis, 71, 73
5-Oxoproline, 10
nomenclature, 6
Oxytocin
carboxy-terminal amide, 218, 221
mobility on HPLC, effect of chemical
modification, 27

P

O^3-[(ω-Pantetheinyl)phospho]serine, 13
k-Papacasein, bovine, NH$_2$-terminal sequence analysis, 38
Penicillinase, lipoprotein membrane
lipophilic modification
nature of, 365, 366
sequence required for, 368
N-terminal glyceride-cysteine modification, 365–368
Peptide
boronate affinity chromatography, 79, 83
carboxy-terminal amides, 218–223
HPLC retention time, and amino acid
modification, 25–29
mass spectrometry
gas chromatographic procedure, 33–35
permethylation procedure, 33, 34, 40
mixtures, mass spectrometry, 33–37
mobility on HPLC
effect of specific chemical modifications, 25–27
as signal of modification, 25–29
nomenclature, 5
Periodate oxidation, 79, 86, 87
Pertussis toxin
assays, 416–418
effect, on adenylate cyclase system, 411, 413
subunits, 416
Phenylalaninamide, 11
Phenylalanine
structure, 8
symbol, 4
systematic name, 4
L-Phenylalanine, modifications, not involving prosthetic group, 11
Phenylthiohydantoin derivative
identification, on HPLC, 22–25, 32
mass spectrometry, 32, 33
Phosphatidyl, symbol, 7
Phosphodiesterase
assay, 507, 508
contaminants, 507
hydrolysis of poly(ADP-ribose) pyrophosphate bonds, 439, 505, 506